AF789776

Farm Animals and the Environment

Farm Animals and the Environment

Edited by

Clive Phillips

School of Agricultural and Forest Sciences
University of Wales
Bangor
UK

and

David Piggins

Department of Psychology
University of Guelph
Canada

C·A·B International

C·A·B International
Wallingford
Oxon OX10 8DE
UK

Tel: Wallingford (0491) 32111
Telex: 847964 (COMAGG G)
Telecom Gold/Dialcom: 84: CAU001
Fax: (0491) 33508

© C·A·B International 1992. All rights reserved. No part of this publication may be reproduced in any form or by any means, electronically, mechanically, by photocopying, recording or otherwise, without the prior permission of the copyright owners.

A catalogue record for this book is available from the British Library

ISBN 0 85198 788 5

VMED
SF
140
.E25
F376
1992

Typeset by Alden Multimedia
Printed and bound in the UK at the University Press, Cambridge

SVM LIBRARY

Contents

Section II
Perception of the Environment by Farm Animals

Section III
Animal Responses to the Environment

Section IV
Effects of Farm Animals on the Human Environment

Contributors

Alcock, M.B., *School of Agricultural and Forest Sciences, University of Wales, Bangor, Gwynedd LL57 2UW, UK.*

Archer, J.R., *Agricultural Development and Advisory Service, Ministry of Agriculture, Fisheries and Food, 17 Smith Square, London, UK.*

Bartle, N.C., *Faculty of Agricultural and Food Sciences, University of Nottingham, Sutton Bonington, Loughborough, Leics. LE12 5RD, UK.*

Baxter, M.R., *Centre for Rural Building, Craibstone, Bucksburn, Aberdeen AB2 9TR, UK.*

Blom, J.Y., *Danish Institute for Fisheries Technology and Aquaculture, DK-9850 Hirtshals, Denmark.*

Boissy, A., *Institut National de la Recherche Agronomique, Laboratoire de la Lactation et de l'Elevage des Ruminants, Theix, 63122 Saint-Genes-Champanelle, France.*

Boivin, X., *Institut National de la Recherche Agronomique, Laboratoire de la Lactation et de l'Elevage des Ruminants. Theix, 63122 Saint-Genes-Champanelle, France.*

Broom, D.M., *Department of Clinical Veterinary Medicine, University of Cambridge, Madingley Road, Cambridge CB3 0ES, UK.*

Ewbank, R., *Universities Federation for Animal Welfare (UFAW), 8 Hamilton Close, South Mimms, Potters Bar, Herts EN6 3QD, UK.*

Geraert, P.A., *INRA, Station de Recherches Avicoles, Nouzilly, 37380 Monnaie, France.*

Habeeb, A.A.M., *Bioclimatology Unit, Radiobiology Department, Nuclear Research Centre, Atomic Energy Authority, Cairo, Egypt.*

Heffner, H.E., *Laboratory of Comparative Hearing, Department of Psychology, 2801 W. Bancroft, University of Toledo, Toledo, OH 43606-3390, USA.*

Heffner, R.S., *Laboratory of Comparative Hearing, Department of Psychology, 2801 W. Bancroft, University of Toledo, Toledo, OH 43606-3390, USA.*

Herpin, P., *INRA, Station de Recherches Porcines, 35590 St-Gilles, France.*

Hurnik, J.F., *Department of Animal and Poultry Science, University of Guelph, Guelph, Ontario, Canada N1G 2W1.*

Kamal, T.H., *Bioclimatology Unit, Radiobiology Department, Nuclear Research Centre, Atomic Energy Authority, Cairo, Egypt.*

Kendrick, K.M., *AFRC Institute of Animal Physiology and Genetics Research, Cambridge Research Station, Babraham, Cambridge, CB2 4AT, UK.*

Lacey, R.W., *Department of Microbiology, University of Leeds, Leeds LS2 9JT, UK.*

Le Dividich, J., *INRA, Station de Recherches Porcines, 35590 St-Gilles, France.*

Le Neindre, P., *Institut National de la Recherche Agronomique, Laboratoire de la Lactation et de l'Elevage des Ruminants, Theix, 63122 Saint-Genes-Champanelle, France.*

Ley, S., *Department of Pharmacology, University of Bristol, University Walk, Bristol BS8 1TD, UK.*

Livingston, A., *Department of Pharmacology, University of Bristol, University Walk, Bristol BS8 1TD, UK.*

Marai, I.F.M., *Department of Animal Production, Faculty of Agriculture, Zagazig University, Zagazig, Egypt.*

Nicholson, R.J., *Agricultural Development and Advisory Service, Bristol, BS10 6YW, UK.*

Nilsson, C., *Department of Farm Buildings, Swedish University of Agricultural Sciences, PO Box 945, S-22009 Lund, Sweden.*

Owen, J.B., *School of Agricultural and Forest Sciences, University of Wales, Bangor, Gwynedd LL57 2UW, UK.*

Perry, G.C., *Department of Animal Husbandry, University of Bristol, Langford House, Langford, Bristol BS18 7DU, UK.*

Phillips, C.J.C., *School of Agricultural and Forest Sciences, University of Wales, Bangor, Gwynedd LL57 2UW, UK.*

Piggins, D., *Department of Psychology, University of Guelph, Guelph, Ontario, Canada N1G 2W1.*

Seabrook, M.F., *Faculty of Agricultural and Food Sciences, University of Nottingham, Sutton Bonington, Loughborough, Leics. LE12 5RD, UK.*

Tamminga, S., *Department of Animal Nutrition, Agricultural University, Haagsteeg 4, 6708 PM Wageningen, The Netherlands.*

Veissier, I., *Institut National de la Recherche Agronomique, Laboratoire de la Lactation et de l'Elevage des Ruminants, Theix, 63122 Saint-Genes-Champanelle, France.*

Vermorel, M., *INRA, Unité de Recherches Métabolismes Enérgetique et Lipidique, Theix, 63122 Ceyrat, France.*

Waterman, A., *Department of Veterinary Surgery, University of Bristol, University Walk, Bristol BS8 1TD, UK.*

Wathes, C.M., *AFRC Silsoe Research Institute, Wrest Park, Silsoe, Beds. MK45 4HS, UK.*

Preface

This book contains edited and revised papers from an international conference on Farm Animals and the Environment held on 4–5 September 1991 at the University College of North Wales, Bangor. There are four major sections in the book: environmental factors affecting farm animals; farm animals' perception of the environment; the effect that the environment has on an animals' welfare and behaviour; and finally the effect that farm animals have on our own environment.

A novel feature of the book is the consideration of the interaction between farm animals and the environment principally from a zoocentric perspective rather than from an anthropocentric point of view (profitability of different production systems, pollution of our environment, etc.). A number of relevant questions may be asked: How does the animal perceive its environment and what effect does it have on the animal's welfare? How can we measure an animal's welfare and its response to its environment? Can we maintain a greater diversity of animal species on our farms and what might be the behavioural result of such an interplay? Another important feature is the acknowledgement that many of the important aspects of the environment to which animals respond are not immediately obvious to the casual observer. The crucial role played in any farming system by the stockman is carefully considered, particularly in relation to the rearing of young animals.

The different sensory capacities of farm animals compared to our own are examined, particularly the greater reliance on smell. Animals often appear not to utilize their maximum sensory capacity, as evidenced by our own sensory spectrum normally being dominated by vision at the expense of, for example, olfaction. It is important therefore to begin to understand how farm animals process and respond to sensory input in their terms rather than assuming that their sensory capabilities reflect their perception of the environment in anthropomorphic terms.

Finally the effect that farm animals have on our own environment is discussed. Pollution can be caused by excessive concentrations of livestock or inappropriate waste disposal. Many of the natural fauna that have existed for centuries on our farms are now threatened with extinction by modern agricultural practices, such as the use of agrochemicals and the intensive land use that allows no room for non-productive animals on the farm. It is an old cliché but 'small is beautiful' is one which usually applies in the conservation of our fauna on livestock farms.

Motivation for short-term expediency and profitability can endanger the lives of humans and livestock alike when conditions prevail that encourage new disease organisms to develop, as in the recent case of the bovine spongiform encephalopathy. Many people have recognized the unsustainability of this and other features of modern production systems, and long-term solutions urgently need to be found.

Some recent research that was presented at the conference in the form of poster papers is contained in the appendix. Though diverse in its subject matter it is heartening to see that there is some excellent work being conducted in this area.

The conference would not have been possible without the support of a great many people and organizations. We are grateful to Dr H. Omed, who organized everything from the advertising to the after dinner entertainments, Angels Varea, Jon Evans and Ishmael Rind for technical management, Phyllis Ellis for typing and the Royal Society for the Prevention of Cruelty to Animals, Pickup, the Meat and Livestock Commission, Gwynedd County Council and the Publicity Office, UCNW for financial support. We would like to thank all the speakers for their contributions, those contributing to the poster sessions and last but by no means least the delegates attending the conference for their attention and searching questioning. We are also indebted to the authors for rapidly and enthusiastically preparing the chapters of this book and hope that it will be of value to many agricultural researchers, extension workers, students and leading farmers.

Clive Phillips
David Piggins
December 1991

Section I

Environmental Factors Influencing the Production and Welfare of Farm Animals

The term environment can mean almost anything to anybody. Indeed the environment of any organism is different from that of every other simply because no two organisms share exactly the same spatial location. Basically an environment is a medium, and a dictionary definition of the term refers to two characteristics: firstly, one of 'surroundings' and, secondly, that of the 'conditions of life or growth'. Thus the first definition refers to the physical aspects of an animal's surroundings with such considerations as temperature, photoperiod and ventilation. These aspects have been the subject of much detailed research with sufficient knowledge now being available to begin to model animal responses to such variables. Other aspects of this first definition include such considerations as floor and space allowance, these being somewhat less well researched, but with currently much interest in their relationship to their effect upon animal comforts and well-being rather than simply productivity.

The second definition may be seen as including the human factor itself, which though not entirely absent under 'surroundings' achieves a particularly special status under the second definition in that the farmer's actions, rather than mere presence, have a profound effect upon the animal at all stages in its life, influencing both its physical and mental well-being. Previously often neglected or even ignored, or taken for granted, the human factor is increasingly seen as being of major significance in intensive husbandry systems, where physical aspects of the environment have been carefully considered and the optimum or economic optimum, provided for the animals. Farm animals now receive much less contact with humans than they used to, but such a trend has not been matched by research determining the attendant effects upon the well-being and productivity of the stock.

This first section, therefore, sees the integration of many aspects, some well researched and others where the information has been newly gathered and not even digested by those with an interest in farm animals, or their productivity, comfort, and welfare. Thus a friendly, cold or hostile environment may be so described for three reasons: its physics, chemistry and attitudes of animals, or a combination of the three.

1 Cold Stress

J. Le Dividich,[1] P. Herpin,[1] P.A. Geraert[2] and M. Vermorel[3]

[1]*INRA, Station de Recherches Porcines, St-Gilles, France;* [2]*INRA, Station de Recherches Avicoles, Monnaie, France;* [3]*INRA, Unité de Recherches Métabolismes Energétique et Lipidique, Ceyrat, France*

Summary

The effects of cold stress on survival of the neonates and on productivity and efficiency of feed conversion by the major farm animals are reviewed. However, most data will concern pigs and to a lesser extent cattle, sheep and poultry. Since the neonates are very susceptible to cold, attention is paid to the development of thermoregulation and factors involved in cold resistance. These include body energy stores, colostrum intake, birthweight, age, genotype and conditions of parturition. The effects of cold stress on production are analysed in relation to the level of feed intake and given in terms of weight gains, feed conversion, carcass composition and meat quality, and milk and egg production. Mechanisms of acclimation including morphological, behavioural, physiological and metabolic adjustments are presented. Emphasis is given to energy substrate production and utilization (especially fat and carbohydrate) and hormonal control of thermogenesis by thyroid hormones, catecholamines and glucagon. Finally, the development of regulatory heat production mechanisms in organs and tissues is considered in terms of replacement of shivering by non-shivering thermogenesis.

Introduction

Environmental temperature has a direct effect on the animal's energy expenditure and voluntary feed intake and therefore on the production performance of farm animals. Hence interest lies in determining the range of environmental temperature over which heat loss from the animal is minimal. This range is called the thermoneutral range with the lower limit called the lower critical temperature (LCT). Below LCT, achievement of homeothermy implies an increase in heat production leading to inevitable reduction in the efficiency of conversion of feed to animal products (growth, fetus, milk, egg). From a nutritional point of view, knowledge of LCT is important since it corresponds to both the lowest limit of the temperature range in which the energy available for production is optimal and to the starting point towards cold environmental conditions. Within classes of livestock, the newborn are very susceptible to cold and the effects are expressed in terms of morbidity and mortality, whereas in heavier animals cold normally represents a smaller problem than heat in as much as they are fed *ad libitum*. However, due to the high stocking densities used for some species (e.g. poultry, pig), a cold indoor climate may lead to a large scale of economic loss.

During the past years the effects of climatic conditions on the energy metabolism of farm animals, including the pathway of heat exchanges, the determination of LCT in relation to animal, nutritional and environmental factors have been extensively studied and have been the subject of several excellent reviews (Mount, 1968; Holmes and Close, 1977; van Kampen, 1981; Young and Degen, 1981; Curtis, 1983; Vermorel, 1987) and will not be specifically discussed in this chapter. Data presented in this chapter will focus on three main points: (i) the effects of cold stress on neonates with emphasis being given to factors involved in cold resistance; (ii) the effect of prolonged cold exposure on production traits; and (iii) the acclimation to cold including mechanisms leading to decreased heat loss and those leading to increased heat production.

Effects of Cold on the Neonate

Survival

The period of 2–3 days following birth represents the most critical period for the survival of newborn farm animals. In the swine industry, early (0–48 h) postnatal mortality of live born piglets accounts for over 50% of total preweaning losses (English and Morrison, 1984). Similarly about 50% of calf mortality and 75% of lamb mortality occur within the first postnatal day (Obst and Day, 1968; Theriez, 1982). Although various factors accounting for this early mortality have been identified, it is usually accepted that cold stress at the time of birth is both directly and indirectly a major cause of death. In this respect, the literature survey of Houston and Maddox (1974) suggests that about 50% of the neonatal mortality of lambs may be attributed to cold and cold-induced starvation. In the very drastic conditions of winter in Canada, Jordan *et al.* (1969) reported a mortality rate of 8% and 50% among calves born indoors and outdoors, respectively. Similarly, a mortality rate of 6% is reported in piglets born at an environmental temperature of 25°C while at 10°C, the mortality rate was 31% (Parker *et al.*, 1980). In fact, at birth the newborn experiences a sudden and dramatic cold stress since the gap between the dam's uterus temperature and the external temperature may vary from 15–20°C for the piglet to 25–35°C for the lamb and the calf. In these conditions, maintenance of homeothermic balance implies an efficient thermogenesis and the availability of body energy reserves until the newborn is consuming sufficient energy-yielding substrates from its diets.

Table 1.1. Basal (BM) and summit (SM) metabolism ($W\,kg^{-1}$) of the newborn lamb, calf and piglet.

	Lamb	Calf	Piglet	Piglet
Age at test (h)	2–4	2.5–15	2	24
BM	3.6	2.0	3.7	4.9
SM	16.5	7.4	10.8	11.3
Theoretical survival time (h) at SM[1]	11	55	11	–

Source: After Eales and Small, 1980; Okamoto *et al.*, 1986; Berthon, personal communication, 1991.
[1]Assuming the neonates being deprived of food.

Thermoregulation capacity of the newborn

At birth, the heat conserving capacity is very low in piglets, lambs and calves because the piglet is virtually hairless and devoid of subcutaneous fat and lambs and calves have their coat saturated in fetal fluid. Their ability to maintain a normal body temperature is then largely dependent on their capacity to produce heat. Despite the fall in rectal temperature (T_r) occurring at birth, this capacity is well developed in normal newborn lambs and calves since their maximum metabolic rate (summit metabolism, SM) is 4–5 times the basal rate (Table 1.1). Also piglets respond quickly to cold conditions by increasing their heat production. For example, the metabolic rate of piglets is 30% higher at 18°C than at 31°C within the first 20 minutes after birth (Noblet and Le Dividich, 1981) and their SM is by 4 times the basal rate (D. Berthon, personal communication, 1991). However, more important than the magnitude of SM itself is its persistence, which is largely dependent on the availability of energy substrates. In this respect carbohydrate (glycogen) and lipids are the most important energy sources in the newborn. Calculated available energy derived from glycogen and lipid stores amounted to 425, 660 and 1470 $kJ\,kg^{-1}$ body weight in normal piglet, lamb and calf, respectively (Mellor and Cockburn, 1986: Robelin, 1986). It is noticeable that the available energy of the newborn calf is 3.5 times that of the piglet. In addition, available fat in piglet is white fat while being brown fat in lamb and calf. Nevertheless, energy available from body stores provided a 'theoretical' starvation survival time of about 11, 11 and 55 h in piglet, lamb and calf, respectively, maintained at ambient temperature corresponding to SM, which emphasizes the importance of colostrum intake.

Factors of cold resistance

Colostrum intake

An early and high intake of colostrum by neonates appears to be vital not only to acquire immunological protection from diseases but also to ensure sufficient supply of energy for heat production. Colostrum provides large amounts of fat, carbohydrate (lactose) and proteins which are readily digested by the neonate. In the

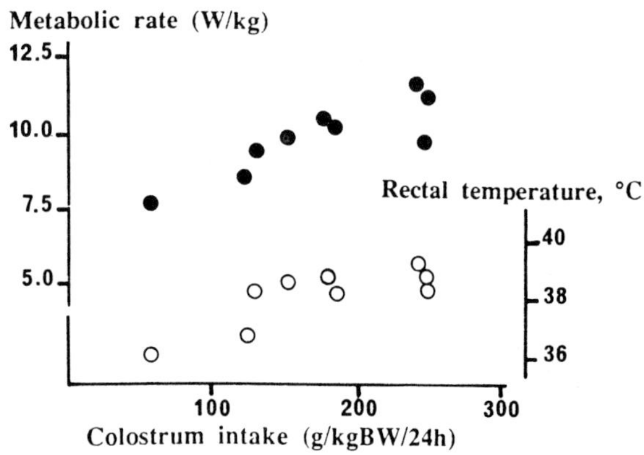

Fig. 1.1. Relationship between metabolic rate (●), rectal temperature (○) and colostrum intake in the neonatal pig maintained in the cold (18°C) (after Noblet and Le Dividich, 1981).

piglet, this is illustrated by the rapid rise in blood glucose following colostrum ingestion while the concomitant elevation of respiratory quotient (RQ) suggests that among the nutrients absorbed, carbohydrates are quickly metabolized (Le Dividich and Noblet, 1984). In fact colostrum consumed within the first 12 postnatal hours (i.e. $\approx 200\,g\,kg^{-1}$ body weight) provides as much available energy (lactose, fat) to the newborn piglet as the total amount of energy reserves at birth.

Evidence for the thermogenic importance of colostrum has been provided for the newborn piglet and lamb (Fig.1.1). In the neonatal pig, this is suggested, first, by the fact that in cold conditions both body temperature and heat production are positively related to the amount of colostrum intake (Noblet and Le Dividich, 1981), and second, by the observation that no improvement in thermostability occurs during the first day of life in piglets which consume very little or no colostrum (Le Dividich *et al.*, 1991). Similarly, ingestion of colostrum by the newborn lamb involves a 46% increase in basal metabolism and a 20% increase in summit metabolism (Eales and Small, 1981). In contrast, although a postprandial increase in heat production is reported by Vermorel *et al.* (1989a) in the newborn calf, Okamoto

Table 1.2. Effect of cold stress on colostrum intake in piglets or that available in lambs.

	Environmental temperature	
	Thermoneutral	Cold
Piglet		
Colostrum intake (g/24 h)	290	212
Lamb		
Colostrum available (ml/h)		
Single	86	76
Twin	76	39

Source: After Le Dividich and Noblet, 1981; Thompson, 1983.

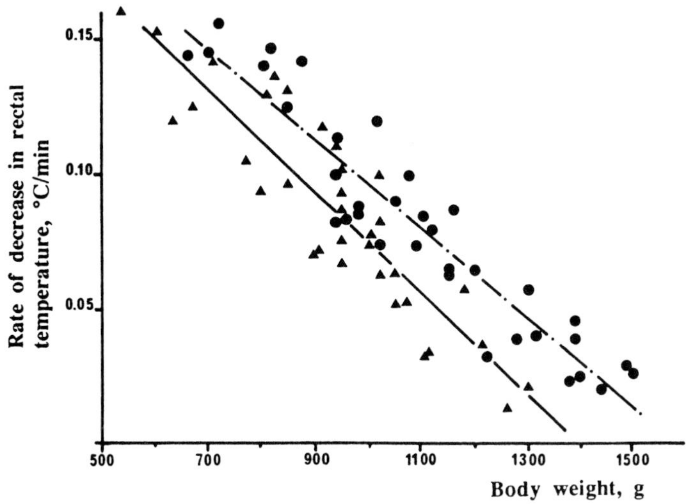

Fig. 1.2. Relationship between the rate of decrease in rectal temperature during exposure to cold (6–7°C) for 45 min and body weight in Large White (●) and Meishan (▲) piglets aged 2 h (after Le Dividich *et al.*, 1991).

et al. (1986) did not find any effect of colostrum feeding on basal and summit metabolism. However, prolonged cold exposure may affect cold resistance through a decreased colostrum consumption. In this respect cold stressed (18–20°C) piglets are found to consume 27% less colostrum than their littermates maintained in thermoneutral conditions (Le Dividich and Noblet, 1981), probably because cold stress results in less aggressive nursing behaviour (Table 1.2). Very similar results are reported in neonatal rabbits reared in incubators (Bernard and Hull, 1964). Also severe cold stress may reduce colostrum intake in the newborn lamb through a reduced teat-seeking activity (Alexander and Williams, 1966) or through a decrease in milk production by the ewe (Thompson, 1983), this being of particular importance when twin or triplet lambs are born.

Birthweight and age

Birthweight (BW) is an important component of cold resistance in the early postnatal period in that the higher the BW is, the more resistant to cold is the neonate (Alexander, 1962; Curtis *et al.*, 1967; Okamoto *et al.*, 1986). The impaired ability of small animals compared to large ones to maintain homeothermic balance during cold exposure is largely explained by their greater surface area to body mass ratio, which results in greater heat loss (Fig. 1.2). However, a low BW is also associated with a reduced thermogenesis as indicated in newborn lambs by the lower SM per unit of surface area (Stott and Slee, 1987). In addition, the cold resistance of lambs from undernourished ewes is seriously impaired by, first, reduce thermal insulation due to a lower pelage population density and shorter hair fibre length, and second, a reduced ability to sustain high metabolic rate due to a proportionally lower amount of energy stores (Alexander, 1974). Nevertheless, the effects of BW on cold resistance decline rapidly after birth. For example, in piglets, a 200 g variation in BW is

Table 1.3. Drop (°C) in rectal temperature (T_r) in piglets or in clocal temperature (T_c) in chicks during exposure to cold as related to age.

	Piglets[1]		Chicks[2]	
Age (h)	Weight (g)	Change in T_r	Weight (g)	Change in T_c
6–8	1440	− 4.4	43	− 14.0
18	1460	− 3.2	–	–
30–32	1440	− 0.9	43	− 7.5
128	–	–	57	− 2.9

Source: After Curtis *et al.*, 1967; Wekstein and Zolman, 1967.
[1]90 min exposure at 4°C
[2]60 min exposure at 10°C.

associated with a change in rectal temperature amounting to 0.036°C min^{-1} during exposure to 6–7°C ambient temperature at 2 h of age whereas at 24 h the corresponding change is 0.021°C min^{-1} (Le Dividich *et al.*, 1991). This underlines that newborns of low BW are at particular risk from acute hypothermia in the very first hours following birth.

Cold resistance is markedly improved in the early postnatal period (Table 1.3). This is attested to by evidence showing the improvement in the ability to resist a fall in body temperature during exposure to a cold environment in piglets during the first day after birth (Curtis *et al.*, 1967) and in chicks during the first week after hatch (Wekstein and Zolman, 1967). This rapid improvement in cold resistance in the piglet is indicative of a marked improvement in thermoregulation, largely due to the increased contribution of fat catabolism as a metabolic fuel (Noblet and Le Dividich, 1981). By contrast, in both lambs and calves, the improvement in cold resistance during the first postnatal day is more notably due to the improvement in heat conserving ability by the evaporation of the fetal fluid. The chick's homeothermic development occurs over a period of about 2 weeks (Freeman, 1988), and is largely dependent on the development of feather cover.

Genotype

Resistance to cold is to some extent dependent on genotype. This has been demonstrated when comparing wild to domestic piglets (Foley *et al.*, 1971) and Chinese (Meishan breed) to Large White piglets (Le Dividich *et al.*, 1991) (Table 1.4). Although being lighter, wild piglets are more resistant to cold than domestic ones because they possess greater pelage insulation and a more pronounced ability to utilize fat. The reasons for the greater resistance of Meishan piglets to cold are less understood. However, it is probably due to a greater physiological maturity at birth. Differences in cold resistance are also found when comparing lambs (Slee, 1981) and calves (Vermorel *et al.*, 1989a) from different breeds and are attributed to differences in capacity to conserve heat and/or to produce heat. In lambs, part of the variation between breeds is associated with coat type and skin thickness with heavy and longer-coated lambs showing greater cold resistance. Recently selection for cold

Table 1.4. Cold resistance of neonatal pig from European (Large White, Lw) and Chinese (Meishan, Ms) breeds.

	Age at cold exposure	
Breed	2 h	24 h
Body weight (g)		
Lw	1135	1195
Ms	961	1005
Rate of fall in T_r (°C min^{-1})		
Non-adjusted data		
Lw	0.081	0.042
Ms	0.087	0.041
Adjusted data		
Lw	0.095	0.053
Ms	0.072	0.032

Source: After Le Dividich *et al.*, 1991. Cold resistance was assessed by the rate of change in rectal temperature (T_r) during exposure to 6–7°C ambient temperature for 45 min and 75 min when aged 2 h and 24 h, respectively. Data were adjusted by covariance to common body weight of 1050 g and 1095 g at 2 h and 24 h of age, respectively.

resistance in Scottish Blackface lambs has been proved to be efficient (Slee and Stott, 1986), the realized heritability of cold resistance being 0.27 for upward selection.

Conditions of parturition

The two major complications encountered at calving and lambing are prolonged delivery due to dystocia and premature birth. Further, dystocia is sometimes associated with partial separation of the placenta from the uterus, and the umbilical cord may be compressed or it might even rupture prematurely leading to severe hypoxia before delivery. These are reported to dramatically increase early mortality of lambs and calves (Menissier and Petit, 1984). In fact dystocial lambs and calves are handicapped in maintenance of homeothermy. Studies conducted on calves by Vermorel *et al.* (1989b) clearly indicate that the initial drop in T_r was higher in dystocial than in eutocial calves (Fig. 1.3). Thereafter, T_r remains some 0.8°C below that of eutocial calves. Also, their heat production was lower because of a severe acidosis, or reduced mobilization of body reserves, and lower plasma thyroid hormone levels. Similarly calves born one month before normal term are unable to combat cold, and this is due to the fact that they are less active after birth and to their reduced ability to produce heat, their heat production averaging half that of high-vitality calves born at term (Vermorel *et al.*, 1983).

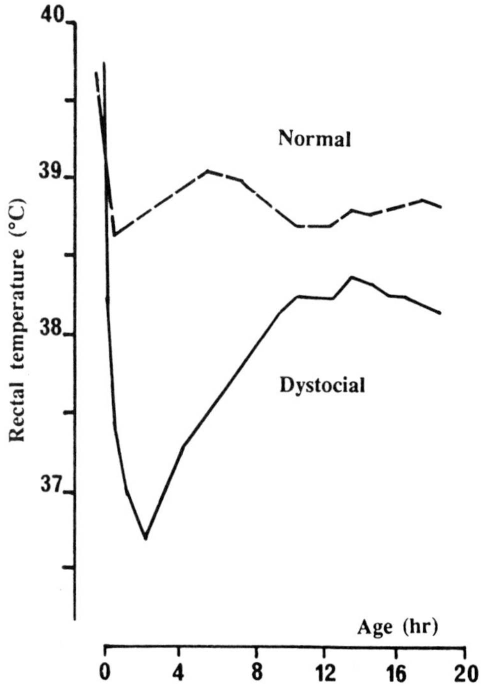

Fig. 1.3. Variations in rectal temperature of eutocial and dystocial calves maintained at an ambient temperature of 10°C (after Vermorel *et al.*, 1989b).

Hypothermia

Hypothermia appears as the ultimate result of excessive heat loss due to intolerable cold or depressed heat production due to starvation, with runt piglets, twin and triplet lambs being of particular risk. In lambs, Eales *et al.* (1982a) identified two periods of greater risk from hypothermia, first, soon after birth because of the intense thermolysis and, second, between 13 and 36 hours corresponding to a reduced thermogenesis due to depletion of body energy reserves. Hypothermia can be classified as moderate (T_r of about 35–36°C for both piglets and lambs) or deep (T_r of about 33°C for piglets) depending on the ability of the neonate to produce sufficient metabolic heat to regain euthermic state without assistance. Hypothermia delays the absorption of immunoglobulin in the calf (Olson *et al.*, 1980) and impairs carbohydrate metabolism in the newborn pig (Close *et al.*, 1981), through a suppression of insulin release leading to a decrease in glucose utilization rate. This impairment is of particular importance since the newborn pig relies almost exclusively on carbohydrate as a source of energy during the first hours of postpartum life. This suppression of insulin release is largely explained by the increased release of catecholamines and can be reversed by rewarming the hypothermic animals. Rewarming hypothermic piglets and lambs has proved to be successful in increasing the survival rate of the newborns. According to Eales *et al.* (1982b), 54% of rewarmed deeply hypothermic lambs (mean T_r of 29.6°C) survived to weaning.

Effects of Cold on Production Traits

Compared to pigs, there is little reliable information on the effects of cold on production in cattle, sheep and poultry. Therefore considerations of cold effects on production will mainly concern data obtained in pigs and to a lesser extent in the above-mentioned species.

Growth performance

Feed intake

An increase in feed intake is an efficient strategy used by cold-exposed animals to cope with the increased energy expenditure. This adjustment is rapid in pigs since feed intake is stabilized within 6 days following cold exposure (Verhagen *et al.*, 1988). According to data obtained by Rinaldo and Le Dividich (1991) in growing pigs (10 to 30 kg live weight (LW)) and by Nienaber and Hahn (1982) in finishing pigs (40 to 85 kg LW), voluntary feed intake (Y) is related to environmental temperature (T, °C) and LW by the following equations:

$$\text{Grower: } Y\ (\text{g/day}) = 1163 + 16.8T - 0.82T^2 \quad (R^2 = 0.73)$$

$$\text{Finisher: } Y\ (\text{kg/week}) = 10.65 + 0.074\text{LW} + 0.382T - 0.018T^2 \quad (R^2 = 0.67)$$

Below the environmental temperature corresponding to optimal performance (25°C for growers, 20°C for finishers) it can be calculated that a 1°C decrease in temperature within the 15–25°C range is associated with a 1.5% increase in feed intake in a 20 kg pig. Within the 10–20°C temperature interval the corresponding increase in a 60 kg pig amounts to 1.0%. Constant cold stress does not change the nycthemeral distribution of feed intake (Herpin *et al.*, 1987a), while a reduction in nocturnal temperature increases the proportion of feed consumed during the night time (Rinaldo *et al.*, 1989). However, the increase in food intake is due to an increased frequency of meals whereas their size is reduced (Nienaber *et al.*, 1990).

The increase in feed intake during cold exposure is reported to be associated with a reduction in feed digestibility in ruminants due to a reduction in retention time (Young and Degen, 1981). In pigs there is also some evidence that dietary energy digestibility is reduced in cold conditions (Phillips *et al.*, 1982). However, it should be noted that heat associated with the utilization of metabolizable energy is, at least partly, used for the thermoregulatory demand in the cold (Close, 1981; Noblet *et al.*, 1989) and this counterbalances to some extent the decline in digestibility.

Growth rate and feed efficiency

In *ad libitum* feeding conditions, it is generally accepted that the performance of growing–finishing pigs is optimal at about 20°C (Verstegen *et al.*, 1978; NRC, 1981).

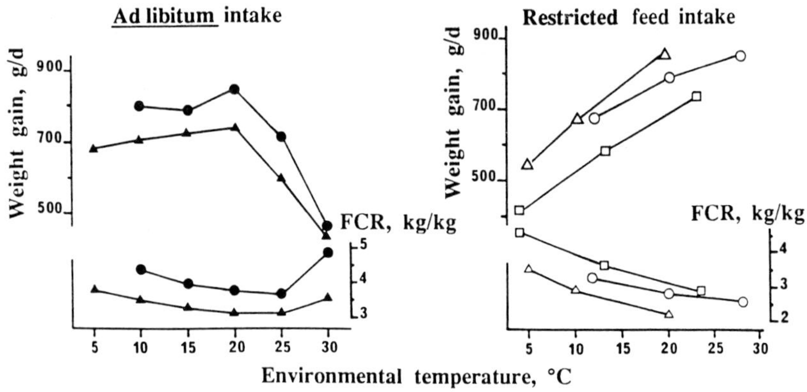

Fig. 1.4. Effect of environmental temperature on weight gain and food conversion ratio (FCR) of growing–finishing pigs in relation to level of feed intake. After Fuller and Boyne, 1971 (□); Bresk and Stolpe, 1982 (△); Nichols *et al.*, 1982 (●), Nienaber and LeRoy Hahn, 1983 (▲) and Le Dividich, 1991 (O).

Below this temperature, the feed conversion ratio decreases by about 0.044 1°C^{-1} decrease in temperature. Nevertheless this decline in feed efficiency is compensated by an increase in feed intake so that between 20 and 8°C growth rate remains practically constant (Fig. 1.4). However in equalized feeding conditions, growth rate decreases in the cold as a result of an increased heat loss leading to a reduction in energy available for growth. This decrease amounts to 11–13 g day^{-1} °C^{-1} coldness in growing pigs (10–30 kg) and to 15–17 g day^{-1} °C^{-1} coldness in heavier animals (for references, see Le Dividich, 1991). Alternatively, the additional daily amount of feed required to sustain similar growth rate in the cold as in the optimum temperature is in the range 11–14 g °C^{-1} coldness in growing pigs and 39–42 g °C^{-1} in heavier ones. Maintenance of performance by extra feed consumption or by heating the animal house depends on the relative costs of feed and heating energy.

Similar to data obtained in pigs, results obtained in *ad libitum* fed broilers (Howlider and Rose, 1987) indicated that below the optimal temperature (22°C) growth rate remains constant within a large temperature range (7–22°C), while feed conversion was increased by 10%. Nevertheless, unlike chickens, in turkeys weight gain was lower at 12°C compared to 18°C (Hurwitz *et al.*, 1980). This species difference may be attributed to differences in carcass fatness, in that broiler carcasses contains about 12% fat while that of the turkey at the age studied contains only 2–3% fat (Hurwitz *et al.*, 1980)

The effects of cold stress on the performance of growing cattle are less obvious. Webster *et al.* (1970) found the rate of gain for calves held outdoors with access to an open front shelter to be very similar to those housed in a warm environment, but feed consumption for the outdoor cattle was 20–30% higher. However, Milligan and Christison (1974) reported that in the very severe climatic conditions of winter in western Canada the rate of gain of feedlot steers was 22–33% lower compared to the other seasons and feed efficiency was reduced by 27–37%. Irrespective of the season a 10°C decline in environmental temperature resulted in a 0.14 kg day^{-1} reduction in weight gain and a 1100 kcal ME (metabolizable energy) increase in energy requirement per kg weight gain.

Abrupt changes in temperature may also affect animal performance. After an abrupt switch to cold, pig performance drops drastically for a while (Curtis, 1983). Similarly, in lambs, a change in temperature from 20°C to −5°C reduces daily weight gain by 30–40% and feed efficiency by 60–70% (Ames and Brink, 1977).

Milk and egg production

Milk production and composition in cows remains remarkably constant over a large range of temperature (0–25°C). However, cold conditions may reduce milk production depending on the availability of feed. According to Young (1981), milk yield begins to decline in dairy cows fed *ad libitum* at −4°C, and there is a marked depression in yield at −23°C. The absolute amount of milk fat produced was also reduced. However, percentage of fat in milk was increased.

In laying hens, most of the literature deals with high environmental temperatures compared to the usual practical conditions (Payne, 1967). Hvidsten and Haugen (1977) investigated the laying performance of Leghorn hens between 10°C and 28°C and concluded that decreasing temperature resulted in increased egg weight and also improved shell quality. Daily egg production was not influenced between 10°C and 22°C, however, as expected the amount of feed required per kg egg was higher at the lower temperature, amounting to 33.9 and 30.6 MJ ME at 10°C and 22°C, respectively.

Carcass composition and meat quality

The effects of cold on the carcass composition of pigs are closely dependent on the level of feed intake. In restricted feeding conditions, prolonged exposure to cold environment results in a leaner carcass (Close, 1981; Le Dividich, 1991). In pigs fed *ad libitum* or fed to achieve the same rate of weight gain in the cold as in thermoneutral conditions, little or no effect is observed (Verstegen *et al.*, 1984). Unlike pigs, broilers fed *ad libitum* are 'leaner' when cold exposed. According to Howlider and Rose (1987), weight of abdominal fat and total fat decreases linearly with the decline in environmental temperature.

Finally, prolonged cold exposure induces changes in the composition of fat depots and in muscle histology and this is clearly evidenced in pigs (Table 1.5). Low ambient temperature produces a high degree of unsaturated fatty acids which can influence the melting point and physical characteristics of the fat (Fuller *et al.*, 1974; Le Dividich *et al.*, 1987). Cold exposure affects also muscle histology. Percentage of type I fibres was dramatically increased in the red muscle (semi-spinalis muscle) while their mean cross-sectional area was slightly decreased (Lefaucheur *et al.*, 1991). Conversely, no effect was recorded in the white muscle (Longissimus dorsi muscle). However, the metabolic significance of change in the red muscle remains to be elucidated.

Table 1.5. Effect of cold on the unsaturation rate of backfat and characteristics of red muscle of pigs.

	Environmental temperature	
	Thermoneutral	Cold
Unsaturation rate of backfat	Control	Cold acclimated
– Iodine value	67.2	64.2
– Unsaturated fatty acids (%)	62.8	60.8
Red Muscle characteristics		
– Fibre types (%)		
I	42.6	58.4
II A	36.7	29.4
II B	12.2	20.8
– Fibre area (μm^2)		
I	3532	3125
II A	4099	3504
II B	4354	3614
– Relative area (%)		
I	38.6	55.4
II A	38.6	31.0
II B	22.8	13.7

Source: After Fuller *et al.*, 1974; Le Dividich *et al.*, 1987; Lefaucheur *et al.*, 1991.

Cold Adaptation

Cold adaptation covers a variety of physiological responses that better enable animals to face the deleterious effects of cold. In fact, cold adaptation results from co-ordinated mechanisms which decrease heat loss and/or increase heat production.

Decrease in heat loss

Decreasing heat loss is mainly achieved by behavioural, morphological and anatomical changes. Behavioural adjustments largely contribute to short- and mid-term adaptation to cold stress. These include mainly huddling, postural changes, piloerection (pigs) and feather fluffing (birds). Huddling of animals by reducing the effective exposed surface results in a substantial decrease in the LCT in pigs (Close *et al.*, 1981). In young chickens maintained at 20–25°C, this behaviour reduces heat production by about 15% (Hillman *et al.*, 1985) compared to individual chicks. Similarly Bouchillon *et al.* (1970) estimated that tripling the feather coat thickness by fluffing would reduce heat loss by 24% in chicken exposed at 10°C. Dairy cows stop grazing and orient their bodies so as to reduce exposure to sudden inclement climatic conditions (Molenat and Micol, 1982).

Prolonged exposure to cold stress induces morphological and anatomical changes in a number of mammalian species. Studies in pigs (Dauncey and Ingram, 1986; Le Dividich *et al.*, 1987; Rinaldo and Le Dividich, 1991) clearly indicated that pigs living in the cold (Fig. 1.5) had a rotund appearance, with a body of shorter

Fig. 1.5. Physical appearance of littermate pigs at 30 kg live weight. Animals were fed *ad libitum* and had been living at environmental temperatures of 12°C (left) and 25°C (right). Photo by J. Chevalier, INRA.

length, short limbs and tail, small ears, a short snout and hairy skin. The weight of internal organs, including the liver, empty digestive tract and kidneys, is increased as a result of an increase in feed intake. Interestingly there is a modification of body fat distribution (Table 1.6) between subcutaneous and deep body sites, in that the percentage of backfat is increased in the cold whereas the percentage of leaf fat is depressed (Le Dividich *et al.*, 1987). This shift of body fat from deep sites to external sites occurring in the cold-exposed pig, together with the above-mentioned mor-

Table 1.6. Effect of environmental temperature on body fat distribution in pigs (expressed in percent of control).

	Environmental temperature (°C)		
	12	20 (control)	28
Similar level of feed intake			
Carcass fat %	91	100	104
Backfat %	92	100	105
Leaf fat %	85	100	117
Similar rate of weight gain			
Carcass fat %	97	100	98
Backfat %	109	100	95
Leaf fat %	83	100	120

Source: After Le Dividich, 1991.

Table 1.7. Effect of environmental temperature on the activity of lipogenic enzymes in pig backfat.

	Environmental temperature	
	Thermoneutral	Cold
Outer layer		
AcCoA carb.	8.86	9.95
ME	430	640
G6 PDH	492	606
Inner layer		
AcCoA carb.	10.60	12.60
ME	489	708
G6 PDH	542	689

Source: After Lefaucheur *et al.*, 1991.
AcCoA carb. = acetylcoenzyme A carboxylase; ME = malic enzyme; G6 PDH = glucose-6-phosphate dehydrogenase. Enzyme activities are expressed as nmol substrate $min^{-1}g^{-1}$ of tissue.

phological changes, represent adaptations for minimizing heat loss in the cold through enhanced insulation and reduced surface exposure, respectively.

Increase in heat production

An increase in heat production can be achieved in three complementary ways: first, energy substrate mobilization and utilization; second, modification of hormonal status; and third, development of regulatory heat production mechanisms at the cellular level.

Energy substrates

The increased heat production in cold-acclimated animals must use up extra substrates for oxidation. This is achieved through an increase in food intake in *ad libitum* fed animals, and some additional adaptations in carbohydrate and free fatty acid (FFA) metabolism.

In pigs, it is well known that fat stores are easily mobilized during the post-weaning period, the mobilization being enhanced by low environmental temperatures (Le Dividich *et al.*, 1980). In addition, investigations performed at the tissue level have clearly shown an overall enhancement of lipid metabolism in cold conditions: stimulation of triglyceride uptake, lipogenesis (Table 1.7) and lipolysis (Fig. 1.6), especially in white adipose tissues (Herpin *et al.*, 1987a, 1991) suggesting an increased rate of FFA turnover. These results confirm the prominent part played by white adipose tissues in the control of substrate supply to the fatty acid utilizing tissues (liver, heart and muscle), according to the energy requirements of the animals. In cattle, the changes in blood lipids occurring during exposure to cold also support the idea that fat is a fuel for heat production (Thompson, 1976). In newborn calves, there is a progressive increase in plasma FFA concentration in the first 2 hours after

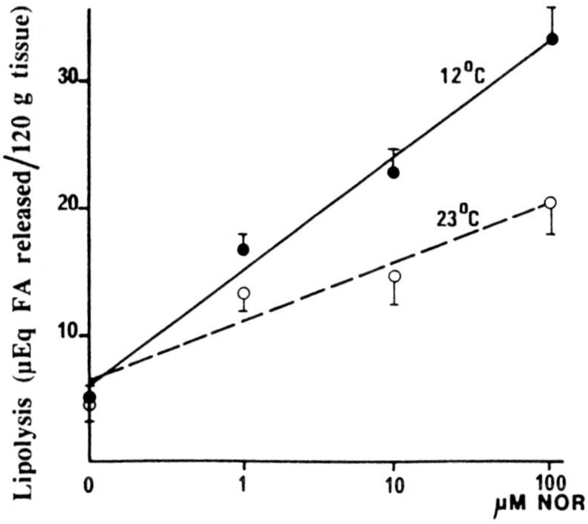

Fig. 1.6. Effect of cold acclimation on *in vitro* noradrenaline-induced lipolysis in white adipose tissue of the weaned piglet (after Herpin *et al.*, 1991).

birth if they are kept at 25°C, the increase being greater at 5°C. In beef cows, high levels of plasma FFA persist through the winter months in Canada (Dietz and Young, 1972). However, to our knowledge, no data are available on the effect of cold on lipid metabolism at the tissue level in cattle. Finally, in birds, the enhancement of heat production induced by cold acclimation is associated with a stimulated lipolysis, a decreased respiratory quotient (Freeman, 1966) and an increased level of FFA (Duchamp, 1990).

It is generally recognized that carbohydrate contributes significantly to the overall increase in heat production induced by cold exposure. This is clearly evidenced in the newborn pig by the mobilization of muscle glycogen and the enhanced glucose turnover (Duée *et al.*, 1988). In older pigs we have also found a 20% increase in glucose turnover in cold acclimation conditions (Herpin *et al.*, 1987b). The effect of cold on plasma glucose concentrations has also been studied in cattle but the results are quite variable. Khan *et al.* (1970) reported no effect of cold exposure on plasma glucose in newborn calves whereas Eales and Small (1986) found increased glucose level during summit metabolism in newborn lambs. Vermorel *et al.* (1983) calculated from RQ that glucose may contribute 15% to 30% of the thermogenesis in newborn calves exposed at 10°C. Finally Dietz and Young (1972) also measured glucose levels of pregnant beef cows during winter and noted an increase only during the first severe cold spell. Following cold exposure domestic and wild birds always show a marked increase in carbohydrate utilization with an enhanced hepatic glycogenolysis (Freeman, 1966). Thus cold exposed birds appear to use readily available glycogen reserves in the skeletal muscles as a major source of fuels (Hissa, 1988).

Hormonal changes

The major endocrine glands implicated in the responses of homeotherms to cold exposure are the adrenal, the thyroid and the pancreas. Response to chronic cold exposure represents the manifestation of their synergistical action. It must be noted that whereas thyroid hormones and catecholamines may be the agents which mediate stimulation of thermogenesis in pigs, cattle and sheep, glucagon has been demonstrated to have a powerful calorigenic effect in birds.

Increased thyroid hormone secretion and thyroid activity during cold acclimation has been reported in pigs (Herpin *et al.*, 1987a), cattle (Young, 1981) and birds (Chaffee and Roberts, 1971). In birds, as in mammals, thyroid hormone involvement in the acclimation process is reflected by an elevated thermoneutral resting metabolic rate and by a considerably greater calorigenic capacity (summit metabolism). This latter effect could be partly explained by a potentiation of the calorigenic effect of catecholamines by thyroxine. But there is still considerable uncertainty as to how thyroid hormones function at the cellular level to increase metabolic rate. However, their most consistent effect in mammals has been reported to be the control of gene expression, and especially the control of key enzymes involved in mitochondrial function (Oppenheimer *et al.*, 1977), and the stimulation of energy breakdown through the Na^+/K^+-ATPase pump (Ismail-Beigi and Edelman, 1971). They have also been suggested to stimulate mitochondria biogenesis (Mutvei *et al.*, 1989). These effects are not yet established in birds and there is no information on the role of thyroid hormones on avian oxidative metabolism *in vitro*. Finally, it must be noticed that changes in thyroid activity in cold-acclimated animals appear also to be interrelated with the observed changes in food intake and the modification of tissue sensitivity to T_3 (number and affinity of nuclear T_3 receptors) (Dauncey, 1990).

Within catecholamines, adrenaline has been generally associated with the short-term responses induced by cold stress or acute cold exposure. But during prolonged cold exposure in mammals, the increased involvement of the sympathetic nervous system will ensure higher levels of circulating noradrenaline. This is well illustrated in cold-acclimated piglets (Fig. 1.7) with a fivefold increase in noradrenaline excretion in urine after 10 days spent in the cold and a much reduced variation for adrenaline (Herpin *et al.*, 1991). More generally, in sheep, in goats and probably in all domestic animals urinary excretion of catecholamines increases in the cold to a degree proportional to the intensity of the stimulus (Webster, 1976). In mammals, catecholamine release during cold exposure is associated partly with cardiovascular adjustments leading to maximum heat production (increased blood flow to heat-producing organs or tissues) and minimum heat loss (peripheral vasoconstriction), and partly with direct actions on thermogenesis. The principal function of these hormones is therefore to mobilize energy-rich substrates (FFA and glucose) for catabolism. Finally, noradrenaline also induces regulatory non-shivering thermogenesis (NST) in those species which possess brown adipose tissue (BAT).

In birds, catecholamines have been shown to have no or very little calorigenic effect (Freeman, 1970), probably because they are not the main hormones involved in substrate mobilization in this class. However, their effects on blood pressure and peripheral vasoconstriction cannot be completely ruled out (Hissa, 1988), since this will ensure adequate blood flow, energy and oxygen supply to active tissues in cold

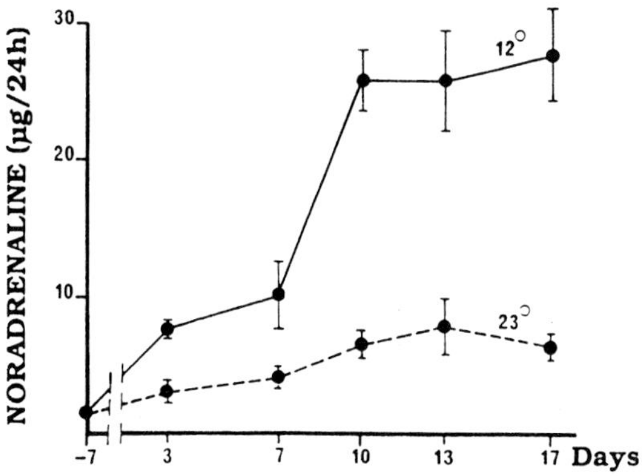

Fig. 1.7. Effect of cold acclimation on urinary noradrenaline excretion in the weaned piglet (after Herpin *et al.*, 1991).

conditions. The main hormone involved in cold acclimation in birds is glucagon. It exerts powerful lipolytic, glycogenolytic and calorigenic effects in ducks, geese and hens (Langslow *et al.*, 1970; Barré, 1984; Duchamp, 1990), increasing heat production up to 70% in quail (Krimphove and Opitz, 1975). Its involvement in cold acclimation mechanisms has been clearly established recently since plasma levels increase in cold-acclimated ducklings (Barré *et al.*, 1987) and since a chronic glucagon treatment is able to reproduce most of the physiological and biochemical changes induced by cold acclimation (Duchamp, 1990).

Location and mechanisms of regulatory heat production

Heat is generated as a product of metabolic reactions but it is often convenient to distinguish between obligatory and regulatory heat production mechanisms. Obligatory thermogenesis refers only to heat generated via reactions necessary to maintain cell integrity and steady state condition of the animal, i.e. basal metabolism. All tissues contribute to obligatory heat production. On the contrary, regulatory thermogenesis, the heat generated when the animal is exposed to various stimuli (e.g. cold) is attributable to only a few tissues, mainly muscle and brown adipose tissue (Foster and Frydman, 1978). The ability to develop regulatory thermogenesis and the relative importance of shivering and non-shivering thermogenesis (NST) in cold acclimation are different between species and are highly affected by the age of the animal.

Shivering thermogenesis is defined as an involuntary rhythmic contraction of skeletal muscle myofibrils involving no voluntary movements or external work. All the major domestic animals (pigs, cattle, sheep and poultry) are able to produce heat by shivering and all substrates (protein, carbohydrate and fat) can support shivering thermogenesis, fats being preferentially utilized. Electromyographic recordings indicate the presence of shivering but estimation of its quantitative importance is still difficult. It is generally accepted that during cold acclimation shivering is progressively replaced by NST. However, it seems now that the thermogenic capacity of shivering could also be adaptable: for a similar electromyographic activity the amount of heat produced by shivering is higher after cold acclimation than during a cold stress in mammals (Ivanov, 1989) and birds (Barré *et al.*, 1985).

Regulatory NST reflects the heat generated in metabolic reactions associated with processes other than shivering. The identity of the site of NST has been a subject of controversy for many years but it seems now that both BAT and muscles contribute to regulatory NST – BAT being the main site in small mammals (Foster and Frydman, 1978). Among domestic animals, BAT has been found in newborn lambs, calves and rabbits which is of great interest since it facilitates neonatal survival in cold conditions. Very soon after birth brown adipocytes are changed into white adipocytes (Casteilla *et al.*, 1988) and lose their heat-producing ability. However, in cold exposed newborn calves, the changes of brown adipose tissue into white adipose tissue is delayed and might be part of an acclimation process. On the other hand, BAT is present neither in pigs (Trayhurn *et al.*, 1989) not in domestic birds (Duchamp, 1990). However, during chronic cold exposure these two species are able to develop some non-shivering heat-producing mechanisms in skeletal muscles: loose coupling between oxidation and phosphorylation in mitochondria, (pigs, Herpin and Barré, 1989; ducklings, Duchamp, 1990), and enhancement of Na^+/K^+-ATPase pump activity (Herpin *et al.*, 1987b).

Conclusion

Cold results in various effects on domestic farm animals. Among classes of livestock, the neonates are the most susceptible to cold but thermoregulation is well developed in most neonates and improves rapidly with age. However, maintenance of high metabolic rate is dependent on the availability of energy substrates, which emphasizes the importance of an early intake of sufficient colostrum in mammalian neonates. Birthweight, genotype and conditions of parturition are important components of cold resistance. In heavier animals, the effects of cold are evident on growth, feed efficiency and carcass composition (pig). All domestic animals exhibit specific metabolic and physiological adaptative responses to chronic cold exposure. However, heat is the end product of oxidative metabolism and hence the energy available for animal products is reduced.

References

Alexander, G. (1962) Temperature regulation in the newborn lamb. V. Summit metabolism. *Australian Journal of Agriculture Research* **13**, 100–121.

Alexander G. (1974) Birth weight of lambs: influences and consequences. In: *Size at Birth.* Ciba Foundation Symposium, Elsevier, Amsterdam, pp. 215–246.

Alexander, G. and Williams, D. (1966) Teat-seeking activity in the new-born lamb: the effects of cold. *Journal of Agricultural Science (Cambridge)* **67**, 181–189.

Ames, D.R. and Brink, D.R. (1977) Effect of temperature on lamb performance and protein efficiency ratio. *Journal of Animal Science* **44**, 136–140.

Barré, H. (1984) *Recherche sur la thermogenèse de régulation chez l'oiseau. Etude expérimentale chez le poussin de Manchot Royal et le caneton de Barbarie*. Thèse d'Etat, Université Pierre et Marie Curie, Paris.

Barré, H., Geloen, A., Chatonnet, J., Dittmar, A. and Rouanet, J.L. (1985) Potential muscular thermogenesis in cold-acclimated Muscovy ducklings. *American Journal of Physiology* **249**, R533–R538.

Barré, H., Cohen-Adad, F. and Rouanet, J.L. (1987) Two daily glucagon injections induce nonshivering thermogenesis in Muscovy ducklings. *American Journal of Physiology* **252**, E616–E620.

Bernard, E. and Hull, D. (1964) The effect of the environmental temperature on growth of new-born rabbits reared in incubators. *Biology of the Neonate* **7**, 172–178.

Bouchillon, C.W., Reece, F.N. and Deaton, J.W. (1970) Mathematical modeling of thermal homeostasis in a chicken. *Transactions of the American Society of Agricultural Engineers* **13**, 648–652.

Bresk, R. and Stolpe, J. (1982) Fattening performance tests for assessment of critical temperature of swine. *Monatshefte für Veterinaermedizin* **37**, 374–380.

Casteilla, L., Champigny, O., Bouillaud, F., Robelin, J. and Ricquier, D. (1988) Sequential changes in the expression of mitochondrial protein mRNA during the development of BAT in bovine and ovine species: sudden occurrence of uncoupling mRNA during embryogenesis and its disappearance after birth. *Biochemical Journal* **257**, 665–671.

Chaffee, R.R.J. and Roberts, J.C. (1971) Temperature acclimation in birds and mammals. *Annual Review of Physiology* **33**, 155–202.

Close, W.H. (1981) The climatic requirements of the pig. In: Clark, J. (ed), *Environmental Aspects of Housing for Animal Production*. Butterworths, London, pp. 149–166.

Close, W.H., Le Dividich J. and Duée, P.H. (1981) Influence of environmental temperature on glucose tolerance and insulin response in the new-born piglet. *Biology of the Neonate* **47**, 84–91.

Curtis, S.E. (1983) *Environmental Management in Animal Agriculture*. Iowa State University Press, Ames.

Curtis, S.E., Heidenreich, C.J. and Harrington, R.B. (1967) Age dependent changes of thermostability in neonatal pigs. *American Journal of Veterinary Research* **31**, 576–587.

Dauncey, M.J. (1990) Thyroid hormones and thermogenesis. *Proceedings of the Nutrition Society* **44**, 203–215.

Dauncey, M.J. and Ingram, D.L. (1986) Acclimatization to warm or cold temperatures and the role of food intake. *Journal of Thermal Biology* **11**, 89–93.

Dietz, W. and Young, B.A. (1972) Plasma metabolites in beef cows over winter. *Proceedings of the Western Section of the American Society of Animal Science* **23**, 265–269.

Duchamp, C. (1990) *Démonstration d'une thermogénèse sans frisson d'origine musculaire chez l'oiseau*. Thèse de doctorat, Université Claude Bernard, Lyon.

Duée, P.H., Pégorier, J.P., Le Dividich, J. and Girard, J. (1988) Metabolic and hormonal response to acute cold exposure in newborn pig. *Journal of Developmental Physiology* **10**, 371–381.

Eales, F.A. and Small, J. (1980) Summit metabolism in newborn lambs. *Research in Veterinary Science* **29**, 211–218.

Eales, F.A. and Small, J. (1981) Effect of colostrum on summit metabolic rate in Scottish Blackface lambs at five hours old. *Research in Veterinary Science* **30**, 266–269.

Eales, F.A. and Small, J. (1986) Plasma hormone concentrations in newborn Scottish Blackface lambs during basal and summit metabolism. *Research in Veterinary Science* **40**, 339–343.

Eales, F.A., Gilmour, J.S., Barlow, R.M. and Small, J. (1982a) Causes of hypothermia in 89 lambs. *Veterinary Record* **110**, 118–120.

Eales, F.A., Small, J. and Gilmour, J.S. (1982b) Resuscitation of hypothermic lambs. *Veterinary Record* **110**, 118–120.

English, P.R. and Morrison, V. (1984) Causes and prevention of piglet mortality. *Pigs News and Information* **5**, 369–376.

Foley, C.W., Seerley, R.W., Hansen, W.J. and Curtis, S.E. (1971) Thermoregulatory responses to cold environment by neonatal wild and domestic piglets. *Journal of Animal Science* **32**, 926–929

Foster, D.O. and Frydman, L. (1978) Brown adipose tissue: the dominant site of nonshivering thermogenesis in the rat. In: Girardier, L. and Seydoux, J. (eds), *Effectors of Thermogenesis*. Birkhaüser Verlag, Basel, pp. 147–151.

Freeman, B.M. (1966). The effects of cold, noradrenaline and adrenaline upon the oxygen consumption and carbohydrate metabolism of the young fowl (*Gallus domesticus*). *Comparative Biochemistry and Physiology* **18**, 369–382.

Freeman, B.M. (1970) Thermoregulatory mechanisms of the neonate fowl. *Comparative Biochemistry and Physiology* **33**, 219–230.

Freeman, B.M. (1988) The domestic fowl in biomedical research: physiological effects of the environment. *World's Poultry Science Journal* **44**, 41–60.

Fuller, M.F. and Boyne A.W. (1971) The effects of environmental temperature on the growth and metabolism of pigs given different amounts of food. I. Nitrogen retention. *British Journal of Nutrition* **25**, 259–272.

Fuller, M.F. Duncan, W.R.H. and Boyne A.W. (1974) Effects of environmental temperature on the degree of unsaturation of depot fats of pigs given different amounts of food. *Journal of the Science of Food and Agriculture* **25**, 205–210.

Herpin, P. and Barré, H. (1989) Loose-coupled subsarcolemmal mitochondria from rhomboideus muscle in cold-acclimated piglets. *Comparative Biochemistry and Physiology* **92**, 59–65.

Herpin, P., Bertin, R., Le Dividich, J. and Portet, R. (1987a) Some regulatory aspects of thermogenesis in cold-exposed piglets. *Comparative Biochemistry and Physiology* **87**, 1073–1081.

Herpin, P., McBride, B.W. and Bayley, H.S. (1987b) Effect of cold exposure on energy metabolism in the young pig. *Canadian Journal of Physiology and Pharmacology* **65**, 236–245.

Herpin, P., Bertin, R., De Marco, F., Portet, R. and Le Dividich J. (1991) Adipose tissue lipolytic activity and urinary catecholamine excretion in cold-acclimated piglets. *Canadian Journal of Physiology and Pharmacology* **69**, 362–368.

Hillman, P.E., Scott, N.R. and Van Tienhoven, A. (1985) Physiological responses and adaptations to hot and cold environments. In: Yousef, M.K. (ed.), *Stress Physiology in Livestock*. Vol. III, Poultry. CRC Press, Boca Raton, Florida, pp. 2–71.

Hissa, R. (1988) Controlling mechanisms in avian temperature regulation: a review. *Acta Physiological Scandinavica* **132** (Suppl. 567), pp. 1–145.

Holmes, C.W. and Close, W.H. (1977) The influence of climatic variables on energy metabolism and associated aspects of productivity in pigs. In: Haresign, W., Swan, H. and Lewis, D. (eds), *Nutrition and the Climatic Environment*. Butterworths, London, pp. 51–74.

Houston, D.C. and Maddox, J.G. (1974) Causes of mortality among young Scottish Blackface lamb. *Veterinary Record* **95**, 575.

Howlider, M.A.R. and Rose, S.P. (1987) Temperature and growth in broilers. *World's Poultry Science Journal* **43**, 228–237.

Hurwitz, S., Weiselberg, M., Eisner, U., Bartov, I., Riesenfeld, G., Sharvit, M., Niv, A. and Bornstein, S. (1980) The energy requirements and performance of growing chickens and turkeys as affected by environmental temperature. *Poultry Science* **59**, 2290–2299.

Hvidsten, H. and Haugen, A.E. (1977) Influence of temperature on laying performance. *Poultry International* **16**, 62–66

Ismail-Beigi, F. and Edelman, I.S. (1971) The mechanism of the calorigenic action of thyroid hormones. Stimulation of Na^+–K^+-activated adenosinetriphosphatase activity. *Journal of General Physiology* **57**, 710–722.

Ivanov, K.P. (1989) Thermoregulatory chemical metabolism and muscle work efficiency. *Journal of Thermal Biology* **14**, 1–18.

Jordan, W.A., Lister, E.E. and Commeau, J.E. (1969) Outdoor versus wintering of fall calving beef cows and their calves. *Canadian Journal of Animal Science* **49**, 127–129.

Kampen, M., van (1981) Thermal influences on poultry. In: Clark, J. (ed.), *Environmental Aspects of Housing and Animal Production*. Butterworths, London, pp. 131–147.

Khan, M.A., Dickson, W.M. and Meyers, K.M. (1970) The effect of low environmental temperature on plasma corticosteroid and glucose concentrations in the newborn calf. *Journal of Endocrinology* **48**, 355–363.

Krimphove, M. and Opitz, K. (1975) Untersuchungen der calorigenen wirkung von glucagon. *Archives Internationales de Pharmacodynamie et de Thérapie* **216**, 328–350.

Langslow, D.R., Butler, E.J., Hales, C.N. and Pearson, A.W. (1970) The response of plasma insulin, glucose and non-esterified plasma fatty acids to various hormones, nutrients and drugs in the domestic fowl. *Journal of Endocrinology* **46**, 243–260.

Le Dividich, J. (1991) Influenza della temperatura ambiente sulles performances del suino. *Selezione Veterinaria* **32**, 191–207.

Le Dividich, J. and Noblet, J. (1981) Colostrum intake and thermoregulation in the neonatal pig in relation to environmental temperature. *Biology of the Neonate* **40**, 167–174.

Le Dividich, J. and Noblet, J. (1984) Effect of colostrum intake on metabolic rate and plasma glucose in the neonatal pig in relation to environmental temperature. *Biology of the Neonate* **46**, 98–104.

Le Dividich, J., Vermorel, M., Noblet, J., Bouvier, J.C. and Aumaitre, A. (1980) Effects of environmental temperature on heat production, energy retention, protein and fat gain in early weaned piglets. *British Journal of Nutrition* **44**, 313–323.

Le Dividich, J., Noblet, J. and Bikawa, T. (1987) Effect of environmental temperature and dietary energy concentration on the performance and carcass characteristics of growing–finishing pigs fed to equal rate of gain. *Livestock Production Science* **17**, 235–246.

Le Dividich, J., Mormède, P., Catheline, M. and Caritez, J.C. (1991) Body composition and cold resistance of the neonatal pig from European (Large White) and Chinese (Meishan) breeds. *Biology of the Neonate* **59**, 268–277.

Lefaucheur, L., Le Dividich, J., Mourot, J., Monin, G., Ecolan, P. and Krauss, D. (1991) Influence of environmental temperature on growth and adipose tissue metabolism and meat quality in swine. *Journal of Animal Science* **69**, 2844–2854.

Mellor, D.J. and Cockburn, F. (1986) A comparison of energy metabolism in the new-born infant, piglet and lamb. *Quarterly Journal of Experimental Physiology* **71**, 361–379.

Menissier, F. and Petit, M. (1984) Poids et vitalité des veaux à la naissance: leurs implications zootechniques. In: Jarrige, R. (ed.). *Physiologie et pathologie périnatales chez les animaux de ferme*. INRA, Paris, pp. 279–308.

Milligan, J.D. and Christison, G.I. (1974) Effects of severe winter conditions on performance of feedlot steers. *Canadian Journal of Animal Science* **54**, 605–610.

Molenat, G. and Micol, D. (1982) Influence directe et indirecte des facteurs climatiques sur le comportement de l'animal et l'ingestion d'herbe. In: *Actions du climat sur l'animal au pâturage*. INRA, pp. 149-157.

Mount, L.E. (1968) *The Climatic Physiology of the Pig*. Edward Arnold, London.

Mutvei, A., Husman, B., Anderson, G. and Nelson, B.D. (1989) Thyroid hormone and not growth hormone is the principal regulator of mammalian mitochondrial biogenesis. *Acta Endocrinologia* **121**, 223–228.

National Research Council. (1981) *Effects of Environment on Nutrient Requirements of Domestic Animals*. National Academy of Sciences, National Research Council, Washington, DC.

Nichols, D.A., Ames, D.R. and Hines, R.A. (1982) Effect of temperature on performance and efficiency of finishing swine. *Proceedings of the 2nd International Livestock Environment Symposium*. Ames, Iowa. pp. 376–379.

Nienaber, J.A. and LeRoy Hahn, G. (1982) Heat production and feed intake of ad-libitum-fed growing swine as affected by temperature. *ASAE Paper No. 82–4065*. St Joseph, Michigan.

Nienaber, J.A. and LeRoy Hahn, G. (1983) Performance of growing finishing swine in response to the thermal environment. *ASAE Paper No. 83–137*. St Joseph, Michigan.

Nienaber, J.A., McDonald, T.P., LeRoy Hahn, G. and Chen, Y.R. (1990) Eating dynamics of growing–finishing swine. *Transactions of the American Society of Agricultural Engineers* **36**, 2011–2018.

Noblet, J. and Le Dividich, J. (1981) Energy metabolism of the new-born pig during the first 24 h of life. *Biology of the Neonate* **40**, 175–182.

Noblet, J., Dourmad, J.Y., Le Dividich, J. and Dubois, S. (1989) Effect of ambient temperature and addition of straw in the diet on energy metabolism in pregnant sow. *Livestock Production Science* **21**, 309–324.

Obst, J.M. and Day, H.R. (1968) The effect of inclement weather on mortality of Merinos and corrieadale lambs on Kangaloo island. *Proceedings of the Australian Society of Animal Production* **7**, 239–242.

Okamoto, M., Robinson, J.B., Christopherson, R.J. and Young, B.A. (1986) Summit metabolism of newborn calves with and without colostrum feeding. *Canadian Journal of Science* **66**, 937–944.

Olson, D.P., Papasian, C.J. and Ritter, R.C. (1980) The effects of cold stress on neonatal calves, II. Absorption of colostral immunoglobulins. *Canadian Journal of Comparative Medicine* **44**, 19–23.

Oppenheimer, J.H., Silva, E., Schwartz, H.C. and Surks, M.I. (1977) Stimulation of hepatic mitochondrial α-glycerophosphate deshydrogenase and malic enzyme by L-triidothyronine. *Journal of Clinical Investigation* **59**, 517–527.

Parker, R.O., Williams, P.E.V., Aherne, F.X. and Young, B.A. (1980) Serum concentration changes in protein, glucose, urea, thyroxine and triidothyronine and thermostability of neonatal pigs farrowed at 25 and 10°C. *Canadian Journal of Animal Science* **60**, 503–509.

Payne, C.G. (1967) The influence of environmental temperature on egg production: a review. In: Carter, T.C. (ed.), *Environmental Control in Egg Production*. Oliver and Boyd, Edinburgh, pp. 40–54.

Phillips, P.A., Young, D.B. and McQuitty, J.R. (1982) Live weight, protein deposition and digestibility responses in growing pigs exposed to low temperature. *Canadian Journal of Animal Science* **62**, 95–108.

Rinaldo, D. and Le Dividich, J. (1991) Assesment of optimal temperature for performance and chemical body composition. *Livestock Production Science* **29**, 61–75.

Rinaldo D., Salaün, M.C. and Le Dividich J. (1989) Influence d'une réduction de la température ambiante ou d'un abaissement nocturne de la température ambiante sur les performances du porcelet sevré. *Journées Recherche Porcine en France* **21**, 239–244.

Robelin, J. (1986) *Composition corporelle des bovins: évolution au cours du développement et différences entre races*. Thése de doctorat, Université de Clermont-Ferrand.

Slee, J. (1981) A review of genetic aspects of survival and resistance to cold in newborn lambs. *Livestock Production Science* **8**, 419–429.

Slee, J. and Stott, A.W. (1986) Genetic selection for cold resistance in scottish Blackface lambs. *Animal Production* **43**, 397–404.

Stott, A.W. and Slee, J. (1987) The effects of litter size, sex, age, body weight, dam age and genetic selection for cold resistance on the physiological responses to cold exposure of

Scottish Blackface lambs in a progressively cooled water bath. *Animal Production* **45**, 477–491.

Theriez (1982) La mortalité des agneaux: point de vue du zootechnicien. *7èmes Journées de la Recherche Ovine et Caprine*. INRA-ITOVIC, Paris. pp. 1–17.

Thompson, G.E. (1976) Effect of cold on energy metabolism of cattle. In: Tromp, S.W. (ed.), *Progress in Biometeorology. Division B: Progress in Animal Biometeorology*. Vol. 1, Part. 1. Swets and Zeitlinger, B.V. Amsterdam, pp. 210–217.

Thompson, G.E. (1983) The intake of milk by suckled, newborn lambs and the effects of twinning and cold exposure. *British Journal of Nutrition* **50**, 151–156.

Trayhurn, P., Temple, N.J. and Van Aerde, J. (1989) Evidence from immunoblotting studies on uncoupling protein that brown adipose tissue is not present in the domestic pig. *Canadian Journal of Physiology and Pharmacology* **67**, 1480–1485.

Verhagen, J.M.F., Geers, R. and Verstegen, M.W.A. (1988) Time taken for growing pigs to acclimate to change in ambient temperature. *Netherlands Journal of Agricultural Science* **36**, 1–10.

Vermorel, M. (1987) Effects of climatic conditions on energy metabolism and performance of calves. In: Verstegen, M.W.A. and Henken, A.M. (eds), *Energy Metabolism in Farm Animals. Effects of housing, stress and disease*. Martinus Nijhoff Publishers, Dordrecht, Boston and Lancaster, pp. 180–198.

Vermorel, M., Dardillat, C., Vernet, J., Saido and Demigne, C. (1983) Energy metabolism in the newborn calf. *Annales de Recherches Vétérinaires* **14**, 382–389.

Vermorel, M., Vernet, J., Dardillat, C., Saido, Demigne, C. and Davicco, M.J. (1989a) Energy metabolism and thermoregulation in the newborn calf; variations during the first day of life and differences between breeds. *Canadian Journal of Animal Science* **69**, 103–111.

Vermorel, M., Vernet, J., Dardillat, C., Saido, Demigne, C. and Davicco, M.J. (1989b) Energy metabolism and thermoregulation in the newborn calf: effect of calving conditions. *Canadian Journal of Animal Science* **69**, 113–122.

Verstegen, M.W.A., Brascamp, E.W. and van der Hel, W. (1978) Growing and fattening of pigs in relation to temperature of housing and feeding level. *Canadian Journal of Animal Science* **58**, 1–13.

Verstegen, M.W.A., Brandsma, H.A., Mateman, G. and van der Hel, W. (1984) Effects of cold thermal environment on feed requirements, growth rate and slaughter quality in pigs. *Archiv für Experimentale Veterinaermedizin* **38**, 431–438.

Webster, A.J.F. (1976) Effect of cold on endocrine responses of ruminants and other domestic animals. In: Tromp, S.W. (ed.), *Progress in Biometeorology. Division B: Progress in Animal Biometeorology*. Vol. 1. Part 1. Swets and Zeitlinger, B.V. Amsterdam. pp. 33–41.

Webster, A.J.F., Chlumecky, J. and Young, B.A. (1970) Effects of cold environments on energy exchanges of young beef. *Canadian Journal of Animal Science* **50**, 89–100.

Wekstein, D.R. and Zolman, J.F. (1967) Homeothermic development of the young chick. *Proceedings of the Society for Experimental Biology and Medicine* **125**, 294–297.

Young, B.A. (1981) Cold stress as it affects animal production. *Journal of Animal Science* **52**, 154–163.

Young, B.A. and Degen, A.A. (1981) Thermal influences on ruminants. In: Clark, J.A. (ed.), *Environmental Aspects of Housing for Animal Production*. Butterworths, London, pp. 167–180.

2 Heat Stress

A. Alnaimy M. Habeeb,[1] I. Fayaz M. Marai[2] and T.H. Kamal[1]

[1]*Bioclimatology Unit, Radiobiology Department, Nuclear Research Centre, Atomic Energy Authority, Cairo, Egypt:* [2]*Department of Animal Production, Faculty of Agriculture, Zagazig University, Zagazig, Egypt*

Summary

When European breeds of farm animals are introduced to tropical and subtropical countries, they are faced with many problems relating to the hot climate, particularly conditions of heat stress. A vast array of physiological and biochemical changes are induced in such animals. Appetite, feed intake, feed efficiency and feed utilization are impaired. Normal thermoregulatory reactions, i.e. respiration, sweating and rectal temperature, are increased causing disturbances in the metabolism of water (increased water intake and body water content), energy, protein and minerals (negative nitrogen, energy and mineral balances). These disturbances also occur in enzymatic reactions, e.g. an increase in transaminase enzyme activities, etc., and hormonal secretion where there is a decrease in insulin, T_4, T_3 and aldosterone and an increase in cortisol. Such disturbances lead to a depression in some of the blood metabolites, i.e. glucose, total protein, total lipids, cholesterol, etc. The final result of all these changes is the impairment of growth, milk yield and reproduction. The possible mechanisms of these changes are reviewed.

Introduction

Heat is the major constraint on animal productivity in arid zones. Growth, milk production and reproduction are impaired as a result of the drastic changes in biological functions caused by heat stress.

This chapter presents a review of productive, reproductive, physiological and biochemical changes under heat stress. At the same time, the possible mechanisms that may take place for inducing the biological changes in heat-stressed animals are illustrated. Most of the research referred to uses ruminants, and particularly European cattle, rather than monogastric animals, as the endogenous heat load is considerably greater in the former. Thus while hot countries more often have fibrous feeds available to feed to ruminants, the additional heat load on the animal often reduces productivity substantially unless the environment is ameliorated.

noregulation Mechanisms and Animal Response

In the wide range of environmental temperatures that comprise the thermoneutral zone, animals maintain their heat balance through vasomotor control by regulating the amount of blood flowing through the cutaneous vessels by either vasodilation or vasoconstriction. Vasodilation stimulates the pilomotor centre to flatten the hair cover to allow better heat dissipation through conduction, convection and radiation (sensible means). In addition, some heat is lost by evaporation through the lungs and also from the surface of the body as a result of diffusion of water through the skin as insensible perspiration.

As the ambient temperature increases, the amount of water lost by insensible perspiration increases slightly until active sweating starts. The circulation of blood transfers heat from the core to the periphery. Excess heat is dissipated by evaporation of sweat as a protective mechanism of the body against overheating.

At the same time, respiration rate or panting increases and becomes much more shallow to permit an efficient ventilation of the upper respiratory tract, without undue overventilation of the lungs themselves (Shafie and Abdelghany, 1978). When the animal cannot sustain homeothermy, it reduces the heat production using internal physiological means to help in re-establishment of the thermal balance. Feed consumption and thermogenic hormone secretion decrease to lower the basal metabolism resulting in a rapid decline in productivity. If all these physiological mechanisms fail to balance the excessive heat load, the body temperature rises and the animal enters the acute phase of heat stress. If these systems still fail to stop the elevation in body temperature, the animal succumbs with heat stroke and dies.

The acute phase normally occurs within a few days of the animal encountering the high temperature and is accompanied by a rapid decline in productivity. If the ambient conditions remain the same and are not severe (or the exposure to stressful heat is intermittent), a rapid acclimation to the conditions takes place and the animal returns to the chronic phase of heat stress. This results in better productivity which eventually stabilizes at a level greater than that observed during the acute phase, but lower than the normal level (Fig. 2.1). Some of these changes may occur rather rapidly (days), whereas others may require a longer period (weeks). When more favourable climatic changes occur, the performance is improved and a compensatory response occurs quite often, resulting in a return to productive levels above the normal ones (Ames and Ray, 1983).

Appetite and Feed Utilization

The most important reaction to heat exposure is the depression in feed consumption. High environmental temperature stimulates the peripheral thermal receptors to transmit suppressive nerve impulses to the appetite centre in the hypothalamus (Fig. 2.2) causing the decrease in feed consumption. Thus fewer substrates become available for enzymatic activities, hormone synthesis and heat production, which minimizes thermal load (Kamal, 1975). Exposure to severe heat suppresses the production by the hypothalamic centre of hormone releasing factors causing a

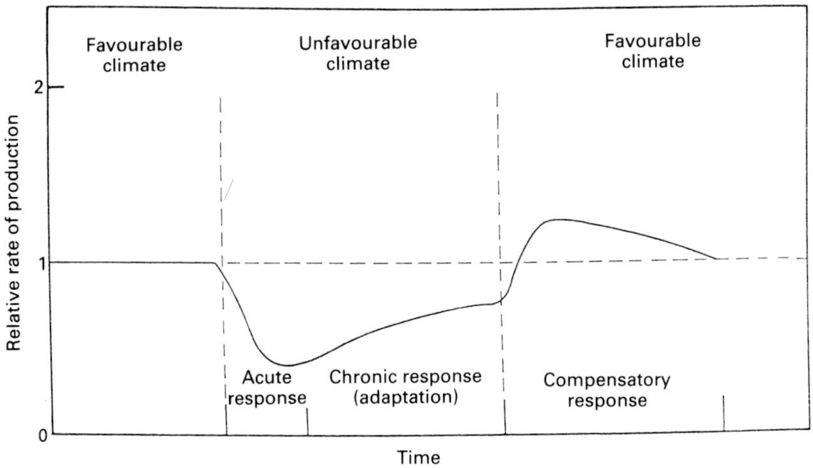

Fig. 2.1. Theoretical animal response to periodic climatic change. Solid line represents the animal's response to changing climatic conditions relative to favourable climate (1 on vertical scale) (after Ames and Ray, 1983).

decrease in pituitary hormonal secretion (Johnson, 1974), insulin and possibly thyroxine (Habeeb, 1987). The metabolic pathways slow down, causing drastic impairment of protein utilization due to shortage of energy substrates, hormones and enzymes, and a dramatic decrease in dry matter intake, apparent digestibility, volatile fatty acid production, rumen pH and electrolyte concentrations in the rumen fluids (Niles *et al.*, 1980). Under these conditions, the protein synthesis becomes unable to counteract the protein catabolism which leads to a negative nitrogen balance. This destruction in protein tissues is due to the increase in glucocorticoid hormones (proteolytic hormones) responsible for protein catabolism. The increase in glucocorticoid hormones may occur through the increase in gluconeogenisis, which delivers the amino acids to their corresponding α-keto acids (Alvarez and Johnson, 1970) or in the hepatic capture of blood amino acids (Noall *et al.*, 1957) or through inhibiting the oxidation of glucose which is essential for providing the energy required for peptide synthesis (Welt *et al.*, 1952). The increase in catecholamines (lipolytic hormones) (Winegrad, 1962) or the decrease in insulin, responsible for protein anabolism (Habeeb, 1987) may also contribute to tissue destruction.

In young animals, the nitrogen balance still decreases significantly under high temperature, but it does not reach the negative nitrogen balance found in older animals. This phenomenon may be due to the fact that heat-induced protein catabolism is not high enough to offset the well-known high rate of protein synthesis in young animals.

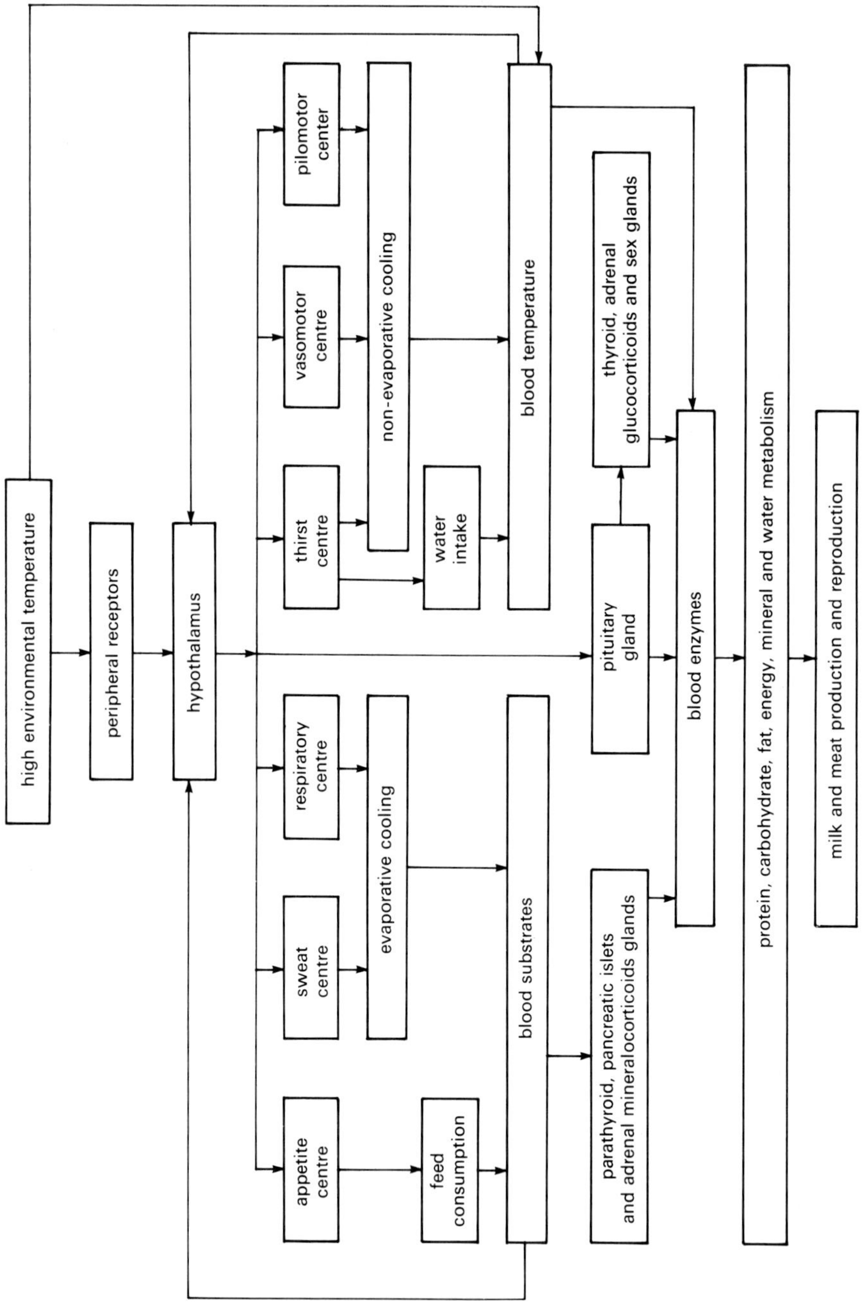

Fig. 2.2. Pathways of heat stress syndrome in European cattle in a hot climate.

Animal Production

Growth

Growth, which may be defined as an increase of live matter or cell multiplication, is controlled genetically and environmentally by well balanced available nutrients, hormones and enzymes. However, the data in the literature regarding the effects of elevated temperatures on growth performance are conflicting.

Some studies on Brown Swiss, Holstein and Jersey calves (Kamal *et al.*, 1962), Holstein and Hereford calves (Thompson *et al.*, 1963), Friesian heifers (Kamal and Seif, 1969) and Friesian calves (Habeeb *et al.*, 1991b,c; Marai *et al.*, 1991) showed that growth performance, i.e. growth rate, daily weight gain, dry body weight (total body solids), solids daily gain and live body weight, are impaired at elevated temperatures. The calculated loss in body solids due to heat stress conditions was found to be 23% in Friesian heifers (Kamal and Seif, 1969), 14–29% in Guernsey cattle (Kamal and Johnson, 1971), 17% in Friesian calves (Habeeb, 1987) and 15% in sheep (Macfarlane, 1968). In solids daily gain, the loss in Friesian calves was 51% (Habeeb, 1987) and 46% (Daader *et al.*, 1989; Marai *et al.*, 1991). The effects of elevated temperature on growth performance are the products of the decrease in anabolic activity and increase of tissue catabolism. The decrease of anabolism is essentially caused by the decrease in voluntary feed intake of essential nutrients (Morrison and Lofgreen, 1979), particularly metabolizable energy for both maintenance and weight gain (Bianca, 1965). This causes loss of production per unit of food under heat stress (Ames and Brink, 1977; Kamwanja *et al.*, 1980). The increase of tissue catabolism occurs mainly in fat depots and/or lean body mass (Kamal and Johnson, 1971). Specifically, there is a reduction in body amino-N (El-Fouly *et al.*, 1978) and endogenous DNA and RNA purine catabolism (El-Fouly and Kamal, 1979) as a result of the increase in catecholamines and glucocorticoids.

The other studies showed either no appreciable change in live body weight (Habeeb, 1981) or with Brahman cattle, a steady increase with rising environmental temperature (Ragsdale *et al.*, 1951).

Kamal and Johnson (1971) attributed the contradictory response of live body weight under heat stress to the interaction between tissue destruction and water retention. The increase in total body water could be less than, equal to, or more than the loss in total body solids, thus resulting in a decrease, no change or increase in live body weight, respectively. The authors found that heat stressed calves lost 10.6 kg body fat and gained 0.6 kg lean mass solids with a net total body solids of 10 kg, in three days. This loss was replaced by extra body water retained during these three days without a significant change in body weight. From another point of view, the conflicting responses of animals' live body weight under hot conditions could be attributed to the differences in the exposure periods and to breed and stage of maturity of the treated animals.

Milk yield and composition

Cattle milk production is significantly impaired in hot climates (Habeeb *et al.*, 1989, 1991a). A rise in the average temperature by 1.6, 3.2 and 8.8°C above normal (21°C) results in the decrease in the average daily milk yield by 4.5, 6.8 and 14% respectively, and a decline in the daily temperature by 7°C below normal resulted in an increase in the daily milk yield by 6.5% in dairy cattle (Petkov, 1971). At 30°C, the high producing animals showed a mean reduction of 2.0 kg day^{-1} compared to a reduction of only 0.65 kg day^{-1} for the low producing animals (Vanjonack and Johnson, 1975). In the hot climate (38°C), the reduction in the average milk yield in Friesian cows was lower by 30% than in the mild climate (18°C) (Kamal *et al.*, 1989b). In imported purebreds from mild climates to the humid tropics, Raun (1976) reported that milk production rarely exceeded 12–15 kg day^{-1} and most usually was less than 10 kg day^{-1}. The reaction of the lactating cows to hot environments seemed to be related to stage of lactation, since Bober *et al.* (1980) reported that the early-, mid- and late-lactation milk production decreased by 25%, 41% and 47%, respectively, at 72 h after the beginning of heat exposure.

Milk constituents are also greatly affected by hyperthermia. Friesian cows maintained under 38°C showed lower averages of total solids, fat, protein, ash and lactose yields than in the same animals when maintained under thermoneutral environmental temperatures (*c.*18°C) (Habeeb *et al.*, 1989). The percentage reductions were 28%, 27%, 7%, 27% and 30%, respectively. Rodriguez *et al.* (1985) demonstrated that fat and protein percentages declined between 8 and 37°C, chloride content increased when the temperature was above 21°C and total solids, lactose and mineral yields and specific gravity decreased above 22°C. In addition the ratio of protein to fat decreased at temperatures above 29°C, in Friesian cows. It may be of relevance that in Egypt, summer calving buffaloes have lower daily averages of fat, protein, total solids and chloride than those calving in winter (Ragab *et al.*, 1958). The average values for phosphorus and magnesium were also found to be less in summer calves. Citric acid and calcium contents decreased during early lactation, while potassium decreased in all lactation stages at high temperatures (Kamal *et al.*, 1962).

The decrease in yield and constituents of milk of dairy cattle as a result of exposure to high environmental temperature might be due to the decline in protein, carbohydrate, lipid mineral and vitamin metabolism which leads to a negative balance in nitrogen (Kamal *et al.*, 1962), energy (McDowell *et al.*, 1969) and minerals (Kamal *et al.*, 1984) resulting in low protein turnover, less heat production and fewer minerals for biosynthesis of milk. The depression in many hormone levels in heat stressed cattle, especially the thermogenic hormones such as insulin (Habeeb, 1987), thyroxine (El-Masry and Habeeb, 1989), and cortisol (Kamal *et al.*, 1989a,b) may also be responsible for the decrease in milk production as well as milk composition. Upper critical temperatures are in the range 21–27°C and 24–30°C for milk production and growth rates of European cattle respectively.

Reproductive Performance

High environmental temperatures or the rapid and sudden fluctuations of temperature such as often occur in many parts of the subtropics, cause unfavourable effects on reproductive function of cattle.

In males, the high environmental temperature exerts negative effects on sexual libido, despite the thermoregulatory mechanism of the testes. Consequently, testicular function is affected, causing reduced semen quality (Van Heerden, 1963) and spermatogenesis, metamorphosis of sperms, semen degeneration processes (Coser *et al.*, 1979), reduced ejaculate volume, live sperm percentage, sperm concentration, viability and motility (Coser *et al.*, 1979; Gamcik *et al.*, 1979). These effects reduce conception and fertility rates of male semen, i.e. reduce the male fitness.

In females, high environmental temperatures may suppress oestrus resulting in periods of anoestrus which interfere with ovulation. Hyperthermia can also interrupt early pregnancy by causing death and resorption of embryos and abortion of well grown fetuses (Van Heerden, 1963). Age at puberty, oestrous cycle length, duration of oestrus, incidence of abnormalities of the ova, embryonic mortality, fetal death rates, gestation length and fetal size (Hafez, 1968), incidence of oestrus with weak signs, percentage of silent heat, ovulation failure, interval from parturition to conception and the number of services per conception may increase (Mohamed, 1974). The frequency of ovulatory oestrus, fertilization rate and neonatal survival may decrease as a function of heat stress (Hafez, 1968). Heat stress also causes loss of libido, ovarian activity and conception rate (Mohamed, 1974).

Comparison between the seasons of the year also shows that in a hot summer ovarian activity decreases, while it increases during winter and spring when ambient temperature decreases (El-Fouly *et al.*, 1976). The percentage of heat with strong symptoms is also lower in the hot season (25%) than in the cold season (48%) (Barkawi, 1981). It may be of interest to note that in Egyptian buffaloes, the highest percentage of silent heat was also observed during summer (6.0%), while the lowest was in winter (4.8%), and the oestrous cycle length was 37 and 25 days during summer and winter, respectively (Borady *et al.*, 1982). Ovarian inactivity is one of the causes of the prolonged ovulatory cycle.

The detrimental effects of the hot climate on a cow's reproductive performance are a result of the drastic changes in its biological functions. These are controlled by the external environment through a chain of reactions involving thermoreceptors, photoreceptors, sensor capacities, the hypothalamus, central nervous system, endocrine glands and gonads.

Insufficient gonadotrophin (FSH and LH) hormone secretion under heat stress conditions, could lead to inadequate oestrogen and/or progesterone production and consequently to reproductive failure. Decreased LH may lead to ovulation failure with subsequent developmental failure of the corpus luteum in females. It also causes atrophy of the interstitial cells in the testis in males. The thyroid gland also plays a role in decreasing the reproductive activity (Farghaly, 1984), with a decrease in the level of triiodothyronine (T_3) and thyroxine (T_4) hormones. Deficiencies in minerals such as phosphorus, copper, cobalt, iodine, zinc and manganese and vitamins (A and D) in heat stressed cattle may also contribute to the impairment of reproductive performance. Specifically, vitamin A deficiency results in anoestrus, frequent returns to service, abortions and birth of weak and stillborn calves in cows. This vitamin is also necessary for sexual development and normal spermatogenesis in the bull (Salisbury and Van Demark, 1961).

Physiological and Biochemical Reactions to Heat Stress

Physiological reactions

Increased respiration rate is the first reaction when animals are exposed to environmental temperatures above the thermoneutral zone (Kamal, 1975; see also Fig. 2.2). This response ensures direct heat stimulation of the peripheral receptors which transmit nervous impulses to the heat centre in the hypothalamus. The cardiorespiratory centre is stimulated to send impulses to the respiratory activity. Specifically, in Holstein and Jersey cows, respiration rate begins to increase at about 16 and 21°C, respectively (Kibler *et al.*, 1949). With increasing environmental temperature, the respiration rate continues to rise linearly until a certain temperature, where the rate of increase in respiration rate slows, then it almost levels off or slightly decreases (Kibler, 1962; Bianca, 1963). At this level, the respiratory muscles reach a threshold of maximum activity and thus the respiration rate cannot increase further. Specifically, the respiration rate of European breeds of cattle never exceeds 50 respirations/minute at all temperatures below 26°C (Dobinson, 1951).

The significance of the increase of the respiration rate under heat stress is that it enables the animals to dissipate the excess of body heat by vaporizing more moisture in the expired air, which accounts for about 30% of the total heat dissipation (McLean, 1963). However, the excessive respiratory activity may cause respiratory alkalosis as a result of excessive CO_2 respiration and consequently loss of blood alkali reserve.

The onset of sweating is the next rapid reaction of animals to heat exposure and this increases linearly with the increase in the ambient temperature (Kamal, 1975). This increase in sweating is also controlled by the hypothalamus which is stimulated by the increase in core body temperature and/or by nervous impulses transmitted by the peripheral receptors. The significance of the increase in sweating is that the heat loss by sweating represents 60% of the total when the ambient temperature reaches 37.8°C, at which point the thirst, vasomotor and pilomotor centres are unable to respond further (McLean, 1963). Moreover, when the surrounding environmental temperature is equal to (or above) the animal's body temperature, the evaporative cooling (respiration and sweating) system represents the only means of heat dissipation. The importance of sweating in dissipating heat load is due to the high thermal capacity of water ($4.2\,J\,g^{-1}\,°C^{-1}$ at 30°C), and its high heat of evaporation ($2.4\,kJ\,g^{-1}$).

The pulse rate reflects primarily the homeostasis of circulation along with the general metabolic level. It increases on exposure to high environmental temperature (Salem, 1980; Aboul-Naga, 1987). This increases blood flow from core to surface to give a chance for more heat to be lost by sensible and insensible ways. At very high temperature, it may decrease (Yousef and Johnson, 1966) due to the decrease in metabolic rate of the animals under heat stress. However, some studies showed that pulse rate does not always change appreciably under high environmental temperatures (Marai *et al.*, 1991).

In a wide range of environmental temperatures within the thermoneutral zone, a balance between heat gain and loss is needed to make the animal's life comfortable (Hafez, 1968). However, rectal temperature rises if heat production exceeds heat dissipation which may be caused by failure in proper response of the peripheral receptors, hypothalamus, neutrohumora, nervous system, endocrine glands or enzymes.

The significance of a body temperature increase is its importance in thermoregulation. It coincides with the impulses transmitted by peripheral thermal receptors which stimulate the central receptors of the hypothalamus to induce thermal polypnoea and sweating to increase heat loss (Kamal, 1975). In general, the rise in rectal temperature occurs above 21–26°C in European cattle and above 32°C in Zebu cattle; younger cattle are affected at lower temperatures than old ones (Kibler *et al.*, 1949; Dobinson, 1951; Kibler, 1962; Bianca, 1963).

Biochemical reactions

Blood haematology

Most research has shown that haemoglobin concentration decreases during heat stress (El-Masry, 1987; Daader *et al.*, 1989; Yousef, 1990) due to the depression of haematopoiesis (Shebaita and Kamal, 1973) and to haemodilution. Blood volume (Shebaita and Kamal, 1975) and total body water (Kamal and Seif, 1969; Habeeb, 1987) increase with increasing evaporative processes in heat stressed animals.

With regard to the packed cell volume (haematocrit percentage) most investigators reported that it also decreases in heat stressed animals (Rowlands *et al.*, 1974; Bond *et al.*, 1984; Kappel *et al.*, 1984; Marai *et al.*, 1991) due to red cell destruction and/or to haemodilution (Shebaita and Kamal, 1975) and to reduction in cellular oxygen requirements to minimize the metabolic heat load (Lee *et al.*, 1976).

The erythrocyte (RBC) count was also found to decrease significantly by 12–20% in cattle under heat stress conditions (38°C) (Salem, 1980; Abdel-Samee, 1987; Habeeb, 1987) due both to the destruction of erythrocytes (Shaffer *et al.*, 1981) and to haemodilution effect.

The leucocyte (WBC) count values have been mostly found to increase by 21–26% in Friesian cattle (Abdel-Samee, 1987; Aboul-Naga, 1987), under heat stress due to thyromolymphatic involution.

Blood glucose

The results of the studies of the effect of the high ambient temperature on plasma glucose content are conflicting. Some studies showed that blood glucose decreased significantly with different percentages in animals exposed to heat stress conditions (Abdel-Samee, 1987; El-Masry, 1987; Habeeb, 1987; Yousef, 1990). The decrease was found to be 19% in Jersey cows (Rick and Lee, 1948), 24% in Friesian cows (Shaffer *et al.*, 1981), 18% in lactating Friesian cows (Kamal *et al.*, 1989b), 13% in

Holstein, Brown Swiss and Jersey heifers (Kamal *et al.*, 1962), 10% in Holstein heifers (Segura *et al.*, 1979) and 8% in Friesian calves (Habeeb, 1987). The decrease in glucose level during heat exposure relates in part to the decrease in concentrations of insulin (Herbein *et al.*, 1985; Habeeb, 1987) and thyroxine (El-Masry, 1987), which are correlated closely to the decrease in energy metabolism during heat exposure. The decrease in plasma glucose could also be due to the marked dilution of blood and body fluids as a whole in the heat stressed animals (Habeeb, 1987) or the increase in glucose utilization to produce more energy for greater muscular expenditure required for high respiratory activity (Habeeb, 1987). Moreover, the decrease in production of propionic acid in the rumen (Kelley *et al.*, 1968) and the decrease in roughage intake (Colditz and Kellaway, 1972; McDowell, 1972), as well as decreased hepatic capacity for gluconeogenesis (Sano *et al.*, 1983), would also be related to a decrease of plasma glucose in heat stressed animals.

Other studies show that glucose concentration can be increased under heat stress conditions (Bianca and Findlay, 1962; Webster, 1976; Collier *et al.*, 1982). This could be due to decreased glucose utilization, depression of both catabolic and anabolic enzyme secretions and the subsequent reduction of metabolic rate (Webster, 1976), or to the rapid panting which results in increased breakdown of glycogen into free glucose by the increase in glucocorticoid hormones (Thompson, 1973). These have hyperglycaemic action by enhancing gluconeogenisis through activating the enzyme glutamic pyruvic transaminase (GPT) (Rosen *et al.*, 1959) or by inhibiting the hypoglycaemic action of insulin (Bornstein and Park, 1953).

The conflict shown above may be because high environmental temperature affects blood glucose content by altering the hormonal balance (i.e. insulin, adrenaline, glucagon, thyroxine and adrenocorticotrophic hormone balance). Thus either the glucose utilization and glucogenesis or glycogenolysis and gluconeogenesis are increased with a consequent decrease or increase in blood glucose concentration respectively.

Blood protein

The plasma protein provides an efficient way of transferring the heat from inside the body to the outer surface in the skin for heat dissipation by non-evaporative processes during heat stress, since it holds an adequate percentage of water in the intravascular fluids and maintains the viscosity of the blood (Kamal *et al.*, 1962).

Serum protein, particularly albumin concentrations, usually decrease under heat stress by about 10% (Kamal *et al.*, 1962, 1989a; Aboul Naga, 1987; El-Masry, 1987; Habeeb, 1987), although some workers have noted increases in globulin concentrations. Kamel *et al.* (1962) confirmed a negative relationship between plasma protein content and environmental temperature.

The significant decline in serum protein with rising temperature seems to be due to dilution of plasma proteins, decreased protein synthesis as a result of the depression of anabolic hormonal secretion (El-Masry and Habeeb, 1989) and the increase in the catabolic hormones such as glucocorticoids and catecholamines (Alvarez and Johnson, 1973). The decrease in serum protein may also be due to the decrease in feed nitrogen and mineral intake which occurs under heat stress conditions.

Total lipids and cholesterol

The serum total lipids concentration decreases significantly in ruminants with prolonged exposure to environmental temperature (Noble *et al.*, 1973; O'Kelly, 1973; Daader *et al.*, 1989; Kamal *et al.*, 1989b). Such a phenomenon may be due to the increase in either body water content or utilization of fatty acids for energy production as a consequence of the decrease in glucose concentration. The cholesterol concentration also decreases markedly with the high environmental temperature (Shaffer *et al.*, 1981; Abdel-Samee, 1987; Aboul-Naga, 1987; El-Masry, 1987). Salem (1980) and Shaffer *et al.* (1981) confirmed that cholesterol concentration in cattle was lower in summer than in winter.

The marked decrease in cholesterol concentration may be due to the increase in total body water or the decrease in acetate concentration which is the primary precursor for the synthesis of cholesterol. The marked increase in progesterone (El-Shafie *et al.*, 1983) and glucocorticoid hormone levels (Thompson, 1973) in heat stressed cattle may be another factor causing the decline in blood cholesterol.

Urea nitrogen

Segura *et al.* (1979), Shaffer *et al.* (1981), El-Masry (1987), Habeeb (1987) and Kamal *et al.* (1989a) found in heat stressed Friesians that blood urea-N level decreased by 16% (El-Masry, 1987; Kamal *et al.*, 1989a) in lactating cows and 28% (Aboul-Naga, 1987) to 30% (Habeeb, 1987) in calves.

The depression in blood urea-N associated with heat stress in animals may be due to more resorption of the urea-N from the blood to the rumen to compensate for the decrease in ruminal ammonia-N as a result of the decrease in feed intake (El-Fouly *et al.*, 1978; Yousef, 1990). In addition, the increase in urinary nitrogen excretion under severe heat stress conditions as indicated by a negative nitrogen balance (Kamal *et al.*, 1962) may also contribute to the decrease of serum urea level under such conditions.

Blood enzyme activities

The blood enzymes are easily and often influenced by the external environment, including feeding practices, type of shelter and many other aspects of herd management, since they are intimately related to metabolism. Accordingly, seasonal changes of the enzymes are very important and must be considered (Bedrek, 1965). It is important to control carefully all experimental conditions, especially environmental ones, when measuring the enzyme activity in any animal (Boots *et al.*, 1969).

Most research shows that the serum transaminase activities increase with increase in environmental temperature. In European cattle, heat stress causes an increase in the activities of serum glutamic oxaloacetic transaminase (SGOT) and serum glutamic pyruvic transaminase (SGPT) due to increased stimulation of gluconeogenesis by corticoids (increase in cortisol, cortisone or adrenocorticotrophic hormone) (Thompson, 1973; Habeeb, 1987; Kamal *et al.*, 1989b).

Alkaline phosphatase enzyme has been mostly found to decrease significantly (Peterson and Waldern, 1981; Shaffer *et al.*, 1981; Aboul-Naga, 1987; and El-Masry,

1987) by heat stress and may be attributed to the reduction in thyroid hormones which takes place under heat stress. With regard to acid phosphatase enzyme level, Roussel and Stallcup (1966) and Aboul-Naga (1987) reported that it was not affected in heat stressed cattle.

Body water

Under thermoneutral conditions, water intake equals water loss in the normal adult animal. Under heat stress, the major physiological reaction is an increase in water intake or requirement and consequently body water content. As water is lost through urine, skin and respiratory vaporization under heat stress, a temporary water deficit ensues, with increased body fluids concentration (blood hypertonicity) which stimulates the hypothalamic thirst centre to increase water consumption. The increase in water consumption causes an increase in blood volume; the hydrostatic pressure in blood vessels and water flux from the blood vessels to the surrounding tissues causes an increased body water content in animals.

The water retained in the animal under heat stress increases the heat dissipation by assisting evaporative cooling via sweating, respiration and non-evaporative cooling via diuresis. However, although the increase in body water content in animals under hot climatic conditions is considered as an adaptive reaction to ameliorate heat stress, it is thought that heat-tolerant animals are those which manifest the least changes in most of the physiological functions, including body water content, when subjected to hot climate (Kamal, 1975).

Research shows that body water content increases by different percentages when different animal species are exposed to hot climatic conditions. The increment values in body water content have been recorded as 26% in Friesian heifers (Kamal and Seif, 1969), 8% in Friesian calves (Habeeb, 1987), 8–23% in camels (Siebert and Macfarlane, 1971), 20% in Merino wethers (Morris *et al.*, 1962), 10% in Merino ewes (Macfarlane, 1974) and 15% in Nubian goats (Kamal and El-Masry, 1984).

Hormonal secretions

Hormonal secretions are known to be of major importance in the body thermoregulation. With prolonged heat exposure, the hypothalamic hormone releasing factors are suppressed. Consequently, the pituitary hormones and other hormones, either autonomous or pituitary hormones and other hormones, either autonomous or pituitary controlled, are affected. It is believed that not only the hormonal concentrations, but also the levels available for cellular metabolic activities and cellular multiplications are altered by high environmental temperature.

The hormones connected with thermoregulation are numerous. Below, the changes in hormonal balance of insulin, thyroxine, cortisol and aldosterone under heat stress are considered.

Insulin

With regard to the effect of high environmental temperatures on insulin concentration in European cattle, the studies on Holstein cows (Herbein *et al.*, 1985; Abdel-

Samee *et al.*, 1989), Friesian heifers (Sejrsen *et al.*, 1980) and Friesian calves (Habeeb, 1987) showed that plasma insulin decreased significantly by 33%, 54% and 30% respectively. This decrease may be due to the decreased heat production. However, McVeigh and Tarrant (1982) noticed no significant difference between the stressed and unstressed Friesian bulls in plasma insulin concentration.

Thyroid hormones

Thyroid hormones, either thyroxine (T_4) or triiodothyronine (T_3) are known to play an important role in the animal's adaptation to environmental changes. However, T_3 is more concerned with thermogenesis. With regard to T_3, many investigators (Bober *et al.*, 1980; Magdub *et al.*, 1982 and Mohamed and Johnson, 1985) reported that it declined significantly in heat stressed cattle. It has been extensively confirmed by numerous workers that thyroxine hormones levels, in the long term, decrease under heat stress conditions by up to 25% (Vanjonack and Johnson, 1975; Collier, *et al.*, 1982; Magdub *et al.*, 1982; El-Masry, 1987). However, the decrease does not occur abruptly, but takes at least 72 hours to reach a peak and may then decline. The reduction has been found to be 15% and 37% after 48 and 72 hours of heat exposure, respectively, in Friesian cows (Kamal and Ibrahim, 1969). In summer, Kamal and Ibrahim (1969) reported that the thyroid activity decreased by 16% in Friesian cows relative to winter. The decrease in thyroid hormones is due to the decrease in basal metabolic rate and muscle activity to decrease heat production.

Cortisol

Activation of the hypothalamic–pituitary–adrenal axis and the consequent increase of plasma glucocorticoid concentrations are perhaps the most important responses of animals to stressful conditions. Adrenal corticoids, mainly cortisol, elicit physiological adjustments which enable animals to tolerate stressful conditions (Christison and Johnson, 1972). Moreover, the literature dealing with the effect of hot climate on plasma cortisol level in cattle is rather conflicting. In some studies, heat stress increased plasma glucocorticoids (Alvarez, 1968; Satterlee *et al.*, 1977) and in others they are either decreased significantly (Lee *et al.*, 1976; Niles *et al.*, 1980; Collier *et al.*, 1982; Kamal *et al.*, 1989a, b) or not significantly affected (El-Nouty *et al.*, 1980; Gwazdauskas and Vinson, 1979). The work of Alvarez and Johnson (1973) may aid interpretation of the results. They found that glucocorticoids increased by 38% after 1 h and 62% after 2 h of exposure of animals to hot conditions, reaching a peak of 120% at 4 h, then declined gradually to values not different from normal at 48 h and remained at or below this level for the rest of the exposure duration. The above contradiction may, therefore, be attributed to the difference in duration of exposure to heat stress. The basal cortisol concentrations vary greatly so that values may not be reliable indicators of an animal's ability to adjust to short (acute) or long (chronic) stressful conditions (Willet and Erb, 1972; Rudson *et al.*, 1975). However, it can be concluded that plasma cortisol level increases during acute heat stress and decreases during the chronic phase. The increase of plasma cortisol level during acute heat stress is attributed to the fact that the glucocorticoid hormones have hyperglycaemic action to increase gluconeogenesis and provide the expected increase in glucose utilization in heat stressed animals. However, it is possible to consider the initial reactions of the animal to acute heat stress as an emotional rather than a thermoregulatory response (Collins and Weiner, 1968). The decline which occurs in chronic heat stress is attributed to the fact that cortisol is thermogenic in animals and, consequently, the reduction of adrenocortical activity under thermal stress is a thermoregulatory protective mechanism preventing metabolic heat production in a hot environment. This indicates the role of the adrenal cortex gland in adaptation to stress (Alvarez and Johnson, 1973).

Aldosterone

Plasma aldosterone concentration decreases significantly in European cattle under high environmental temperature (El-Nouty *et al.*, 1980; Niles *et al.*, 1980; Aboul-Naga, 1987) due to the large decrease in potassium retention in heat stress (Kamal *et al.*, 1962). With prolonged heat exposure, mineralocorticoids seem to decrease due to the change in blood electrolytes. The increase in body fluids which occurs in heat stressed cattle may also be partly responsible for this decrease (Aboul-Naga, 1987), since the increase in the extracellular fluids volume decreases the aldosterone secretion.

Minerals

The maintenance of mineral balance in animals is of profound importance for their milk production and reproduction due to their role in bone and teeth formation, blood clotting, proper functioning of nerve tissue, regulation of osmotic pressure in body fluids, maintenance of homeostasis in the acid–base balance and acting as co-factors of enzymes or as catalysts in enzymatic reactions.

Thermal stress induces alterations in electrolytes and a negative mineral balance. High environmental temperatures (32 to 39°C) induced significant decreases in the retention of Na (23%), K (37%), Ca (20%), P (20%) and Zn (24%) in European cattle (Kamal and Johnson, 1977; Aboul-Naga, 1983). At 39°C, Aboul-Naga (1983) also reported a decrease in the retention of Mg, Fe and Co in heat stressed Friesian heifers of 28%, 43% and 22% respectively. The total excretion of all minerals increases significantly. On the other hand, high environmental temperatures induce a decrease in the intake of most minerals (Aboul-Naga, 1983).

Regarding the concentration of blood electrolytes, significantly lower values were found in serum Na, K, Ca and Zn (4%, 10%, 7% and 12%, respectively) in lactating Friesian cows at 38°C (Kamal *et al.*, 1989a) and in serum Na, K, Ca, P, Mg, Fe, Zn and Co (3%, 10%, 7%, 13%, 13%, 13%, 11% and 16%, respectively) in Friesian calves at 36°C (Aboul-Naga, 1987). However, other studies showed that cattle maintain a fairly normal electrolyte balance at high temperature. Some studies showed no marked changes in plasma Na and K (Blincoe and Brody, 1951), and in Ca (Kamal and Abdelaal, 1972; Shaffer *et al.*, 1981; Kiatoko *et al.*, 1982) and others showed plasma P (Kamal and Abdelaal, 1972) and serum Na, K and Ca (Shebaita and Pfau, 1982) increased significantly in cattle exposed to heat stress. These different effects may depend on whether the heat stressed animals can select a diet rich in minerals, especially sodium.

The decrease in aldosterone and parathyroid hormone secretion in heat stressed cattle is probably associated with both the rise in urinary mineral excretion and the increase in the body fluids and water turnover rate, which dilute the absolute quantities of plasma minerals and help in washing out these minerals during heat stress. These hormones may contribute to the reduction of blood electrolytes. In addition, the decrease in feed intake, especially roughages, and the catabolic processes may also contribute to the phenomenon. The increase of glucocorticoid hormones leads to tissue destruction, thereby eliminating minerals from the body in the urine and faeces. The increase in glomerular filtration rate in the kidney may also contribute.

References

Abdel-Samee, A.M. (1987) The role of cortisol in improving productivity of heat stressed farm animals with different techniques. PhD Thesis, Faculty of Agriculture, Zagazig University, Zagazig, Egypt.

Abdel-Samee, A.M., Habeeb, A.A., Kamal, T.H. and Abdel-Razik, M.A. (1989) The role of urea and mineral mixture supplementation in improving productivity of heat stressed Friesian calves in the sub-tropics. *Proceedings of 3rd Egyptian–British Conference on Animal, Fish and Poultry Production*, Vol 2. Alexandria, Egypt, pp. 637–641.

Aboul-Naga, A.I. (1983) The role of heat induced physiological changes of mineral metabolism in the heat stress syndrome in cattle. MSc Thesis, Faculty of Agriculture, Mansoura University, Mansoura, Egypt.

Aboul-Naga, A.I. (1987) The role of aldosterone in improving productivity of heat stressed farm animals with different techniques. PhD Thesis, Faculty of Agriculture, Zagazig University, Zagazig, Egypt.

Alvarez, M.B. (1968) Relation of environmental temperature to the activity of the adreno-sympathetic system of cattle. PhD Thesis, University of Missouri, USA.

Alvarez, M.B. and Johnson, H.D. (1970) Urinary excretion of adrenaline and noradrenaline in cattle during heat and cold exposure. *Journal of Dairy Science* **53**, 928–930.

Alvarez, M.B. and Johnson, H.D. (1973) Environmental heat exposure on cattle plasma catecholamine and glucocorticoids. *Journal of Dairy Science* **56**, 189–194.

Ames, D.R. and Brink, D.R. (1977) Effect of temperature on lamb performance and protein efficiency ratio. *Journal of Animal Science* **44**, 136–140.

Ames, D.R. and Ray, D.E. (1983) Environmental manipulation to improve animal productivity. *Journal of Animal Science* **57** (Suppl.), 209–220.

Barkawi, A.H. (1981) Ovarian activity and estrus detection in Egyptian buffaloes. MSc Thesis, Faculty of Agriculture, Cairo University, Egypt.

Bedrek, A. (1965) Seasonal changes in plasma cholesterol and serum alkaline phosphatase and transaminases activities in cross-bred cattle. *Journal of Applied Physiology* **20**, 287–290.

Bianca, W. (1963) Rectal temperature and respiratory rate as indicators of heat tolerance in cattle. *Journal of Agricultural Science* **60,** 113–120.

Bianca, W. (1965) Reviews of the progress of dairy science. A- Physiology of cattle in a hot environment. *Journal of Dairy Research* **32,** 291–345.

Bianca, W. and Findlay, J.D. (1962) The effect of thermally induced hyperpnoea on the acid base status of the calves. *Research in Veterinary Science* **3**, 38–49.

Blincoe, C. and Brody, S. (1951) The influence of temperature on blood composition of cattle. *Missouri Agricultural Experiment Station, Research Bulletin*, No. 488.

Bober, M.A., Becker, B.A., Valtorta, S.E., Katt, P., Mertsching, H., Johnson, H.D. and

Shanklin, M.D. (1980) The relationship of growth hormone and thyroxine to milk production under heat in Holstein cows. *Journal of Animal Science* **51** (Suppl. 1), 261–268.

Bond, J., Powell, J.B., Undersander, D.J., More, P.M., Tyrrel, H.F. and Ottjen, R.R. (1984) Forage composition and growth and physiological characteristics of cattle grazing several varieties of tall fescues during summer conditions. *Journal of Animal Science* **59**, 584–593.

Boots, L.R., Crist, W.L., Davis, D.R., Brum, E.W. and Ludwick, T.M. (1969) Effect of age, body weight, stage of gestation and sex on plasma glutamic-oxaloacetic and glutamic-pyruvic transaminase activities in immature cattle. *Journal of Dairy Science* **52**, 211–216.

Borady, A.M.A., Morad, H.M., El-Shafie, M.M., Bedeir, L.H. and Khattab, R.H. (1982) Seasonal variations in the reproductive activity of Egyptian buffaloes. 6th International Conference on Animal and Poultry Production. Zagazig University, Zagazig, Egypt, pp. 210–230.

Bornstein, J. and Park, C.R. (1953) Inhibition of glucose uptake by the serum of diabetic rats. *Journal of Biological Chemistry* **205**, 503–511.

Brody, S. (1949) *Bioenergetics and Growth.* Reinhold Publishing Company, New York.

Christison, G.I. and Johnson, H.D. (1972) Cortisol turnover in heat-stressed cows. *Journal of Animal Science* **53**, 1005–1010.

Colditz, P.J. and Kellaway, R.C. (1972) The effect of diet and heat stress on feed intake, growth and nitrogen metabolism in Friesian F_1, Brahman × Friesian, and Brahman heifers. *Australian Journal of Agricultural Research* **23**, 717–725.

Collier, R.J., Breede, D.K., Thatcher, W.W., Israel, L.A. and Wilcox, C.J. (1982) Influence of environment and its modification on dairy animal health and production. *Journal of Dairy Science* **65**, 2213–2227.

Collins, K.H. and Weiner, H.S. (1968) Endocrinological aspects of exposure to high environmental temperature. *Physiological Reviews* **48,** 785–794.

Coser, A.M.L., Godinho, H.P. and Fonseca, V.O. (1979) Effect of high temperatures on spermatogenesis in Brazilian woolless rams under experimental conditions. *Arquivos da Escola de Veterinaria da Univ. Federal de Minas Gerais* **31**, 147–154.

Daader, A.H., Marai, I.F.M., Habeeb, A.A. and Yousef, H.M. (1989) Improvement of growth performance of Friesian calves under Egyptian sub-tropical conditions. 1. Internal cooling technique using diuretics and drinking cool water. *Proceedings of 3rd Egyptian–British Conference on Animal, Fish and Poultry Production*, Vol 2, Alexandria, Egypt, pp. 595–605.

Dobinson, J. (1951) Heat tolerance in European breeds of cattle exposed to high environmental temperature. *Nature* **168**, 882.

El-Fouly, H.A. and Kamal, T.H. (1979) Effect of short-term heat exposure on urinary allantoin-N in Friesian calves. *World Review of Animal Production* **15**, 61–64.

El-Fouly, H.A., Kamal, T.H., Abou-Akkada, A.R. and Abou-Raya, A.K. (1978) Protein catabolism in heat-stressed sheep. *Isotope and Radiation Research* **10** (Suppl.), 145 (Abstract).

El-Fouly, M.A., Kotby, E.A. and El-Soubhy, H.E. (1976) The functional reproductive peak in Egyptian buffalo cow is related to day length and ambient temperature. *Archivio Veterinario Italiano* **27**, 123–129.

El-Masry, K.A. (1987) The role of thyroxine in improving productivity of heat stressed animals with different techniques. PhD Thesis, Faculty of Agriculture, Zagazig University, Zagazig, Egypt.

El-Masry, K.A. and Habeeb, A.A. (1989) Thyroid function in lactating Friesian cows and water buffalos under winter and summer Egyptian conditions. *Proceedings of 3rd Egyptian–British Conference on Animal, Fish and Poultry Production*, Vol. 2, Alexandria, Egypt, pp. 613–620.

El-Nouty, F.D., Elbanna, I.M., Davis, T.P. and Johnson, H.D. (1980) Aldosterone and ADH response to heat and dehydration in cattle. *Journal of Applied Physiology* **48**, 249–255.

El-Shafie, M.M., Borady, M.A., Mourad, H.M. and Khattab, R.M. (1983) Physiological and seasonal factors affecting reproductive performance of Egyptian buffalo heifers. *Journal of Animal Production, Egypt* **23**, 1–4.

Farghaly, H.A. (1984) Thyroid and some ovarian hormones and their relation to reproduction in Egyptian water buffalo. MSc Thesis, Faculty of Agriculture, Cairo University, Cairo.

Gamcik, P., Mesaros, P. and Schvarc, F. (1979) The effect of season on some semen characters in Slovakian Merino rams. *Zivocisna Vyrobo* **24,** 625–630.

Gwazdauskas, F.C. and Vinson, W.E. (1979) Adrenal response to adrenocorticotropin in Holstein heifers exposed to a cool environment. *Journal of Dairy Science* **62**, 1811–1813.

Habeeb, A.A. (1981) Comparison of different tracer techniques of assessment of body fluids in Animals. MSc Thesis, Faculty of Agriculture, Cairo University, Egypt.

Habeeb, A.A. (1987) The role of insulin in improving productivity of heat stressed farm animals with different techniques. PhD Thesis, Faculty of Agriculture, Zagazig University, Zagazig, Egypt.

Habeeb, A.A., Abdel-Samee, A.M. and Kamal, T.H. (1989) Effect of heat stress, feed supplementation and cooling technique on milk yield, milk composition and some blood constituents in Friesian cows, under Egyptian conditions. *Proceedings of 3rd Egyptian–British Conference on Animal Fish and Poultry Production,* Vol. 2. Alexandria, Egypt, pp. 629–635.

Habeeb, A.A., Ibrahim, M.Kh. and Hiekal, A.H. (1991) Environmental heat exposure effect on biosynthesis of milk components and some hormones in Friesian cows. *Egyptian Journal of Dairy Science* (in press).

Habeeb, A.A., Marai, I.F.M., Daader, A.H., Yousef, H.N. and Johnson, H.D. (1991) Effects of Egyptian subtropical summer conditions and heat stress alleviation technique of water spray and diaphoretic on growth and physiological functions of Friesian calves. *Annual Meeting of The American Society of Animal Science*, University of Wyoming, Wyoming State, USA.

Hafez, E.S.E. (1968) Environmental effects on animal productivity. In: Hafez, E.S.E. (ed.), *Adaptation of Domestic Animals.* Lea & Febiger, Philadelphia, PA.

Herbein, J.H., Aiello, R.J., Eckler, L.I., Pearson, R.E. and Akers, R.M. (1985) Glucagon, insulin, growth hormone, and glucose concentrations in blood plasma of lactating dairy cows. *Journal of Dairy Science* **68,** 320–325.

Johnson, H.D. (1974) Tracer studies in environmental adaptation. In: *Tracer Techniques in Tropical Animal Production. Proceedings of International Atomic Energy Agency, Vienna*, pp. 89–97.

Kamal, T.H. (1975) Heat stress concept and new tracer methods for heat tolerance in domestic animals. 1st Science Conference on Peaceful Uses of Atomic Energy for Scientific and Economic Development, Baghdad, Iraq. *Proceedings of IAEA/FAO*, Vienna.

Kamal, T.H. and Abdelaal, A.E. (1972) Seasonal changes in P-32 and Ca-45 metabolism in Friesians and water buffaloes. Symposium on Isotope Studies on the Physiology of Domestic Animals, Athens, Greece. *Proceedings of IAEA*, Vienna, pp. 95–102.

Kamal, T.H. and El-Masry, K.A. (1984) Effect of hot climate on water turnover rate, total body water and total body solids in Goats. *Proceedings of the 1st Egyptian–British Conference on Animal and Poultry Production.* Zagazig University, Zagazig, Egypt.

Kamal, T.H. and Ibrahim, I.I. (1969) The effect of the natural climate of the Sahara and controlled climate on thyroid gland activity in Friesian cattle and water buffaloes. *International Journal of Biometeorology* **13**, 275–285.

Kamal, T.H. and Johnson, H.D. (1971) Total body solids as a measure of a short-term heat stress in cattle. *Journal of Animal Science* **32,** 306–311.

Kamal, T.H. and Johnson, H.D. (1977) Effect of high environmental temperature and age on trace element metabolism in cattle. Symposium on *Trace Elements in Drinking Water, Agriculture and Human Life*, Goethe Institute, Cairo, Egypt, pp. 54–68.

Kamal, T.H. and Seif, S.M. (1969) Changes in total body water and dry body weight with age and body weight in Friesians and water buffaloes. *Journal of Dairy Science* **52**, 1650–1656.

Kamal, T.H., Johnson, H.D. and Ragsdale, R.C. (1962) Metabolic reactions during thermal stress (35 to 95°F) in dairy animals acclimated at 50° and 80°F. *Missouri Agricultural Experiment Station, Research Bulletin* No. 785.

Kamal, T.H., Mehrez, A.Z., El-Shinnawy, M.M. and Aboul-Naga, A.I. (1984) Effect of high environmental temperature on mineral metabolism in Friesian cattle. *Proceedings of 1st Egyptian–British Conference on Animal and Poultry Production*. Zagazig University, Zagazig, Egypt.

Kamal, T.H., Habeeb, A.A., Abdel-Samee, A.M. and Abdel-Razik, M.A. (1989a) Supplementation of heat-stressed Friesian cow, with urea and mineral mixture and its effect on the milk production, in subtropics. *Proceedings of International Symposium on the Constraints and Possibilities of Ruminant Production in the Dry Subtropics*. Cairo, Egypt, pp. 183–185.

Kamal, T.H., Habeeb, A.A., Abdel-Samee, A.M. and Marai, I.F.M. (1989b) Milk production of heat-stressed Friesian cows and its improvement in the subtropics. *Proceedings of International Symposium on the Constraints and Possibilities of Ruminant Production in the Dry Subtropics*. Cairo, Egypt, pp. 156–158.

Kamwanja, L.A., Schillo, K.K., Hansen, P.J., Hauser, E.R. and Dierschke, D.J. (1980) The influence of season on body weight, growth rate and feed consumption as related to puberty in heifers. *Journal of Animal Science* **51** (Suppl. 1), 289 (Abstract).

Kappel, L.C., Ingraham, R.H., Morgan, E.B. and Babcock, S.E. (1984) Plasma copper concentration and backed cell volume and their relationship to fertility and milk production in Holstein cows. *American Journal of Veterinary Research* **45**, 346–357.

Kelley, R.O., Martz, F.A. and Johnson, H.D. (1968) Effect of environmental temperature on ruminal volatile fatty acids levels with controlled feed intake. *Journal of Dairy Science* **50**, 531–533.

Kiatoko, M., McDowell, L.R., Bertrand, J.E., Chapman, H.L., Pate, F.M., Martin, F.G. and Conrad, J.H. (1982) Evaluating the nutritional status of beef cattle herds from four soil order regions of Florida. I-Macroelements, protein, carotene, vitamins A and E, haemoglobin and haematocrit. *Journal of Animal Science* **55**, 28–37.

Kibler, H.H. (1962) Energy metabolism and related thermoregulatory reactions to thermal stress in 50 and 80°F acclimated dairy heifers. *Missouri Agricultural Experimental Station, Research Bulletin* No. 793.

Kibler, H.H., Brody, S. and Worstell, D.M. (1949) Influence of temperature, 50 to 105°F, on heat production and cardiorespiratory activities in dairy cattle. *Missouri Agricultural Experimental Station Research Bulletin* No. 435.

Lee, J.A., Roussell, J.D. and Beaty, J.F. (1976) Effect of temperature-season on bovine adrenal cortical function blood cell profile and milk production. *Journal of Dairy Science* **59**, 104–114.

Macfarlane, W.V. (1968) Adaptation of ruminants to tropics and deserts. In: Hafez, E.S.E. (ed.), *Adaptation of Domestic Animals*. Lea & Febiger, Philadelphia, PA.

Macfarlane, W.V. (1974) Tracer studies in environmental adaptation. In: *Tracer Techniques in Tropical Animal Production. Proceedings of IAEA/FAO*, Vienna.

Magdub, A.B., Johnson, H.D. and Belyea, R.L. (1982) Effect of environmental heat and dietary fiber on thyroid physiology of lactating cows. *Journal of Dairy Science* **65**, 2323–2329.

Mallone, P.G., Beede, D.K., Collier, R.J. and Wilcox, C.J. (1985) Production and physiologi-

cal responses of dairy cows to varying dietary potassium during heat-stress. *Journal of Dairy Science* **68**, 1479–1487.
Marai, I.F.M., Daader, A.H., Habeeb, A.A. and Yousef, H.M. (1991) Effect of winter and summer environmental conditions and their amelioration by dietary supplementation and cooling treatment on Friesian calves, in Egypt. *Journal of Arid Environments* (in press).
McDowell, R.E. (1972) *Improvement of Livestock Production in Warm Climates.* W.H. Freeman & Company, San Francisco, CA, USA.
McDowell, R.E., Moody, E.G., Van Scoest, P.J., Lehmann, R.P. and Ford, G.L. (1969) Effect of heat stress on energy and water utilization of lactating cows. *Journal of Dairy Science* **52**, 188–194.
McLean, J.A. (1963) The regional distribution of cutaneous moisture vaporization in the Ayrshire calf. *Journal of Agricultural Science* **61,** 275–280.
McVeigh, J.M. and Tarrant, P.V. (1982) Behavioral stress and skeletal muscle glycogen metabolism in young bulls. *Journal of Animal Science* **54**, 790–795.
Mohamed, A.A. (1974) Physiological changes in the reproductive organs of buffaloes from parturition to conception. PhD Thesis, Faculty of Agriculture, Al-Azhar University, Egypt.
Mohamed, M.E. and Johnson, H.D. (1985) Effect of growth hormone on milk yield and related physiological functions of Holstein cows exposed to heat stress. *Journal of Dairy Science* **68**, 1123–1130.
Morris, R.J.H., Howard, B. and Macfarlane, W.V. (1962) Interaction of nutrition and air temperature with water metabolism of Merino wethers shorn in winter. *Australian Journal of Agricultural Research* **13,** 320–334.
Morrison, S.R. and Lofgreen, G.P. (1979) Beef cattle response to air temperature. *Transactions of the American Society of Agricultural Engineering* **22**, 861–872.
Niles, M.A., Collier, R.J. and Croom, W.J. (1980) Effects of heat stress on rumen and plasma metabolite and plasma hormone concentration of Holstein cows. *Journal of Animal Science* **5** (Suppl. 1), 152 (Abstract).
Noall, M.W., Riggs, T.R., Walker, L.M. and Christensen, H.N. (1957) Endocrine control of amino acid transfer. *Science* **126**, 1002–1005.
Noble, R.C., O'Kelly, J.C. and Moore, J.H. (1973) The effects of exposure to warm environments on bovine plasma lipid composition. *Lipids* **8,** 216–220.
O'Kelly, T.C. (1973) Seasonal variations in the plasma lipids of genetically different types of cattle: Steers on different diets. *Comparative Biochemistry and Physiology* **44A**, 303–312.
Peterson, R.G. and Waldern, D.E. (1981) Repeatabilities of serum constituents in Holstein Friesians affected by feeding, age, lactation and pregnancy. *Journal of Dairy Science* **64**, 822–831.
Petkov, G. (1971) Environmental temperature and milk production of cows. *Veterinaria Sbirka* **75**, 23–28.
Ragab, M.T., Asker, A.A. and Kamal, T.H. (1958) The effect of age and season of calving on the composition of Egyptian buffalo milk. *Journal of Indian Dairy Science* **11**, 18–23.
Ragsdale, A.C., Thompson, H.J., Worstell, D.M. and Brody, S. (1951) Influence of increasing temperature 4°F to 105°F on milk production in Brown Swiss and Brahman cows and heifers. *Missouri Agriculture Experimental Station Research Bulletin* No. 471.
Raun, N.S. (1976) Beef and cattle production practices in the lowland American tropics. *Animal Review* **19**, 18–21.
Riek, R.F. and Lee, D.H.K. (1948) Reactions to hot atmospheres of Jersey cows in milk. *Journal of Dairy Research* **15**, 219–226.
Rodriguez, L.R., Mekonnen, G., Wilcox, C.J., Martin, F.G. and Krienke, W.A. (1985) Effect of relative humidity, maximum and minimum temperature, pregnancy and stage of lactation on milk composition and yield. *Journal of Dairy Science* **68**, 973–978.

Rosen, F.N., Roberts, N.R. and Nichol, C.A. (1959) Glucocorticoids and transaminase activity: Increased activity of glutamic-pyruvic transaminase in four conditions associated with glucogenesis. *Journal of Biological Chemistry* **234**, 476–480.

Roussel, J.D. and Stallcup, O.T. (1966) Influence of age and season on phosphatase and transaminase activities in blood serum of bulls. *American Journal of Veterinary Research* **27**, 1527–1530.

Rowlands, G.J., Manston, R., Pocock, R.M. and Dew, S.M. (1974) Relationships between stage of lactation and pregnancy and blood composition in a herd of dairy cows and the influence of seasonal changes in management on these relationships. *Journal of Dairy Research* **42**, 349–362.

Rudson, S., Mullord, M., Whittlestone, W.G. and Payne, E. (1975) Diurnal variations in blood cortisol in dairy cows. *Journal of Dairy Science* **58**, 30–36.

Salem, I.A. (1980) Seasonal variations in some body reactions and blood constituents in lactating buffaloes and Friesian cows. *Journal of Egyptian Veterinary Medicine Association* **40**, 63.

Salisbury, G.W. and Van Demark, N.L. (1961) *Physiology of Reproduction and Artificial Insemination of Cattle.* W.H. Freeman, San Francisco.

Sano, H., Takahashi, K., Ambo, K. and Tsuda, T. (1983) Turnover and oxidation rates of blood glucose and heat production in sheep exposed to heat. *Journal of Dairy Science* **66**, 856–861.

Satterlee, D., Roussel, J.D., Gomila, L.F. and Segura, E.T. (1977) Effect of exogenous corticotropin and climatic conditions on bovine adrenal cortical function. *Journal of Dairy Science* **60**, 1612–1616.

Segura, E.T., Roussel, J.D., Satterlee, D.G., Gomila, L.F., Shaffer, L. and Bergeron, J.C. (1979) Interaction of exogenous corticotropin and environment on protein bound iodine and other plasma biochemical parameters. *Journal of Dairy Science* **62**, 278–283.

Sejrsen, K., Fitzgerald, E.M., Tucker, H.A. and Huber, J.T. (1980) Effect of plane of nutrition on serum prolactin and insulin in pre- and postpubertal heifers. *Journal of Dairy Science* **53** (Suppl. 1), 326–327 (Abstract).

Shaffer, L., Roussel, J.D. and Koonce, K.L. (1981) Effect of age, temperature, season and breed on blood characteristics of dairy cattle. *Journal of Dairy Science* **64**, 62–70.

Shafie, M.M. and Abdelghany, M. (1978) Structure of respiratory system of sheep as related to heat tolerance. *Acta Anatomica* **100**, 441–460.

Shebaita, M.K. and Kamal, T.H. (1973) *In vivo* body composition in ruminants. 1- Blood volume in Friesian and water buffaloes. *Alexandria Journal of Agricultural Research* **21**, 339–350.

Shebaita, M.K. and Kamal, T.H. (1975) Lean body mass and body fat changes in lactating animals under Sahara climate. *World Review of Animal Production* **11**, 50–57.

Shebaita, M.K. and Pfau, A. (1982) Changes in the electrolytes of serum and urine with heat exposure in the bovine. In: *Proceedings of the 6th International Conference on Animal and Poultry Production*, Vol 1. Zagazig University, Zagazig, Egypt, pp. 261–271.

Siebert, B.D. and Macfarlane, W.V. (1971) Water turnover and renal function of dromedaries in the desert. *Physiological Zoology* **44**, 225–240.

Thompson, G.E. (1973) Review of the progress of dairy science. Climatic physiology of cattle. *Journal of Dairy Research* **40**, 441–473.

Thompson, R.D., Johnston, J.E., Breidenstein, C.P., Guidry, A.J., Banerjee, M.R. and Burnett, W.T. (1963) Effect of heat conditions on adrenal cortical, thyroid and other metabolic reponses of dairy heifers. *Journal of Dairy Science* **46**, 227–231.

Van Heerden, S. (1963) Livestock infertility. Aspects of tropical and subtropical infertility in livestock. *Animal Health Branch, Monograph, No. 5*, FAO Rome.

Vanjonack, W.J. and Johnson, H.D. (1975) Effects of moderate heat and yield on plasma thyroxine in cattle. *Journal of Dairy Science* **58,** 507–516.

Webster, A.J.F. (1976) The influence of the climatic environment on metabolism in cattle. In: Swan, H. and Broster, W.H. (eds.) *Principles of Cattle Production* , Butterworths, London.

Welt, I.D., Stretten, D. Jr., Ingle, D.J. and Morely, E.H. (1952) Effect of cortisone upon rates of glucose production and oxidation in the rat. *Journal of Biological Chemistry* **197**, 57–67.

Willet, L.B. and Erb, R.E. (1972) Short-term changes in plasma corticoids in dairy cattle. *Journal of Animal Science* **34**, 103–109.

Winegrad, A.I. (1962) Endocrine effects on adipose tissue metabolism. *Vitamins and Hormones* **20**, 141–153.

Yousef, H.M. (1990) Studies on adaptation of Friesian cattle, in Egypt. PhD Thesis, Faculty of Agriculture, Zagazig University, Zagazig, Egypt.

Yousef, M.K. and Johnson, H.D. (1966) Calorigenesis of dairy cattle as influenced by thyroxine and environmental temperature. *Journal of Animal Science* **25**, 150–156.

3 Photoperiod

C.J.C. PHILLIPS

School of Agricultural and Forest Sciences, University College of North Wales, Bangor, UK

Summary

Photoperiod is used by animals to time many events, in particular their parturition in relation to the period of optimum feed availability. For most animals this occurs in spring/early summer and parturition is timed to occur in early spring so that peak feed requirements of dam plus offspring are in early summer.

In farm animals, these photoperiod effects are stongest in animals that have evolved furthest from the Equator or whose breeding patterns have not been modified by man's selection programmes.

Many methods have been developed by man to reorientate the breeding pattern of farm animals and overcome seasonality effects. Photoperiod may be artificially extended in controlled environments, and this is widely practised for poultry, increasingly in skeletal light sequences to minimize the cost of the additional lighting. In breeding ewes, short daylength may be simulated by implants that release melatonin, the hormone released by the pineal gland during the dark period of the 24 hour cycle.

Photoperiod can also affect the intake, productivity and partitioning of nutrients in farm animals. There is evidence that intake, and hence productivity, is decreased during short photoperiods in most ungulate species. In winter, short daylength encourages the growth of body fat reserves at the expense of lean tissue growth or milk production. It is probable that animals perceive photoperiod as a relative rather than an absolute phenomenon and there are critical parts of the day that regulate photoperiod perception. This rhythm of perceived photoperiod is ingrained from an early age, probably *in utero*, and can have considerable residual effects after photoperiodic manipulation.

Finally, evidence from cattle suggests that the photoperiod may influence hedonistic environmental perception. In the dark, social stresses are more evident and the welfare and productivity of animals in a crowded building may be adversely affected by short daylength.

Introduction

Photoperiod is the most reliable indicator of season and as such is extensively used to control the important productive functions of animals, particularly growth and reproduction. This achieves both the synchronization of peak nutrient demands with peak availability and the preparation and synchronization of the animal's body for reproduction at the optimum time, usually to produce a spring parturition. The synchronization of parturition is particularly important in species whose neonatal offspring 'follow' their parents rather than 'hide' (Rutberg, 1987) to avoid predation. Sheep may be classified as 'followers' and are strongly photosensitive, whereas cattle and pigs are 'hiders' and are weakly photosensitive. Chicks, like pigs, have nests but usually spend a large proportion of the day 'following' the hen. Other possible reasons for the greater photosensitivity of sheep and goats than pigs and cattle are the greater seasonal variation in food availability for the smaller ruminants, which evolved in harsher conditions, and the increased selective breeding by man that cattle and pigs have been exposed to. There are strong genetic influences on photosensitivity. These are more pronounced between species and genera, but within a population variation is less pronounced so that synchrony of parturition can be achieved. In the wild this synchrony would produce a limited availability of neonates for predation, thus reducing the predator population.

Photoperiod can in essence be considered as a biphasic light intensity change. However, this change is relative rather than absolute, since the lunar cycle during the scotophase can have additional effects on reproduction. In artificially manipulated photoperiods aspects of the photophase such as light intensity colour (wavelength) and length of the transition between the photophase and scotophase (artificial dawn/dusk) affect the animal's response. Artificial biphasic sequences of more or less than the normal 24 hour day (ahemeral cycles) can also be used, as well as the replacement of part of the photophase with darkness (skeletal photoperiods).

In this review the effects of photoperiod on the production and reproduction of farm animals, the mechanisms involved and methods of artificial manipulation are considered. For more comprehensive reviews the reader is referred to papers by Terqui *et al.* (1984) and Ortavant *et al.* (1988) for sheep, Morris (1967) and Cavalchini *et al.* (1983) for poultry, Ortavant *et al.* (1985), Forbes (1982) and Tucker (1985) for growth, lactation and feed intake in farm animals, particularly cattle.

Photoperiod Perception

Photoperiod is perceived by the eyes in mammals and additionally also through the skull in poultry. The photic information is relayed to the hypothalamus and the pineal gland via the superior cervical ganglion – the pineal gland encoding the information by producing melatonin in higher concentrations in the dark than the light. Research in sheep has demonstrated that daylength is measured not by the total exposure to light but by the illumination of two special set points during the day about 16 hours apart (Ortavant *et al.*, 1988). An animal can therefore be 'tricked' into giving a physiological long day response by being given, for example, a one hour burst of light 16 hours after a 7 hour short day. This elementary mechanism may yet prove to be too simple, as it does not explain why 24 hours of light do not usually give a long day physiological response.

This circadian rhythm of melatonin production entrains the animal to photoperiodism from an early age, in fact influence of the maternal melatonin rhythm begins the process *in utero*. As the rhythm becomes more firmly entrained into the animal's biorhythm the animal builds up a resistance to photomanipulation, and residual effects lasting up to several months maintain the physiological response to natural photoperiod.

Perception of light intensity depends on the ratio of rods to cones in the eye and most farm animals appear rather more adapted to low intensity vision and less to visual acuity than man. For example, in cattle the ratio of rods to cones is 15 : 1 compared with 20 : 1 in humans, giving cattle optimum *visual acuity* at a light intensity as low as 120 lux (Dannenmann *et al.*, 1984/5). Further details of the visual perception capabilities of various animals are given in Chapter 8.

Milk Production

Cattle

Several long-term studies comparing cows given natural daylength of 8–10 hours and a period of supplementary light of 4–8 hours with cows receiving the natural daylength only have demonstrated that milk yield is increased with supplementary light by 7–16% (Peters *et al.*, 1978, 1981; Bodurov, 1979; Stanisiewski *et al.*, 1985; Phillips and Schofield, 1989). Level of milk yield of the cows does not affect this response (Peters *et al.*, 1981; Stanisiewski *et al.*, 1985), but in the application to commercial herds it has been found that the response is not uniform (Stanisiewski *et al.*, 1985). Where the effects of supplementary light have been examined using a short-term *changeover* experimental design, little or no response has been obtained (Murrill *et al.*, 1969; Peters *et al.*, 1978; Phillips and Schofield, 1989). The reason for this is not clear but may be related to temporary inhibition of prolactin elevation when supplementary light is applied (Eisenman and Chew, 1982) or may arise from a relative rather than absolute perception of photoperiod by the cows.

Perera and Hacker (1982) have reported that 6 hours' light during the day plus a 2 hour pulse at night administered during the last two months of gestation will increase yield persistency, compared to cows given 12–14 hours continuous light daily during two months. To be effective the light pulse had to begin more than 2 hours after the end of the daylight. Further research on this technique would seem worth while.

Other species

Bocquier (1985) reported that ewes subjected to 16 hours light (16L) and 8 hours dark (8D) from day 100 of gestation to day 60 of lactation had heavier lambs and increased milk yield than ewes on 8L, 16D. Some of the increased milk yield could have been due to the ewes having heavier lambs at birth. However, photoperiod *per se* appears to have had the dominant effect since crossing over photoperiod treatments at 60 days resulted in milk production levels crossing over within 20 days. The magnitude of the milk yield response to photoperiod (20%) appears greater than that usually recorded in cattle. This has implications for frequent lambing systems where ewes may be lactating in short daylength. Similar responses to long days have also been recorded in goats (Linzell, 1973; Terqui *et al.*, 1984). However, some cyclicity is maintained in the absence of light and temperature cues, demonstrating the existence of a photoperiodic history.

In pigs, Mabry *et al.* (1983) reported increased nursing frequency and increased litter weights at weaning when sows were exposed to 16L, 8D compared with 8D, 16L. However, Jones (1984) found no effect of these photoperiods, although a creep light was provided for the piglets in both treatments and the supplementary light intensity (100 lux) was low.

Milk Composition

Peters *et al.* (1978, 1981) reported that milk fat content of dairy cows was not affected by supplementary light, but in a more robust design of experiment Stanisiewski *et al* (1985) reported that milk fat content was reduced by 1.6 g kg^{-1}. A larger reduction was reported by Phillips and Schofield (1989), but this was not statistically significant. The apparent reduction may not be related to the increase in milk yield, since in a separate experiment Phillips and Schofield (1989) found milk fat content to be reduced by supplementary light in the absence of any effect on milk yield. In the only report on the effect of light on milk protein content the same workers found it to be reduced where the light was of high intensity (529 lux at cow eye level), but not at lower intensities (101 or 191 lux). In goats' milk the protein content was also found to be reduced in long daylength, but milk fat content was unaffected (Linzell, 1973).

Growth and Carcass Composition

Cattle

Several experiments have demonstrated that an extended photoperiod of approximately 16L, 8D increases the growth rate of heifers by up to 10% compared to natural photoperiods of 8–10L (Petitclerc, 1985; Zinn *et al.*, 1986a,b). Others have found no effect (Hansen *et al.*, 1983; Bourne *et al.*, 1984; Petitclerc *et al.*, 1984; Roche and Enright, 1988; Zinn *et al.*, 1989). With steers similar results have been obtained, some researchers finding that an extended photoperiod increased growth rate by up to 10% (Zinn *et al.*, 1988; Johnson *et al.*, personal communication), but others finding no effect (Roche and Enright, 1988; Zinn *et al.*, 1989; Weiguo and Phillips, 1991). One reason for this dichotomy may be that the extended photoperiod reduces fat accretion relative to that of protein in both heifers (Petitclerc *et al.*, 1984; Zinn *et al.*, 1986b) and steers (Johnson *et al.*, personal communication), especially on a high plane of nutrition (Petitclerc *et al.*, 1984). The concomitant reduction in the energetic content of carcass gain would be expected to increase the growth rate slightly. In the wild it is expected that animals entering a period of short days in the autumn will divert nutrients from protein accretion (an increase in size) to fat accretion (an increase in body reserves to sustain them through the winter).

Skeletal long (6L, 8D, 2L, 8D) photoperiods have been tested with growing cattle but appear to have less benefit than a conventional long photoperiod (Zinn *et al.*, 1986a). The major economic benefit of long photoperiods for growing cattle is the reduction in carcass fatness, allowing the producer to rear cattle to a heavier weight before slaughter. To a certain extent this could be offset in heifers by earlier puberty in light-supplemented cattle (Fauconneau and Gauthier, 1984; Hansen *et al.*, 1983).

Pigs

In pigs no benefit of 16L, 8D compared with 8L, 16D has been found in the feed intake or weight gain of growing pigs (McGlone *et al.*, 1988; Gooneratne and Thacker, 1990). Similarly an increasing daylength through the rearing period does not alter performance compared with a declining daylength (Wright *et al.*, 1984). However, compared with 8L, 16D a skeletal photoperiod of 2L, 5D, 2L, 15D has been shown to increase growth rate by 3% and feed conversion efficiency by 2%, and reduce the electricity consumption for lighting by 48% (Melhorn and Dorn, 1985).

Poultry

Continuous light or 8L, 16D retards growth in the male compared to 14L, 10D (Leeson and Summers, 1985; Legare *et al.*, 1986; Okamoto *et al.*, 1989). The beneficial effect of 14L, 10D is most pronounced in the first 12 weeks and extending daylength above 8L after 12 weeks has little effect on growth rate (Leeson and Summers, 1985; Classen, 1990). However, increasing the photoperiod in the first 2–6 weeks of age from 6–14L to 23L has been shown to almost double the incidence of leg disease in broilers (Classen, 1990) and also to increase the incidence of sudden death syndrome.

Light intensity is usually kept low to reduce the activity level of the birds, counteract aggression and reduce the electricity cost. In this respect a blue light, which is perceived by the birds as less bright than a red or white light, is preferable to a white light to stimulate growth, at least in the first 16–18 weeks (Lavenick and Leighton, 1988), while the birds are establishing their dominance order. Thereafter white or red light acts as a better photostimulant than blue light (Lavenick and Leighton, 1988), conceivably because the birds are settled in a dominance order but have a greater number of meals under the perceived brighter light.

Ahemeral skeletal photoperiods of 2L, 2D have been shown to increase growth rate, compared with conventional diurnal, 12L, 12D, cycles (Lavenick and Leighton, 1988). A skeletal photoperiod of 1L, 3D is possibly too extreme initially because it does not allow sufficient feeding time in the light. However, the consequent reduction in initial growth rate has advantages of reducing leg disorders and other fast

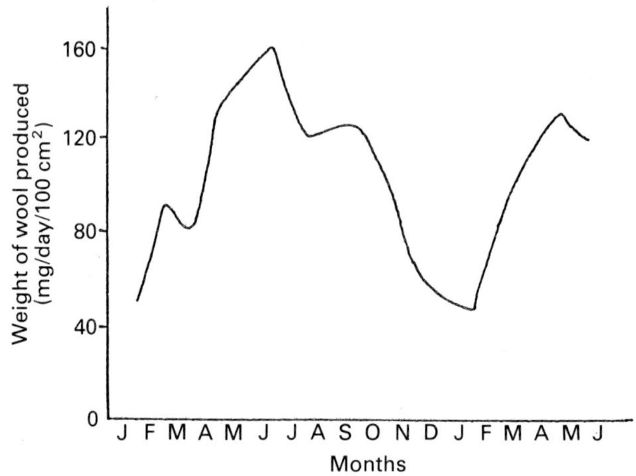

Fig. 3.1. Annual pattern of wool growth (after Story and Ross, 1960).

growth problems, and the birds compensate in growth later on. Such a regime may therefore improve the health of the birds without compromising productivity.

Sheep and deer

In growing lambs a long photoperiod, e.g. 16L, 8D has been shown to significantly increase live weight gain compared with 8L, 16D (Forbes *et al.*, 1979, 1981). However, at least 50% of the increase is due to an increase in gut contents (Forbes *et al.*, 1979, 1981), even in lambs fed at restricted intakes. Continuous light, as with cattle and poultry, is detrimental to the growth of lambs (Hulet *et al.*, 1968). Like cattle, the carcass composition of lambs on long daylength tends to show increased protein and reduced fat content (Forbes *et al.*, 1979, 1981).

As with most other species skeletal long photoperiods, e.g. 7L, 9D, 1L, 7D have been successfully used to replace long photoperiod (Foster *et al.*, 1988) and achieve the same increase in weight gain.

In deer there is also clear evidence for effects of daylength on growth rate. Abott

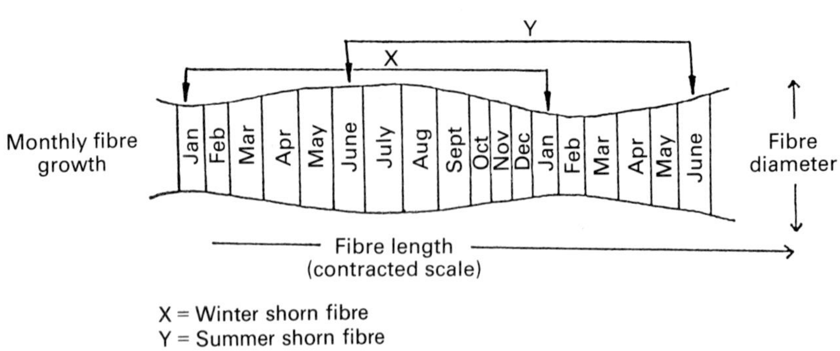

Fig. 3.2. Relation of time of shearing to wool fibre diameter (after Story and Ross, 1960).

et al. (1984) reported faster weight gains with doe fawns exposed to 8L, 16D than 16L, 8D. The increased weight gain appears to be largely accounted for by increased lipogenesis (Verme, 1988).

Body Covering

Wool

In sheep photoperiod controls the timing of the moult, now confined largely to unimproved breeds, and the growth of wool, this being minimal in winter and maximal in summer (Fig. 3.1) (Story and Ross, 1960). This is mainly a difference in wool fibre diameter (Fig. 3.2) (Story and Ross, 1960), which creates a weak point in the middle of the fibre in summer shorn sheep. Quality tests have shown that the single thread strength test is an average of 20% greater for winter shorn fibres, with an increase in the hank test of 24%. Much of the summer shorn wool therefore has to be used for carding in woollen manufacture, whereas winter shorn fibres can more often be used for combing for worsted production.

If the natural annual photoperiodic sequence is reversed, to give long days in winter and short days in summer, the periodicity of wool growth is reversed within two years. This gives further evidence for the strong influence of photoperiodic inertia in mature animals, seen also in milk production experiments.

Hair

In cattle the coat is normally thick and long in winter but short and sleek in summer. Extending the photoperiod artificially in winter prevents the growth of the winter coat (Bourne *et al.*, 1984).

Feathers

In poultry the annual moult is photoperiodically determined but is reduced in birds intensively bred for high egg production (Meyer and Millam, 1986). In these animals the moult can be abruptly induced by reducing the photoperiod to 8L, 16D. After the moult the birds go through a photorefractory period of limited reproductive potential. Photosensitivity can be restored by subjecting photorefractory birds to several weeks of non-stimulatory short photoperiod (Meyer and Millam, 1986).

Antlers

In deer, antler growth is seasonally controlled, with most rapid growth during long photoperiods in summer and least growth in winter. The cycle can be accelerated to give up to three cycles per year with artificial photoperiod control (Goss, 1969).

Puberty

Photoperiod has a strong influence on the age at puberty of strongly seasonal breeders, such as sheep and deer, in order to time the first parturition to the rest of the adult population. In farm animals without strong seasonal breeding patterns there are still some photoperiodic influences on age at puberty. Evidence suggests that in many species there is a photosensitive phase, usually at about one-half of the earliest pubertal age, when long days will stimulate early puberty.

Sheep

In the ewe lamb the timing of puberty is determined primarily by photoperiod and nutrition. The lamb must reach an adequate size before she is receptive to photostimulation. Normally the spring-born lamb will enter puberty at 6 months of age in the following autumn, whereas lambs born by ewes induced to give birth in the autumn will not enter puberty until 12 months of age, also in the following autumn.

The hormonal control of these events is shown in Fig. 3.3. The external cues are photoperiod and nutrition; if nutrition is adequate the GnRH (gonadotroph-releasing hormone) pulse generator operates at a low frequency. Photoperiod is regulated by melatonin release from the pineal gland during the dark cycle. This release pattern takes up to 20 weeks to become established in the neonate, and it is the photosensitization to a short day period by the long day melatonin secretion pattern that triggers an increase in GnRH pulse frequency and the subsequent LH release that causes ovulation. The photosensitization, or refractoriness to long days, develops soon after the long day melatonin secretion pattern is established, as a long day period of as little as one week at 20 weeks of age is sufficient to induce short day receptiveness (Yellon and Foster, 1985). However, a lamb kept constantly in short or long days from birth will not enter puberty at the normal age of 6 months. Similarly autumn-born lambs do not enter puberty until they have experienced both the long days in the following summer and the shortening days in the autumn (Foster *et al.*, 1985). The evidence is that the lamb develops a photoperiodic reference, using long days to photosensitize the lamb to enter puberty in the short days of autumn.

Cattle

In cattle, although reproduction is not limited to a particular time of year as it is in sheep, there is evidence of photoperiodic effects on puberty. Spring-born heifers are usually older than autumn-born heifers at puberty (Kamwanja, 1984). This is because the gonadotrophin production develops later in the former than in the latter. In spring-born heifers extending the photoperiod artificially to 16L, 8D during the first winter reduces the time to puberty by about 20 days and reduces the weight at puberty by about 10 kg (Petitclerc *et al.*, 1983) to 35 kg (Hansen *et al.*, 1983) but may reduce conception rate (Hansen *et al.*, 1983). Long photoperiod at this time also stimulates mammary parenchymal growth, and in postpubertal virgin heifers (Petitclerc, 1985), but has no effect on the mammary growth of pregnant heifers (Newbold *et al.*, 1989).

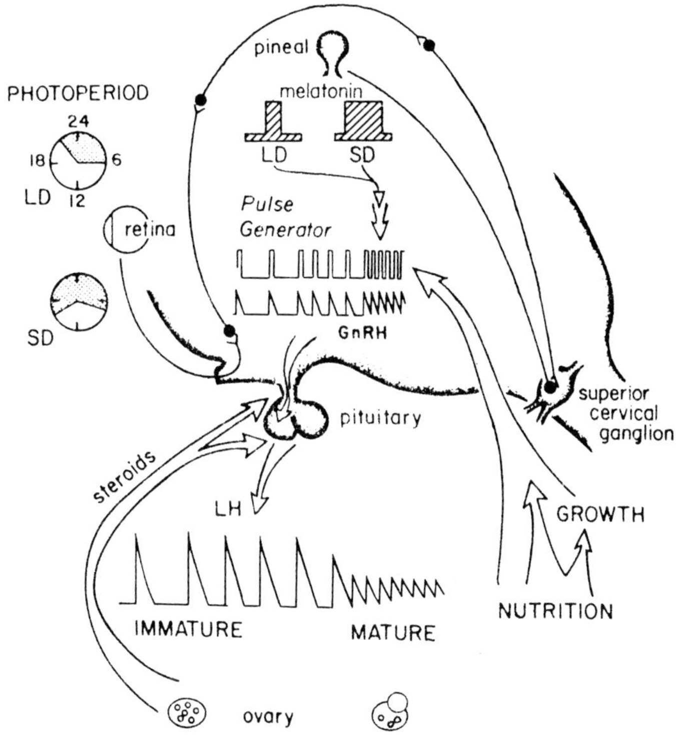

Fig. 3.3. Hypothetical control of pulsatile LH secretion in the developing female sheep through modulation of the GnRH pulse generator by internal cues (melatonin, metabolic signals). The internal cues mediate the effects of external factors (photoperiod, nutrition). Having experienced the appropriate photoperiod sequence (long day melatonin then short day melatonin) and attained a sufficient physiological size (assessed by metabolic cues) for pregnancy, the LH pulse frequency increases to produce a preovulatory follicle (after Foster *et al.*, 1985).

Pig

In the pig, spring-born gilts have delayed puberty, which occurs at a later age and greater weight than pigs born in other seasons (Mavrogenis and Robison, 1976), but they have a reduced incidence of silent oestrus compared to gilts born at other times of the year. It is likely, therefore, that long days retard reproductive development. Most researchers agree that rearing gilts in the complete dark retards reproductive development (e.g. Ntunde *et al.*, 1979), although Dufour and Bernard (1968) report accelerated puberty by 11 days under darkened conditions compared with natural light. The differences in age to puberty between artificial photoperiods of 10L and 18L appear small (Ntunde *et al.*, 1979), although Klotchov *et al.* (1971) found an increased number and size of corpora lutea in pubertal gilts with long days. In spring-born boars extension of natural photoperiod (10L) to 15L in the first winter has been shown to advance puberty (Berger *et al.*, 1980) but has little effect on semen quality (Mahone *et al.*, 1979).

Poultry

In egg strain pullets it is generally established that long daylengths up to 16–17L will advance age at first lay up to 2 weeks (Morris, 1967) and result in smaller eggs (Leeson and Summers, 1980). Continuous light is detrimental in that it results in an unacceptable incidence of blindness (Payne, 1975). Not all researchers have found large effects of photoperiod on age at maturity (e.g. Leeson and Summers, 1985), while it has even been found to be delayed by long photoperiod with a corresponding increase in egg size in meat strain pullets (Payne, 1975).

In female turkeys a short photoperiod is required for 3–6 weeks prior to lay as a conditioning period to maximize subsequent egg production (McCartney *et al.*, 1961). In male turkeys more rapid development of semen production occurs in similar short photoperiod conditions (Hocking, 1988).

Male and Female Reproduction

Sheep

Ewes normally confine their breeding season to the late summer/early autumn period, but there is a large variation between breeds, and less effect on sheep at low latitudes. The ovulation rate peaks in the middle of the breeding season but oestrous behaviour accompanies all ovulations equally. Reversing the annual photoperiod pattern will reverse the seasonality of ewe reproduction. It is interesting that the onset of oestrus in the autumn typically occurs at a considerably longer daylength (e.g. 14L in Suffolk ewes) than the onset of anoestrus in the spring (e.g. 11.5 h in Suffolk ewes; Robinson and Karsch, 1984). This suggests that the ewes cease breeding because they become photorefractory to short daylength, rather than a limiting photoperiod for reproductive activity. Using contracted photoperiodic cycles the ewe can be induced to ovulate frequently, with ovulation starting 50 days after the transition from long to short days (Ortavant *et al.*, 1988).

In the ram seasonal variation in reproductive activity is less pronounced than in the ewe, and the peak of activity precedes that in the ewe by 1.5 to 2 months, presumably to allow the rams time to find the ewes and stimulate them to enter oestrus early by pheromonal induction. In peak activity during late summer/early autumn rams normally have greater testes weight, greater spermatozoan output, better quality semen and increased libido (Ortavant *et al.*, 1988), although low forage availability at this time and high seasonal activity rates can reduce testes weight (Masters and Fels, 1984).

The control of gonadal activity is both by photoperiod and gonadal steroid production, since long daylength is inhibitory to luteinizing hormone (LH) release in intact but not in castrated or ovariectomized sheep. Thus the gonadal steroids inhibit LH release in long, increasing daylength but not in short, declining daylength. The hormonal mediator to monitor daylength from the photic cues and to induce changes in gonadotrophin production is melatonin. Exogenous melatonin administration by oral or subcutaneous implant methods will induce reproductive cyclicity in ewes and spermatogenesis in rams with increasing photoperiod (English *et al.*, 1986). However, constant cyclicity cannot be induced by repeated melatonin or short daylength administration since, after a period, gonadal steroid production increases and LH declines. Alternating long days with melatonin-stimulated short days in two month cycles can induce almost constant fertility as the gonads are inactive. This obviates the need for lightproof buildings and is particularly useful to maintain functional rams for semen production for artificial insemination, or for a frequent lambing flock (Schanbacher, 1988). The administration of long days to sheep has also been shown to reduce the gestation length of sheep by about 1 day, independent of any effect on lamb birthweight.

Cattle

In cattle there are seasonality effects on reproductive activity in the female, but seasonal anoestrus does not occur. In autumn-calving cows the interval from calving to first ovulation is less than in spring-calving cows (Roche and Boland, 1980), this arising from photoperiod rather than nutritional cues (King and Macleod, 1983). However, the delayed period to first ovulation only occurs for early spring calvers, who traditionally would have been producing their calves during the most severe feed shortage in midwinter (King and Macleod, 1983). There is also evidence that the reproductive performance of dairy and beef autumn calving cows can be improved by extending the photoperiod during the winter breeding season to 16–18L (Sweetman, 1950). This is probably a combination of several factors, none of which has been extensively researched. The interval from calving to first ovarian activity is probably reduced under long daylength (Hansen and Hauser, 1984; Garel *et al.*, 1987), oestrous exhibition may be reduced (Phillips and Schofield, 1989) but the herdsman can see cows displaying oestrus better in an extended lighting pattern.

Pigs

In the wild, pig reproduction is basically biphasic, with a primary breeding season in autumn and a secondary one in spring (Claus and Weiler, 1985). These differences occur in both the boar and sow with the latter being anoestrous in summer. They are present in the domesticated pig too, but with less pronounced variation in gonadal activity than in the wild pig.

In the boar changes in ejaculate volumes and the number of motile spermatozoa per ejaculate show a maximum in November, a decrease in February, followed by a slight increase in April/May, and a nadir in June. In the sow, weaning to oestrus interval is longest in the summer and the oestrus is prolonged, making it difficult to time accurately mating or AI. Ovulation and conception rates are reduced, and the proportion of the litter lost either as embryos or aborted fetuses is greatest for sows mated in June to September.

Artificially advancing the declining autumnal daylength to the summer has been effective in increasing boars' libido and spermatazoa production and reducing the sows' weaning to oestrus interval from 24 to 6 days (Claus and Weiler, 1985).

Behaviour

Many of the effects on production and reproduction previously described are often assumed to arise from changes in the animals' physiology, stimulated by photoreception. However, particularly in the case of effects on production, the photic stimuli change the animals' environmental perception so dramatically that there are marked effects on behaviour, which could be partially responsible for observed effects. In dairy cattle a small increase in lying and a large reduction in walking has been recorded in 18L compared with natural (10L) daylength (Phillips and Schofield, 1989). The change in locomotion may be due to the ability of subordinate cattle to see dominant cattle for most of the day and avoid unexpected confrontation. Normally subordinate cattle have to move about frequently to avoid confrontation with dominant cattle. Similar, though less pronounced effects have also been observed with group-housed calves (Weiguo and Phillips, 1991) but no effect of photoperiod on daily lying time has been observed in bulls (Dechamps *et al.*, 1989) or grazing cows (Phillips and Hecheimi, 1989). The reduction in activity observed for loose housed dairy cows, and hence their maintenance requirements, could account for at least one-half of the increased milk production observed under long daylength.

Additional benefit to production or welfare might accrue from a reduction in stress or fear in light supplemented, group housed animals. Reduced concentrations of cortisol in bulls (Leining *et al.*, 1980) and β-endorphins in sheep (Ebling and Lincoln, 1987) suggest that stress is reduced under long days. Some of the reduced stress could be a conditioned reflex as these animals are normally kept in a more benign environment (nutrition, space, etc.) in the long days of summer than the short days of winter. There is ample evidence from photoperiodic manipulation that animals develop a photoperiod memory, which can take up to two years to eradicate by photoperiodic manipulation, and the ability of a remembered visual stimulus to alleviate stress by conditioned response has been demonstrated by Willis and Mein (1983).

Within the changes in daily behaviour patterns supplementary light evokes specific responses during the period of supplementation. As farm animals are by choice diurnal, any increase in photoperiod extends the period of activity and lengthens the rests in between activity bouts. Housed dairy cattle under 18L have five meals a day compared to four in 10L (Phillips and Schofield, 1989), and in grazing cattle the number of grazing bouts is directly related, and the length of the grazing bouts inversely related, to photoperiod (Phillips and Hecheimi, 1989). Total daily feeding time is not significantly affected by photoperiod in housed or grazing cattle, at least in the short term. However, if production is increased under extended photoperiod it would be expected that feed intake might ultimately increase unless the maintenance requirement is reduced (by reduced activity) or the efficiency of digestion is increased. The latter has been observed in sheep (Forbes *et al.*, 1979, 1981). An increase in digestive efficiency could arise from a more even flow of feed to the rumen. Normally cattle spread out their feeding bouts to attempt to cover the photophase and the most intensive feeding periods are seen at dawn and dusk. The reason for this is that on the one hand an even supply of feed to the rumen will allow more efficient fermentation, as the micro-organisms will have a more constant supply of substrates (particularly the rapidly soluble cell solutes), and on the other hand they prefer not to feed in the dark, probably because of a strong vestigial anti-predator instinct.

Light intensity also affects behaviour, with high intensities stimulating activity, especially social interactions, in calves (Dannenmann *et al.*, 1984/5). The optimum light supplementation to stimulate production and reduce activity would therefore appear to be a long daylength (*c.* 16–18L) of low intensity, which is the pattern adopted in most commercial controlled environments for pigs and poultry.

References

Abott, M.F., Ullrey, D.E., Ku, P.K., Schmitt, S.M., Ronsos, D.R. and Tucker, H.A. (1984) Effect of photoperiod on growth and fat accretion in white-tailed doe fawns. *Journal of Wildlife Management* **48**, 776–787.

Berger, T., Mahone, J.P., Svoboda, G.S., Metz, K.W. and Clegg, E.D. (1980) Sexual maturation of boars and growth of swine exposed to extended photoperiod during decreasing natural photoperiod. *Journal of Animal Science* **51**, 672–678.

Bocquier, F. (1985) Influence de la photopériode et de la température sur certains équilibres hormonaux et sur les performances zootechniques de la brebis en gestation et en lactation. Thèse Doctorat Ingénieur, Institut National Agronomique Paris-Grignon.

Bodurov, N. (1979). Effect of supplementary artificial illumination with visible rays on some biochemical parameters in the blood serum, milk production and fertility in cows during lactation. *Veterinarnomeditsinki Nauki* **16**, 58–65.

Bourne, R.A., Trump, P.R. and Kirk, J.A. (1984) The effect of extended photoperiod in winter on the growth and coat characteristics of heifer calves. *Animal Production* **38** , 558 (Abstract).

Cavalchini, L.G., Cerolini, S. and Giardini, A. (1983) Light in poultry breeding. *Avicoltura* **52**, 13–17.

Classen, H.L. (1990) Effects of increasing photoperiod length on broiler productivity and health. *Proceedings of the Australian Poultry Science Symposium*, pp. 55–61.

Claus, R. and Weiler, U. (1985) Influence of light and photoperiodicity on pig prolificacy. *Journal of Reproduction and Fertility* (Suppl. 33), 185–197.

Dannenmann, K., Buchenauer, D. and Fliegner, H. (1984/5) The behaviour of calves under four levels of lighting. *Applied Animal Behaviour Science* **13**, 243–258.

Dechamps, P., Nicks, B., Canart, B. and Istasse, L. (1989) Influence of supplemental lighting on the resting behaviour of fattening bulls kept in a stanchion barn. *Applied Animal Behaviour Science* **22**, 303–311.

Dufour, J. and Bernard, C. (1968) Effects of light on the development of market pigs and breeding gilts. *Canadian Journal of Animal Science* **48**, 425–430.

Ebling, F.J.P. and Lincoln, G.A. (1987) Beta-endorphin secretion in rams related to season and photoperiod. *Endocrinology* **120**, 809–818.

Eisenman, J.R. and Chew, B.P. (1982) The effect of extended photoperiod on prepartum plasma prolactin levels and subsequent milk production in late pregnant cows. *Journal of Animal Science* **55** (Suppl. 1), 500 (Abstract).

English, J., Poulton, A.L., Arendt, J. and Symons, A.M. (1986) A comparison of the efficiency of melatonin treatments in advancing oestrus in ewes. *Journal of Reproduction and Fertility* **77**, 321–327.

Fauconneau, B. and Gauthier, D. (1984) Influence of 4 hours' supplemental lighting on growth and occurrence of puberty in Creole heifers grown in a tropical country. *Tropical Animal Production* **9**, 81–85.

Forbes, J.M. (1982) Effects of lighting pattern on growth, lactation and food intake of sheep, cattle and deer. *Livestock Production Science* **9**, 361–374.

Forbes, J.M., Driver, P.M., Brown, W.B. and Scanes, C.F. (1979) The effect of daylength on the growth of lambs. 1. Comparisons of sex, level of feeding, shearing and breed of sire. *Animal Production* **29**, 33–42.

Forbes, J.M., Brown, W.B., Al-Banna, A.G.M. and Jones, R. (1981) The effect of daylength on the growth of lambs. 3. Level of feeding, age of lamb and speed of gut-fill response. *Animal Production* **32**, 23–28.

Foster, D.L., Yellon, S.M. and Olster, D.H. (1985) Internal and external determinants of the timing of puberty in the female. *Journal of Reproduction and Fertility* **75**, 327–344.

Foster, D.L., Ebling, F.J.P. and Claypool, L.E. (1988) Timing of puberty by photoperiod. *Reproduction, Nutrition, Dévelopement* **28**, 349–364.

Garel, J.P., Gauthier, D., Petit, M. and Thimonier, J. (1987) Effect of photoperiod on changes in live weight and the development of ovarian activity postpartum in the lactating cow. *Reproduction, Nutrition, Dévelopement* **27**, 305–306.

Gooneratne, A.D. and Thacker, P.A. (1990) Influence of an extended photoperiod on sow and litter performance. *Livestock Production Science* **24**, 83–88.

Goss (1969) Cited by Linzell (1973).

Hansen, P.J. and Hauser, E.R. (1984) Photoperiodic alteration of postpartum reproductive function in suckled cows. *Theriogenology* **22**, 1–14.

Hansen, P.J., Kamwanja, L.A. and Houser, E.R. (1983) Photoperiod influences age at puberty of heifers. *Journal of Animal Science* **57**, 985–992.

Hocking, P.M. (1988) Effect of restricting body weight gain or limiting the photoperiod during rearing on sexual maturity, viability and semen production in Large White turkey males. *British Poultry Science* **29**, 531–544.

Hulet, C.V., Price, D.A. and Foote, W.C. (1968) Effects of variation in ewes during the breeding season. *Journal of Animal Science* **27**, 684–690.

Jones, D.W. (1984) NAC pig unit photoperiod trial. Report of the Meat and Livestock Commission Planning and Development Group. 22 June 1984.

Kamwanja, L.A. (1984) Regulation of onset of puberty in the bovine female: the role of season. *Dissertation Abstracts International* **45**, 1332 (Abstract).

King, G.J. and Macleod, G.K. (1983) Reproductive function in beef cows calving in the spring or fall. *Animal Reproduction Science* **6**, 255–266.

Klotchkov, P.V.A., Klotchkova, A., Kim, A.A. and Belyaev, D. (1971) Cited in Berger *et al.* (1980).

Lavenick, C.K. and Leighton, A.T. (1988) Effects of photoperiod and filtered light on growth, reproduction and mating behaviour of turkeys. 1. Growth performance of two lines of males and females. *Poultry Science* **67**, 1505–1513.

Leeson, S. and Summers, J.D. (1980) Effect of early light treatment and diet self selection on laying performance. *Poultry Science* **59**, 11–15.

Leeson, S. and Summers, J.D. (1985) Response of growing Leghorn pullets to long or increasing photoperiods. *Poultry Science* **64**, 1617–1622.

Legare, E.T., Birrenkott, G.P. and Hale, K.K. (1986) Effect of photoperiod on quail processing yields. *Poultry Science* **65**, 631–635.

Leining, K.B., Tucker, H.A. and Kesner, J.S. (1980). Growth hormone, glucocorticoid and thyroxine response to duration, intensity and wavelength of light in prepubertal bulls. *Journal of Animal Science* **51**, 932–942.

Linzell, J.L. (1973) Innate seasonal oscillations in the rate of milk secretion in goats. *Journal of Physiology* **230**, 225–233.

Mabry, J.W., Coffey, M.T. and Seerley, R.W. (1983) A comparison of an 8- versus 16-hour photoperiod during lactation on suckling frequency of the baby pig and maternal performance of the sow. *Journal of Animal Science* **57**, 292–295.

McCartney, M.G., Sanger, V.L., Brown, K.I. and Chamberlain, V.D. (1961) Photoperiodisms as a factor in the reproduction of the turkey. *Poultry Science* **40**, 368–376.

McGlone, J.J., Stansbury, W.F., Tribble, L.F. and Morrow, J.L. (1988) Photoperiod and heat stress influence on lactating sow performance and photoperiodic effects on nursery pig performance. *Journal of Animal Science* **66**, 1915–1919.

Mahone, J.P., Berger, T., Clegg, E.D. and Singleton, W.L. (1979) Photoinduction of puberty in boars during naturally occurring short day lengths. *Journal of Animal Science* **48**, 1159–1164.

Masters, D.G. and Fels, H.E. (1984) Seasonal changes in the testicular size of grazing rams. *Proceedings of the Australian Society of Animal Production* **15**, 444–447.

Mavrogenis, A.P. and Robison, O.W. (1976) Factors affecting puberty in swine. *Journal of Animal Science* **42**, 1251–1252.

Melhorn, G. and Dorn, W. (1985) Visible light regime and live-weight development in pigs. In: Hilliger, H.G. (ed.), *Proceedings 5th International Congress for Animal Health, Hannover*, pp. 10–13.

Meyer, W.E. and Millam, J.R. (1986) Primary feather moult and serum luteinizing hormone concentration in chukar partidge (*Alectoris chukar*) during a photoperiodically induced moult with or without fasting. *Poultry Science* **65**, 1615–1622.

Morris, T.R. (1967) Light requirements of the fowl. In: Carter, T.C. (ed.), *Environmental Control in Poultry Production.* Oliver and Boyd, Edinburgh, pp. 15–39.

Murrill, F.D., Eide, R.N., Leonard, R.O. and Bath, D.L. (1969) Effects of corral lighting on feed intake and milk production of dairy cattle. *Journal of Dairy Science* **52**, 947 (Abstract).

Newbold, J.A., Chapin, L.T., Zinn, S.A., Hughes, T.L. and Tucker, H.A. (1989) Effects of photoperiod on mammary growth in pregnant Holstein heifers. *Journal of Dairy Science* **72** (Suppl. 1), 310–311 (Abstract).

Ntunde, B.N., Hacker, R.R. and King, G.J. (1979) Influence of photoperiod on growth, puberty and plasma LH levels in gilts. *Journal of Animal Science* **48**, 1401–1406.

Okamoto, S., Nagata, S., Kobayashi, S. and Matsuo, I. (1989) Effects of photoperiod and cage density on growth and feed conversion in large and small quail lines selected for body weight. *Japanese Poultry Science* **26**, 150–156.

Ortavant, R., Pelletier, J., Ravault, J.P., Thimonier, J. and Volland-Nail, P. (1985) Photoperiod: main proximal and distal factor of the circannual cycle of reproduction in farm mammals. *Oxford Review of Reproductive Biology* **7**, 305–345.

Ortavant, R., Bocquier, F., Pelletier, J., Ravault, J.P., Thimonier, J. and Volland-Nail, P. (1988) Seasonality of reproduction in sheep and its control by photoperiod. *Australian Journal of Biological Science* **41**, 69–85.

Payne, C.G. (1975) Daylength during rearing and the subsequent egg production of meat-strain pullets. *British Poultry Science* **16**, 559–563.

Perera, A.N.M. and Hacker, R.R. (1982) The influence of light pulses on bovine lactation. *Journal of Animal Science* **55** (Suppl. 1), 197–198.

Peters, R.R., Chapin, L.T., Leining, K.B. and Tucker, H.A. (1978) Supplemental lighting stimulates growth and lactation in cattle. *Science* **199**, 911–912.

Peters, R.R., Chapin, L.T., Emery, R.S. and Tucker, H.A. (1981) Milk yield, feed intake, prolactin, growth hormone and glucocorticoid response of cows to supplemental light. *Journal of Dairy Science* **64**, 1671–1678.

Petitclerc, D. (1985) Influence of diet and photoperiod on growth of dairy heifers and mammary gland development. Bulletin No. 8. Station de Recherches de Lennoxville, 19–23.

Petitclerc, D., Chapin, L.T., Emery, R.S. and Tucker, H.A. (1983) Body growth, growth hormone, prolactin and puberty response to photoperiod and plane of nutrition in Holstein heifers. *Journal of Animal Sciences* **57**, 892–898.

Petitclerc, D., Chapin, L.T. and Tucker, H.A. (1984) Carcass composition and mammary development responses to photoperiod and phase of nutrition in Holstein heifers. *Journal of Animal Science* **58**, 913–919.

Phillips, C.J.C. and Hecheimi, K. (1989) The effect of forage supplementation, herbage and season on the ingestive behaviour of dairy cows. *Applied Animal Behaviour Science* **24**, 203–216.

Phillips, C.J.C. and Schofield, S.A. (1989) The effect of supplementary light on the production and behaviour of dairy cows. *Animal Production* **48**, 293–303.

Robinson, F.E. and Karsch, F.J. (1984) Refractoriness to inductive day lengths terminates the breeding season of the Suffolk ewe. *Biology of Reproduction* **31**, 656–663.

Roche, J.F. and Boland, M.P. (1980) Effect of extended photoperiod in winter on growth rate of Friesian male cattle. *Irish Journal of Agricultural Research* **19**, 85–90.

Roche, J.F. and Enright, W.J. (1988) Photoperiodic effects on cattle growth and reproduction. Technical report No. 4. Dairy Research Unit, University of Wales, Bangor, pp. 16–19.

Rutberg, A.T. (1987) Adaptive hypotheses of birth synchrony in ruminants; an interspecific test. *American Naturalist* **130**, 692–710.

Schanbacher, B.D. (1988) Responses of market lambs and Suffolk rams to a stimulatory skeleton photoperiod. *Reproduction, Nutrition, Dévelopement* **28**, 431–441.

Stanisiewski, E.P., Mellenberger, R.W., Anderson, C.R. and Tucker, H.A. (1985) Effect of photoperiod on milk yield and milk fat in commercial dairy herds. *Journal of Dairy Science* **68**, 1134–1140.

Story, L.F. and Ross, D.A. (1960) Effect of shearing time on wool. VI. Rate of growth of wool and its relation to time of shearing. *New Zealand Journal of Agricultural Research* **3**, 113–124.

Sweetman, W.J. (1950) Artificial breeding in Alaska and the effect of extra light during the winter days. *Journal of Dairy Science* **33,** 391 (Abstract).

Terqui, M., Delouis, C. and Ortavant, R. (1984) Photoperiodism and hormones in sheep and goats. In: Roche, J.F. and O'Callaghan, D. (eds.) *Manipulation of Growth in Farm Animals*. Martinus Nijhoff, The Hague, pp. 246–259.

Tucker, H.A. (1985) Photoperiodic influences on milk production in dairy cows. In: Haresign, W. and Cole, D.J.A. (eds), *Recent Advances in Animal Nutrition*. Butterworths, London, pp. 211–221.

Verme, L.F. (1988) Lipogenesis in buck fawn white-tailed deer: photoperiod effects. *Journal of Mammology* **69**, 67–70.

Weiguo, L. and Phillips, C.J.C. (1991) The effects of supplementary light on the behaviour and performance of calves. *Applied Animal Behaviour Science* **30**, 27–34.

Willis, G.L. and Mein, G. (1983) Classical conditioning of milk ejection using a novel conditioned stimulus. *Applied Animal Ethology* **9,** 231–237.

Wright, I.A., English, P.R. and Robertson, A.M. (1984) Daylength and reproductive development in pigs. *Farm Building Progress* **75**, 21–23.

Yellon, S.M. and Foster, D.L. (1985) Alternate photoperiods time puberty in the female lamb. *Endocrinology* **116**, 2090–2097.

Zinn, S.A., Chapin, L.T. and Tucker, H.A. (1986a) Response of body weight and clearance and secretion rates of growth hormone to photoperiod in Holstein heifers. *Journal of Animal Science* **62**, 1273–1278.

Zinn, S.A., Purchas, R.W., Chapin, L.T., Petitclerc, D., Merkel, R.A., Bergen, W.G. and Tucker, H.A. (1986b) Effects of photoperiod on growth, carcase composition, prolactin, growth hormone and cortisol in prepubertal and post-pubertal Holstein heifers. *Journal of Animal Science* **63,** 1804–1815.

Zinn, S.A., Chapin, L.T. and Tucker, H.A. (1988) A note on the effects of previous photoperiod exposure and gradual transitions of light intensity at dawn and dusk on growth in Holstein heifers. *Animal Production* **46,** 300–303.

Zinn, S.A., Chapin, L.T., Enright, W.J. and Tucker, H.A. (1989) Growth, carcass composition and serum hormone responses to photoperiod and ovariectomy in heifers. *Animal Production* **49**, 365–373.

4 The Space Requirements of Housed Livestock

M.R. Baxter

Centre for Rural Building, Aberdeen, UK

Summary

Space allocation is one of the most important single variables affecting the welfare of farm livestock. In addition, insufficient space allocation has been shown to affect growth and reproductive performance and has been strongly argued to affect health.

Space is not a simple homogeneous variable but can be categorized into a number of functional subdivisions: resting space, feeding space, social space, etc. The concept is further complicated by the fact that space requirements increase as animals grow.

This chapter describes a biological approach to the study of space requirements, where space is expressed as functions of animal body size. These allometric expressions are then used to describe the space required by livestock for the performance of different activities. Finally, the concept of a universal space requirement will be explored where one allometric expression could represent the minimum space requirement of all livestock.

Introduction

From research studies, space allocation is known to affect both the performance and welfare of livestock. An analysis to date from 29 published papers on pig performance in relation to space allocation, found a significant improvement in both growth and feed conversion efficiency as space was increased (Kornegay and Notter, 1984). Similar effects have been shown from single experiments with poultry (Hughes, 1975; Hill, 1977) and cattle (Arave *et al.*, 1974). In terms of welfare, reduced space allocation has been related to frustration (Nicol, 1987; Nicol and Guildford, 1991), social instability (Jensen, 1982), greater stress (Ladewig *et al.*, 1985), greater aggression (Bryant and Ewbank, 1972) and greater restlessness. In practice, therefore, the decision about how much space to give livestock is an important one. It is made all the more important because the space in any given pen is determined

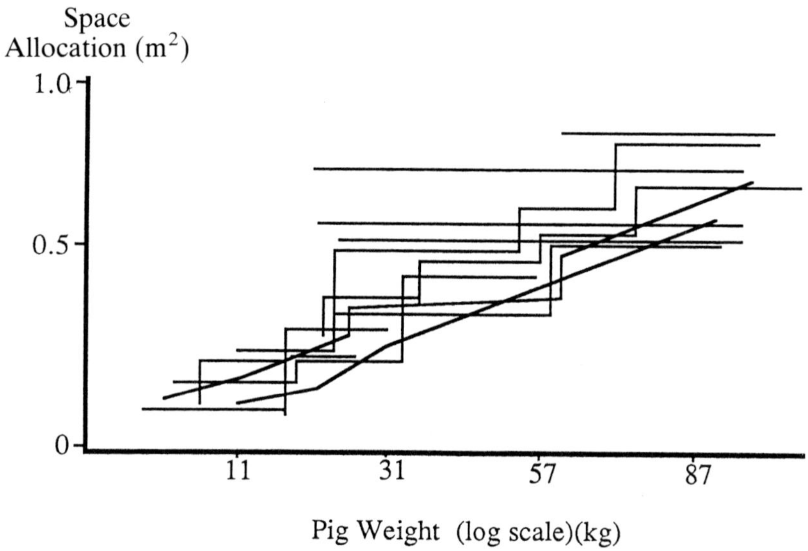

Fig. 4.1. Recommended space allocations for pigs published in the advisory literature (after Petherick, 1982).

by structural features – walls and pen gates – and cannot readily be changed, once built.

Unfortunately, the decision is not an easy one. Certainly, scientific studies suggest that, within limits, more space gives better performance and improved welfare. That, however, is very different from establishing minimum space requirements. The concept of minimum space allocation is confounded by differences between animals and differences in their environmental circumstances. What size of animal? Hot or cold weather? Bedded, concrete or slatted floors? Minimum in what respect – productivity, return on capital investment or welfare? The problem with overgeneralizing becomes clear looking at recommended space allocations for pigs from the published advisory literature (Fig. 4.1). Clearly, for the pig farmer who has to decide what size to build his weaner pens or whether to put an extra pig in his existing pens, these recommendations are less than ideal.

A general problem here appears to be that we don't understand what determines an animal's requirement for space. Looking at the natural space requirement of livestock species does not appear to help much. Pigs in a natural environment have home ranges of up to 800 ha (Kurz and Marchington, 1972), a space requirement which Baxter (1986a) estimated to be 15 million times the current recommendation for intensively housed pigs. It might conceivably be possible to reduce this figure by subtracting space which is not needed under intensive husbandry conditions (e.g. the spatial distribution of food sources over which the pig must forage). With such a large difference, though, there is little prospect of this being an accurate way of assessing space requirements for confined livestock.

A more attractive alternative, and one which has now been adopted in many studies, is to identify a baseline in space requirements and add to that as different spatial needs are identified. The minimum space requirement of an animal under any circumstances will be its body size: it must have space to physically contain its body mass. If body size could be accurately estimated, taking into account breed differences and animals of different weights, then considerable progress would have been made towards overcoming the difficulties with space requirements outlined above: the baseline would have been established.

The Principle of Allometry

A fundamental problem in estimating the body size of animals is finding some way of expressing size in relation to live weight. Allometry provides the solution to precisely this problem. In terms of simple geometry, a piglet or a calf is, roughly speaking, just a scaled down version of a mature pig or a mature cow. This means that, within species, animals of all sizes can be considered as geometrically similar figures. As a consequence of this, animal body dimensions should vary according to geometric scaling rules.

For ease of illustration, an example is given using these rules and employing a cube as an example (big cubes and little cubes also being geometrically similar figures).

The volume of a cube (V) is equal to its length (L) to the power of three and the area (A) of one surface of that cube is equal to its length squared. From these two basic principles, several others can be derived.

$$\text{If } V = L^3 \text{ then } L = V^{0.33}$$

$$\text{If } A = L^2 \text{ then } L = A^{0.5}$$

$$\text{If } L = V^{0.33} \text{ and } A = L^2 \text{ then } A = V^{0.67}$$

To deal with less regular shapes than cubes we need to add another term to the expressions. Consider changing our cube into a cuboid of the same volume but twice the height and half the width.

$$\text{Height} = 2 \times V^{0.33}$$

$$\text{Width} = 0.5 \times V^{0.33}$$

The general expression for this is $L = k \times V^{0.33}$ where k is a coefficient expressing the degree of departure from a perfect cube.

Reverting to livestock, we would expect all the above principles to hold, with an accuracy dependent on their geometric similarity. If, in addition, we can assume there are no significant differences in the specific gravity of livestock as they grow, then body dimensions can be related to weight rather than volume. We would, therefore, expect livestock body dimensions to vary according to the following expression: $L = kW^{0.33}$ where L is the body dimension, W is live weight and k is an empirically derived coefficient. Expressions of this type are not new in biology. Many biologists from the 1920s onwards have studied the effects of scale on animal biology (e.g. Huxley, 1932; Pedley, 1977). They used the term allometry to describe these relationships with body weight.

Table 4.1. Allometric relationship between live weight and body dimensions in pigs (from Petherick, 1983b).

Body dimension	Best-fit regression	0.33 power constraint
Breadth at shoulders	$0.059W^{0.34}$	$0.061W^{0.33}$
Length tail to scapula	$0.172W^{0.36}$	$0.185W^{0.33}$
Height at shoulders	$0.166W^{0.29}$	$0.150W^{0.33}$

Body dimensions are in metres and live weight in kilograms.

Livestock Size and Body Space Requirements

Most of the research work in this area has been on pigs so they will be reviewed first.

Pigs

Petherick (1983b) presented data on the allometric relationships between body dimensions of pigs and live weight. The best fit regressions were found to be very close to the expected expression $L = kW^{0.33}$ and it was concluded that this adequately described the data (Table 4.1).

The simplest applications of these allometric expressions is where a single body dimension is related to some unidimensional space requirement. Baxter (1986b), for example, plotted several recommendations for feeding trough space found in advisory literature and national standards (Fig. 4.2). The lower end of these recommendations was found to fit quite well between 1 and 1.1 times pig shoulder width.

Baxter and Schwaller (1983) related the three main body dimensions of sows (length, breadth and height) to the corresponding dimensions of commercially available farrowing crates (Fig. 4.3). From this, several crates were found to be too short to accommodate the 95th percentile of sows. This conclusion has been supported by reports from pig producers that they cannot fit their large sows into certain proprietary crates. One producer had to go to the trouble of welding 300 mm extensions to the back of all his crates to accommodate his older sows.

Taking allometric expressions a stage further, we can use pairs of body dimensions to estimate floor area requirements. Petherick and Baxter (1981) used data on the length, breadth and height dimensions of the body to represent a plan view of pigs lying down either fully recumbent (Fig. 4.4b) or in a more upright posture with all four legs tucked under their bodies (Fig. 4.4a). In an attempt to validate these estimates Baxter and Zappavigna (unpublished study) photographed pigs in thermoneutral conditions and measured their space occupancy. This was found to range from 0.024 to $0.29W^{0.67}$ depending on the extent to which their hind legs were tucked under their body. Edwards *et al.* (1987) conducted a study of pig performance in which space was varied to maintain allometric space allocations (i.e. the space increased as the pigs grew, according to a two-thirds power of live weight). They concluded that $0.027W^{0.67}$ was the minimum space requirement for pigs on fully slatted floors, without reducing growth rates and food conversion efficiency. From

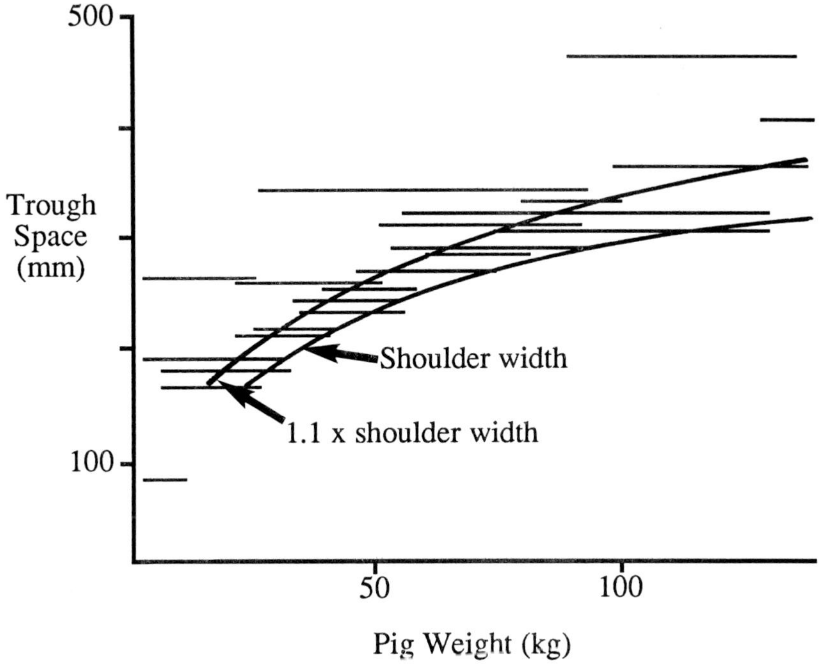

Fig. 4.2. Trough space allocations for pigs published in the advisory literature (after Baxter, 1986b).

the photographic study cited above, this would appear to be the space required to lie down in thermoneutral conditions.

Beyond Body Space

Allometric expressions of animal body dimensions appear to give useful estimates of both linear space requirements and resting space requirements, but can they be used to estimate overall space requirements? In principle they should. It is difficult to imagine any component of space requirement which would not increase as the animal grows. An argument can even be made in this respect for social space requirements.

The aggressive interactions between pigs consist of many behavioural elements (Jensen, 1980, 1982). Different elements have been shown by Jensen (1982) to serve different functions throughout a fight. For example, 'head-to-head' and 'head-to-body' knocks are aggressive behaviours whereas 'head-tilt' and 'retreat' are submissive behaviours. The importance of providing space for the pigs to perform such behaviours is suggested from Jensen's (1982) finding that pigs in the smaller of two space allocations formed a less stable hierarchy. The behaviour which is most demanding of space is 'parallel/inverse pressing', a frequently observed behavioural element during dominance formation. This appeared to be the pigs' way of assessing their opponents and hence its disruption might well interfere with the pigs' assessment of each other and hence with dominance formation. Baxter (1985) explored the

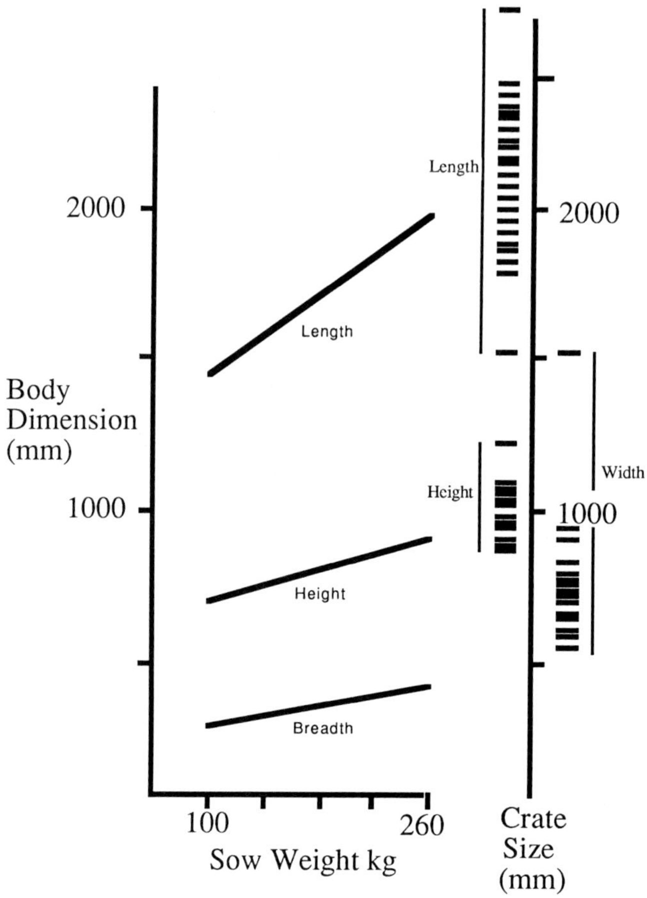

Fig. 4.3. The size of sows compared to the size of farrowing crates (after Baxter and Schwaller, 1983).

implications of this behaviour for space requirements. As shown in Fig. 4.5 this behaviour will require a clear space of two pig body lengths (tail to shoulder). If the circling behaviour, which is commonly seen while pigs are in this parallel/inverse pressing position, is an important part of the behaviour then a circle with a diameter of two pig body lengths will be the minimal space requirement. Clearly bigger pigs will require more space than smaller pigs for this behaviour and again allometry would seem to give us a way of estimating this space requirement.

Other Livestock Species

Poultry

Studies by Bognor *et al.* (1979), Freeman (1983) and Dawkins and Hardie (1989) have all measured the space occupied by hens. Their results (Table 4.2) show some

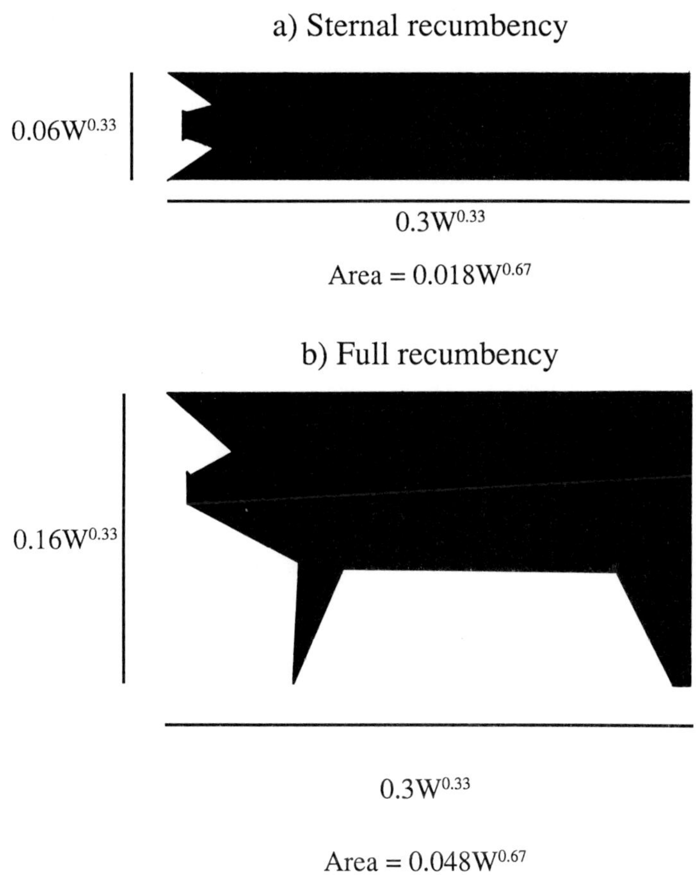

Fig. 4.4. Estimated floor area requirements of pigs based on pairs of body dimensions (after Petherick and Baxter, 1981).

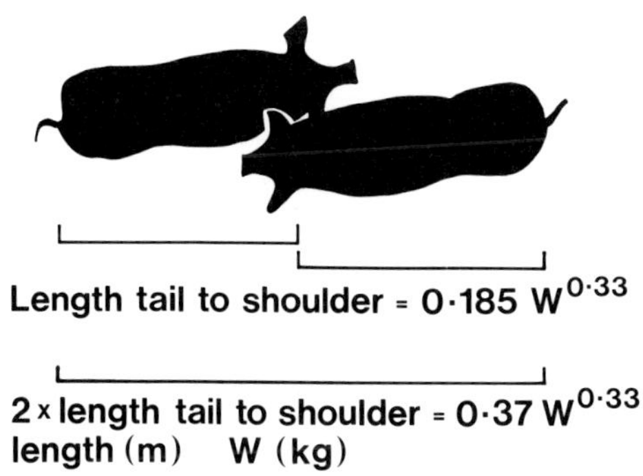

Fig. 4.5. Social space requirements based on allometry – the space required for 'parallel-inverse pressing' during dominance formation (after Baxter, 1985).

Table 4.2. Published estimates of space occupied by resting hens.

Source	Average live weight, *W* (range)	Average area occupied, *A* (range)	Allometric expression
Bognor *et al.*, 1979	1.8 kg (unknown)	0.0424 m^2 (unknown)	$A = 0.029W^{0.67}$
Freeman, 1983	2.1 kg (1.0–4.0 kg)	0.0569 m^2 (0.0384–0.0992 m^2)	$A = 0.035W^{0.67}$
Dawkins and Hardie, 1989	2.1 kg (1.9–2.6 kg)	0.0475 m^2 (0.0428–0.0592 m^2)	$A = 0.029W^{0.67}$

measure of agreement, although Freeman's data suggest considerably greater areas than both Bognor *et al.* (1979) and Dawkins and Hardie (1989). In his paper Freeman urges caution in interpreting his data since the birds he studied were pure bred rather than modern commercial hybrids.

The extra space needed for active behaviour was measured by both Bognor *et al.* (1979) and Dawkins and Hardie (1989) (Table 4.3) but with a considerable measure of disagreement in their results. It should be noted that the methods of Bognor *et al.* (1979) raise doubt about the accuracy of their measurements – their camera was fitted with a 50 mm lens which will lead to substantial parallax errors resulting from movements of the birds away from and towards the camera. It is, however, clear that all the empirical evidence suggests that hens need substantially more space than the 450 cm^2 provided in the current EC directive (European Community Directive 86/113/EEC).

Cattle

From the recommendations in BSI (1981) both the length and width of cow cubicles appear to vary with the expected one-third power of body weight. Using data on cattle ranging from 75 to 600 kg, cubicle length was found to vary with $0.31W^{0.31}$ ($r^2 = 0.993$) and width varied with $0.15W^{0.32}$ ($r^2 = 0.996$). Again constraining the

Table 4.3. Extra space needed by hens for active behaviour.

	Source	
	Bognor *et al.*, 1979	Dawkins and Hardie, 1989
Resting space (baseline)	$0.029W^{0.67}$	$0.029W^{0.67}$
Feather ruffling	Baseline + 17%	Baseline + 84%
Preening	Baseline + 19%	Baseline + 142%
Body Shake	Baseline + 24%	No data
Wing spreading/ stretching	Baseline + 27%	Baseline + 88%
Turning	No data	Baseline + 166%
Wing flapping	No data	Baseline + 295%

Table 4.4. Cattle body dimensions and recommended cubicle dimensions.

	Length (m)	Width (m)
Body dimensions (recalculated from Enomoto *et al.*, 1976)	$0.28W^{0.33}$	$0.13W^{0.33}$
Cubicle dimensions (recalculated from BSI, 1981)	$0.28W^{0.33}$	$0.14W^{0.33}$

exponent to 0.33 reduces the coefficient for cubicle length to 0.28 and for cubicle width to 0.14. Enomoto *et al.* (1976) published data on the body dimensions of cattle, but unfortunately analysed them using linear regressions rather than the logarithmic regressions required for allometry. Reanalysing the data (estimated by eye from the published graphs), logarithmic regressions appear to fit the data equally well. The best fit expressions for body length and width while lying were $0.238W^{0.36}$ and $0.134W^{0.33}$ respectively (the r^2 values were 0.97 and 0.85 but some caution is required in their interpretation since error was introduced by estimating the data from the published graphs). Constraining the length expression to an exponent of 0.33 changes the coefficient to 0.28. This means that the length and width of cattle while lying down, as estimated from body dimension data, is identical to the length and width of cubicles, as estimated from published recommendations (Table 4.4). This suggests that cattle cubicles are intended to be only just big enough to accommodate the cattle while lying. Clearly, from this, problems would be expected from cubicles designed on an underestimate of the weight of the cattle or in failing to accommodate the larger animals within the naturally occurring variation in live weight.

Sheep

There are many recommended space allocations for sheep in the advisory literature. One such set of recommendations were originally prepared by Cunningham and Soutar (1967), and have been widely used subsequently (e.g. Williams, 1970; Robertson, 1978; BSI, 1981; Bryson, 1984). After conversion to metric units, Cunningham and Soutar's recommended space allocation for slatted floors regresses against live weight according to the expression $A = 0.063W^{0.64}$ (where A is floor area (m^2) and W is live weight (kg), $r^2 = 0.997$), a remarkably close fit to the two-thirds relationship expected from allometric principles. Indeed, considering the few data available, it was considered best to constrain the regression to a 0.67 exponent, a change which reduces the coefficient to 0.055. It should be remembered that these recommendations are for total space allocation, not just resting space, as has been the case for the other species so far discussed. A similar analysis of the data on trough space gives a best fit regression of $TS = 0.095W^{0.37}$ (where TS is trough space (m) and W is live weight (kg), $r^2 = 0.999$). Again, constraining the regression to the expected one-third exponent gives a coefficient of 0.112. Treating these expressions as a baseline, we can now consider space requirements under different circumstances. Again from the data in Cunningham and Soutar (1967), keeping sheep on straw bedding appears

Table 4.5. Allometric expressions estimating the space requirements of sheep.*

Housing conditions	Floor area (m^2)	Trough space (m)
Unshorn, fully slatted pens (baseline)	$0.055W^{0.67}$	$0.112W^{0.33}$
Straw bedded pens	Baseline + 30%	Baseline
Ewe with lambs	Baseline + 30%	Baseline
Shorn ewe	Baseline − 15%	Baseline − 15%
Horned breeds	Baseline + 17%	Baseline + 17%

*See text for data sources.

to increase their space requirements by 30%. It appears that this increase in space allocation is not so much for the sheep as to keep the straw from getting too dirty. A similar increase is required to provide room for the lambs once they arrive. From other sources (Loynes, 1983; MAFF, 1983), shearing the sheep is estimated to reduce their floor area requirement by 15%. It would be expected that the main effect of shearing would be to alter the body width of the sheep and that their trough space requirement would be correspondingly reduced. From recommendations given by Dickson and Stephenson (1979) both the floor area and trough space requirements can be calculated to increase by 17% for horned breeds (although some authorities dispute that horned breeds need extra space; A.M. Robertson, personal communication). All of this can be summarized as a set of allometric expressions as shown in Table 4.5.

An Estimate of Space Use Common to All Livestock?

Interestingly, all the estimates of space occupied by livestock bear remarkable similarities (see Table 4.6). Bognor *et al.* (1979) and Dawkins and Hardie (1989) estimates of the area occupied by birds while resting are $0.029W^{0.67}$. This is at the high end of the range measured for pigs, which can perhaps be explained by the space needed for the birds' plumage. This generalization can only be continued to the cattle measurements (Enomoto *et al.*, 1976), by firstly estimating the area of a rectangle described by body length and body width and then estimating what proportion of

Table 4.6. An estimate of resting space requirement common to all livestock?

Species	Origin of estimate	Allometric expression
Poultry (1–4 kg)	Bognor *et al.*, 1979, Dawkins and Hardie, 1989; measurements of space occupied	$0.029W^{0.67}$
Pigs (1–300 kg)	Baxter and Zappavigna (unpublished); measurements of space occupied	$0.24W^{0.67}$ up to $0.29\ W^{0.67}$
Cattle (50–600 kg)	Enomoto *et al.*, 1976; fraction of measured length × breadth	$0.24W^{0.67}$

that rectangle is occupied by the cow's body. The length × breadth rectangle is $0.0375W^{0.67}$ and of this, 0.64 is estimated (from a plan drawing provided in Enomoto *et al.*, 1979) to be occupied by the cow. This gives $0.024W^{0.67}$ for the space occupied by a cow while lying down, a value which coincides with the lower end of the measured range for pigs. Considering that the hens are smaller than the heavier cattle by a factor of 400, it is remarkable that their estimated resting space requirements vary by only 16%. This suggests that $0.027W^{0.67}$ would probably be a reasonable estimate of the space occupied by any animal while resting, with the possible exception of animals of very different geometry (e.g. snakes).

Context Dependent Space Requirements

So far, space requirements have been discussed as if they were fixed and immutable, but in reality they are often much more flexible and dependent upon a variety of factors. A simple study by Oester (1977) recorded the flight distance of cows before and after dehorning and demonstrated a marked reduction after dehorning. This would suggest that the overall space requirements may be greater for horned than unhorned cows (see below for the effect of horns on space requirements of sheep). In a similar vein, the space required to accommodate aggressive behaviour in pigs also appears to be variable. When pigs are mixed, at weaning, considerable fighting occurs until a stable dominance hierarchy is established and this can often result in the pigs becoming heavily scratched and cut. McGlone and Curtis (1981) argued that this aggression was exacerbated by the fact that pigs could not escape from their aggressor when they wanted to submit. To test this they provided a series of holes in the pen walls in which the pigs could hide their head and shoulders. This was reported to reduce the duration of fighting by 40% and reduce the area of wounding resulting from fights by 50%, as compared to a similar pen without the hides. Space allocation was not varied in this experiment and hence it is impossible to say whether extra space would have reduced the fighting even further. However, the important principle from this study is that space requirements are not independent of the environmental conditions provided within that space. Allowing the submissive pig to get away from its attacker could presumably have also been achieved by providing a lot of extra space. This experiment of McGlone and Curtis could be envisaged as demonstrating a trade-off between space and environmental resources.

Potter and Broom (1987) reported a similar finding with dairy cattle. Studying cubicle use, they found that low ranking cows spent significantly longer standing with their front feet in the cubicles than did high or mid-ranking cows. The cubicles they chose to use for this were the central cubicles in a row which were furthest from areas of commotion near feeders and drinking bowls. They argue that the cubicles are being used 'as a zone where the effective personal distance is increased by the bars of the cubicle'.

This trade-off between space and environmental resources was examined more directly in a study of feeder space requirements (Baxter, 1986b). Six experimental treatments were studied as shown in Fig. 4.6. Three of these treatments had different feeder space allocations (1.1, 2.2 and 4.4 times shoulder width), whereas the other three had different trough barriers between the pigs. Aggressive behaviour was not

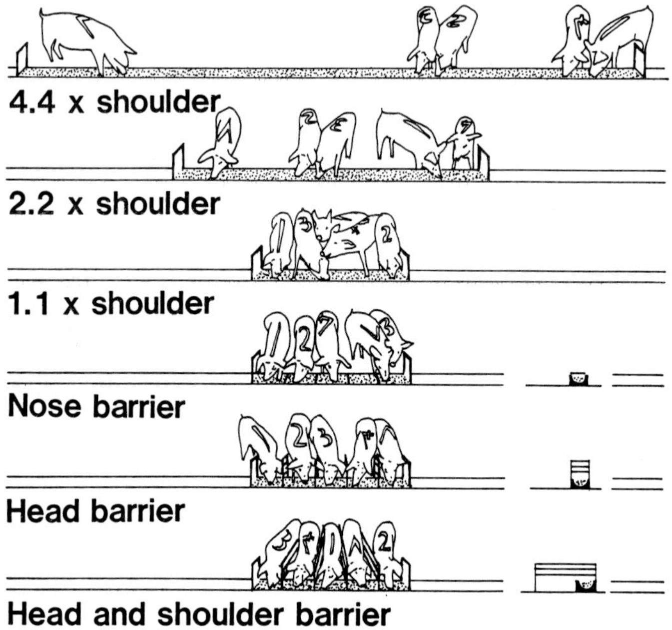

Fig. 4.6. An experiment to study the trade-off between space and environmental resources in pigs (after Baxter, 1986b).

significantly affected by the different space allocations although the trend was for less aggression with more feeder space. The head barrier treatment did, however, reduce aggression and the head and shoulder barrier virtually eliminated it (Fig. 4.7). So long as there is sufficient space to give all the pigs access to the trough, there seems to be little advantage in providing extra space. Improving the feeding conditions by adding trough barriers, on the other hand, makes a substantial difference.

Another factor which can have a large influence on space requirements is the number of animals in the social group. Bryant and Ewbank (1972) first pointed out that increasing the group size has the effect of increasing the amount of free space available to the animals. The space allocated to any animal will have to accommodate its body size and leave a certain amount of space for movement. In group housing, this movement space can be shared which means that when all but one of the animals is lying down, the active animal has free use of a very much larger space than if it was housed individually. Published recommendations on space allocation have occasionally recognized this. For example, Soutar (1965) proposed that sheep housed in slatted pens in groups of 50 require 65 cm^2 per animal but in groups of 200 this drops to 46 cm^2. Also, the EC regulations on space allocations in battery cages for laying hens, which were mentioned in the introduction, give different recommendations for different numbers of hens. A single hen must be given 1000 cm^2, two hens must have 750 cm^2 per bird, three hens 550 cm^2 per bird and four or more hens 450 cm^2 per bird.

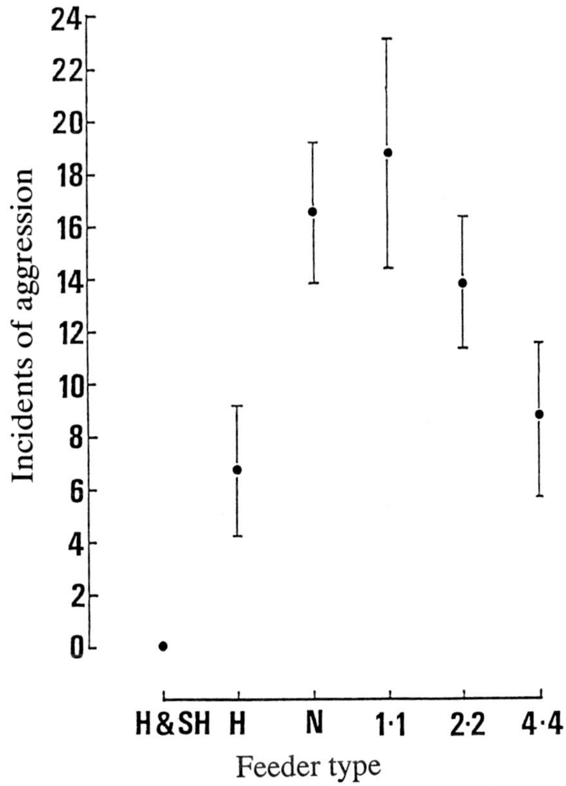

Fig. 4.7. Results of space versus environmental resources experiment (after Baxter, 1986b).

Conclusions

Space requirements for housed livestock remain poorly understood and we are still some way from being able to allocate space accurately on the basis of the animal's needs. Describing space requirements in terms of allometry appears to at least provide a sensible foundation to build upon. This is a parallel concept to the use of metabolic live weight (weight raised to the three-quarters power) to describe animal energetics. Both take account of variability due to animal size and expose size-independent sources of variation. It is hoped that the prospect of raising the study of livestock space requirements to a thorough and empirical scientific discipline will substantially improve our understanding of spatial needs in years to come.

References

Arave, C.W., Albright, J.L. and Sinclair, C.L. (1974) Behaviour, milk yield and leucocytes of dairy cows in reduced space and isolation. *Journal of Dairy Science* **57**, 1497–1501.

Baxter, M.R. (1985) Social space requirements of pigs. In Zayan, R. (ed.), *Social Space for Domestic Animals.* Martinus Nijhoff, The Hague, pp. 116–127.

Baxter, M.R. (1986a) Pig space requirements. *Pig Veterinary Society Proceedings* **15**, 56–65.

Baxter, M.R. (1986b) The design of the feeding environment for the pig. PhD Thesis, University of Aberdeen.

Baxter, M.R. and Schwaller, C.E. (1983) Space requirements for sows in confinement. In: Baxter, S.H., Baxter, M.R. and MacCormack, J.A.D. (eds), *Farm Animal Housing and Welfare.* Martinus Nijhoff, The Hague, pp. 181–195.

Bognor, H., Peschke, V., Seda, V. and Popp, K. (1979) Studie Zum Flachenbedarf von Legehennen in Kafigen bei bestimmten Aktivaten. *Berl. Munch. Tierarztl. Wschr.*, **92,** 340–343.

BSI (1981) British standard code of practice for the design of buildings and structures for agriculture. BS 5502: Section 2.2 (Livestock buildings). British Standards Institution, London.

Bryant, M.J. and Ewbank, R. (1972) Some effects of stocking rate and group size upon agonistic behaviour in groups of growing pigs, *British Veterinary Journal* **128**, 64–70.

Bryson, T. (1984) *The Sheep Housing Handbook.* Farming Press, Ipswich, UK.

Cunningham, J.M.M. and Soutar, D.S. (1967) Sheep housing. *Scottish Agriculture*, winter.

Dawkins, M.S. and Hardie, S. (1989) Space needs of laying hens. *British Poultry Science* **30**, 413–416.

Dickson, I.A. and Stephenson, D.E. (1979) The housing of ewes. West of Scotland Agricultural College Technical Note No. 63.

Edwards, S.A., Armsby, A.W. and Spechter, H.H. (1987) Space allowances for growing pigs on fully slatted floors. *Animal Production* **44,** 490–491.

Enomoto, H., Iketaki, T., Hidari, H. and Suzuki, S. (1976) Dimensions of dairy cattle for use in the design of dairy buildings and equipment. *Research Bulletin of Obihiro University* **10**, 129–140.

Ewbank, R. and Bryant, M.J. (1972) Aggressive behaviour amongst groups of domesticated pigs kept at various stocking rates. *Animal Behaviour* **20**, 21–28.

Freeman (1983) *Veterinary Record* **113**, 562–563.

Hill, A.T. (1977) The effects of space allowance and group size on egg production traits and profitability. *British Poultry Science* **18,** 483–492.

Hughes, B.O. (1975) The concept of an optimum stocking density and its selection for egg production. In: Freeman, B.M. and Boorman, K.N. (eds), *Economic Factors Affecting Egg Production. British Poultry Science*, Edinburgh, pp. 271–298.

Huxley, J.S. (1932) *Problems of Relative Growth.* Methuen, London.

Jensen, P. (1980) An ethogram of social interaction patterns in group-housed dry sows. *Applications of Animal Ethology* **6**, 341–350.

Jensen, P. (1982) An analysis of agonistic interaction patterns in group-housed dry sows – aggression regulation through an 'avoidance order'. *Applications of Animal Ethology* **9**, 47–61.

Kiley-Worthington, M. (1977) *Behavioral Problems in Farm Animals.* Oriel Press, Stocksfield, UK.

Kornegay, E.T. and Notter D.R. (1984) Effects of floor space and number of pigs per pen on performance. *Pig News and Information,* **5**, 23–33.

Kurz, J.C. and Marchington, R.L. (1972) Radiotelemetry studies of feral hogs in South Carolina. *Journal of Wildlife Management* **36**, 1240–1248.

Ladewig, J., Schlichting, M.C., Beneke, B., von Borell, E., Stuhec, I. and Smidt, D. (1985). Physiological aspects of social space in heifers and pigs. In Zayan, R. (ed.), *Social Space for Domestic Animals.* Martinus Nijhoff, The Hague, pp. 151–159.

Loynes, I.J. (1983) Sheep house design. In: *Housing Sheep.* Farm Buildings Information Centre, pp. 4–9.

McGlone, J.J. and Curtis, S.E. (1981) A behaviour/performance study to evaluate an alternative nursery pen for swine. *Journal of Animal Science* **51**, 129–130.

MAFF (1983) Winter shearing of the housed flock. In: *Housing Sheep*. Farm Buildings Information Centre, pp. 12–13.

MAFF (1987) Codes of recommendations for the welfare of domestic fowls. Ministry of Agriculture, Fisheries and Food Leaflet No. 703.

Nicol, C.J. (1987) Behavioural responses of laying hens following a period of spatial restriction. *Animal Behaviour* **35**, 1709–1719.

Nicol, C.J. and Guildford, T. (1991) Exploratory behaviour as a measure of motivation in deprived hens. *Animal Behaviour* **41**, 333–341.

Oester, H. (1977) Auswirkungen der enthornung bei schweizer braunvieh milchkuhen im tiefstreulaufstall auf allgemeine und spezielle activitaten. *Liz. Arbeit. Zool. Institut*, Bern. (Cited in Metz, J.H.M. and Wierenga H.K (1987) Behavioural criteria for the design of housing systems for cattle. In: Wierenga, H.K. and Peterse, D.J. (eds), *Cattle Housing Systems, Lameness and Behaviour*. Martinus Nijhoff, The Hague, pp. 14–25.)

Pedley, T.J. (1977) *Scale Effects in Animal Locomotion*. Academic Press, London.

Petherick, J.C. (1982) Biological basis of space in pig housing. MSc Thesis, University of Aberdeen.

Petherick, J.C. (1983) A note on allometric relationships in Large White × Landrace pigs. *Animal Production* **36**, 497–500.

Petherick, J.C. and Baxter, S.H. (1981) Modelling the static spatial requirements of livestock. In MacCormack, J.A.D. (ed.), *Modelling, Design and Evaluation of Agricultural Buildings*. Scottish Farm Buildings Investigation Unit, Aberdeen, pp. 75–82.

Potter, M.J. and Broom, D.M. (1987) The behaviour and welfare of cows in relation to cubicle house design. In Wierenga, H.K. and Peterse, D.J. (eds), *Cattle Housing Systems, Lameness and Behaviour*. Martinus Nijhoff, The Hague, pp. 129–147.

Robertson, A.M. (1978) Sheep housing. *Farm Building Progress* **51,** 1–4.

Soutar, D.S. (1965) Contemporary livestock housing systems. *Journal of the Institute of Agricultural Engineering* **21**, 112–121.

Williams, L.J. (1970) Design, construction and costs. In: *Housing Sheep*. Farm Buildings Centre Report No. 13, pp. 3–10.

5 Ventilation

C.M. Wathes
Agricultural and Food Research Council, Silsoe Research Institute, Silsoe, UK

Summary

'Good' ventilation of livestock buildings has long been thought of as one of the Ten Commandments of intensive animal husbandry. Although thermal criteria for ventilation can be readily described for animal performance in terms of air velocity and temperature, there is much less information on the role of the aerial environment in animal health. Recent epidemiological and laboratory studies are beginning to define critical burdens of exposure to the common aerial pollutants of dust and ammonia through their effects on the incidence and severity of respiratory diseases, such as atrophic rhinitis in pigs and chronic pulmonary disease in horses. Evidence is also accumulating for the airborne transmission of primary pathogens, not only of the respiratory tract but also of the gut. These criteria can then be translated into specifications for air change rate, flow pattern and cubic capacity for those major domesticated species which are housed intensively. Given such specifications, ventilation systems can, in turn, be designed according to the aims of 'good' health, welfare and performance.

Introduction

'Good' ventilation of animal houses has long been considered one of the Ten Commandments of animal husbandry. Indeed, it is enshrined in the current UK Codes of Livestock Welfare. The main aim of ventilation is to provide an aerial environment in which an animal's health can be maintained and its productivity sustained. This simple statement of good intent conceals at least three primary mechanisms by which ventilation affects health and performance, namely aerial spread of pathogens and pheromones, regulation of thermal exchanges, and maintenance of air quality. Analysis of these mechanisms in turn, leads to specifications for airflow pattern, air exchange rate and cubic capacity of the air space. These three characteristics of a ventilation system are almost inextricably linked.

Air flow through an animal house can be driven by either natural or mechanical means. This chapter is not concerned with details of either system of ventilation since the main principles of operation are now well established in theory and proven in practice (e.g. Carpenter, 1981; Randall, 1981; Randall and Boon, in press). Instead, ventilation is considered with regard to its biological consequences. Furthermore, a distinction is often made between so-called 'controlled environment' and climatic houses, which are conventionally suited to pigs and poultry and to ruminants and horses respectively. The former description implies complete definition of an animal's environment: in practice, among the atmospheric variables, only air temperature and perhaps air speed and airflow pattern are fixed while humidity and the concentration of aerial pollutants are uncontrolled. An attempt is also usually made to eliminate spatial and temporal variations, which thereby purposely deny animals the opportunity to express their individual preferences, e.g. for heat.

Influence of Ventilation upon Health and Performance

Aerial spread of pathogens and pheromones

Many diseases of farm animals are known to be transmitted through the air. Table 5.1 gives common examples for pigs and poultry (Wathes, 1987). For some micro-organisms, the aerial pathway is the normal route, e.g. the hardy spores of fungi and some bacteria, in contrast to the majority of bacteria and viruses which are ill-equipped to survive the rigours of aerial suspension. Nevertheless, sufficient numbers of the latter can survive to initiate infection, especially where the shedding rate is high, the atmospheric conditions are (relatively) favourable for survival, and the infective dose is low. Foot-and-mouth disease virus provides the best example of a disease which can be transmitted aerially, even over long distances of up to several kilometres (Donaldson, 1983).

Aerial speed is not confined to respiratory pathogens nor does it imply that the respiratory tract is the sole route of inoculation. Other portals include the conjunctiva (Moore, 1957), while ingestion of microbes from surfaces contaminated by deposition may provide the necessary dose of infection. Trillat and Kaneko (1921) showed that mice could be infected by aerosol with *Salmonella typhimurium* (Danysz strain) with a lower infection dose than needed orally. This finding was extended to calves by Wathes *et al.* (1988). Further evidence for the importance of non-contagious modes of transmission (including the aerial route) comes from an epidemiological study by Hardman *et al.* (1991), who showed that contagion accounts for about 40% of secondary cases of infection with *S. typhimurium* among calves penned individually. The conclusion to be drawn from these findings is that air mixing should be minimal in animal houses.

Contrariwise, thorough mixing of air within a livestock building is most desirable where animals are kept for breeding. For example, olfactory stimulation by sex pheromones will help induce oestrus in prepubertal gilts, encourage sexual development in immature boars and increase copulatory performance in mature boars (Baxter, 1984). It is likely that similar responses occur in other farm mammals.

Table 5.1. Common pathogens of pigs and poultry known to be transmitted aerially (Wathes 1987).

Bacteria	
Bordetella bronchiseptica	*Mycobacterium tuberculosis*
Brucella suis	*Mycoplasma gallisepticum*
Corynebacterium equi	*Mycoplasma hyorhinus*
Erysipelothrix rhusiopathiae	*Mycoplasma suipneumoniae*
Escherichia coli	*Pasteurella multocida*
Haemophilus gallinarus	*Pasteurella pseudotuberculosis*
Haemophilus parasuis	*Salmonella pullorum*
Haemophilus pleuropneumoniae	*Salmonella typhimurium*
Listeria monocytogenes	*Staphylococcus aureus*
Leptospira pomona	*Streptococcus suis* type II
Mycobacterium arium	
Fungi	
Aspergillus flavus	*Coccidioides immitis*
Aspergillus fumigatus	*Cryptococcus neoformans*
Aspergillus nidulans	*Histoplasma farcinorum*
Aspergillus niger	*Rhinosporidium seeberi*
Rickettsia	
Coxiella burnetii	
Protozoa	
Toxoplasma gondii	
Viruses	
African swine fever	Infectious nephrosis of fowls
Avian encephalomyelitis	Infectious porcine encephalomyelitis
Avian leukosis	Marek's disease
Foot-and-mouth disease	Newcastle disease
Fowl plague	Ornithosis
Hog cholera	Porcine enterovirus
Inclusion body rhinitis	Swine influenza
Infectious bronchitis of fowls	Transmissible gastroenteritis of swine
Infectious laryngotracheitis of fowls	

Thermal environment

The predominant objective over the past 30 years of environmental manipulation in pig and poultry housing has been control of the thermal environment. This has been achieved primarily by design and operation of ventilation systems which regulate air change rate and provide a stable airflow pattern, though other factors such as insulation, feeding level and stocking density are also influential. Except where heating is provided for neonates, environmental temperature within livestock buildings will be coupled closely with air temperature, with the corollary that adjustments to building ventilation rate will control effectively heat losses from individual animals. Airflow patterns affect thermal exchanges directly via air speeds over animals, with a draught conventionally defined as a speed in excess of 0.3 m s^{-1}, and indirectly via deliberate creation of temperature gradients within a building, e.g. to discourage pigs from defecating in the lying area (Randall and Armsby, 1983).

Regulation of environmental temperature within the narrow limits of thermoneutrality is a sound policy for pig and poultry production and, in the absence of other criteria, is used routinely to determine air change rate (see Bruce, 1981; Charles, 1981). The 'minimum' ventilation rate under cold conditions is in fact a misnomer since the rate is the fastest which can be permitted without lowering air temperature below a set point. Where the purpose of ventilation is to dilute aerial contaminants, then the rate adopted represents a true minimum since it is the slowest needed to maintain acceptable concentrations. The 'maximum' ventilation rate in hot weather is the slowest rate consistent with limiting the rise of building air temperature to 2–3°C above ambient.

Manipulation of environmental temperature for ruminants and horses kept indoors and sheltered from extremes of wind and rain is unnecessary and undesirable, exept for the young or sick (Webster, 1981). There is therefore no need to restrict air flow through cattle sheds or stables and other factors, such as stocking density, assume greater importance in the maintenance of health (see below).

Air quality

Air quality is almost inevitably compromised once animals are housed at the high stocking densities typical of intensive production. The design and operation of a ventilation system can play an important role in determining air quality in livestock buildings. The other major factors are stocking density, method and type of feeding and bedding, and the system of waste management. The most common aerial pollutants are fine organic dusts from raw or partially digested feed, desquamated skin cells or faeces, noxious or irritant gases such as NH_3 and H_2S and commensal micro-organisms (as distinct from primary pathogens). Chronic exposure to these pollutants may increase host susceptibility to respiratory diseases, such as atrophic rhinitis of pigs and enzootic pneumonia of calves, but at present there is insufficient knowledge to set tolerance limits based on the criterion of disease incidence. Instead, human occupational limits are adopted (see Donham, 1987). These are based on weekly exposures to single pollutants lasting 40 h: animals may be exposed continuously without remission to several or more pollutants simultaneously.

Inhalation of organic dusts elicits specific immune responses, which may induce airway hypersensitivity following repeated exposure (Rylander, 1986). Alternatively, these dusts may behave like some inorganic dusts and cause mechanical blockage of macrophages (Gilmour *et al.*, 1989). Ammonia irritates reversibly the mucous membranes of the respiratory tract and has been shown in pigs to reduce bacterial clearance from the lungs (Drummond *et al.*, 1978), depress growth rate (Drummond *et al.*, 1980) and exacerbate turbinate atrophy (Drummond *et al.*, 1981). In poultry, acute exposure to 20 ppm NH_3 increased the susceptibility of fowls to infection with Newcastle disease virus (Anderson *et al.*, 1964). These laboratory experiments on animal responses to aerial pollutants have been supported by several epidemiological studies of the benefits of air filtration in calf (Pritchard *et al.*, 1981) and pig (Carpenter *et al.*, 1986) houses. More recently, detailed observations of pig health from weaning to slaughter convincingly implicated exposure to aerial pollutants as a major risk factor for atrophic rhinitis (Robertson *et al.*, 1990). Taken together the above evidence suggests that poor air quality in animal houses raises the incidence and severity of infectious respiratory diseases of young farm animals. Unfortunately, tolerance limits have yet to be defined.

There have been numerous studies of the natural history of air hygiene in animal houses (see review by Wathes, 1987). The overwhelming conclusion is that the air teems with a microbial miasma, whose composition and numbers vary greatly according to the husbandry system and climate. Commensal skin bacteria will be carried aloft on rafts of skin squamae, e.g. *Staphylococcus* spp., while fungal propagules of *Aspergillus* spp., *Penicillium* spp. and *Scopulariopsis* will also be found (Madelin and Wathes, 1989) and will dominate the air spora where poor quality hay or straw is used. The concentrations of airborne microbes will be up to 10^6 times higher than those found outdoors. Whether or not these burdens are harmful to animal health is a matter for speculation.

Strategies for Ventilation

Pollutant dynamics

The concept of a balance sheet of sources and sinks provides a useful framework within which various strategies for air change rate and cubic capacity can be discussed. Figure 5.1 (after Wathes *et al.*, 1983), shows the influence of air change rate upon the concentration of an aerial contaminant, measured in arbitrary units. The calculations are based on a simple steady-state model of air quality in a calf house and assume perfect mixing from a uniform internal source in a constant climate. The concentration of gases falls rapidly with faster ventilation rate and well illustrates the law of diminishing returns. Particulate matter, such as airborne dusts, is also cleared by sedimentation at an overall rate dependent on aerodynamic size. No account is taken here of the influence of turbulence structure of the airflow pattern on either a microscopic or macroscopic scale, which may keep particles in suspension for longer periods than would otherwise occur. Assuming a particle diameter of 6 μm, then the equivalent air change rate for settling is five air changes per hour (ach). Moreover, losses due to inactivation of viable microbes in aerosol can also be considered in ach equivalents, typically 10 ach for those airborne micro-organisms common in animal houses. However, the most astonishing conclusion to be drawn from these simple calculations concerns the relative benefits of doubling air space allowance from 6 to 12 m^3 per head. For viable microbes, the reduction in concentration achieved by doubling cubic capacity can only be reproduced by raising ventilation rates from 6 to 30 ach. Such a rapid rate could only be realized in practice by creating large openings in the building or by using fans of larger capacity, both of which would then be uncomfortably draughty. Clearly, the maintenance of satisfactory air quality involves attention to both ventilation rate and stocking density.

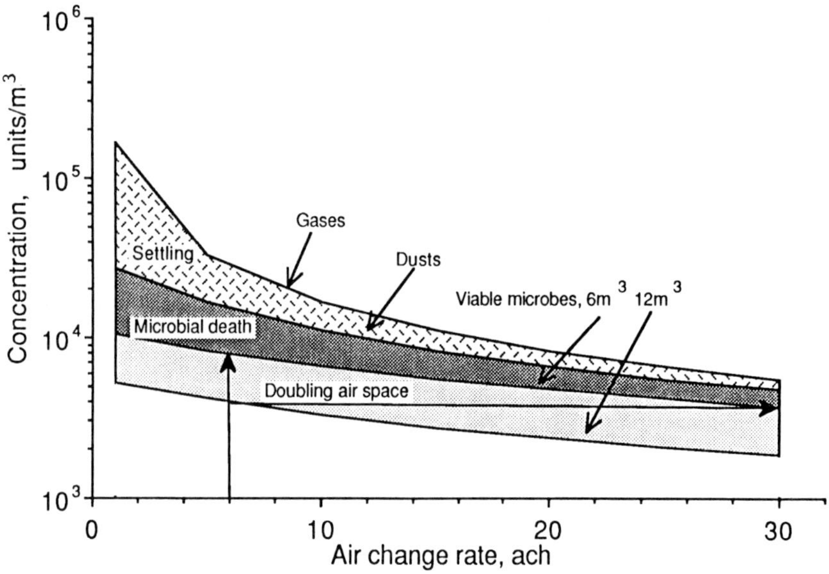

Fig. 5.1. Model relationship between air change rate and aerial concentration of gases, respirable dusts and viable microbes in a calf house stocked at 6 m^3 per calf. The clearance rates for settling and microbial death are equivalent to 5 and 10 air changes per hour respectively. The additional benefit of increasing air space allowance to 12 m^3 per calf is shown for viable microbes alone. The solid-headed arrows indicate the equivalence of doubling air space compared to raising air change rate from 6 to 30 ach (after Wathes *et al.*, 1983).

Ventilation rate, cubic capacity and airflow pattern

A strategy for ventilation of livestock buildings must specify the (minimum) ventilation rate and cubic capacity, both of which must be related to airflow pattern. Sometimes independent specifications can be made, as shown above, for the concentration of a gas generated internally, which increases *pro rata* with emission rate irrespective of cubic capacity. In this case ventilation rates can be expressed as m^3 per h per head or per kg (Wathes *et al.*, 1983). However, where other mechanisms of clearance are involved, such as sedimentation or inactivation of viable microorganisms, then ventilation rate is more properly expressed in ach and must be related to cubic capacity. The above model of air quality illustrates quite simply the relative benefits of increasing air change or cubic capacity.

A major flaw in the design and operation of ventilation systems in livestock (and other) buildings arises from the assumption of perfect mixing of incoming with ambient air. All practical designs are invariably based on air flows at the building's inlets and exhausts. The ideal flow pattern of plug or displacement flow is difficult to establish within the large airspace of an animal house, though it has been produced effectively within the confines of a negative-pressure isolator for laboratory animals

(Wathes and Johnson, 1991). A pressurized ventilation system with incoming air delivered via a glass-fibre ceiling and exhausted via a perforated floor, as practised in many deep-pit laying houses, provides the nearest approximation to plug flow in livestock buildings. Recently Breum *et al.* (1990) quantified the effectiveness of air distribution in a piggery and observed a range of up to 15% in mixing efficiency at different locations. A theory of air mixing is described by Barber and Ogilvie (1982).

Further departures from theory may follow from spatial inhomogeneities in the distribution of sources of heat, pollutants and other atmospheric constituents, which may thereby cause corresponding vertical and horizontal gradients of concentration. Large, open-span buildings, such as a modern deep litter house for 30 000 broilers, are essentially uniform in source strengths as shown by the spatial homogeneity of most pollutant concentrations (Conceiçào *et al.*, 1989). This finding has importance for the random siting of samplers for bioaerosols and other pollutants, though there will still be a requirement to consider the number of replicates needed for statistical accuracy. Conversely, in a complex building, e.g. a farrowing house for sows and piglets with partially slatted floors, the spatial inhomogeneity of pollutants can be expected to be greater.

Practical recommendations for ventilation rate, cubic capacity and sometimes airflow pattern have been made for poultry (Charles, 1981), pigs (Bruce, 1981), cattle (Webster, 1988) and horses (Wathes, 1989). Surveys of ventilation rates are described for cattle by MacCormack *et al.* (1984) and horses by Jones *et al.* (1987).

Conclusions

The link between ventilation and animal health has long been recognized by veterinarians and farmers alike. While the financial consequences of over- or under-ventilation are quite clear in terms of temperature control and hence production, similar cost/benefit relationships are not yet available for aerial pollutants. Once tolerance limits have become established, then reliable sensors, which are capable of working in the harsh environment of an animal house, will need to be developed to control ventilation systems.

References

Anderson, D.P., Beard, C.W. and Hanson, R.P. (1964) The adverse effects of ammonia on chickens including resistance to infections with Newcastle disease virus. *Avian Diseases* **8**, 369–379.

Barber, E.M. and Ogilvie, J.R. (1982) Incomplete mixing in ventilated airspaces. Part 1 – Theoretical considerations. *Canadian Agricultural Engineering* **24**, 25–29.

Baxter, S. (1984) *Intensive Pig Production.* Granada, London.

Breum, N.O., Takai, H. and Rom, H.B. (1990) Upward vs. downward ventilation air flow in a swine house. *Transactions of the American Society of Agricultural Engineers* **33**, 1693–1699.

Bruce, J.M. (1981) Ventilation and temperature control criteria for pigs. In: Clark, J.A. (ed.), *Environmental Aspects of Housing for Animal Production.* Butterworths, London, pp. 197–216.

Carpenter, G.A. (1981) Ventilation systems. In: Clark, J.A. (ed.), *Environmental Aspects of Housing for Animal Production.* Butterworths, London, pp. 331–350.

Carpenter, G.A., Cooper, A.W. and Wheeler, G.E. (1986) The effect of air filtration on air hygiene and pig performance in early-weaner accommodation. *Animal Production* **43**, 505–515.

Charles, D.R. (1981) Practical ventilation and temperature control for poultry. In: Clark, J.A. (ed.), *Environmental Aspects of Housing for Animal Production.* Butterworths, London, pp. 183–195.

Conceiçào, M.A.P., Johnson, H.E. and Wathes, C.M. (1989) Air hygiene in a pullet house: spatial homogeneity of aerial pollutants. *British Poultry Science* **30**, 765–776.

Donaldson, A.I. (1983) Quantitative data on airborne foot-and-mouth disease virus: its production, carriage and deposition. *Philosophical Transactions of the Royal Society, London B* **302**, 529–534.

Donham, K.J. (1987) Human health and safety for workers in livestock housing. In *Latest Developments in Livestock Housing. Proceedings of Commission Internationale du Génie Rural,* Section 2 Seminar. Illinois, pp. 86–95.

Drummond, J.G., Curtis, S.E. and Simon, J. (1978) Effects of atmospheric ammonia on pulmonary clearance in the young pig. *American Journal of Veterinary Research* **39**, 211–212.

Drummond, J.G., Curtis, S.E., Simon, J. and Norton, H.W. (1980) Effects of aerial ammonia on growth and health of young pigs. *Journal of Animal Science* **50**, 1085–1091.

Drummond, J.G., Curtis, S.E., Meyer, R.C., Simon J. and Norton, H.W. (1981) Effects of atmospheric ammonia on young pigs experimentally infected with *Bordetella bronchiseptica. American Journal of Veterinary Research* **42**, 963–968.

Gilmour, M.I., Baskerville, A., Taylor, F.G.R. and Wathes, C.M. (1989) The effect of titanium dioxide inhalation on the pulmonary clearance of *Pasteurella haemolytica* in the mouse. *Environmental Research* **50**, 157–172.

Hardman, P.M., Wathes, C.M. and Wray, C. (1991) Salmonella transmission among calves penned individually. *Veterinary Record* **129**, 327–329.

Jones, R.D., McGreevy, P.D., Robertson, A., Clarke, A.F. and Wathes, C.M. (1987) A survey of the designs of racehorse stables in the South West of England. *Equine Veterinary Journal* **19**, 454–457.

MacCormack, J.A.D., Clarke, J.J. and Knowles, L.C. (1984) Survey of naturally ventilated buildings for beef cattle. *Farm Building Progress* **78,** 31–35.

Madelin, T.M. and Wathes, C.M. (1989) Air hygiene in a broiler house: comparison of deep litter with raised netting floors. *British Poultry Science* **30**, 23–37.

Moore, B. (1957) Observations pointing to the conjunctiva as the portal of entry in salmonella infection of guinea-pigs. *Journal of Hygiene* **55**, 414–433.

Pritchard, D.G., Carpenter, G.A., Morzaria, S.P., Harkness, J.W., Richards, M.S. and Brewer, J.I. (1981) Effect of air filtration on respiratory disease in intensively housed veal calves. *Veterinary Record* **109**, 5–9.

Randall, J.M. (1981) Ventilation system design. In: Clark, J.A. (ed.), *Environmental Aspects of Housing for Animal Production.* Butterworths, London, pp. 351–369.

Randall, J.M. and Armsby, A.W. (1983) Cooling gradients across pens in a finishing piggery. 1. Measured cooling gradients. *Journal of Agricultural Engineering Research* **28**, 235–245.

Randall, J.M. and Boon, C.R. Ventilation control and systems. In: Charles, D.R. and Wathes, C.M. (eds), *Livestock Housing.* CAB International, Wallingford, UK (in press).

Robertson, J.F., Wilson, D. and Smith, W.J. (1990) Atrophic rhinitis: the influence of the aerial environment. *Animal Production* **50,** 173–182.

Rylander, R. (1986) Lung diseases caused by organic dusts in the farm environment. *American Journal of Industrial Medicine* **10**, 221–227.

Trillat, A. and Kaneko, R. (1921) Activité de l'infection par voie aerienne. *Comptes Rendus de l'Academie des Sciences, Paris* **173**, 109–111.

Wathes, C.M. (1987) Airborne microorganisms in pig and poultry houses. In Bruce, J.M. and Sommer, M. (eds), *Environmental Aspects of Respiratory Disease in Intensive Pig and Poultry Houses including the Implications for Human Health.* Commission of the European Communities, Luxembourg, pp. 57–71.

Wathes, C.M. (1989) Ventilation of stables. *Farm Buildings and Engineering* **6**, 21–25.

Wathes, C.M. and Johnson, H.E. (1991) Physical protection against airborne pathogens and pollutants by a novel animal isolator in a level 3 containment laboratory. *Epidemiology and Infection* **107**, 157–170.

Wathes, C.M., Jones, C.D.R. and Webster, A.J.F. (1983) Ventilation, air hygiene and animal health. *Veterinary Record* **113**, 554–559.

Wathes, C.M., Zaidan, W.A.R., Pearson, G.R., Hinton, M. and Todd, J.N. (1988) Aerosol infection of calves and mice with *Salmonella typhimurium*. *Veterinary Record* **123**, 590–594.

Webster, A.J.F. (1981) Optimal housing criteria for ruminants. In: Clark, J.A. (ed.), *Environmental Aspects of Housing for Animal Production.* Butterworths, London, pp. 217–232.

Webster, A.J.F. (1988) Beef cattle and veal calves. In: *Management and Welfare of Farm Animals*, 3rd edition. Universities Federation for Animal Welfare. Baillière Tindall, London, pp. 47–79.

6 Walking and Lying Surfaces in Livestock Houses

C. NILSSON

Department of Farm Buildings, Swedish University of Agricultural Sciences, Lund, Sweden

Summary

The biological requirements of domestic animals with respect to the floors in livestock houses are discussed. The requirements are transformed into technical terms, and suitable design solutions that fulfil these requirements are indicated. The animals dealt with are dairy cows, sows and their litters and growing–finishing pigs. The analysed floor properties are mainly thermal comfort, softness, friction and abrasiveness.

The requirements of cattle and pigs with respect to the thermal characteristics of the lying area are analysed with thermodynamic models. The requirements of animals differ widely depending on the animal species, body weight, food intake, production, air temperature and radiation. The design of the floor must consider all these factors.

Using preference tests with dairy cows, it was observed that a bed thickness of 10–15 cm (sawdust) is considered by the animals as optimal. On the basis of choice studies and softness tests of the floor material, graphs can be drawn describing the range of softness for floor materials.

In order to avoid risk of slipping, floors in livestock houses must have a certain friction. The requirements set by cattle and pigs with respect to the floor friction have been analysed and recommendations are given.

In order to get a non-slip floor the surface can be made too rough. A floor with a high degree of abrasiveness may injure the animal. The abrasiveness can be measured by a technical method and using injury studies of animals a recommended roughness value can be set.

Mechanical and chemical attacks on the surface may wear down the floor. Special requirements must be observed concerning the floor materials, for example concrete, in order to give a long lasting construction.

Introduction

The general trend in modern animal production is, in view of the present economic situation, to reach optimum production at the lowest possible cost. This goal has been attained by an extensive mechanization of farm labour, e.g. animal care and feed handling.

Housing systems, primarily designed to cut down labour requirements, have been introduced. Building new animal houses and rebuilding old ones using a wide variety of new building materials, fittings and equipment which are different from the environment in the traditional animal building, requires an increased knowledge of environment–animal interaction.

When changing or developing new floor constructions or surfaces for animals, it is important to determine the requirements set by the animals. The floor is of particular interest since this is the part of the building with which the animal will come into closest contact.

What requirements do the animals have with regard to the surfaces where they both stand and lie? This is a complex question and impossible to answer unequivocally, although it is often answered in biological terms. However, for a designer or for a floor manufacturer it is of interest to have most of these requirements expressed in objective technical terms. This will enable new floors to be developed on the basis of theoretical calculations, and the manufacturing process to be monitored using appropriate techniques, to be certain that it is correct.

When developing new floors, consideration must be given to many factors, the most important of which is to maintain and improve animal health (Jongebreur, 1983). The most important physical properties of the walking and lying surfaces in livestock buildings are:

- thermal comfort
- softness
- friction
- abrasiveness.

Thermal Characteristics

Temperature extremes may cause injuries directly, such as burns or frostbite. One of the reasons for housing domestic animals is to control the climate they are exposed to, especially in extreme climates. An animal is also in a position to control its own heat loss, for instance, to the floor, by varying its lying position, behaviour and physiology.

The requirements of animals with respect to thermal factors is to a great extent determined by their heat production. Mount (1974) described temperature regulation in warm-blooded animals and the production of sensible and latent heat. The amount of heat production and loss via the skin primarily depends on the species, size, production stage and ambient climate. In the thermoneutral zone, just above the lower critical temperature, a 'zone of thermal comfort' has been defined, which is

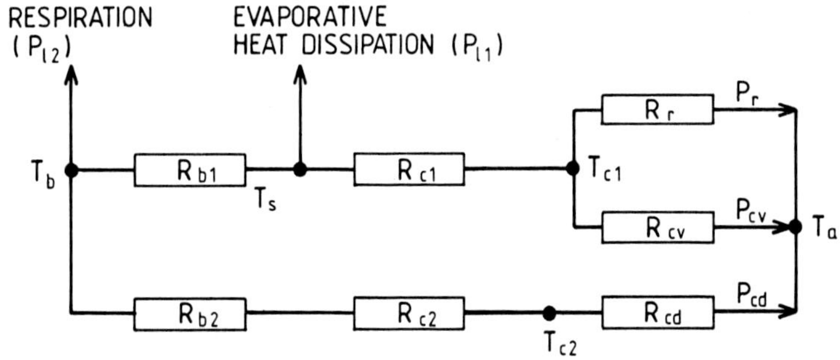

Fig. 6.1. Model of thermal network showing heat losses from the animal's body: T_a, air temperature; T_b, deep body temperature; T_s, skin temperature; T_{c1}, coat outside temperature (air contact); T_{c2}, coat outside temperature (floor contact); R_{b1}, body thermal resistance (air contact); R_{b2}, body thermal resistance (floor contact); R_{c1}, coat thermal resistance (air contact); R_{c2}, coat thermal resistance (floor contact); R_{cd}, floor effective thermal resistance; R_{cv}, convective heat transfer thermal resistance; R_r, radiant heat transfer thermal resistance; P_{cd}, conductive heat loss; P_{cv}, convective heat loss; P_r, radiant heat loss (after Nilsson, 1988).

considered to be the area preferred by an animal (Mount, 1974; Webster, 1974). Therefore, the goal must be to maintain animals under these conditions.

Many investigations have been made to determine the heat production of animals and to develop appropriate models. A summary of this work is found in, among others, Bruce and Clark (1979) for growing–finishing pigs and Ehrlemark (1988) for dairy cows. Reviews by the CIGR (1984) and Gustafsson (1988) describe models of heat production for the most common livestock. An animal's requirement with respect to heat flow to the floor varies greatly. Factors in the thermal ambient environment, such as temperature, air speed and radiation conditions will interact with the animal's own thermal regulation, which makes it impossible to define the requirements in general terms. Therefore, they must be specified for different conditions.

The sum of the heat loss to the floor, the surrounding air and other contact surfaces in the ambient environment should be equivalent to the heat produced. From this, mathematical models may be constructed whereby the requisite heat flow from an animal to the floor may be calculated (P_{cd} in Fig. 6.1). Such a model has developed by Bruce and Clark (1979) originally for pigs. It has also been adapted for cattle (Bruce *et al.*, 1984; Bruce, 1986). In the models all conductive heat from the animal to the floor is assumed to lead to the surrounding air. No heat flow is supposed to occur downwards into the ground.

In order to evaluate the suitability of a floor, its thermal properties (R_{cd}) must be known. When an animal lies down, an unsteady three-dimensional heat flow to the floor occurs. The mathematical solutions to describe this flow have, with the help of the Finite Element Method (FEM), been presented by Andersen (1981). The results of the calculations are greatly dependent upon the model used, and the input data. For example, assumptions must be made with respect to the heat conductivity of the bedding and manure and urine residues present on the lying area.

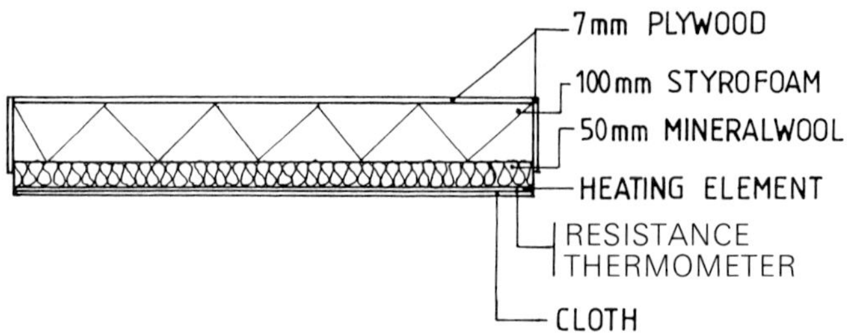

Fig. 6.2. The construction of the simulator used by Nilsson (1988).

Alternatively, the thermal resistance of the lying area may be evaluated using either an artificial animal, a simulator (Hannusch, 1970; Graee, 1971; Bergschöld and Ohlén, 1973; Bruce, 1977; Nilsson, 1988, etc.), or a heat flow meter placed between an animal and the floor (Mount, 1967; Spillman and Hinkle, 1971; Stanzel, 1972; Rist and Mathys, 1973, etc.).

The heat loss to the floor from an animal depends on the size of the contact area. The relationships between body surface area and the lying area have been studied by Grommers *et al.* (1970), Graee (1971), Webster (1974), Esmay (1977) and others. Mount (1967) concludes that the contact area of an animal with the floor while resting is 20% of its total area. Bruce (1979) suggests in his model that this is only true if the floor thermal resistance is greater than the sum of the heat transfer resistances between the body surface and the surrounding air. If it is less, an animal will reduce its contact area to only 10% of the total area.

In a series of investigations by Nilsson (1988) the effective thermal resistance (R_{cd}) of a number of the floor constructions commonly used in farming was determined. These were carried out partly in the laboratory, and partly in the field. The observations were made using the simulator shown in Fig. 6.2 having a floor contact area of 0.447 m^2; equivalent to a 124 kg cow or a 120 kg pig. Since the effective thermal resistance varies directly with the size of the contact area, the observed values must be corrected for animals of other sizes as shown in Nilsson, 1988. The calculated thermal resistance for a 100 kg cow varied from 0.066 m^2·K/W on 100 mm unbedded concrete to 0.78 m^2·K/W on a 40 mm mattress (Table 6.1).

Using a heat balance model developed by Svensson (1988), the observed values were used to study how the lower critical temperature will be affected when floors of differing thermal characteristics are used for growing–finishing pigs of varying weight. For 60 kg pigs in a group of eight, fed according to a Swedish standard level (Scan, 1984), the lower critical temperature may be reduced from +13°C to +6.5°C if the concrete floor received 2.5 cm straw bedding. If the floor is built up with a 15 mm concrete screed on 45 mm styrofoam insulation, the thermal characteristics will be improved so the lower critical temperature under similar conditions may be lowered to +6.2°C (Fig. 6.3). These calculations were made using an outside temperature of −10°C, air speed of 0.2 m/s and a *U*-value of the walls and roof of 0.25 W/(m^2·K). By altering these conditions the requirements for the floor construction may be determined for other situations.

The model is not applicable for calculations concerning deep litter in the present state. In this case the animal–straw contact area will be more than 20%. Wejfeldt

Table 6.1. Laboratory measurements of floor thermal resistance R_{cd}, for a 100 kg animal measured after lying for 60 minutes on different floor constructions.

Floor construction	R_{cd} ($m^2 \cdot K/W$) Cattle	R_{cd} ($m^2 \cdot K/W$) Pigs
100 mm clean concrete	0.066	0.067
Concrete with 17 mm rubber cow mat	0.12	0.12
Concrete with 17 mm rubber mat plus 2 l/m² sawdust	0.15	0.15
Concrete with 15 mm dry sawdust	0.27	0.27
Concrete with 25 mm dry straw	0.39	0.39
Concrete with 47 mm dry straw	0.63	0.64
100 mm concrete + 45 mm styrofoam + 15 mm concrete screed	0.47	0.48
Concrete with 40 mm mattress	0.78	0.79

Source: Nilsson, 1988.

(1991) has made laboratory measurements with a modified artificial pig and determined the effective thermal resistance for pigs with different areas in contact with the straw. After that the lower critical temperature was calculated. It was shown that a 34 kg pig fed *ad libitum* embedded with a 70% contact with the straw had a lower critical temperature of −7°C. If the same pig in addition to the 70% contact with the straw also huddles together with 16 pen mates, the critical temperature is −22°C.

In the heat balance model for pigs, steady-state conditions have been assumed, that is, the animals have been lying for so long that it is assumed a constant heat flow between the animals and the floor has been reached. This simplification may actually be realistic, since pigs spend about 80% or more of the time lying (Baxter, 1984). Cattle spend a greater part of their time standing (Hedrén, 1971; Bergschöld and Ohlén, 1973; Metz and Wierenga, 1987), with lying periods of 1–2 hours (Hedrén, 1971) although there are large individual variations. This means that for cattle the heat flow to the floor may never be considered as steady. Thus, to evaluate the suitability of the floors in a lying area for cattle, from the viewpoint of the thermal properties, the energy loss from an animal during the first hour of the lying period has been used as the reference figure. This amount of energy must not exceed the mean sensible heat loss through the skin under the climate conditions used.

An evaluation of different floors in lying areas for dairy cows has been made (Table 6.2). Since heat production varies greatly with weight, yield (if lactating), and length of gestation, the requirements for the floor will also vary. The results of such studies have been evaluated with respect to the differing demands. For example, a high-yielding cow or one late in gestation needs to lose so much heat that an unbedded concrete floor will be adequate, at least from the aspect of the thermal characteristics. On the other hand, non-productive cattle need well-bedded lying areas, at least equivalent to 1 cm straw on concrete or a rubber mat with sawdust bedding (Table 6.3).

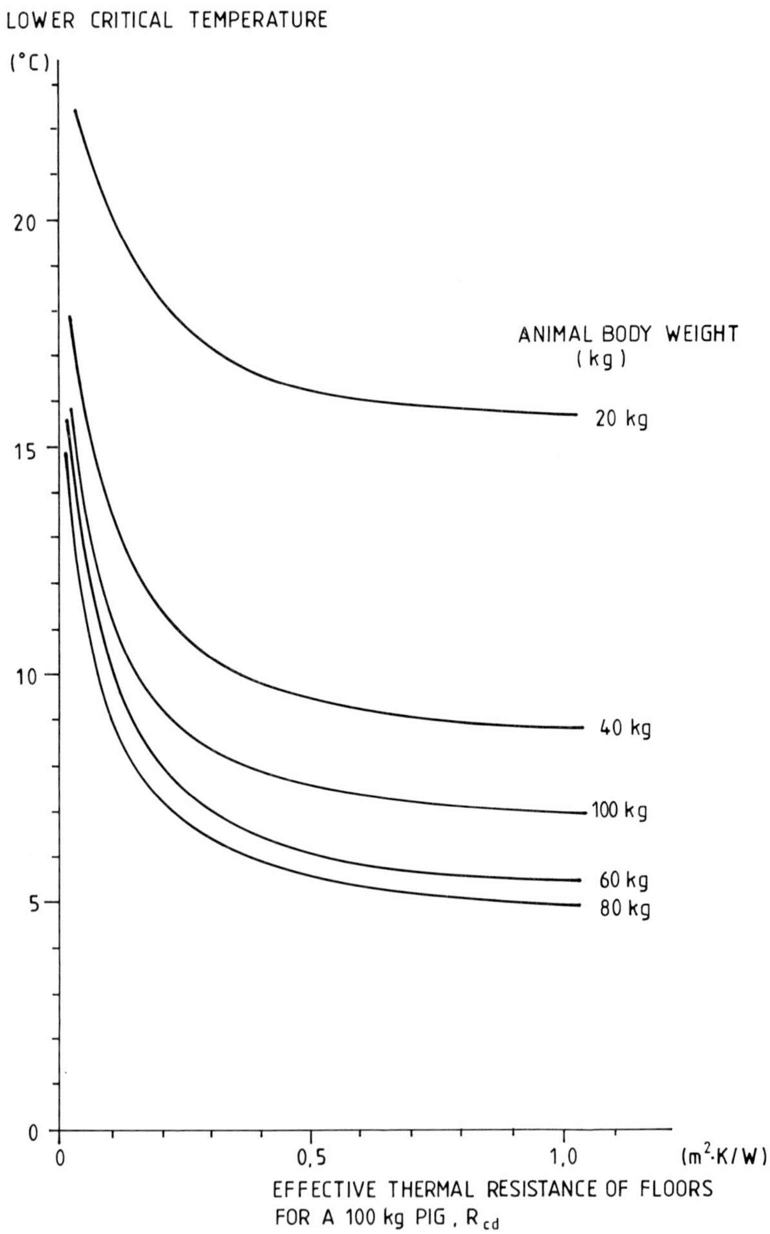

Fig. 6.3. Calculated lower critical temperatures for a group of eight pigs on floors with various effective thermal resistance (see Table 6.1). Outside temperature: -10°C. *U*-value of roof and walls of the building: 0.25 W/(m^2·K). Indoor air velocity: 0.2 m/s (after Nilsson, 1988).

Table 6.2. Heat losses to the floor from cattle of different sizes during the first hour after lying down on different floor constructions (W • h/m^2). Room temperature + 12°C

Floor material	Animal weight (kg) 500	600	700
Concrete, strawless	187	176	167
Concrete + 2 mm sawdust	131	123	117
Concrete + 17 mm rubber mat	113	107	101
Concrete + 17 mm rubber mat + 2 mm sawdust	87	82	78
Concrete + 15 mm sawdust	53	50	47
Concrete + 25 mm straw	32	30	29
Concrete + 40 mm mattress	19	18	17

Source: After Nilsson, 1988

Table 6.3. Calculated heat losses from dairy cows of different weights, length of gestation and milk yields. Room temperature + 12°C.

a) Animal weight = 600 kg

	Pregnancy (days)							
	0			140			280	
	Milk yield (kg/day)			Milk yield (kg/day)			Milk yield (kg/day)	
	0	20	40	0	10	20	0	10
Total heat loss (W)	679	1119	1559	723	943	1163	1030	1250
Sensible heat flow through the skin (W m^{-2})	95	157	219	101	132	163	145	175

b) Animal weight = 700 kg

	Pregnancy (days)							
	0			140			280	
	Milk yield (kg/day)			Milk yield (kg/day)			Milk yield (kg/day)	
	0	20	40	0	10	20	0	10
Total heat loss (W)	762	1202	1642	806	1026	1246	1113	1333
Sensible heat flow through the skin (W m^{-2})	98	155	211	103	132	160	143	172

Source: After CIGR, 1984.

Softness

A great many methods for measuring the hardness (or softness) of materials have been developed with penetration depth or calculated contact pressure given as a value of softness. Common industrial methods, where a test body is pressed against the material, include the Brinell test using steel balls of 1.0, 2.0, 2.5, 5.0 or 10.0 mm diameter, the Rockwell test using 1/16 (16 mm) or 1/8 (32 mm) inch diameter balls, or a cone with a top angle of 120°, and the Vickers test using a diamond pyramid. These are all laboratory methods. A small hand apparatus for evaluating softness is used in the rubber and plastic industry. This is the Shore Durometer, which presses a needle-shaped test body against the material by means of a spring in a handle. In order to study the resistance of floor materials to penetration under point load, Bring (1968) proposed a method resembling the Rockwell test, but using a 20 mm diameter steel ball. A ball of this diameter and one of 50 mm were used in softness tests by Løken (1978).

If the suitability of the softness of a floor material in animal houses is to be evaluated, a test ball with dimensions equivalent to the animal, or some portion thereof, should be used. This will require larger test balls, with dimensions depending on the species and age concerned. A 100 mm diameter steel ball has been used, which should be equivalent to the knee of a small bovine (Nilsson, 1988). Lasson and Boxberger (1976) used a 240 mm steel ball. When testing floor material, it should be remembered that softness is not a constant characteristic. It depends upon, for example, the thickness of the material, temperature, humidity, speed of loading, size of force and the time between loading and reading (Løken, 1978). Therefore, it is necessary to specify the test conditions when citing softness values.

The softness of the floor is of greater importance for a larger animal than a smaller one, as the contact pressure to the lying area is relatively greater, with the animal's surface theoretically proportional to its weight raised to the power of 2/3. For this reason, dairy cows are especially interesting to study.

Lying areas for dairy cows must have a certain degree of softness to lessen the incidence of injuries and to provide comfort. Studies of this problem (Nilsson, 1988) showed that the incidence of injuries decreases with an increase in the softness of the floor in standings for tied dairy cows (Fig. 6.4). However, the surface must not be too soft. Using preference tests, it was observed that a bed thickness of 10–15 cm (sawdust) is considered as optimal for the animals, within the limits of the test conditions (Wander and Fricke, 1974). Wander and Fricke (1974) and Lasson and Boxberger (1976) explained this observation by suggesting that animals experience a very soft underlay as being unstable for standing on. They must have a firm footing for the hooves.

In order to study the preference of dairy cows with respect to softness, tests were carried out in a loose housing barn, where they were permitted to choose between lying cubicles which varied in the softness of the floors (Nilsson, 1988). The softest cubicles, which contained about 15 cm sawdust, were preferred. Of the artificial floorings tested, all of which were generally significantly harder than sawdust, the softest of these were also preferred. On the basis of these results and softness tests

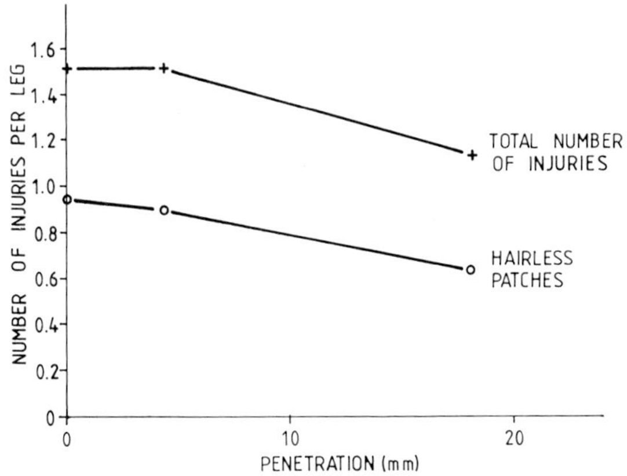

Fig. 6.4. Number of observed injuries (hairless patches, swellings and wounds) according to the softness of the floor material (measured with a 100 mm diameter steel sphere). Investigation of 44 tied dairy cows during three winter housing seasons (after Nilsson, 1988).

on the materials, graphs were drawn describing the range of softness for stall floor materials suitable for dairy cows (Fig. 6.5).

These graphs were originally based on results obtained using a 100 mm test ball, and may not directly be used in terms of other diameters. The corresponding softness limits have been calculated for 20 and 240 mm diameter balls, respectively.

Friction

The friction properties of the floor have been studied in many countries over a long period of time. Two comprehensive summaries of the work, primarily in connection with residences and roads, have been published by Bring (1964) and Brungraber (1976). To describe friction characteristics, the coefficient of friction (or friction number) is used. This is the ratio between the frictional force parallel to the surface, a comprehensive expression of the overall effect of the forces that cause friction, and the normal force. It may be determined either when the body (the foot) is at rest, i.e. the coefficient of static friction, or moving, the coefficient of sliding friction. The most commonly used methods of measuring friction are:

- the drag test
- the inclined plane
- the inclined leg
- the pendulum apparatus.

To these methods could be added measurements made with rotating wheels, measurement of sliding distance, friction wheels, etc. Bring (1978) suggested, after intensive studies of the human gait, an apparatus having a foot moving against the material and pushing from the heel. If the friction is adequate, the foot will 'walk',

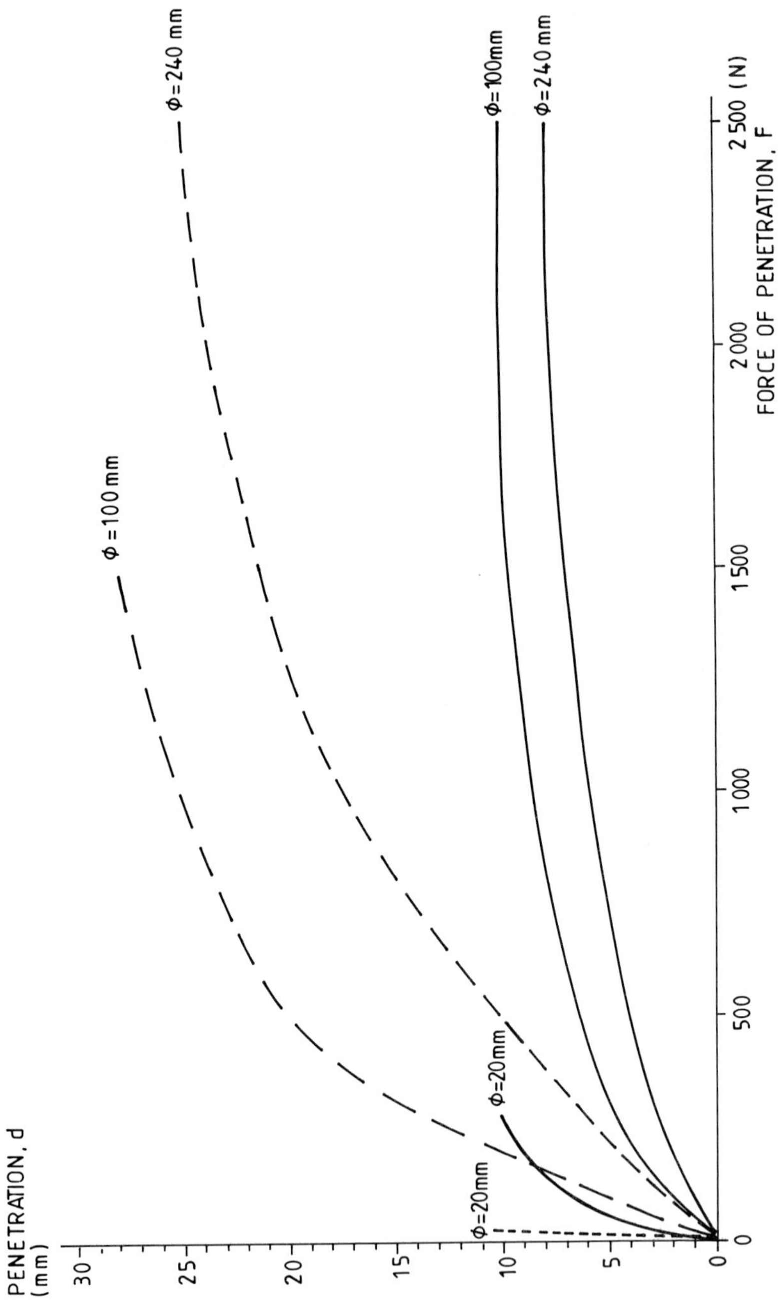

Fig. 6.5. Softness of floor materials; minimum and maximum requirements for lying areas used by dairy cows, using test balls of 20, 100 and 240 mm diameters respectively: ——, minimum requirements; – – –, maximum requirements; Ø, ball diameter (after Nilsson, 1988).

if it is slippery, it slips. The method was developed using the principle that a test method should be related to the true conditions, which is a very important guideline.

Special studies of the friction and sliding problem in animal houses have been made. The drag test has been used by Beer (1976), Bähr and Türpitz (1976), Takai and Møller (1981) and Nilsson (1988). The coefficient of friction was given by these workers to describe friction characteristics, with the exception of Takai and Møller (1981). They only used the horizontal force values, which make their results difficult to compare with the others.

The inclined leg method has been used by Løken (1979). In this test, the force from a specific weight loads the friction foot (sliding body) via a hinged leg. By gradual displacement of the frame the leg is forced from a vertical to an inclined position until the foot starts to slide. The tangent of the leg angle in this position equals the coefficient of static friction. The coefficient of sliding friction cannot be evaluated using this method.

Some of the earliest studies of the relationships between animal injuries and the physical characteristics of the floor used a pendulum apparatus called a skid-resistance tester (Penny *et al.*, 1965; Mitchell, 1974). It consisted of a friction foot connected to a pendulum where the foot is adjusted to sweep a path across the floor surface and the loss of energy by the pendulum is a measure of the sliding friction.

An apparatus for measuring the friction characteristics of floors in animal houses must fulfil four main demands:

- it must be portable for field studies
- it should have sufficient precision and repeatability
- it should give values in agreement with other friction testers
- it should rank the materials as they are experienced by the animals.

With these in mind a friction tester was designed and built as described by Nilsson (1988). This worked according to the drag method, and the original sliding body, having the same size and appearance of a cow hoof, was loaded so the normal force reached a maximum of 2287 N, which is in the order of the load from a large domestic animal. The relatively great load was necessary for testing floors with soft surfaces. The penetration of a hoof in the material will increase the friction coefficient.

Beer (1976) used real hoof material in his first laboratory model for evaluating friction, as did Bähr and Türpitz (1976) in theirs. In both tests the workers found that a dry hoof slips more easily than a moistened hoof, since the latter is softer and adapts itself to the underlying surface.

In order to have a test material with more constant properties, Beer (1976) in later experiments used a synthetic material which showed approximately the same coefficient of friction as a dry hoof, namely polythene. This material was also used in studies by Løken (1979) and Nilsson (1988). However, since grooves will appear in the plastic sliding body when it is tested, Takai and Møller (1981) chose a steel brush as a sliding body since it was soft enough to adapt to small irregularities in the surface.

By studying the movements of the animals or by measuring the gait force, it is possible theoretically to calculate the coefficient of friction required. Wander (1970) stated that when a cow walks, the ratio between the horizontal and vertical forces

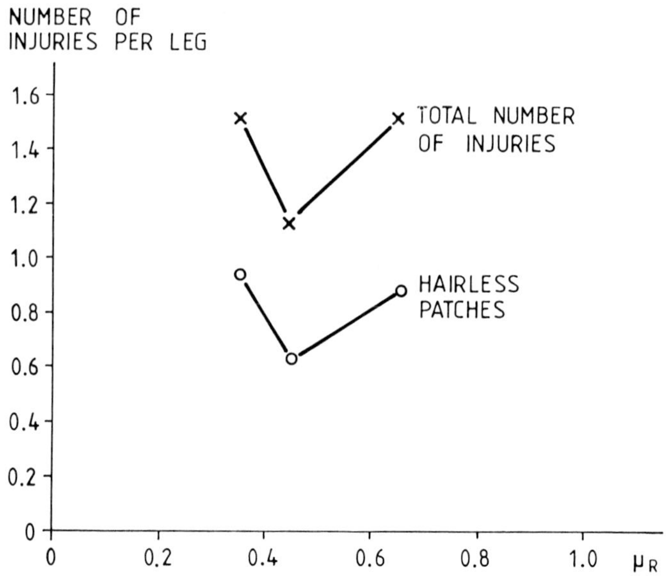

Fig. 6.6. Number of observed injuries (hairless patches, swellings and wounds) according to the coefficient of static friction. Investigation of 44 tied dairy cows during three winter housing seasons (after Nilsson, 1988).

on the hoof lies between 1 : 2 and 1 : 3. This means that a coefficient of friction of 0.33–0.50 is required. Webb and Clark (1981) give friction coefficients lying between 0.2 and 0.6 for a cow with extreme values observed during the gait (static friction). Bähr and Türpitz (1976) note values between 0.2 and 0.4 depending upon species, age and housing system. In studies of injuries on tied dairy cows Nilsson (1988) reported the optimal static friction coefficient as lying between 0.4 and 0.5 (Fig. 6.6). Analysis of the data from Gjestang and Løken (1980), indicates that the incidence of slipping for tied dairy cows is acceptably low when the static friction coefficient is about 0.4 or more (Fig. 6.7).

Studies carried out by Wander (1970) show that the front leg of a cow forms an angle of about 45° to the floor when she rises. If the force to the floor follows the direction of the leg, a friction coefficient of 1.0 would be required. Observations of the relationships between the horizontal and vertical force in this case have not been reported in the literature. There is nothing to disagree with the assertion that an animal, with the help of the muscles, exerts more vertical force to the floor, and thus the requisite coefficient of friction may be presumed to be lower. Therefore, the friction coefficient for stalls and standings should be at least 0.35–0.40, as measured with our friction meter.

Many of the available floor materials for stalls have been tested in the laboratory in the dry, wet and manure-covered states. Examples from the tests are given in Table 6.4. For hard materials, the static friction was always higher than the sliding friction, whereas the reverse was true for soft materials, such as rubber mats and mattresses. It was often difficult to determine a coefficient of static friction for the soft materials. Due to elastic deformation, the sliding started with a transition

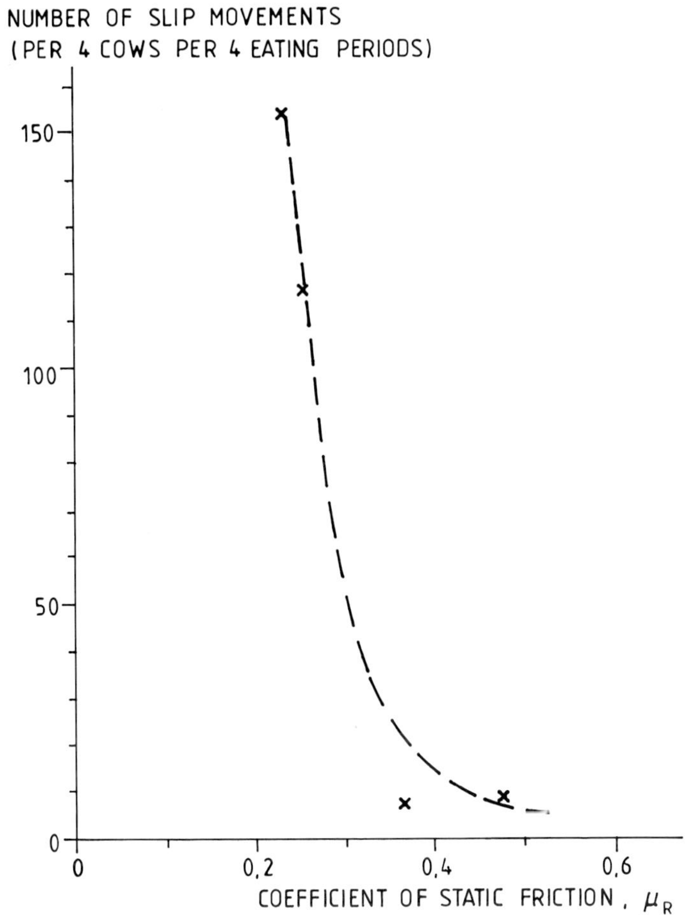

Fig. 6.7. Observed slip movements in tie stalls (×) according to the coefficient of static friction (dry materials) as measured with a Norwegian friction tester. The broken line shows a possible relationship (after Gjestang and Løken, 1980).

between rest and movement that was difficult to define. Hard materials had the lowest friction when moist and the highest when dung covered, with an intermediate value observed when they were dry. The sliding friction for soft materials was highest when wet, and lowest when covered with manure.

For hard materials such as wood, concrete and plastic compounds, the observed coefficient of friction was lower than that required for cow standings. However, the friction properties of the rubber mats and most types of mattresses fulfilled this requirement.

Table 6.4. Friction coefficients measured on different floor materials using a drag test method. Dry materials, 2287 N normal force.

Material	Coefficient of static friction	Coefficient of sliding friction
Particle board	0.28	0.27
Sawed pinewood	0.24	0.20
Steel-trowelled concrete	0.32	0.28
Floated concrete	0.38	0.34
Hard rubber mat	0.47	0.91
Soft rubber mat	0.77	0.86
20 mm mattress covered with PVC-coated fabric	0.41	0.67

Source: After Nilsson, 1988.

Abrasiveness

A certain amount of wear is necessary for animal foot health; underworn hooves may be just as great a problem as overworn ones (Camara, 1970; Beer, 1976; Gravert, 1977; Clark, 1983). The type of floor may also cause injuries on the body, for example on the legs of piglets (Penny *et al.*, 1965, 1971; Gravås, 1979; Bengtsson *et al.*, 1982, etc.). By giving bedding materials and thereby reducing direct contact with a rough floor surface, the incidence of injuries can be reduced (Maton *et al.*, 1978).

Camara (1970) and Camara and Gravert (1971) determined the effect of different types of concrete floors on hoof wear in cattle, and Kovács (1975) studied the effect of concrete, asphalt, tile and wood on pig claw wear. From the results of these studies, recommendations with respect to floor design were presented. According to Kovács (1975), the wear of the horn is greatly affected by its moisture content. Even where the specimen has been taken from the hoof has significance.

A method for testing abrasiveness using a plaster test body was developed in order to define more clearly the test body and obtain more precise values (Nilsson, 1988). By dragging a weighed plaster block over a floor surface, the weight lost may be used as a value of the abrasive properties of the floor. During the investigations four types of floors were tested with the results in good agreement with studies of injuries on the pigs using these floors (Table 6.5). This study showed that by grinding a floated concrete floor before setting pigs in the pen, the disadvantages of such a rough floor may be lessened. A surface structure obtained by treating a concrete surface with varnish or a plastic-based compound lasted significantly longer than that of untreated concrete. Treating the floor surface also strengthened it so the wear and tear due to the animals was reduced. However, if these treatments are to be successful, the surface must be sufficiently smooth from the beginning, although there should be no risk of slipping.

The plaster method has subsequently been used by Christison and Farmer (1981) and Farmer (1982) after modifications. However, plaster has the disadvantage of absorbing moisture, which will present problems if used in animal houses with wet floors. Thus, Takai and Møller (1981) constructed a test apparatus using a welding

Table 6.5. Abrasiveness measurements with a plaster block and number of injuries to the legs of piglets for the seven first farrowings on each floor type (seven pens in each).

Floor surface	Measured abrasiveness (kg/m^2)	No. of injuries per animal (mean) at piglet age (days)		
		3	10	21
Floated concrete	20.4×10^{-2}	6.0	2.9	0.2
Machine ground concrete	11.5×10^{-2}	3.6	1.6	0.2
Floated concrete painted with chlorinated rubber	13.2×10^{-2}	4.3	3.7	0.2
Polyurethane with rubber particles	15.1×10^{-2}	3.6	2.7	0.4

Source: After Nilsson, 1988.

carbon rod as a test body. Clark (1983) improved the method used by Nilsson (1988) by using an india-rubber test body loaded with a 2 kg weight during testing. He concluded that the method gave results agreeing with the observed animal injuries, and that it was suitable to use in future studies.

Regular and effective cleaning of the floor in animal houses is necessary to maintain a hygiene level high enough for an economic production. However, if high pressure cleaning systems are used, the floor surface will be exposed to greater wear, increasing the roughness of the surface. In addition, holes may also appear in the floor, increasing the risk of injury to the animals.

In order to determine the level of quality necessary for a concrete floor cleaned using a high pressure cleaner, a study of the effect of compression strength, composition and crack formation in concrete on resistance to erosion was carried out (Nilsson, 1971). It was found that the resistance increased with the quality of the concrete. If pea gravel (0–16 mm) was used instead of gravel with a grain size of 0–8 mm the resistance was also improved. Concrete with cracks showed about twice as much erosion after high pressure cleaning than the same concrete with a perfect surface. The amount of erosion was independent of the width of the cracks (0.1–1.0 mm).

Conclusion

It is obvious that the optimal values of all the floor properties are difficult if not impossible to fulfil simultaneously. Instead the aim must be to work out a compromise which as much as possible fulfils the different demands. The parameters of thermal comfort, softness, friction and abrasiveness of the walking and lying surfaces have been discussed. Other important properties of the floors in livestock houses are mechanical strength, cleaning ability, toxicity, pathogenicity, etc.

In the end it is the economical consideration that is most important. The net profit resulting from better animal health, higher output and reduced labour – and straw – costs must be compared to the cost of a better floor construction.

References

Andersen, K.T. (1981) The concrete top layer on insulated lying areas. In: MacCormack, J.A.D. (ed.), *Modelling, Design and Evaluation of Agricultural Buildings.* Commission Internationale du Génie Rural (CIGR), Seminar of Section II, 29–31 August, Theme 3, pp. 205–212. Scottish Farm Buildings Investigation Unit, Aberdeen.

Bähr, H. and Türpitz, L. (1976) Die Trittsicherheit von Stallfussböden und der Einflussfaktor Reibwiderstand (Walking safety on floors in animal houses and the importance of friction). *Agrartechnik* **5**, 241–243.

Baxter, S. (1984) *Intensive Pig Production: Environment, Management and Design.* Granada, London.

Beer, G. (1976) *Einige neue Prüfungsmethoden für die Tierstallböden* (Some new test methods for animal house floors). Report from the Working Session of the 2nd Technical Section of Commission Internationale du Génie Rural (CIGR), Budapest, pp. 473–487.

Bengtsson, A.-C., Fajersson, P. and Svendsen, J. (1982) Benskador på smågrisar-jämförande undersökning på tre olika golvtyper (Leg injuries of piglets – a comparative study using three different floor types). *Sveriges Lantbruksuniversitet, Institutionen för lantbrukets byggnadsteknik, Report No. 26.* Lund.

Bergschöld, B. and Ohlén, P. (1973) Båspallars värmeegenskaper (Heat characteristics of standings for dairy cows). *Lantbrukshögskolan Meddelande A* **182**. Uppsala.

Bring, C. (1964) Friktion och halkning (Friction and slipping). *Statens råd för byggnadsforskning, Report No. 112.* Stockholm.

Bring, C. (1968) Provningsmetoder för golvmaterial och golvkonstruktioner (Test methods for floor materials and floor constructions). *Statens råd för byggnadsforskning, Report No. 20.* Stockholm.

Bring, C. (1978) Provning av halksäkerhet (Testing slip safety). *Kungliga Tekniska Högskolan, Institutionen för byggnadsteknik, Meddelande* **115**. Stockholm.

Bruce, J.M. (1977) Conductive heat loss from the recumbent animal. *Farm Building R & D Studies* **8**, 9–15.

Bruce, J.M. (1979) Heat loss from animals to floors. *Farm Building Progress* **55**, 1–4.

Bruce, J.M. (1986) Lower critical temperatures for housed beef cattle. *Farm Building Progress* **84**, 23–28.

Bruce, J.M. and Clark, J.J. (1979) Models of heat production and critical temperature for growing pigs. *Animal Production* **28**, 353–369.

Bruce, J.M., Broadbent, P.J. and Topps, J.H. (1984) A model of the energy system of lactating and pregnant cows. *Animal Production* **38**, 351–362.

Brungraber, R. (1976) An overview of floor slip-resistance research with annotated bibliography. *National Bureau of Standards, Technical note* 895. Washington DC.

Camara, S. (1970) *Untersuchungen über den Klauenabrieb bei Rindern* (Investigations of the hoof abrasion of cattle). Christian-Albrechts-Universität, Dissertation. Kiel.

Camara, S. and Gravert, H.O. (1971) Untersuchungen über den Klauenabrieb bei Rindern (Investigations of the hoof abrasion of cattle). *Züchtungskunde* **43**, 111–124.

Christison, G.I. and Farmer, C. (1981) *Progress in the Evaluation of Raised Floors for Young Pigs at the Prairie Swine Centre.* Paper presented at the Saskatchewan Swine Symposium, University of Saskatchewan. Saskatoon.

Clark, M. (1983) *Biomechanical Aspects of Piglet Foot Development.* Scottish Farm Buildings Investigation Unit, Aberdeen.

Commission Internationale du Génie Rural (CIGR) (1984) *Report of Working Group on Climatization of Animal Houses.* Scottish Farm Buildings Investigation Unit, Aberdeen.

Ehrlemark, A. (1988) Calculation of sensible heat and moisture loss from housed cattle using a heat balance model. *Sveriges Lantbruksuniversitet, Report No. 60.* Uppsala.

Esmay, M.L. (1977) *Principles of Animal Environment.* AVI Publishing Company, Inc. Westport. Connecticut.

Farmer, C. (1982) Preference and floor characteristics as criteria in the selection of perforated floors by young swine. University of Saskatchewan, Department of Animal and Poultry Science. MSc Thesis. Saskatoon.

Gjestang, K.E. and Løken, K.A. (1980) Friksjonsforhold på båsgolv av betong og gummi (Slipperiness of concrete floors and rubber mats in tie stalls). *Norges Landbrukshøgskole, Institutt for bygningsteknikk Stensiltrykk* No.161. Ås.

Graee, T. (1971) Heat exchange through animal–floor interface. *Norges Landbrukshøgskole, Institutt for bygningsteknikk, Særtrykk No. 171.* Ås.

Gravås, L. (1979) Behavioural and physical effects of flooring on piglets and sows. *Applied Animal Ethology* **5**, 333–345.

Gravert, H.O. (1977) Klauenschäden durch erhöhte Abnutzung (Hoof injuries by increased abrasion). *Die Milchpraxis* November, 6–7.

Grommers, F.J., Christison, G.I. and Curtis, S.E. (1970) Estimating animal–floor contact areas. *Journal of Animal Science* **30** (4), 552–555.

Gustafsson, G. (1988) Luft- och värmebalanser i djurstallar (Air- and heat balances in animal houses). *Sveriges Lantbruksuniversitet, Institutionen för lantbrukets byggnadsteknik, Report 59.* Lund.

Hannusch, D. (1970) Vergleichende Messungen zur Wärmeabführung auf Spaltenböden (Comparative measurements of the heat abduct on slatted floors). *Bauen auf dem Lande* **21**, 49–51.

Hedrén, A. (1971) Mjölkkon och båsplatsen – beteendestudier i båsladugårdar (The milk cow and the standing. Studies of the behaviour of milk cows in tying stalls). *Lantbrukshögskolan, Meddelande A* **153**. Uppsala.

Jongebreur, A.A. (1983) Housing design and welfare in livestock production. In: Baxter, S.H., Baxter, M.R. and MacCormack, J.A.D. (eds), *Farm Animal Housing and Welfare.* Martinus Nijhoff Publishers, The Hague, pp. 265–269.

Kovács, A.B. (1975) Studies on roughness and endurance of floor surfaces in animal houses. *Acta Veterinaria Academiæ Scientiarum Hungaricæ*, **25**, 1–12.

Lasson, E. and Boxberger, J. (1976) Untersuchungen über die Anforderungen von Rinder an die Wärme- und Härteeigenschaften von Stand- und Liegeflächen (Research into cattle requirements in respect to warmth and surface hardness of stands and lying places). *Schriftenreihe der Landtechnik* 2. Weihenstephan.

Løken, K.A. (1978) Methods for testing floor surface properties. *Acta Agriculture Scandinavica* **28**, 342–347.

Løken, K.A. (1979) *Friction Measuring Technique for Floors for Livestock.* Report to the 9th International Congress of Commission Internationale du Génie Rural (CIGR), 8–13 July, 1979. Technical Section, Paper No. II–3–6, East Lansing. Michigan.

Maton, A., Daelemans, J and Hoorens, J. (1978) Claw and carpal lesions in suckling pigs in connection with their housing conditions. *Vlaams Diergeneeskundig Tijdschrift* **47**, 16–21.

Metz, J.H.M. and Wierenga, H.K. (1987) Behavioural criteria for the design of housing systems for cattle. In: Wierenga, H.K. and Peterse, D.J. (eds), *Cattle Housing Systems, Lameness and Behaviour. Proceedings from a Commission of the European Communities Seminar*, 3–4 June 1986. Martinus Nijhoff Publishers, The Hague, pp. 14–25.

Mitchell, C.D. (1974) Are the passageways in your cubicle buildings too slippery? *Farm Building Progress* **37**, 17–20.

Mitchell, D. and Smith, B. (1976) Flooring in farrowing pens. *Pig Farming* (Suppl., December 70–71, 73, 75

Mount, L.E. (1967) The heat loss from new-born pigs to the floor. *Research in Veterinary Science* (8), 175–186.

Mount, L.E. (1974) The concept of thermal neutrality. In: Monteith, J.L. and Mount, L.E. (eds), *Heat Loss from Animals and Man*. Butterworths, London, pp. 425–439.

Nilsson, C. (1971) Krav på betonggolv vid högtrycksrengöring (Requirements of concrete floors with reference to high-pressure cleaning). *Sveriges Lantbruksuniversitet, Aktuellt* 162. Uppsala.

Nilsson, C. (1988). Floors in animal houses – Technical design with respect to the biological needs of animals in reference to the thermal, friction and abrasive characteristics and the softness of the flooring material. *Sveriges Lantbruksuniversitet, Institutionen för lantbrukets byggnadsteknik, Report No. 61.* Lund.

Penny, R.H.C., Osborne, A.D., Wright, A.I. and Stephans, T.K. (1965) Foot-rot in pigs. Observations on the clinical disease. *Veterinary Record* **77**, 1101–1107.

Penny, R.H.C., Edwards, M.J. and Mulley, R. (1971) Clinical observations of necrosis of the skin of suckling piglets. *Australian Veterinary Journal* **47**, 529–537.

Rist, M. and Mathys, H. (1973) Zur Wärmeableitung von Tierbelägen (Of measuring the heat abduct from animal floor coverings). *Schweizerische Landwirtschaftliche Forschung* (1), 81–102.

Scan (1984) Scan Utfodringsnorm (Scan feeding norm). Slakteriförbundet, Stockholm.

Spillman, C.K. and Hinkle, C.N. (1971) Conduction heat transfer from swine to controlled temperature floors. *Transactions of the Americal Society of Agricultural Engineers* **14**, 301–303.

Stanzel, H. (1972) Zur Messung der Wärmeableitung von Tierliegeflächen (Of measuring the heat abduct of animal lying surfaces). *Tierzüchter*, (12) 350–352.

Svensson, L. (1988) Klimat- och produktionsfaktorers inverkan på undre kritiska temperaturen för slaktsvin (The influence of climate and production factors on the lower critical temperature for fattening pigs). *Sveriges Lantbruksuniversitet, Institutionen för lantbrukets byggnadsteknik, Internal report.* Lund.

Takai, H. and Møller, F. (1981) Metoder for karakterisering af staldgulve (Methods for characterizing floors in animal houses). *Statens Jordbrugstekniske Forsøg, Orientering* (8), Horsens.

Wander, J.F. (1970) *Einige Ansprüche der Rinder an den Stallfussboden* (Some demands made by cattle on stall flooring). Commission Internationale du Génie Rural (CIGR), Section 2 Conference, Ghent, vol 1, pp. 13.1–13.9.

Wander, J.F. and Fricke, W. (1974) Zur Einrichtung von Liegeboxenställen für Milchkühe (Equipping cubicles for dairy cows). *Bauen auf dem Lande* (3), 138, 140–141.

Webb, N.G. and Clark, M. (1981) Livestock foot–floor interactions measured by force and pressure plate. *Farm Building Progress* **66**, 23–36.

Webster, A.J.F. (1974) Heat loss from cattle with particular emphasis on the effects of cold. In: Monteith, J.L. and Mount, L.E. (eds), *Heat Loss from Animals and Man*. Butterworths, London, pp. 205–231.

Wejfeldt, B. (1991) Kritisk temperatur för slaktsvin på djupströbädd (Critical temperature for fattening pigs on deep litter). *Sveriges Lantbruksuniversitet, Institutionen för lantbrukets byggnadsteknik, Examensarbete No. 74.* Uppsala.

7 Human Factors

M.F. SEABROOK and N.C. BARTLE
Faculty of Agricultural and Food Sciences, University of Nottingham, UK

Summary

This chapter examines the human factor by considering the role model of the human as an interactor with animals. Current knowledge enables three major human interaction facets to be examined: *hand and arm interaction*; *vocal interaction*; and *holistic empathetic interaction*. Data are presented which provide an insight into the human factor influence on both the production and welfare of animals of economic importance.

The effects of different patterns of hand and arm interaction, described as 'pleasant' and 'aversive', are shown by the presentation of experimental results. The results for the animals include differences in behaviour, growth rates, production levels, hormone levels and heart rates.

Vocal interaction is discussed together with data from limited research on dairy cattle systems. These results suggest an association between stockperson vocal behaviour and animal performance.

The crucial area of holistic empathetic interaction is considered in some detail. Data are presented on the possible role that the stockperson's personality might play in the empathetic component. The evidence from both experimental data and observations on commercial units clearly suggests that correct interaction by the stockperson increases the performance of the animal and probably has beneficial welfare effects. There is also evidence to suggest that the behaviour of the animal is significantly affected by the actions of the stockperson.

The concept of 'individuality' as a feature expressed by animals is highlighted. The need for the identification by stockpersons of different behaviours by animals in response to human interaction is suggested. This is vital in order for stockpersons to adapt their behaviour to match the individual requirements of individual animals.

Introduction

The human factor

Various role models for humans in livestock systems have been identified (Hemsworth and Barnett, 1987; Metz, 1987; Seabrook, 1987). Within the context of this chapter the human as an *interactor* is the most appropriate role model. Research epistemologies in this area are fraught with difficulties as it is virtually impossible to isolate individual factors. For example, while it is possible to devise 'pleasant' and 'unpleasant' handling, as defined by level and nature of the physical interaction of the human body with that of the animal, it is difficult experimentally to separate physical handling from aspects such as voice tones, gestures, smell, sound of foot movements, or other 'messages' from human to animal. It is also of relevance that the quality of treatment is a perceived construct and that an animal may perceive both the applied 'pleasant' and 'unpleasant' stimuli as undesirable. Research may thus enable only a consideration of relative effects and outcomes. Current knowledge enables three major human interaction facets to be examined:

- Hand and arm interaction
- Vocal interaction
- Holistic empathetic interaction.

The last category, holistic empathetic interaction, is an attempt to include the full range of factors that current research does not enable one fully to isolate. For example, 'aversive' handling in a stable and consistent environment from a confident personality type stockperson is likely to have a different response from the apparently similar interaction in an unstable, inconsistent environment from a non-confident stockperson.

In spite of the research difficulties there are data available which enable an insight into the human factor influence on both the production and welfare of animals of economic importance. Production data can be obtained by direct measurement, such as growth rate or milk yield. Animal welfare is more difficult to quantify and may only be assessed through indirect indicators which may be used in isolation, or, ideally, in combination. Such indicators include behavioural and physiological responses.

The animal's role

There is evidence that individual animals will react in different ways to similar environments and stimulation. N.C. Bartle (unpublished data) found, in a group of sows, that reactions to a given set of stimuli are *not* the same for every animal. Heart rate, reaction and behaviour are different depending on the 'individuality' of the sow. An analogous problem arises because one is imposing behaviours on animal groups; for example 'aversive' handling of a pig in a group may well influence the behaviour of other animals in that group. Likewise the space available to the animal may influence both the effect and the behaviour. In a comparable way other factors in the environment, e.g. outdoor compared to indoor housing of pigs, may affect the way the human is perceived. Time delays in reaction may also be crucial factors, as will previous experience and perceptions. Similarly habituation may occur where an animal adjusts its stimulation threshold level. What may once have been perceived as 'aversive' may with habituation become a perceived normality and acceptable. The complex perceptual and response process needs to be considered in detail. It is important, for example, to take into account the behaviour and degree of interaction given by any persons responsible for the maintenance of the animals, unless this is undertaken on each and every occasion throughout the entire experiment by one and the same experimenter. This is possible in a small experiment (e.g. Dryden and Seabrook, 1986), but much more difficult in a large experiment (e.g. Hemsworth and Barnett, 1991). This component is crucial in view of the apparent effects on behaviour of the sight and smell of food, and the concurrent association of humans with the provision of a most positive stimulus: that of food.

Hand and Arm Interaction

Background

Hand and arm interaction can be identified into two patterns:

1. Pleasant, involving patting, stroking and touching of the animal, particularly the head area and the ears;
2. Aversive, involving slapping of the animal around the head and snout, light hitting with a stick or hose but *not* overt aggression.

Research results

It is certainly likely that animals in such handling situations are responding to a range of stimuli some of which one may not be able to isolate and describe. It is probable therefore that some results may not be conclusive and be contradictory simply because of the many complex intervening variables. A further confounding factor is that in some of the research work, e.g. Hemsworth and Barnett (1991), the space allowances in experiments may be larger than some commercial situations. Thus, there may be a larger behavioural effect because of the space enabling the animal to move away. However, there does seem to be a general trend of results. Hemsworth *et al.* (1981a, 1987), Gonyou *et al.* (1986), Dryden and Seabrook (1986), Hemsworth and Barnett (1987, 1991) and Seabrook (1988, 1991) have all produced results that show that different hand and arm interaction treatments of pigs adversely affect performance. These results are summarized in Table 7.1.

These results produced may in fact mask a number of factors, for example Dryden and Seabrook (1986) showed that an 'aversive' handled group in some performance areas had better performance than a control group. Some other experimenters made similar findings. Dryden and Seabrook (1986) suggested that under certain intensive conditions interaction and contact, even if 'unpleasant', may help reduce boredom in the pigs. Consistency of handling may be as crucial as the nature of the handling. Pearce *et al.* (1989) have suggested that group housing may minimize some of the adverse effects, but Hemsworth and Barnett (1991) suggest this

Table 7.1. The effects of handling treatments on performance of pigs in experimental conditions.

	Handling treatments		
Performance criteria	Pleasant	Aversive	Reference
Growth rate 0–4 weeks (kg day^{-1})	0.205	0.192	Dryden and Seabrook (1986)
Growth rate 0–5 weeks (kg day^{-1})	0.533	0.509	Hemsworth and Barnett (1991)
Growth rate 0–10 weeks (kg day^{-1})	0.656	0.641	Hemsworth and Barnett (1991)
Growth rate 5–10 weeks (kg day^{-1})	0.779	0.773	Hemsworth and Barnett (1991)
Growth rate 7–13 weeks (kg day^{-1})	0.455	0.404	Hemsworth *et al.* (1987)
Growth rate 8–18 weeks (kg day^{-1})	0.897	0.837	Gonyou *et al.* (1986)
Growth rate 11–22 weeks (kg day^{-1})	0.709	0.669	Hemsworth *et al.* (1981a)
Age of first detected oestrus (days)	169.0	171.8	Hemsworth and Barnett (1991)
Pregnancy rate (%)	88	33	Hemsworth *et al.* (1987)
Age for fully co-ordinated male mating response (days)	161	193	Hemsworth *et al.* (1987)
Piglets born alive per farrowing	10.1	9.3	Seabrook (1991)
Piglet mortality at 3 weeks (%)	11.1	15.2	Seabrook (1991)

is not necessarily true. However, the intervening variable of space per pig may confound the results of different experiments. The ability of the animal to move away from an 'aversive' handling treatment may be influential. N.C. Bartle (unpublished data) has shown in outdoor sows that the major response to 'aversive' handling is in fact to move away to a perceived safe area, e.g. a wallow hole. Hence 'aversive' handling of constrained sows may be more significant than the same treatment on loose housed animals that have much space to move away.

Performance results with other species suggest similar findings. Seabrook (1984) presented results from a study of 24 very similar one-person dairy units where feed inputs per cow were almost identical. The results are shown in Table 7.2. Within these results there is an association of treatment and performance, although the causal relationship cannot be defined. The results are indicative of a human factor effect and show, for example, that in the higher yielding herds the stockperson touched the cows far more often than in the lower yielding herds. In the higher yielding herds each cow on average was touched 2.1 times per minute while in the parlour compared to 0.5 times per minute on the lower yielding herds. Albright (1984) has suggested that if patting and stroking are carried out at milking time cows may be encouraged to enter the parlour and exhibit less stress and hence less inhibition of milk let-down.

Behavioural response

Analysis of the experimental data in terms of behaviour of the animals is of importance. This must be done in the correct context. A knowledge and understanding of an animal's existing behaviour pattern is required in order to identify any changes in these patterns brought about during the period of the handling treatment. Particularly important are the development of new behaviours and especially the performance of stereotypies. In work at the University of Nottingham, Seabrook and Bartle have recognized the importance of disguising, hiding or totally removing the

Table 7.2. Behavioural observations and herd results from one-person dairy units.

	Observation of touching cows by stockperson (average number of times min^{-1} cow^{-1})	
	2.1	0.5
Dairy unit results	Higher yielding	Lower yielding

Source: After Seabrook, 1984.

observer from the view of the animals, when attempting to monitor normal behaviour patterns. One of the major aspects is that 'pleasantly' handled animals may actively seek contact with the observer. The mean time to interact with humans has been suggested as an indicator of an animal's motivation to interact with humans. Changes in this 'approach' behaviour may reflect the response to treatments (Table 7.3).

Free approach times as used by Hemsworth *et al.* involve moving the pig to an 'experimental arena' in which the opportunity arises to approach a stationary observer. The approach time may well be influenced by other human messages/signals and environmental influences. Thus, while the results are of interest they must be seen to have limitations. On commercial units under the same management system and identical buildings, M.F. Seabrook (1991, unpublished data) has also shown differences in results for approach behaviour. These data, however, indicate a rather smaller divergence in times which may be due to the level of inquisitiveness of the pigs. In experimental situations the environment may well be 'richer', and with more humans present, than in commercial unit situations. On commercial unit situations any change of environment may stimulate exploratory behaviour in approaching. Thus differences in approach behaviour of 'aversive' treatments may

Table 7.3. The effects of handling treatments on interaction behaviour. Results show average time for pigs freely to approach the observer.

	Handling treatments		
	Pleasant	Aversive	Reference
Average time (s)	5[1]	80[1]	Dryden and Seabrook (1986)
Average time (s)	119	157	Hemsworth *et al.* (1987)
Average time (s)	73	147	Gonyou *et al.* (1986)
Average time (s)	10	160	Hemsworth *et al.* (1981a)
Average time (s)	48	120	Hemsworth *et al.* (1987)
Average time (s)	55	165	Hemsworth and Barnett (1991)
Average time (s)	16[2]	46[2]	M.F. Seabrook (unpublished data)
Average time (s)	195[3]	274[3]	N.C. Bartle (unpublished data)

[1] A slightly different technique was used.
[2] Recordings on commercial units.
[3] Sows kept outdoors – measurement is for successful approaches by sows; some 'aversive' treatment sows were less likely to approach humans.

Table 7.4. Handling treatments and the observed behaviour of sows.

Action of stockperson	Reaction	Reference
Stroking and scratching sow's head	Sows were less restless	Hemsworth *et al.* (1981b) Dryden and Seabrook (1986)
	Sows were quieter	Hemsworth *et al.* (1981b) Dryden and Seabrook (1986)
	Sows were more approachable	Hemsworth *et al.* (1981b) Dryden and Seabrook (1986)
	Larger number of piglets reared	Hemsworth *et al.* (1981b) Seabrook (1991)
	Increased piglet growth	Dryden and Seabrook (1986)
Slapping sow's head and shoulders	Sows were more restless	Hemsworth *et al.* (1981b) Dryden and Seabrook (1986)
	Sows were less approachable	Hemsworth *et al.* (1981b) Dryden and Seabrook (1986)
	Reduced sexual activity	Hemsworth and Barnett (1987)

arise from factors other than a fear of the stockperson. Hemsworth and Barnett (1991) have argued that through the process of stimulus generalization the response to the association of 'aversive' treatment by one person will extend to all humans. However, there is certainly good anecdotal evidence to suggest animals can discriminate between humans and may in fact be effective at isolating 'which person does unpleasant things and which one does not'. Pigs have been shown to move away from a 'disliked' person while approaching a 'liked' person. The evidence for generalization is thus far from conclusive. This may explain why some experiments, e.g. Paterson and Pearce (1989) have not shown an adverse reaction to 'aversive' handling.

Many of the observations of pigs in experimental and commercial situations have shown that 'pleasant' handling of pigs alters their behaviour. In a number of studies the sows' behaviour was observed at random times during both day and night, at the entry of the observer (Table 7.4). The conclusions show the stockperson is having an effect on animal behaviour by his or her actions. This change in behaviour of the animals may be one of the factors influencing performance. The results in Table 7.4 show that the sows in the 'pleasant' treatment groups behaved by getting up slowly with some quiet grunting, appearing to welcome the observer by standing or sitting up slowly into a more accessible position for contact. The 'aversive' or 'unpleasant' treatment groups behaved very differently; standing or sitting quickly and violently with loud grunting. The sows in the 'aversive' treatments appeared very restless, preparing either to fight or to run away, and generally were more difficult to deal with. However, Metz (1987) and N.C. Bartle (unpublished data) have indicated problems for the stockperson with animals in 'pleasant' treatment situations. This arises due to the animals' rapid approach behaviour, their 'willingness' to stay around the stockperson and attempts to force interaction from the human.

Physiological response

One has to accept that measuring physiological responses to humans is not easy. Any interaction involving blood sampling is likely to influence the behavioural and physiological responses. It is difficult to sample from cannuli without a human effect, since even hiding behind a screen will not eliminate the effect of the sound and smell of the human. This will initiate a response in the animal. Similarly the process of anaesthesia is likely to establish certain responses in the animal to human interaction. The use of some non-invasive techniques, e.g. saliva collection, will have similar problems. The technique of telemetric heart rate perhaps offers a positive opportunity to overcome certain problem areas.

Hormonal

The research results and other physiological tests suggest that in 'aversive' handling, pigs experience a chronic physiological stress response possibly responsible for the depression in growth and reproduction. Chronic stress indicates poor welfare, and is a prolonged response as opposed to acute stress, where there is a sudden but not long-lasting physiological response. Hemsworth *et al.* (1987) have shown that daytime mean corticosteroid concentrations are higher for pigs in 'aversive' treatments than for those in the 'pleasant' treatments. The continuous elevation of corticosteroids decreases the incorporation of amino acids into tissues and mobilizes energy reserves reducing growth. The hormonal evidence is not totally conclusive: while some experiments have shown that pigs with 'aversive' handling have an increase in their free corticosteroid response to humans, others have not (Hemsworth and Barnett, 1991).

Hemsworth and Barnett (1987) conclude that there is a substantial range in the quality of the human–animal interaction, and at the lower end of the range where the pigs are fearful of humans then the productivity and the welfare of the animals are affected. They propose that the level of fear of humans is equated to the degree to which the pigs will freely approach the observer in a standard test. They provide evidence to support the view that approach to humans is a useful index of the level of fear and is linked with physiological responses. The fear-provoking stimuli result in a range of physiological responses in the animal, one of the most consistent being elevated corticosteroid levels. Their work provides evidence of a wide range in the level of fear shown by pigs in commercial units. The fear response of the pigs is clearly one cause of the chronic stress situation. In another species Duncan (1990) indicated a significant fear response to human beings in poultry. Duncan highlighted the need for research to identify the stimulus that prompts the fearful response to human beings.

Heart rate

Recent interest has focused on heart rate responses of animals to human interaction. Work at the University of Nottingham, using telemetric heart rate monitoring, on both indoor and outdoor housed sows is being undertaken by Bartle and Seabrook. Before the effects of human interaction on heart rate can be determined, it is necessary to have a concise knowledge of the 'normal' heart rate patterns for the species and, perhaps more importantly, a knowledge of each individual's heart rate pattern. It is also important to be aware of the external and internal factors that may influence heart rate, for example circadian rhythms, age, weight, level of activity, stage of pregnancy, time from feeding and ambient temperature. It is possible to eliminate the effects of certain factors by conducting the experiment on a group of animals which is homogeneous for age, weight, etc. To eliminate the heart rate response due to motor activity Baldock and Sibly (1990) used 'general linear modelling' techniques on the heart rate results of sheep. This approach is currently being used by Bartle and Seabrook at the University of Nottingham to determine the effects of hand and arm interaction on the heart rate response of both indoor and outdoor kept sows. Preliminary results suggest that 'aversive' hand and arm interaction with sows elicits a marked, spontaneous increase in heart rate. In outdoor sows, the partitioning of the response between *motor activity* (i.e. running away) and the response to *physical* treatment on the animal (i.e. hand and arm interaction on the head or rump) has not yet been fully elucidated. It is, however, evident that the rise in heart rate is not entirely due to motor activity as the same treatment given to tethered sows generally produces an increase in heart rate, although significantly, *not* for all sows.

Vocal Interaction

Background

Animals respond to sound and human and animal vocalizations. The use of the voice is an effective means of interaction with animals. In spite of its importance there is at present very little thorough research on the effect of aspects of tone and amount and level of vocalization. Tone of voice would appear to be crucial in the animal's discrimination as to the nature and purpose of the interaction. A learnt response by the animal occurs as a result of operant conditioning and the association of negative stimulation occurring as it learns to adjust to the routine required.

Research results

Seabrook (1984) presented results from a study of 24 very similar one-person dairy units where feed inputs per cow were almost identical (Table 7.5). These results show, for example, that on the higher yielding herds the stockperson talked to the cows far more often than on the lower yielding herds. Seabrook has also indicated that the vocal interaction occurs at different levels: 'talking to' the animal as opposed to the sound and reaction of 'talking with' the animal. With another species, pigs, Hemsworth *et al.* (1989) conducted attitudinal profiles of the stockpersons. Their results show that the attitude to the use of verbal effort in moving pigs was highly ($P < 0.01$) correlated to the total litter size.

Table 7.5. Behavioural observations and herd results from one-person dairy units.

	Frequency of activity (average number min^{-1} cow^{-1})		Number of words (average number min^{-1} cow^{-1})	
Stockperson talking 'with' cows*	2.8	0.0	12.4	0.0
Stockperson talking 'to' cows†	1.3	0.6	5.7	4.1
Dairy unit results	Higher yielding	Lower yielding	Higher yielding	Lower yielding

Source: After Seabrook, 1984.
*'with' implies that the stockperson was almost 'expecting' a response; e.g., 'how are you old girl this afternoon?'
†'to' implies that the stockperson was merely issuing 'commands' or 'comments'.

Holistic Empathetic Interaction

Background

Metz (1987), in a review of the experimental work on stockmanship, concluded that many negative effects of the human–animal interaction are related to the fact that the stockperson induces fear and adverse reactions in the animals simply by being in and imposing himself or herself on the environment of the animals. There is similar evidence (Seabrook, 1987) that positive components from the stockperson can either ameliorate these effects or induce positive responses. The environment created by the human will include a number of factors that have received only limited systematic research; equally some factors will be perceived as being at the 'fringe' of science and even outside human sensory perception. The following, while not a definitive list, will enable this aspect of interaction to be discussed in a systematic way:

- Smells and other chemicals secreted by the stockperson
- Patterns of movement and behaviour perceived by the animal
- Perception of changes from 'normal' behaviour of stockperson
- Manner of hand movements, e.g. degree of shaking
- Level of electromagnetic 'forces' created by the stockperson
- Nature of other messages and signals transmitted by stockperson, e.g. creating environmental stability and an air of 'calm and confidence'.

Research results

Seabrook (1984) presented the results of a major survey of a group of one-person dairy units under exactly the same management regime. The analysis of the data attempted to quantify in an unbiased and objective manner the subjective view held by the owner after over 40 years of observation. The analysis shows that 'there was a variation in performance of identical units' and that 'there were measurable changes in performance on the units with a change of stockperson'. The herds were also investigated by a longitudinal participant observation study, which together with the data provides a conclusion to support the view that the stockperson's holistic

Table 7.6. Index of yield (average yield cow^{-1} $year^{-1}$) obtained from a six year study of 12 identical one-person dairy units.

	Year 1	2	3	4	5	6
Mean index for 8 herds with no change of stockperson	100	99	102	89	98	97
Herds with a change of stockperson						
1	100	102	91*	92	74*	74
2	100	98	104	108	72*	74
3	100	102	118*	135	118	120
4	100	101	102	125*	124	119

Source: After Seabrook, 1984.
Base year 1966 (100) Average yield 1966 = 2917 litres cow^{-1} $year^{-1}$ (low input/low output system)
*Indicates a change of stockperson.

empathetic interaction influences the yield of dairy cows. The analysis of the herd results is shown in Table 7.6.

These herds with an identical management and very similar operative skills provide evidence of the role of an empathetic component. Seabrook (1984) highlighted the role that the stockperson's personality might play in the empathetic component and indicated differences in the nature of the interaction process. While it is not always possible to isolate the degree of cause and effect, observation of differences in stockperson behaviour can be related to yield levels.

Recent work with pigs by Seabrook (1991) confirms earlier observations by Hemsworth *et al.* (1981b), and shows that in a longitudinal study of stockpersons working in identical farrowing houses different behaviour patterns of the sows were established (Table 7.7). The classification of the level of interaction is concerned with both the quantity and the quality of the interaction. These results clearly highlight behavioural differences and the variations in the different houses appear to relate well to the nature of the interaction process.

The evidence from both experimental data and observations on commercial units clearly suggests that correct interaction by the stockperson increases the performance of the animal and probably has beneficial welfare effects. There is also evidence to suggest that the behaviour of the animal is significantly affected by the actions of the stockperson. Table 7.8 sets out a summary of observations with dairy cows; these are based on research by Seabrook (1984) and confirm the observations of Albright (1984), especially with regard to flight distances and free approaches of the cows to the observer. These results, while not enabling an exact isolation of cause and effect, clearly indicate behavioural differences created by the stockperson on these one-person units.

Table 7.7. Observations of the behaviour patterns of sows, with different stock-persons.

Activity	Stockperson 'High' interaction	 'Low' interaction
	Average (%) of time spent on activity by sows*	
Bar biting	8	31
Resting	49	15
Standing in trough	0	5
Chewing air/chomping	15	15

Source: After Seabrook, 1991.
*Observation period for 20 minutes after evening feed.

The stockperson

Behaviour of the stockperson

A number of signals intentionally or unintentionally released by the stockperson when attempting to impose routines may be identified by the animal as threatening in nature and so delay or interfere with the development of a good relationship. There are many routine husbandry practices which rely on a close approach, even contact, and inspection of the animal. These include such things as oestrus detection, supervised mating, artificial insemination, vaccinations and health inspections. There are also a number of processes, for example castration and vaccination, which involve negative stimuli. The research results show that unfamiliarity with the stockperson and adverse handling by the stockperson can lead to physiological and behavioural responses by animals, adversely affecting their welfare. While the cause and effect is not conclusively developed the evidence supports the view that positive interaction improves performance.

Personality of the stockperson

Some of the differences in behaviour of the stockpersons can be related to personality attributes (Seabrook, 1990, 1991); these are summarized in Table 7.9. Detailed studies of stockpersons showing high levels of positive interaction have been

Table 7.8. The variation in behavioural components of dairy cows.

Behaviour	Dairy unit results Higher yielding	 Lower yielding
Mean entry time to parlour (s cow^{-1})	9.9	16.1
Field flight distance (metres)	0.5	2.5
Approaches to observer (no. min^{-1})	10.2	3.0
Dunging in the parlour (no. hour^{-1})	3.0	18.2

Source: After Seabrook, 1984.

Table 7.9. Personality traits and the influence on behaviour of the stockperson and performance of the animals.

Traits of stockperson	Behaviour	Reference
Confidence and introversion	More positive interactions (higher yield per cow)	Seabrook (1984)
Low confidence	Interaction inconsistent (lower growth rates)	Seabrook (1991) Hemsworth and Barnett (1987)
High aggression level	More aversive handling (fewer piglets born per sow)	Seabrook (1991)
High aggression level	More aversive handling (higher piglet mortality)	Seabrook (1991)

established by Seabrook, using a modified assessment procedure based on Cattell's 16 Factor Inventory (Cattell, 1963). A summary of the work is shown in Fig. 7.1. This profile is for a group of stockpersons seen to exert high empathetic interaction, as defined by much 'pleasant' handling, vocal contact and regular interaction at close quarters.

It can be argued that personality traits are fixed but it is possible to isolate behaviours to adopt; these are summarized in Table 7.10. Metz (1987) reinforces the importance of adaption of the animal to humans. Stroking may help the animal to cope with stressful situations. Dantzer and Mormède (1983) emphasize differences between species and individuals of the same species. The stockperson must be able to interpret the individual reactions of the animals in order to judge the correct behaviour to adopt.

Conclusions

The conclusions from the research are that humans as a factor of an animal's environment are able to influence both production and welfare through interactive processes. This may be through hand and arm interaction, vocal interaction or a combination of, as yet not totally identifiable, factors which have been summarized as holistic empathetic interactions.

It is becoming more apparent that individuality is not a phenomenon unique to the human species, but is also a feature expressed by animals, which might be termed 'temperament'. A given stimulus or interaction will not necessarily elicit the same response from one animal to another, even of the same species and being genetically similar. The expression of behaviours in response to interactive stimuli is a function which may relate closely to 'temperament' as considered in the context of the perceptive capacity of the animal. In an attempt to improve the production and welfare of animals of economic importance through the human factors in the environment it is becoming evident that distinguishing between the quality of the interaction is insufficient. It is necessary to improve not only the quality of the interaction as a whole but also the perceptive skills of stockpersons to enable them to adapt and modify behaviour to match the individual requirements of individual animals. To perform this successfully, skill is required in 'animal temperament'

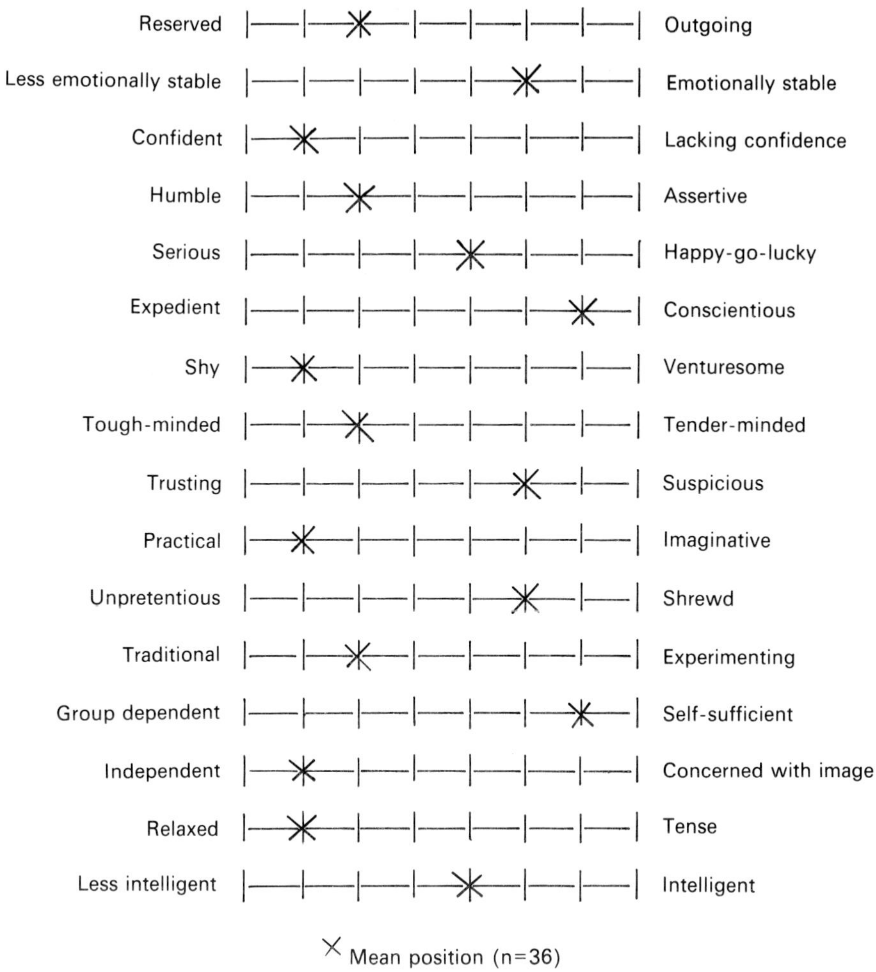

Fig. 7.1. Personality profile of a group of high empathetic pig stockpersons (after Seabrook, 1990, 1991).

Table 7.10. Developing interaction factors associated with the development of good empathetic stockmanship.

Factor	Action
Operant conditioning	Reward *favourable* behaviour of animal
	Use of food and other positive stimuli as a distraction for negative interaction
Physical contact	Stroking and patting
	Scratching animal's head
Social identification	Use of voice and social gestures
Stability	Consistent action
	Confident action
Handling	Non-aggressive behaviour

assessment, accompanied by selective interaction techniques. This merely highlights the importance of the employment of skilled, knowledgeable and adaptive stockpersons in order to meet the changing perceptions, growing expectations and developing knowledge of the consuming population.

References

Albright, J.L. (1984) Human–animal interaction. *Symposium 1 Applied Ethology Meeting.* Carbondale, Illinois, pp. 1–15.

Baldock, N.M. and Sibly, R.M. (1990) Effects of handling and transportation on the heart rate and behaviour of sheep. *Applied Animal Behaviour Science* **28**, 15–39.

Cattell, R.B. (1963) *The Scientific Analysis of Personality*. Penguin, Baltimore.

Dantzer, R. and Mormède, P. (1983) Stress in farm animals; a need for re-evaluation. *Journal of Animal Science* **57**, 6–18.

Dryden, A.L. and Seabrook, M.F. (1986) An investigation into some components of the behaviour of the pigstockman and their influence on pig behaviour and performance. *Journal of Agricultural Manpower Society* **1**, 44–52.

Duncan, I.J.H. (1990) Reactions of poultry to human beings. In: Zayan, R. and Dantzer, R. (eds.), *Social Stress in Domestic Animals.* Kluwer Academic Publishers, Dordrecht, pp. 121–131.

Gonyou, H.W., Hemsworth, P.H. and Barnett, J.L. (1986) Effects of frequent interactions with humans on growing pigs, *Applied Animal Behaviour Science* **16**, 269–278.

Hemsworth, P.H. and Barnett, J.L. (1987) The human–animal relationship and its importance in pig production. *Pig News and Information* (CAB International) **8**, 133–136.

Hemsworth, P.H. and Barnett, J.L. (1991) The effect of aversively handling pigs, either individually or in groups, on their behaviour, growth and corticosteroids. *Applied Animal Behaviour Science* **30**, 61–72.

Hemsworth, P.H., Barnett, J.L. and Hansen, C. (1981a) The influence of handling by humans on the behaviour, growth and corticosteroids in the juvenile female pig. *Hormones and Behaviour* **15**, 396–403.

Hemsworth, P.H., Brand, A., and Willems, P.J. (1981b) The behavioural response of sows to the presence of human beings and their productivity. *Livestock Production Science* **8**, 67–74.

Hemsworth, P.H., Barnett, J.L. and Hansen, C. (1987) The influence of inconsistent handling by humans on the behaviour, growth and corticosteroids of young pigs. *Applied Animal Behaviour Science* **17**, 245–252.

Hemsworth, P.H., Barnett, J.L., Coleman, G.J. and Hansen, C. (1989) A study of the relationships between the attitudinal and behavioural profiles of stockpersons and the level of fear of humans and reproductive performance of commercial pigs. *Applied Animal Behaviour Science* **23**, 301–314.

Metz, J.H.M. (1987) The response of farm animals to humans – examples of the epistemology of experimental research. In: Seabrook, M.F. (ed.), *The Role of the Stockman in Livestock Productivity and Management*, CEC, Luxembourg, pp. 23–37.

Paterson, A.M. and Pearce, G.P. (1989) Boar-induced puberty in gilts handled pleasantly or unpleasantly during rearing. *Applied Animal Behaviour Science* **22**, 225–233.

Pearce, G.P., Paterson, A.M. and Pearce, A.N. (1989) The influence of pleasant and unpleasant handling and the provision of toys on the growth and behaviour of male pigs. *Applied Animal Behaviour Science* **23**, 27–37.

Seabrook, M.F. (1984) The psychological interaction between the stockman and his animals and its influence on performance of pigs and dairy cows. *Veterinary Record* **115**, 84–87.

Seabrook, M.F. (1987) The role of the stockman in livestock productivity and management; Research epistemology – the holistic approach. In: Seabrook, M.F. (ed.), *The Role of the Stockman in Livestock Productivity and Management*, CEC, Luxembourg, pp. 39–51.

Seabrook, M.F. (1988) The behaviour of the pigstockman and its influence on pig performance and behaviour – a review. *Pig News and Information* (CAB International) **9**, 403–406.

Seabrook, M.F. (1990) Reactions of dairy cattle and pigs to human beings, In: Zayan, R. and Dantzer, R. (eds), *Social Stress in Domestic Animals*. Kluwer Academic Publishers, Dordrecht, pp. 110–120.

Seabrook, M.F. (1991) The human factor – the benefits of humane and skilled stockmanship. In: Carruthers, S.P. (ed.), *Farm Animals – it pays to be humane*. Centre for Agricultural Strategy, University of Reading, pp. 60–72.

Section II

Perception of the Environment by Farm Animals

Information about the world in the form of various potential stimuli such as patterns of light, sound or complex molecules in aerosols are received by specific modifying organs such as eyes, ears or nostrils and then transduced by specialist cells such as the retinal photoreceptors, the basilar membranes, hair cells or the olfactory epithelium. Such transducers are the basic interface between an animal and its environment. Information may be casually defined as 'that which reduces uncertainty' and may serve to alert an animal to a predator, signal the onset for lessened activity or appraise an animal of its spatial position. Such information ultimately ends up in the brain and serves as the basis for a decision to fight or flee, chase or be chased, eat or avoid eating. The response may be immediate as in the case of simple reflex activity or may occur much later as in the case of problem solving requiring much 'thought' or 'consideration'. The information enters a brain which, depending upon the species and its level of development, is to some extent prepared to receive the input, and/or of itself may modify the microstructure of the brain, a description indeed of the old nature–nurture problem.

Generally the senses may be regarded as receiving information about distal events, those taking place at a distance from the animal and not in direct contact with it, and those events which are proximal, i.e. in direct contact with it, e.g. the skin. Distal senses include vision and audition; proximal include tactile, kinesthetic and haptic. The initial stimulus, distal or proximal, its modification and transduction and then subsequent processing, and how this affects short- and long-term behaviour are questions addressed in this section. An understanding of the sensory, perceptual and cognitive characteristics of an animal offers at least an intellectual challenge. However, since such processes are vital to a complete organism such knowledge must provide insight into how the farmer, the veterinarian, and animal might enjoy a better relation in terms of treatment, management, economy, yield and comfort and well-being. These factors can range from a simple consideration of optimal, ambient light illumination or temperature levels for an animal's comfort, to the increasing awareness involved in comfort, ethics and welfare. Such a range of concerns was echoed at the conference (see Phillips's chapter on photoperiod, the Heffners' on basic audition and Hurnik's discussion of the broader more complex issues in animal welfare and the behavioural response).

Although we will probably never be fully aware of the experience of another species, much work has been carried out to give at least a general appreciation of how much the sensory input from the eye, ear, skin, etc., are reflected in brain processing and the phenomena which may be observed from such sensory inputs to the brain. The terms sensory, perceptual and cognitive ought not to be considered as describing mutually exclusive domains of behaviour since there can be considerable overlap. For example colour vision may be considered at all levels within the visual process, providing the animal possesses the requisite cone photoreceptors in its retina, while it may also play an important part in more complex pattern recognition involving feeding, as with the perception of shape, intensity and motion; words however mean something, thus for the sake of clarity these terms are defined, the definitions are representative and are taken from Gregory (1987).

Sensation or sensory can be defined as processes or events occurring as a direct result of the transducers or modifying organ such as an ear or skin and the neural processes immediately serving the transducer, such as the retina's 'neural networks' in the eye. Classically, there are five senses, though more contemporary considerations would list some 20 to 30. The purpose and design of a sensory process is 'to detect change' (Evans and Piggins, 1963), thus changes in the position, intensity and frequency of stimuli may be signalled without perhaps detailed recognition of exactly what is sensed. Change then, rather than recognition, is the essential characteristic of sensation. The chapters by Perry, the Heffners, Livingstone, and Piggins all discuss sensory problems drawn from various modalities.

Perception is the awareness of what an event is, the recognition of a pattern or a face, of something specifically in motion rather than simply sensing that motion has occurred. Kendrick's report of cells in the sheep's temporal cortex, which primarily respond to faces rather than to other related stimuli, reveals a physiological substrate of perception, compared for example to receptor cells in the chicken retina responding to a specific waveband of light thus revealing a sensory substrate for colour vision. Perception then is a more complex characteristic of behaviour than sensation. Hurnick, Kendrick and Piggins all refer to such problems.

Finally, cognition refers to the inference or meaning a stimulus might possess and thus involves some degree of memory. The recognition of a shape which represents friend or foe, prompting a response to flee, fight or stay put; which is a preferred food source, and what is to be rejected. Problem- solving is *ipso facto* a cognitive process wherein, for example, a sheep learns to escape from a barn or field. Both Hurnick and Kendrick in rather different ways consider such processing, and reveal what farmers have known for some while – that we underestimate the abilities of such animals as sheep. This author is reminded of a popular article on the species read while in a doctor's waiting room; it was entitled 'Sheep are not dim'!

The methods generally employed to investigate animal behaviour, in this section those of sensory, perceptual and cognitive characteristics, are well represented here. They fall under three headings: physiology, psychophysics, and ethology or naturalistic observation. The intent here is to consider them briefly since specific techniques and instrumentation are described in the chapters where necessary and appropriate.

As the name suggests a *physiological* approach seeks to investigate the wetware of the animal, in the papers described here the wetware comprises such systems as the eye–brain, ear–brain, skin–brain, etc., and their relationships with behaviour. Within this general area a number of major techniques of investigation exist. Firstly, the animals' sensory surface may be stimulated and recordings made from some site in the animals' nervous system, in this case the brain. Many techniques exist for this since many forms of energy may be recorded from the brain itself ranging from that of electrical recording to magnetic, neurochemical or glucose uptake. Additionally of course many morphologic or histologic techniques are available. The number and variation of techniques has increased dramatically over the last two decades, with single unit recording from cortical cells to the EEG where the firing of many neurones is recorded; these represent invasive and non-invasive techniques respectively. On the other hand the brain itself may be stimulated and the response of the animal noted. Here for example, central (i.e. brain) mechanisms may be triggered directly using electrical or chemical stimulation and the receptor surface left unstimulated. Finally, removal (ablation), severing connections (lesioning) or temporarily, chemically altering brain circuits, and viewing the animal's behaviour before or after the operation adds a further general method.

The second major method, that of *psychophysics* basically explores the relationship between stimulus and response. Changes (or otherwise) in the animal's response contingent upon some change in stimulus, e.g. the intensity of light, pressure on the skin or that of more complex behaviour following a change in housing or management all fall under the general purview of psychophysics. It should be noted that questions about the animal's physiology are not being considered directly here, though its action and mechanism may be inferred. As with the physiological method, many techniques exist, most of which were developed in the experimental psychologist's laboratory. A standard technique here rests upon operant or instrumental conditioning where an animal is rewarded, contingent upon its performing a specific action or series of actions; such a technique has proven particularly useful in investigating some of the sensory, perceptual and cognitive behaviours mentioned here. Other psychological techniques typically employed depend upon an animal associating a reward with one particular type of stimulus and then taking part in a controlled environment study where basic parameters of the stimulus are slightly varied (e.g. light or tone intensity, or shape and pattern). Such a learning–discrimination technique has provided much information about the sensory and perceptual characteristics of lower animals over a wide range of species. An animal's cognitive life may be explored using a variety of problem-solving situations, perhaps the most famous (or infamous) of which are the various forms of maze where errors, time and frequency of particular behaviour may all be measured.

The final general method, that of naturalistic observation or *ethology* has been particularly developed by zoologists and to some extent by psychologists. It will be evident that in the psychophysical method mentioned above, it is essential to have much control over the environmental conditions surrounding the animal; in the ethological method the reverse is true, the animal is not only viewed as far as possible in its natural environment, but additionally (and essentially) its behaviour ought not to be influenced by the presence of an observer (hence the use of a 'hide' in observing birds).

Each of these methods has its strengths and weaknesses and for many reasons, some ethical, some economic, it may only be possible to investigate behaviour using one technique under one general method. Obviously it would be more satisfactory and rewarding to examine the same behaviour using techniques from all three methods, on at least the same species, if not on the same animals themselves.

It is evident that the progression from sensory to cognitive function is one of increasing complexity of both event and purpose and occurring at increasingly higher levels in the brain. Considerations of animal consciousness, self-consciousness, awareness of future intentions were discussed at this conference by Livingstone and Hurnik. Such considerations would naturally follow a discussion on cognition. Time constraints alone would have unfortunately restricted much discussion of such topics, though they were touched upon. Perhaps a future conference on awareness in farm animals is planned?

References

Evans, C.A. and Piggins, D.A. (1963) A comparison of the behaviour of geometrical shapes when viewed under conditions of steady fixation, and with apparatus for producing a stabilised retinal image. *British Journal of Physiological Optics* **20**, 261–274.

Gregory, R.L. (1987) *The Oxford Companion to the Mind.* Oxford University Press, Oxford.

8 Visual Perception

D. PIGGINS*

Department of Psychology, University of Guelph, Ontario, Canada

Summary

Although we may never entirely appreciate the experience another species has of its environment, much time and effort has been spent by workers in a variety of disciplines attempting to have an appreciation of the sensory, perceptual and cognitive experience of other animals. Such work has however often been confined to relatively few species of necessity; much of the experimental work in the psychological laboratory on learning and memory, for example, has involved the use of such species as the ubiquitous white rat, various monkeys, pigeons and university undergraduates.

This talk briefly reviews primary visual sensory and perceptual function from eye to brain, together with some of the major methods, psychophysical and physiological, commonly employed to investigate such functions. Such considerations include the evaluation of ocular, retinal and higher visual system functions as contributing to an animal's visual behaviour. The species mainly considered and compared here are the Amazonian Manatee and the sheep, representing aquaculture and agriculture respectively, taken from contrasting environments in Brazil and the United Kingdom. A review of visual function in both species includes a discussion of such sensory matters as physiological optics, colour vision, sensitivity and acuity; whilst more perceptual attributes such as binocular vision, stereopsis, form and object recognition are also considered. Mention is also made of specific attributes in other mammals such as the problem of identifying an accommodative mechanism in the horse eye, and colour vision in cattle.

Introduction

Relatively little is known about most animal species, indeed we know much about a few. Such a statement is true regarding a knowledge of, for example, a species' eating habits, diurnal–nocturnal activity cycle and its sensory, perceptual and

*Temporary address: Department of Psychology, University of North Wales, Bangor, UK.

cognitive abilities. For various reasons, some ethical, others circumstantial, the few species which have been investigated intensively have often acted as models in an attempt to explain aspects of human behaviour and thus most have been close to man in either a mammalian or evolutionary sense. For matters visual, regarded here in a broad sense, this has included various primates such as the chimpanzee, monkeys such as the rhesus or macaque, the domestic cat, laboratory rat and somewhat further afield the pigeon and *Limulus*, the horseshoe crab. With many domestic and farm species interest has been until quite recently 'neck down' rather than 'neck up' where in the latter case the more complex behaviour and processing which is manifest from higher neural centres has been of little interest. The more sensory–cognitive papers delivered at this conference particularly refer to complex behaviour and reflect a contemporarily healthy interest in activities at the animal's head-end.

We will probably never know how another species actually experiences the environment – how it thinks, 'sees' or emotes. However, in some of the species specifically mentioned above we may feel at least anthropomorphically closer to some restricted aspects of behaviour such as pattern recognition or problem-solving. In a number of cases, because of the long-term interest in vision and visual behaviour, relevant sensory, perceptual and cognitive aspects have received much attention. Here the same sensory and perceptual characteristics of two somewhat differing farm species are considered. Not only are the two species common and widely different, but they have been or are being intensively regarded from both a scientific and an agricultural point of view.

The sensory information from the environment and ultimately received by the brain is modified in a number of ways before the appropriate brain centres are reached. The information may be about variations in light intensity, wavelength differences, the shape, size and position of objects in the visual field – a wide variety of information, some of which it is essential to know about currently, some of which may be stored for further retrieval and use, and some of which may be safely ignored. Modifications include those which are effected optically in the production of a well-focused retinal image for objects at varying distances from the animal, the adjustment of the amount of light forming the retinal image by variation in pupil size, the placing of the retinal image on a receptor area to optimize the resolution of detail (or in the case of simple detection, one of optimal sensitivity). There is also the involvement of two eyes, rather than one, requiring the co-ordination of two visual axes to cross at the object under inspection. Although much processing takes place retinally for intensity and wavelength, initially by the transduction process in the photoreceptors, the more complex processing about shape, size, pattern and depth for which decisions may be required, are brain related tasks and require not only access to current information but that which has already been stored and now needs to be retrieved. Some of these parameters will be considered in the rest of this chapter.

Eyes do a number of things and there are usually two of them. They focus, transduce, regulate light intensity and move. Basically vertebrate eyes serve to produce as well focused an image as possible on the retina. This image is then transduced by the photoreceptors and, with some prior processing in the neural retina, is relayed to the many various parts of the brain serving vision. The focusing is accomplished by correlation of eye length and optical power in the cornea and crystalline lens which latter structure enables some eyes to have a variable focus (termed accommodation). There is also control over the amount of light which constitutes the retinal image. This is effected by a sphincter muscle often pigmented, which encloses the pupil aperture; such an aperture may be variable in size and shape and when small can also contribute to the focusing of an image rather like a pin-hole camera. Each eye also possesses a field of vision (termed 'monocular') with usually some overlap between the visual fields of each eye even with quite laterally placed eyes. Objects within this overlapped (termed 'binocular') field will be imaged on each retina at the same time. A possible consequence of a large binocular field is stereopsis, the ability accurately to locate objects in relative depth near and in front of the animal, useful in a predator for example. As mentioned, the photoreceptors (i.e. the transducing cells) represent the interface for the animal between its external and internal environments; the focusing, light control and binocular overlap are important modifiers for the succeeding transduction and further processing. The types of photoreceptor, their relative densities and positions, give rise to such vital aspects of behaviour as the degree of visual acuity, sensitivity, colour vision, visual field extent and capability to function optimally in either day or night or both. Finally, eyes move, the small eye-movements serving to retain a retinal image in awareness by constantly moving the retinal image to adjacent parts of the retina where there is a different state of adaptation in the photoreceptors. Larger eye-movements along with head movements serve to orient the animal and place the retinal images of objects of special interest on appropriate parts of the retina, for example the fovea, the area of highest spatial resolution.

It will be evident that knowledge of the eye portion of the eye–brain system is important in appraising what visual information the system is capable of receiving and transmitting to the brain. This is a major governance in a visual animal's behaviour since it is of interest to know not only what it can, but what it cannot receive, and thus not be able to process.

What follows is a brief discussion of some major ocular and sensory characteristics of the two species sheep and manatee, with comments regarding their consequences for visual behaviour. Following this discussion will be one touching upon more complex visual behaviours such as shape and pattern recognition, and stereopsis, characteristics comprising perception.

The farm species chosen for discussion in some detail in this chapter are a terrestrial mammal the sheep, and an aquatic mammal the manatee. Sheep do not need an introduction, they have been farmed, bred and cross-bred for centuries, and whole economies depend upon them for meat and wool. The manatee on the other hand is a much more recent and as yet unproven aquacultural mammal which probably does need an introduction. The manatee is a purely aquatic herbivorous mammal, one of several species belonging to the order Sirenia which includes also the Dugong and the recently extinct Stellar's Sea Cow, and distantly related to the elephant. The colloquial name for the manatee is indeed 'sea cow'. The species considered here are the Amazonian manatee and the West Indian manatee since they have been the most extensively investigated. In the last 15 years the Amazonian variety has attracted much attention from Brazilian aquaculturalists and much work, mainly of an information gathering nature (colloquially 'stamp collecting'), has been carried out on the species, since very little was known. Practically speaking, it is palatable and being a voracious herbivore enables channels to be kept open in the small river tributaries of the Amazon. It was felt in choosing these species that they would represent a great contrast – terrestrial : aquatic – one for which we have much knowledge and the other very little knowledge; also proven aquacultural species : unproven aquacultural species, and so forth. Where applicable, highly specific examples of sensory and perceptual processing will be introduced using other farm animals as examples: the horse and chicken will be considered for particular comparisons.

The Eye – Physiological Optics

Light refraction by an eye is primarily by a combination of the curved surfaces of the cornea and crystalline lens. It is the cornea which contributes most of the overall ocular power in air since greatest refraction occurs at the surface of greatest refractive index change: the cornea–air boundary. Both cornea and lens have higher refractive indices than the ocular media in contact with them. If the eye's axial length (cornea to retina) correlates accurately with the eye's overall optical power then a well focused retinal image results. If this is for an object at infinity, practically speaking at 6 metres or more, then the eye is said to be emmetropic, with deviations from this being described as ametropic. If the image is formed before the retina is reached the condition is known as myopia; if the image would be formed beyond the retinal plane, then hyperopia results. Colloquially these ametropias are known as 'short sight' and 'long sight' respectively. The common condition of astigmatism refers to differences in overall power of the eye depending upon which anterior–posterior meridian is measured, e.g. there may be a difference between the horizontal power of the eye compared to its power in the vertical meridian. As the name suggests, points on the object are not imaged as points, but in terms of lines or ellipses depending upon where the refracted beam is sampled. Corrective lenses for the ametropias basically reduce the overall ocular power in myopia (they are designated as 'minus'), increase the power in hyperopia (designated as 'plus'), and all combinations of these can occur in astigmatism, where the corrective lens may be described as being combined of two cylinders with axes 90° apart. Ametropia is measured as a deviation from emmetropia (designated quantatively as '0'). The measure is in dioptres or reciprocal metres, for example a hyperopic correction by a lens of 50 cm focal length will be expressed as $+100/50 = +2.0$ D (dioptres), and that of a myopic corrective lens of 20 cm focal length will be $-100/20 = -5.0$ D. Astigmatic corrective lenses will show different powers along different meridians and expressed as the greatest and least power respectively along with their associated axes.

In mammalian eyes variation in the overall ocular power is affected by changes in the curvature of the crystalline lens which is pliable in nature and connected to the ciliary muscle inside the eye. Although ciliary muscular strength is not affected by age, the lens's pliability is, rather like that of the skin. This variation in optical power is described as accommodation and with a lessening in lens pliability and hence variation in power with age, so the nearest point of well focused vision recedes from the eye. When this interferes with reading a book for example (and the arms are not long enough!) the condition is known as presbyopia and 'reading' glasses are prescribed, the power of which to some degree replaces that lost by the now less pliable crystalline lens.

Ametropia and presbyopia may be measured in a number of ways; that employed here for sheep and manatee and other non-literate species is the common optometric and ophthalmologic clinical technique of retinoscopy where basically the characteristics of a beam of light are evaluated as it is reflected from the back of the subject's eye, thus obviating the need for subjective reply! The technique is discussed briefly in the appendix to this chapter. Ametropia is evaluated for a distant object and variation in the refraction while retinoscopy is being carried out can reveal the presence, and give an idea of accommodative ability though in practice a rough estimate only is obtained.

The retinoscope is also useful in estimating the angular subtense of an animal's field of vision. Thus with a stationary eye (more easily said than done!) the extent over which the light reflex from the rear of the eye may be viewed can be measured, and compared to that from the other eye, and any overlap between the monocular fields may also be determined.

There is much variation in eye size across different species and it should be noted that in much smaller eyes not all the inherent structures are reduced by the same scale – retinal thickness for example is not as significantly reduced as that of other structures. Retinoscopic refraction considers reflection from the front of the retina, thus the relevant axial length of much smaller eyes will be relatively much shorter than for bigger eyes when the difference in scale is considered. A consequence of this is that the increase in hyperopia found in smaller eyes is more apparent than real, a point made by Glickstein and Millodot (1970) who plotted hyperopia determined retinoscopically against the axial length of the eye. Their function for a number of species is shown in Fig. 8.1.

This finding is of particular relevance for the manatee and refraction. Reasonably Swift (1726) could not have known of this when he described the vision of Lilliputians (Piggins, 1978) and indeed he is wrong in his conclusions regarding their eye sight. Thus, although there is much variation in eye size there is not in ametropia, indeed a number of workers have strongly suggested that eyes tend towards emmetropia (Sorsby *et al.*, 1957; Glickstein and Millodot, 1970; Hughes, 1977; Martin, 1988).

Hughes (1977) writes '. . . the almost universal hypermetropia established by retinoscopy results from a systematic error of method, and that the true mean refractive state of most, if not all, world vertebrate species is close to emmetropia in the effective image plane', while Sorsby *et al.* (1957) (as cited by Martin, 1988) conclude 'the emmetropic eye is not the end result of a haphazard combination of variable components, but a coordinated organ, the individual emmetropic eye . . . is not accidental'.

Figures 8.2–8.4 show the diagrammed and labelled eyes of the human, sheep (Brewster, 1860) and Amazonian manatee (Piggins *et al.*, 1983), respectively.

Only the structures directly referred to in the context of this chapter are labelled. All the eyes above are vertebrate eyes, the first two being terrestrial, the last adapted for a purely aquatic environment. The eyes in overall structure are closely similar, though the detail makes the difference. The manatee eye shown here is roughly half the size of the human eye. No less a person than Isaac Newton (1680) was the first

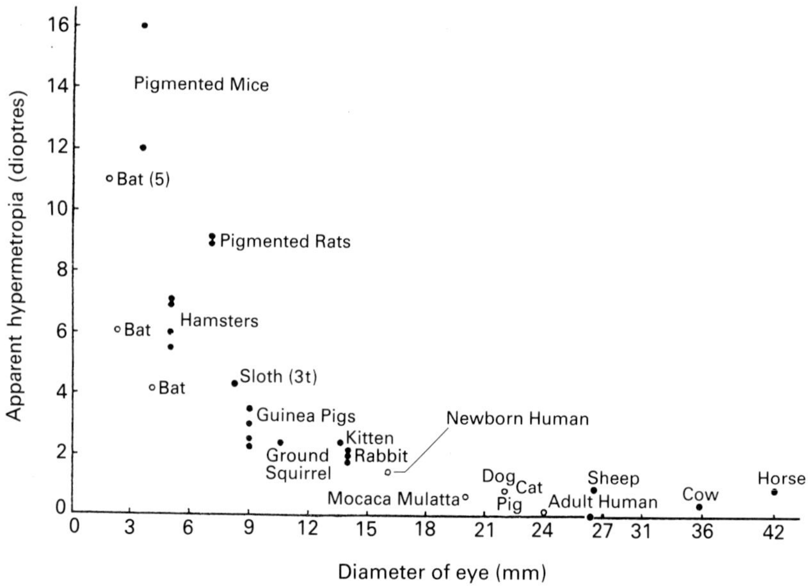

Fig. 8.1. Glickstein and Millodot's (1970) relation for apparent hyperopia and eye size diameter (with additions for dog, sheep, pig, cow and horse).

to completely describe the gross physiological objects of an eye, in this case the sheep eye shown opposite (reprinted in Brewster, 1860) and cited therein. For comparison the overall ocular power of the human eye is given as 58.64 D, that of the sheep at 61.3 D, the cow 47.7 D, the horse 38.5 D and the pig at 78.2 D (Coile and O'Keefe 1988). These are figures calculated from optical parameters in the physiological optical literature and expressed by the authors in producing schematic eyes for a number of common species. The optical powers given also reflect the axial depth of the eyes, cornea to retina; for example the sizes of the human and sheep eyes are similar.

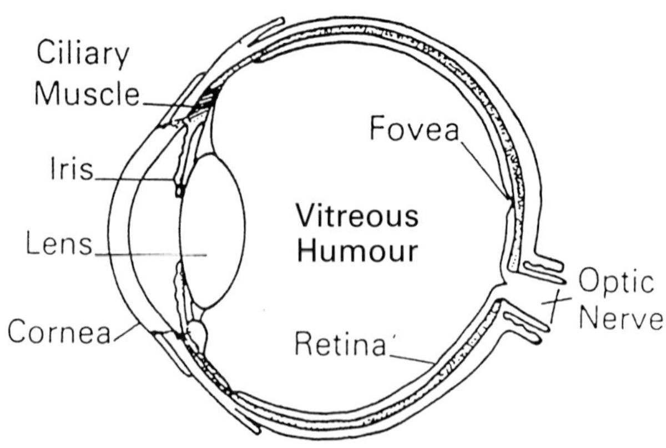

Fig. 8.2. Diagram of human eye.

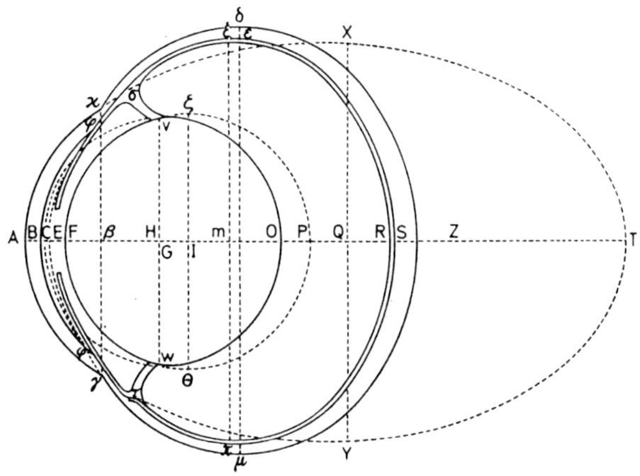

Fig. 8.3. Diagram of sheep eye (Newton (1680), printed in Brewster, 1860).

Refraction

Sheep

The refraction of the sheep eye is close to emmetropia, with negligible astigmatism. Piggins and Phillips (in preparation) report the retinoscopically determined findings for 36 sheep eyes (20 Cambridge and 16 Welsh Hill) with an overall mean of less than 1.0 D of hyperopia with a very narrow range. No myopic eyes were found, and the range of astigmatism was minimal. No correction for the Glickstein and Millodot (1970) apparent hyperopia–eye-size relation was necessary since the sheep eye is of comparable size to that of the human and falls within similar axial length. The statements regarding eyes refractively tending towards emmetropia are thus confirmed for sheep, at least for the 36 mentioned above, while there was no significant support for the suggestion that sheep are astigmatic. Such a suggestion could have been mooted on the basis of the species' large oval pupil, with the long axis horizontal.

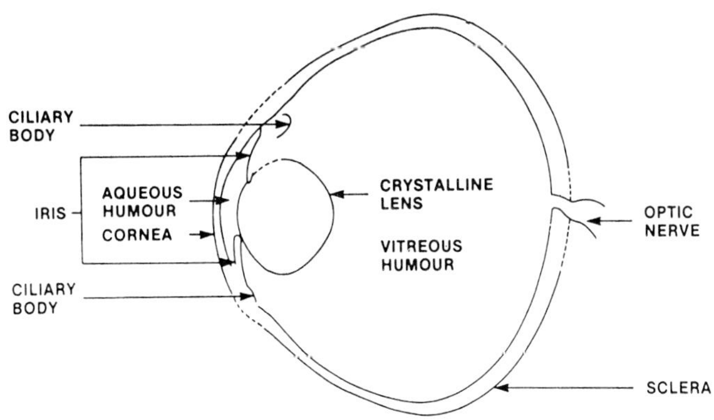

Fig. 8.4. Diagram of Amazonian manatee eye (Piggins *et al.*, 1983).

Table 8.1. Ocular refraction of Amazonian manatee.

	In air[1]		In water[2]	
	Right eye (Dioptres)	Left eye (D)	Right eye (D)	Left eye (D)
	–	+ 18.00	–	+ 0.75
	+ 16.00	+ 16.00	+ 0.75	+ 0.75
	+ 19.00	+ 19.00	+ 0.75	+ 0.75
	+ 17.00	+ 17.00	–	+ 2.25
	+ 17.00	+ 18.50	+ 0.375	+ 1.50
$\bar{X}$	+ 17.25	+ 17.70	+ 0.625	+ 1.20
Range	16.00 to 19.00	16.00 to 19.00	0.375 to 0.75	0.75 to 2.25

Source: Piggins *et al.*, 1983.

[1]These findings are probably influenced by the presence of a soft eyeball observed in all animals refracted in air.

[2]These findings result from a corrected retinoscopic working distance, which adjusts for refraction being carried out through an air–water interface (see Piggins, 1970) when $n'_{air}/n'_{water} = \frac{3}{4}$.

Interestingly, a common farm animal, the chicken, has been used recently in studies of developmental myopia, where the 'importance of patterned visual input to the eye for its co-ordinated development during its sensitive period' (review in Martin, 1988) has been shown to be crucial. Basically it has been found that high myopia can result when even just the width of a developing chick's visual field is constrained from normal, similar results being obtained by preventing patterned images being focused upon the retina, though still allowing light stimulation. In reviewing such recent work, Martin (1988) notes 'these new results suggest an intriguing developmental feedback mechanism between the neural retina, where a patterned image is detected, the optical system which produces that image, and the growth of the eye.'

Manatee

Table 8.1 gives the ocular refraction of five Amazonian manatee. Nine eyes were refracted in air, eight in water. The refractive values in air have been corrected for the distance at which retinoscopy was carried out. Because the refractive findings in water were made essentially through an air–water interface, the under water findings have been adjusted.

The figures represent refractive error, that is the amount (expressed in dioptres), which the eye deviates from emmetropia, the optical condition under which a distant object is imaged sharply upon the retina without the benefit of accommodation. A difference in refractive error along different meridians of the same eye would indicate the presence of astigmatism. This was tested for but not found in these animals. The figures given here represent findings *uncorrected* for eye size. A consideration of refractive error and eye size is important as mentioned earlier by Glickstein and Millodot (1970) and others.

It is possible then to apply a correction factor, initially formulated by Hirschberg

(1882) and utilized more recently by Glickstein and Millodot (1970), to calculate the effective refraction of an eye of small dimension whose refraction has been determined by retinoscopy. Glickstein and Millodot (1970) have plotted this correction for a number of species of animals over a wide range of eye sizes. However, in the case of *Trichechus inunguis* the ocular refraction in air is probably contaminated by the soft bulbous condition of the eyeball observed for all eyes when refraction was attempted in air: all eyes showed very high hyperopia under these conditions. Consequently, it is not reasonable to apply a correction factor to such findings obtained from 'soft' eyes. The overall anterior–posterior length of the eye in Fig. 8.4 was 1.34 cm. Glickstein and Millodot's function would suggest that an apparent hyperopia in air for such an eye size should be about 2.0 dioptres. Although the contribution of the cornea to the overall ocular refraction may not be considerable due to its relative flatness, it is likely to produce myopia since under water, where the corneal contribution to refraction would be minimal, all eight eyes exhibited hyperopia of a little less than 2.0 dioptres (Table 8.1). Consequently, the actual refraction in air of the eye should be that of low to moderate myopia.

The anomalous finding of a moderate degree of hyperopia in air may be explained by the position of the eye when refraction in air was being carried out. It was expedient to carry out retinoscopy in air with the eye pointing in general upwards. A soft bulbous condition would mean that the eye would sag equatorially which would have the effect of further flattening the cornea and reducing the overall refractive power of the eye. Such testing conditions would produce a marked hyperopic shift in refraction.

Accommodation

Sheep and manatee

Neither sheep, nor sirenians in general appear to be able to accommodate, though in the case of the former a well-developed ciliary muscle is present (Walls, 1942). Kendrick (1990) notes that 'sheep lack the accommodative reflex' as does Walls (1956) while Prince (1956) referring to dugong and Walls (1942) to sirenians in general note the lack of such an ability. Certainly in retinoscopically based refractive studies on 36 sheep eyes (Piggins and Phillips, in preparation) and the Amazonian manatee eyes (Piggins *et al.*, 1983) no refractive variation was noted over the extended periods of time taken with refraction, such measurements being made regularly over 10–15 minute periods at least and since accommodation usually takes place both readily and quickly, such a characteristic would have been easily detected. It is probable that such species of herbivore do not need accommodation for non-moving food sources, particularly given the nature of the tactile information they would receive from highly sensitive lips. However, Tenniel's depiction of a presumably presbyopic sheep in Carroll (1871), shown in Fig. 8.5, while basically incorrect could have some merit in producing a clear retinal image at rear, given the low degree of hyperopia present for distance vision.

An interesting, potentially accommodative, mechanism has been posited for the horse eye. Nicolas (1930) noted that the horse eye in sagittal section was asymmetri-

Fig. 8.5. Tenniel's depiction of a presbyopic sheep (Carrol, 1871).

cal in that the lens–retina distance was greater in the upper half of the eye than in the lower. Walls (1942) extrapolates that such a 'ramp' retina is 'the equivalent of slanting the plate of a camera so as to have in simultaneous sharp focus objects at different distances'. He mentions the horse specifically and 'probably many other unstudied ungulates'. Further 'the horse has only to tilt the eye slightly up or down to have a sharp retinal image of any object over a considerable range of distance'. Wisely, he ended the discussion with 'this is, however, only a suspicion which awaits experimental justification', since Sivak and Allen (1975) showed using retinoscopy, *in vivo*, and a rapid freezing technique with eyes, *in vitro*, that there was 'no indication of a ramp retina that could serve accommodation', there being close agreement between the results found from two different measuring methods. Thus they note that evidence for a static accommodative mechanism does not exist, while there is some evidence for a dynamic one.

The Retina

Photoreceptors

As mentioned earlier, the basic interface between an animal's light environment and its brain or mental life is the transduction mechanism itself: the photoreceptors or the rods and cones, so called because of their general shape. Depending upon the type of retina, rods, cones or both may be present. There are functional differences between these cells, which are important in considering an animal's visually guided behaviour.

The rod photochemical rhodopsin is more sensitive to light than the iodopsin of the cones. This is of particular importance in the signalling of light intensity. Rods can signal both the presence of very low light intensities and differences between such intensities while cones do so less well. However, at higher levels of intensity rod action becomes relatively insensitive because of the number of rods which have already become activated and no longer signal such differences, while cones being less sensitive continue to do so. Thus rods are associated with high sensitivity at low light levels and cones with relatively speaking moderate to low sensitivity. In environmental terms the difference between nocturnal and diurnal vision is termed scotopic and photopic vision respectively. High sensitivity and high neural convergence (i.e. many receptors inputting to many fewer ganglion cells later in the retinal process) give rise to a system which detects signals rather than one which resolves detail about them. High acuity, the ability to resolve detail, is associated more with higher levels of illumination and with lower sensitivity cones with low neural convergence at high density in the receptor matrix. Such an increase in cone density is the process of foveation. The presence of both rods and cones in the same retina (termed a duplex retina) will give rise to the best of visual worlds, high sensitivity and high acuity, and particularly so if their inventive owners have produced artificial lighting! Animals possessing only rods, for example many owls, will function best under lower light levels and prefer nocturnal living; those with a purely cone retina, such as the chicken, will best function under higher light levels and be diurnal in nature.

The presence of these two general types of photoreceptor may be revealed by a number of techniques: retinal histology, determining the spectral absorption curves for the photopigment and by psychophysical experiments using operant condition techniques for generating a dark adaptation curve. Here only the last mentioned phenomenon will be discussed, the other techniques being mentioned later where appropriate. Dark adaptation is the process of increasing visual sensitivity over time, experienced by most humans going from a relatively high intensity situation to one that is substantially lower, basically a change from photopic to scotopic vision if the intensities are far enough apart. The dark adaptation curve for a duplex retina is shown in Fig. 8.6a. The curve results from the action of both types of photoreceptor, initially the cone action and latterly the rods, thus a dark adaptation curve which is smooth and without a 'kink' in it represents the action of only one type of photoreceptor, the presence of a 'kink' reveals all; indeed even in the absence of

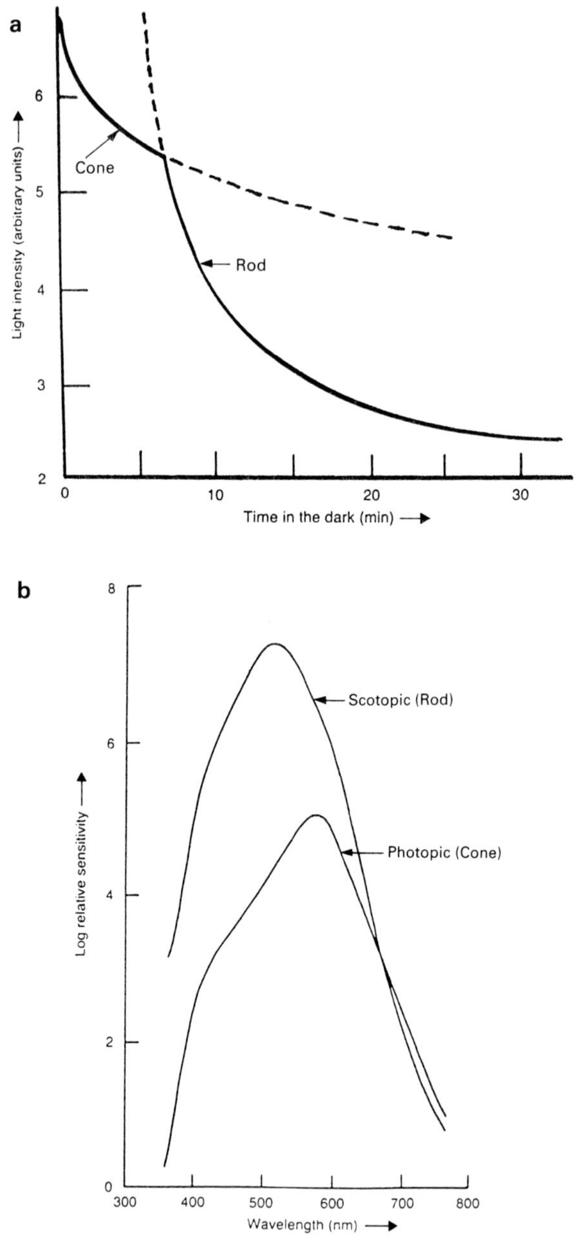

Fig. 8.6 a Dark adaptation curve; **b** relative sensitivities for rods and cones.

histological identification two types of receptor can be inferred from such a curve. Figure 8.6b shows relative sensitivities for rods and cones.

While only one type of photopigment is associated with the rods, namely rhodopsin, three types have been associated with the cone's variants of iodopsin – the spectral absorption curves are shown in Fig. 8.7. The overlap of these curves forms the basis for trichromatic vision, namely that any one wavelength on the

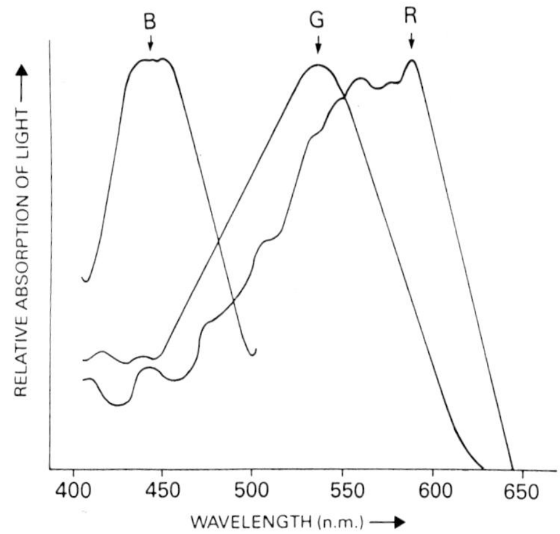

Fig. 8.7. Photochemical basis for trichromatic vision.

visible spectrum can be matched by the judicious mixing of three others when they are suitably distanced apart in wavelength. Colour vision then is a result of cone action which codes for intensity of light and wavelength, while rods give rise to achromatic vision and code for intensity alone.

Sensitivity and acuity

Retinal action can be often considered a compromise between acuity and sensitivity, the resolution of detail and detection of a stimulus respectively. Such major characteristics reflect to some extent (though many possibilities occur; see Walls, 1942), the type(s) of photoreceptor present, the retinal neural connections, receptor density, the presence of a tapetum, pupil size, its variation in size, and the shape of the crystalline lens. A consideration of such elements can indicate an eye and hence a visual system adapted more for acuity or sensitivity where the perception of detail or of sensing/detecting an event has priority. However, this is in some regard to only consider the system a little superficially, since (as is not uncommon) respectable visual acuity and well-developed sensitivity can coexist physiologically with comfort and with little compromise.

The role of photoreceptors has already been considered. What is also important is the manner in which the receptors are wired into the post-receptor neural matrix. At its simplest level, this can be considered *vis-à-vis* neural convergence, the number of receptors inputting to cells further in the retina, usual candidates for such cells being the retinal ganglion cells which send their axons through the optic nerve to the higher mid-brain centres, for example the lateral geniculate nucleus. If a ganglion cell count is low, compared to a much larger receptor cell population, high neural convergence is postulated, and sensitivity rather than acuity is assumed because a ganglion cell will not discriminate among the many inputs it receives and additionally requires much spatial summation to fire. Resolution of detail on the other hand would result from a low receptor–ganglion cell ratio, a few receptors inputting to cells further in the retina; thus closely adjacent receptor cells would signal different events, for example the detail in a stimulus.

With regard to visual sensitivity, Clarke and Whitteridge (1976) note, 'we know nothing so far of peripheral acuity in sheep'. However, the presence of rod photoreceptors and a tapetum would initially lead to a condition of a developed visual sensitivity at least. A note here is apt perhaps regarding the tapetum since it is present in various forms in a number of species and popularly associated with sensitivity, though as will be discussed this is not always the case. It should be noted that 'an area centralis (for acuity) and a tapetum (for sensitivity) are of course not incompatible as is obvious from the situation in the ungulates' (Walls, 1942) and 'a tapetum is perfectly compatible with an area centralis (though it is never associated with a fovea)' (Walls 1942). Walls in his classical book *The Vertebrate Eye and its Adaptive Radiations* (1942) discusses the tapetum at some length. Briefly, in some species a mirroring of incoming light to the eye takes place somewhere behind the photoreceptors which are located at the back of the retina in mammals. This mirror (generically) termed the *tapetum lucidum* gives rise colloquially to 'eye-shine', obvious when the light from a car headlamp is reflected back along a closely similar optical pathway into the driver's eyes. Commercially the 'cat's eye' reflectors found along many UK roads are based upon this principle. 'Functionally, reflectors from the tapetum will increase the absolute and relative differential between object and ground (thus increasing discriminability) under low levels of light intensity' (Walls 1942). The fundi of all eyes will of course reflect light incident upon them, even those lacking a tapetum, as shown by unfortunate choice of flash angle in indoor photography where the pupils appear as orange–red in the resulting photograph! However, the amount of reflection by a tapetum much exceeds that of a non-tapetal fundus. Although a number of different forms of tapeta exist, that found in ungulates is termed the tapetum fibrosum. Walls (1942) notes 'nearly all hoofed animals have this kind . . . none of them has any other kind'. Such a structure has a tendinous reflective layer located choroidally.

Sheep

Piggins and Phillips (in preparation) note that all the eyes of the sheep sampled (20 Cambridge and 16 Welsh Hill) possessed tapetal areas in the superior fundus roughly in the position suggested by Walls (1942). Interestingly, Walls also notes that the structure is 'not always restricted in usefulness to the dimmest of illuminations' and nicely puts the point: 'the all-round capacity of 24-hour eyes is not due to a fence straddling avoidance of specialization, but to a mosaic of compatible specializations for both scotopic and photopic vision'. Such a statement should be borne in mind regarding his earlier mentioned comments regarding the non-mutual exclusivity of an area centralis and a tapetum, not only in the same eye but also in sharing a similar retinal locus. Generally, ungulates have large eyes (as generations of anatomy and physiology students will remember from ocular dissection exercises), and moderate-sized retinal images. Thus a compensation for the 'dimness' of such a sized image, where it is spread over many photoreceptors may be made by a tapetum – the size of image enabling them to attain 'keen' vision despite a low ratio of cones to rods. The tapetum then can possess the dual purpose of rendering sensitivity in low illumination and being at least functional at higher illumination levels regarding large retinal image size and maintaining respectable acuity. Walls (1942) is again succinct here: 'few ungulates . . . have any approach to a slit pupil' and 'it is probably no accident that in these animals the area centralis falls within the confines of the tapetum' and 'If cones are indeed concentrated within these imperfectly known area centrales, the lowered sensitivity of these regions is nicely compensated by the tapetum behind them'. His next statement might need some modification in the light of Clarke and Whitteridge's (1976) considerations of sheep acuity, '. . . we do know that the visual acuity of ungulates is far higher than we would expect to find in the average tapetum bearing *nocturnal* animal with its rod-rich retina'. Certainly, the mention by Ensor in Bazeley and Ensor (1989) of some cones being present and the many reports of large oval pupils with a poorly defined direct reflex to light (Piggins and Phillips, in preparation), to some degree support Walls.

Sheep have been described as possessing 'good' vision on a number of criteria and by a number of observers both anecdotally and experimentally. The term itself has meant considering a wide assortment of visual capabilities including acuity, motion detection, shape recognition and depth perception – both monocular and binocular. Colour vision is one of the omissions. Geist (1971) writes 'their eyesight is marvelled at by hunters'. Kendrick (1990) mentions Geist reporting that sheep can see humans and coyotes at distances exceeding 900 metres, even when they are partially hidden from the sheep. Such reports may consider both acuity and shape discrimination. Baldwin (1981) mentions that 'observation of goats, sheep and calves under field conditions clearly indicates that they possess excellent vision and readily respond to visual stimuli'. Again both acuity and shape may be considered here, the quotation being taken from his paper on shape discrimination in sheep and calves. Kendrick (1990) in a semi-popular review on sheep vision writes, 'they rely heavily on their keen eyesight'.

Here acuity is discussed, other sensory and perceptual characteristics will be discussed where appropriate. Clarke and Whitteridge (1976) in discussing the cortical representation of vision of sheep mention the 'curious discrepancy' between the expected visual acuity using retinal ganglion cell density and the psychophysically determined value. Backhaus (1959) found a visual acuity value of 11′ arc. However, Clarke and Whitteridge quote A. Hughes's (personal communication) retinal ganglion cell density value at 500 cells/degree^2, yielding an acuity of 3–4′ 'if this is related to' . . . cell density. This is relative to the cat with an acuity of 5′ and ganglion cell density of 300/degree^2. The authors continue the discussion in considering that better acuity might be evoked by moving stimuli in that 'sheep are similarly (to red deer) not good at identifying stationary patterns, but readily perceive movement'. Thus, while fair static acuity is possible, kinetic acuity is, at least anecdotally, regarded as much better.

Histological evidence for respectable sheep acuity is evidenced by the presence of both a visual streak and an area centralis (Hughes and Whitteridge, 1973). As the terms suggest, these are areas of increased cell densities, in the first instance horizontally elongated and to some extent matching the long axis of the oval pupil, while in the latter case a much smaller roughly circular area is involved subscribing to the area of greatest visual acuity. Depending upon the size, shape and density of cells such areas as the latter are variously described as being centrale, maculae or foveae, the last mentioned being the smallest and only a degree or so across, as present in humans.

Manatee

In general two assessments of sirenian vision have been made; one of which describes their vision as 'weak' at best, while the other is more optimistic. Thus a number of authors have described the vision of dugong and manatee as 'weak', 'feeble', 'mediocre' or 'poor' (see Piggins *et al.* 1983 for details), and collectively imply that their sense of smell is more important than that of sight. This view is summarized rather strongly by Walls (1942) when he writes, 'all observers agree that the vision of the sirenians is wretched . . .' and further comments upon the adaptation of the sirenian eye to an aquatic environment: 'if we had only the sea-cows to go by we should be forced to conclude that the mammalian eye is too set in its ways to depart far enough from them to give a passable imitation of the eye of a fish'. Prince (1956) echoes these statements with respect to the dugong. Petit and Rochon-Duvigneaud's (1929) evidence and conclusions are supported by Wall's and Prince's views, though they present their own work and in a somewhat more considered manner.

These views of sirenian vision have been based upon a few observations of the animals' visually guided behaviour in the wild and in captivity, the intermediate form of the crystalline lens between a purely terrestrial and aquatic form, and the presence of a cone-free retina with high neural convergence. However, the implied degree of hyperopia for underwater refraction (the dugong eye in air was estimated at 5.00 dioptres of myopia by Dexler and Freund, 1906, which Walls, 1942, suggests would convert to a 'fearsomely hyperopic' eye under water) would appear to be in error, both from an observed and extrapolated point of view. Petit and Rochon-Duvigneaud (1929) suggest on the basis of ocular axial length, crystalline lens dimensions and shape that the dugong eye would be near emmetropic or slightly myopic in air and hyperopic when diving. Vosseler (1924) comments as have others (e.g. Muntz, 1978) later, that vision must fail after a short distance due to muddy water.

In contrast, Hartman (1979) reports more perceptual characteristics from his extensive observations of the behaviour of the Florida manatee (now termed West Indian), that 'visual inspection is the manatee's preferred method of environmental exploration. The animal's typical reaction to an unfamiliar auditory stimulus is to confirm its source visually' in adopting a 'head-on approach'. Hartman suggests that 'the manatee is functionally binocular and may even possess stereoscopic vision. It is obvious, at least, that the visual fields of the two eyes overlap.' Later he also comments: 'visual signals alone suffice to alert manatees', and 'their visual range underwater is as great as a diver's (through a face mask). Under exceptionally clear conditions animals saw and approached me from distances of 35 m. This contradicts the supposition that manatees are myopic. In fact, poor depth perception at close range is evidenced by the animals' tendency to bump their heads and eyes . . .'

The present study shows that the near emmetropic eye of *T. inunguis* (Amazonian manatee) is capable of focusing a clear retinal image for objects situated underwater a metre or so away and further. Hartman's comment regarding myopia in sirenians appears to stem from the *in air* refraction of the dugong as determined by Dexler and Freund (1906). Their findings would convert to hyperopia underwater. The eye of

T. inunguis would certainly appear adapted for underwater vision with its relatively flat cornea and low refractive error. It is also more adapted for sensitivity than acuity, as shown by the rod-retina and high neural convergence. Although *T. inunguis* is arrhythmic by nature (Hartman, 1979; Montgomery *et al.*, 1981 and Reynolds, 1977) the available light for an animal living in muddy or turbid water could be considered the aquatic equivalent of a scotopic environment: in parts of the Amazonian river system, for example, light is attenuated so rapidly that it is unlikely that vision is possible at all at depths greater than about 2 m (Muntz, 1978). The eye is not completely adapted for high sensitivity, however, in that no tapetum was observed for *T. inunguis* (though reported for other species), while the crystalline lens possesses neither a characteristic terrestrial nor an aquatic shape.

Piggins *et al.* (1983) do not report tapeta in the Amazonian manatee, confirming Petit and Rochon-Duvigneaud (1929) for the dugong and Rochon-Duvigneaud (1943). 'Eye-shine' is noted by Vosseler (1924) for the manatee and by Walls (1942) for sirenians in general, though the difference between 'eye-shine' and actual tapetal reflection is not made clear. Dexler and Freund (1906) record tapeta for dugong and manatee, though as with a number of earlier papers, it is sometimes not clear to which species of manatee the authors refer. Hartman (1971) for the Florida (now termed West Indian) manatee, Munie (1872) for manatee Americanis and Pütter (1903) for *T. inunguis* and the dugong all note tapeta.

Colour Vision

Together with high visual acuity and thus a foveate retina, the ability to make wavelength discriminations and hence the possession of at least two types of cone photoreceptor has frequently been regarded as one of the evolutionary peaks in vision. Indeed casual conversation regarding uncommon species about which little may be known often begets the question 'can they see in colour?' It should be stressed that 'seeing colour' is a purely subjective statement in that 'colour' does not exist outside the human brain. Colour is primarily, though not completely, the result of different wavelengths of light being transduced differentially by cone photoreceptors. The title of Wright's book on optics and colour, *The Rays Are Not Coloured*, (1975), puts it succinctly. It is very likely that a large majority of humans can transduce different wavebands of light from the visible wavelength spectrum and perceive colour, though colour naming itself is different across cultures. It would of course be not only brave but scientifically foolhardy to infer that simply because a species, even close to man, possessed a full complement of three types of cone with different spectral absorptions and overlaps between them, that the species would perceive as colourful a world as does man, and that the use to which such information would be put would be in any way similar.

Sheep

Bazeley and Ensor (1989), discussing sheep foraging behaviour, mention that they discriminate between patches of vegetation on either a brightness or a hue criterion, these being the psychophysical correlates of intensity and wavelength respectively.

It is an old problem in vision explaining how such judgements are made, indeed careful colour vision testing in humans relies upon keeping intensity constant and equating it across stimuli, while varying only wavelength. Bazeley and Ensor used a learning discrimination situation to examine brightness and hue discrimination in Border Leicester × Swaledale sheep; stimulus shape (a triangle) was kept constant while in one study intensity was varied, and in the other peak wavelength (550–575 mm – 'green', 575–600 mm – 'yellow'). Brightness differences were discriminated, while hue differences were not, the latter finding confirming Tribe and Gordon (1949). However, these latter authors do not make it clear if the blue and red bulbs used as stimuli in their study were equated for intensity or not. In Bazeley and Ensor's study a possible stimulus problem could have been the close adjacency of peak wavelength values in the reflected light from the painted triangles, while a further consideration could be the spectral distribution of the reflected light. Wide wavelength band reflection could mean that spectrally the green and yellow stimuli were more similar than casual observation, albeit by humans, would suggest. Narrow waveband spectral reflection with little overlap would produce more satisfactory hue-related stimuli. No waveband characteristics were available.

Intriguingly Ensor is mentioned as finding (unpublished work cited in Bazeley and Ensor, 1989), cones in histological preparations from the sheep, whilst Kendrick (1900) mentions 'sheep have some form of colour vision, although no one has yet proved that they actually use colour as opposed to differences in luminance'. At best it can be said that the jury is still out on sheep colour vision but that the verdict with current evidence suggests it is unlikely. Both Bazeley and Ensor (1989) and Kendrick (1990) mention sheep discriminating between dark and light green vegetation and in the latter case a reference to sheep following of a dark green line across light green vegetation, though whether this is by hue or brightness judgement is again unclear. A consideration in a future study could be in equating the spectral reflectance of the stimuli (peak wavelength and waveband) with that of the vegetation regularly encountered by the sheep naturally.

Manatee

Piggins *et al.* (1983) show the Amazonian manatee not to possess the physiological basis for colour vision, retinal histology revealing only rods, and confirmed by photopigment analysis showing a visual pigment difference spectrum of vitamin A_1 based photopigment with maximal absorption at about 505 mm. Such a finding together with an eye leaning towards a design for sensitivity rather than acuity would militate against the species possessing colour vision and suggests only rods being present. Walls (1942), discussing sirenia in general, mentions that they possess at best few cones. Such comments based upon physiology are mentioned in abstracts of conference papers by Woodyard (1985) and Woodyard *et al.* (1987) who carried out a number of learning discrimination studies on the West Indian manatee finding shape and size discrimination present but no convincing evidence for colour vision, only half their subjects showing colour discrimination (no details of stimuli were given in the abstracts).

Cattle

A variant, possibly a perceptual one, in considering cattle colour vision refers to the relationship between 'red' and 'anger'. Much has been made of the relationship between 'colour' and behaviour in humans, though relatively little has been actually established (e.g. see Piggins and Nichols, 1982). Stratton (1923) in examining anger in man extended his investigation to examining the 'popularly held belief that red rouses anger in cattle and especially bulls'. He concludes after a somewhat ethological study that '(1) there was no strong excitement from the colours as a whole or from any one of them, (2) none of the colours used (white, black, red and green) caused anything which could be recognised as anger and (3) brightness and motion, rather than hue were effective'. Riol *et al.* (1989) using discrimination conditioning have examined responses between different grey samples, and those between seven wavelength bands from the visible spectrum and greys matched for brightness (though *not* between different wavelength bands). They conclude that long and medium wavelengths (550–700 nm) are differentiated easily, but shorter wavelengths with difficulty, to some extent confirming previous work. Due to the discrimination between the longer wavelengths the authors suggest that this would act as a mechanism in triggering appropriate responses in fighting cattle. However, no direct inter-wavelength responses were measured, only between wavelength bands and greys adjusted for brightness equality. Gilbert and Arave (1986) conclude from a study comparing responses to green (535 nm), red (610 nm) and blue (450 nm), that the 'results strongly indicate cattle discriminate amongst colours, though less well between blue and green'. Soffie *et al.* (1980) also mention the poor discrimination of blue, with best chromatic discrimination within the region 574–589 nm, while Mander *et al.* (1989) show only a problem for one subject discriminating between red and green, in tests involving red, green, blue and grey. As usual perhaps Walls (1942) has the final word: 'a whole herd of European stud bulls were once provided with red veils, which entirely failed to disturb their equanimity'. The mind boggles!

Visual Perception

Visual perception is a product of processing at levels usually higher than the retina and may be defined as the awareness of the characteristics of a stimulus, some important, some not, depending upon the species and circumstance. They may include the recognition of shape and form, from simple geometric shapes to faces and complex patterns and scenes, binocular vision and stereopsis. Some aspects of these characteristics will be discussed here. Compared to cognition, perception is considered to take place rapidly if not instantaneously from an experiential point of view. The ambiguous or anomalous appearances of Figs 8.8 and 8.9 can visually define for man at least the difference between perception and cognition. In Fig. 8.8, the Necker cube (1832), there occurs an abrupt change in apparent viewing position and in depth when the cube undergoes perceptual reversal, while in Fig. 8.9, Penrose's impossible tribar (1958), time is required to understand why the tribar is anomalous and impossible in three dimensions, though possible in two. Gregory's description of cognition as 'intelligent perception' is particularly apt in this latter

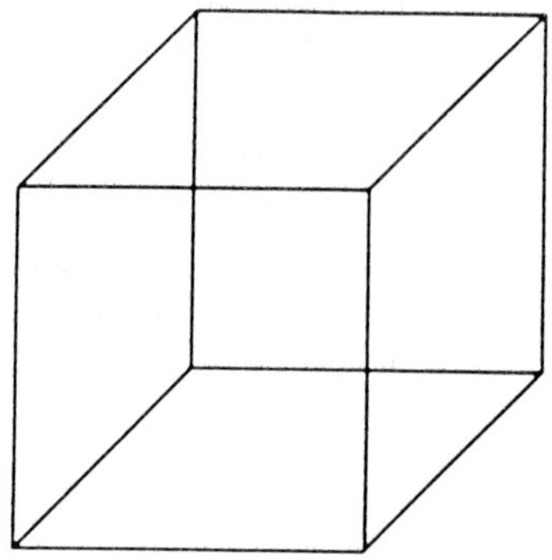

Fig. 8.8. Necker cube (1832).

case. It is not suggested that the farm animals under discussion here would perceive the anomalies in either the cube or the tribar, but for humans they pictorially define the terms.

Some phenomena rely upon memory and much higher processing, others, though considered perceptual in nature, may still have a strong and evident sensory component present. The two figures mentioned above represent examples of the first category. The second category may be represented by an example taken from the visually guided behaviour of laying hens where a lower process can be seen as influencing a much higher one. Hurnik *et al.* (1973) show that feeder 'complexity' has a differential effect upon the hens' food consumption, the greater the 'complexity' the greater the food intake. Complexity was loosely defined as the number of stimulus elements presented with the actual feeder, thus simple achromatic association produced less consumption than a coloured one. The stimulus which elicited most consumption and was the most 'complex' was randomly dotted using four colours. Since the multicoloured stimulus was randomly dotted with four colours, it is unlikely that some global or Gestalt phenomenon was operating as an attractor to the hens, and since hens possess all cone retinas and have trichomatic vision, the direct influence of a lower (sensory and chromatic) process upon a higher one

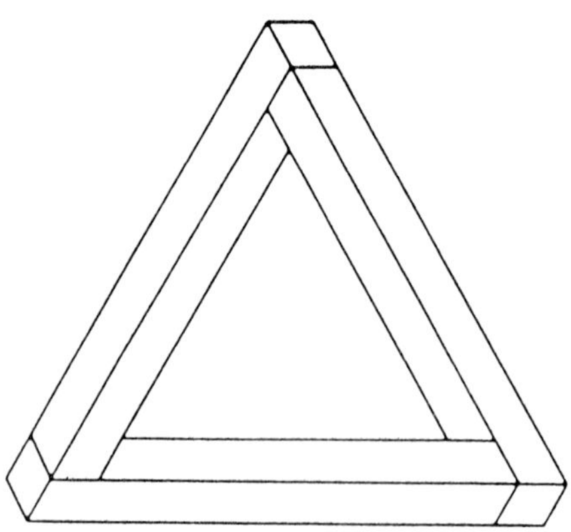

Fig. 8.9. Penrose tribar (1958).

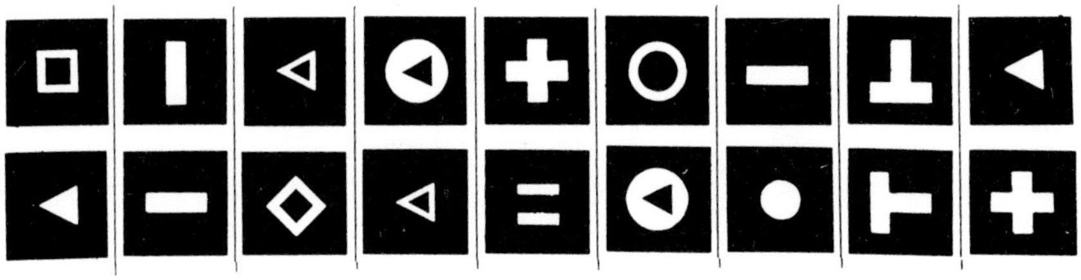

Fig. 8.10. Typical stimuli presented to sheep (from Baldwin, 1981).

(perceptual and complex) is notable. A reasonable question here would be where is the limit of both complexity and consumption?

The perceptually driven behaviour of most animals has received even less attention than that of sensory characteristics, as will be obvious from the discussion of the few better known perceptual characteristics of sheep and manatee given below.

Sheep – shape

Along with goats (Baldwin, 1979), sheep and calves (Baldwin, 1981) show simple shape discrimination under experimental conditions of operant conditioning. Similar stimuli were employed in both studies: simple geometrical shapes presented in pairs, varying in orientation for identical pair presentation, or as outlined, filled-in shapes, or combinations of simple shapes such as a triangle surrounded by a circle, presented with an outlined circled as the other half of the pair. Typical stimuli are shown in Fig. 8.10. Such stimuli are very commonly used in the shape perception studies with a variety of species.

Manatee – shape/motion

Little study has been made of the Manatee's perceptual (as opposed to its sensory) capacities. Woodyard (1985) and Woodyard *et al.* (1987) mention the West Indian manatee as making underwater visual discriminations of brightness, size, shape (simple geometrical figures), horizontal–vertical movement, colour and horizontal–vertical stripe orientation. Since the reports are in conference abstracts, no details were available. Piggins *et al.* (1983) note the Amazonian manatee regularly tracked a moving moderately contrasting bar of light focused on the side of the holding tank. Tracking took place only for underwater presentation and not when the stimulus was above water level. Hartman (1979) makes a similar statement for the West Indian manatee, while Anderson (1979) mentions that dugongs will visually examine stimuli presented above the water level. Moreover D. Woodyard (personal communication) mentions that the West Indian manatee could make size discriminations when the eyes were underwater and the stimuli were in the air.

Depth Perception

Sheep – depth

Many ungulates possess good depth perception while sheep as mentioned previously have been described as possessing excellent vision, including depth judgement; for example Walk and Gibson (1961) have investigated sheep using the 'visual cliff'. The most likely depth cue used upon the visual cliff apparatus has been considered to be motion parallax and can be available to monocular vision alone, upon head or eye movement. A number of species have been shown to employ such a monocular cue to make depth judgements. However, sheep with their relatively large binocular visual field, estimated by Walls (1942) at between 40° and 60°, combined with an area centralis (Hughes and Whitteridge, 1973), would appear to be respectable candidates for stereopsis, although their static visual acuity reported by Backhaus (1959) is 11′. (Human acuity, clinically determined is regarded as normal at 1′, though this depends upon the type of visual task, laboratory studies showing acuity of only a few seconds of arc.) However, a consideration of retinal ganglion cell density (A. Hughes, personal communication, in Clarke and Whitteridge, 1976) converts to an expected acuity of 3–4′, i.e. between that of the cat and rhesus monkey, though closer to the former. Clarke *et al.* (1976) conclude from cortical unit recording studies that disparity is processed in the sheep cortex, though the sheep's stereo vision may be much coarser than man's. However, though coarser (i.e. binocular depth judgements are less accurate in the animal's immediate proximal space), the sheep's stereoscopic space is likely to extend a bit further from the sheep than it is in man, in view of the sheep's inter-ocular distance being approximately twice that of man. Such an increase in ocular baseline will mean that disparity information about objects relatively further away than for man could in principle be stereoscopically processed. Casual observation of sheep suggests that initially their visual response for much information appearing in their visual field is to adopt a 'head on', that is at least a binocular if not stereoscopic, attitude to the novel stimulus, even when it is at some distance away. Figure 8.11 shows a photographic demonstration of binocular overlap in the sheep.

A final note regarding the perception of movement by sheep has been made by Clarke and Whitteridge (1976) in connection with visual acuity. The behaviourally measured acuity by Backhaus (1959) for static stimuli is some three times coarser than would be expected on the basis of retinal ganglion cell density. Clarke and Whitteridge mention that 'possibly much lower figures would be given by moving targets' and state further that field experience of red deer suggests that they are much better at detecting moving objects and that sheep may 'readily perceive movement' as well.

Manatee – depth

The overall visual field (i.e. monocular and binocular combined) of the manatee's near relative the dugong has an estimated value of 265° (Petit and Rochon-Duvigneaud, 1929), close to that of sheep at 270° (Walls, 1942). However, the overlap of the monocular fields is very different, being estimated at 15° for the West Indian

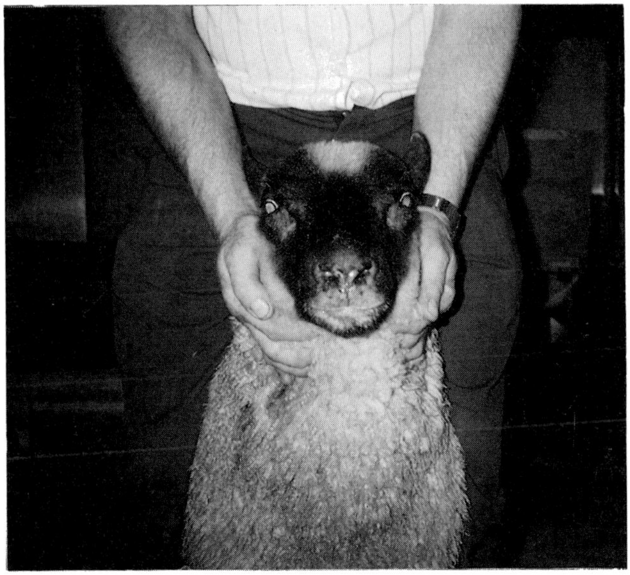

Fig. 8.11. Binocular overlap for sheep visual field, shown by 'eye-shine'.

manatee (Hartman, 1979) and for the Amazonian manatee (Piggins *et al.*, 1983), while the estimated overlap for the sheep is considerably more being estimated at between 40° and 60°. It is thus unlikely that the manatee would possess stereopsis, or indeed very little binocular vision at all. However, in an extensive ethological study, Hartman (1979) mentions that the West Indian manatee prefers to adopt a 'head on' approach when confirming an auditory signal, visually. Hartman further suggests binocularity and even stereopsis for the species. However, poor depth perception is evidenced at short range since the animals tend to bump their heads and eyes. Piggins *et al.* (1983) suggest that the Amazonian species may possess some binocularity for objects 1 metre away and further, though such a visual capability would not be very important in a herbivore relying upon non-moving food sources. Depth perception within a metre or so would be made on the basis of monocular cues alone; support for such a view is alluded to by Hartman's comment that the West Indian manatee repeatedly shifts position so that first one eye is exposed then the other. However, the lack of ocular axes convergence and accommodation, and no visual field overlap nearer than 1 m or so, in addition to the high neural convergence and reports of poor depth perception at near distances (contrasting as mentioned with Hartman's own, earlier view of the possible presence of stereopsis) would indicate that stereopsis was unlikely and that binocular vision was not a primary concern for the animal. As mentioned earlier, casual observation of the visually guided behaviour of *T. inunguis* indicated that the animal did considerably more than 'avoid a bright light' stimulus as is suggested by some workers, since the animal regularly tracked a moderately contrasting bar of light focused upon the side of the tank under water.

General Discussion

Baldwin (1979) writes: 'the behaviour of an animal is greatly influenced by the information it receives from its environment and thus the comparative study of a major sense such as vision is an important part of behavioural research'. Obviously it would be ideal to report the findings on any single perceptual, or sensory or cognitive characteristic from studies on the same species, perhaps even the same animal, using all three major approaches – physiological, psychophysical and ethological. For a number of reasons, not the least being those of an ethical nature, this has only been accomplished on a very few occasions. Most of the work reported in this, and other chapters in this section, has involved circumstances and stimuli which the animal would seldom chance upon in its natural world, even in the farmyard – unless it were run by a psychologist of course! The presentation of squares, triangles and so forth, under controlled laboratory conditions is an important aspect, albeit reductionist in examining an animal's shape perception. However, such a reductionist approach under some circumstances is necessary, but tells us little of what might be important to the animal in its natural setting – that is to say the ecological validity of such studies to some extent goes begging. Laboratory studies tell us little about what an animal would naturally pay attention to, habituate to or regard as novel in its normal environment. Kendrick's cortical unit recording studies do appear to have made a beginning, providing a vital and necessary link between the ethology and physiology of sheep visual perception and cognition.

It would of course be desirable to proceed further. Dunbar (1991) in a popular review of Cheney and Seyforth's book *How Monkeys See the World: Inside the mind of another species* (1991) begins 'We humans are aware of our own states of mind . . . . But are we unique in this or do other animals have similar experiences?' The authors of the book '. . . took the bold step of transferring the techniques of the laboratory psychologist into the field' when specifically examining the social role which vocalization plays in Kenyan Vervet monkeys. Dunbar writes '. . . that monkeys possess very much more sophisticated cognitive abilities than was previously assumed inevitably raises questions about the extent to which they are able to see the world from another individual's point of view'. Although at a guess awareness and self-awareness are likely to be less in ungulates than in monkeys, we have as always underestimated the capabilities and characteristics of lower animals, thus perhaps an attempt is worth while and at least laboratory psychology and field ethology will benefit mutually. To some extent, the fields have always overlapped: another recent approach has been in the consideration of animal signals, voluntary or otherwise. Guilford and Dawkins (1991) suggest that signal design has two components: 'strategic design' and 'efficacy'. The former considering 'how a signal is constructed by natural selection to provide the information necessary to make a receiver respond' and the latter on the design of signals such that the information gets to the receiver. Thus the 'psychology' of the signal receiver is vital.

The above two examples, the Cheney and Seyforth book and the paper by Guilford and Dawkins are reflected in considerations somewhat closer to the farm by Duncan and Petherick (1989) and Wiepkema (1989) in abstracts for papers presented at the Society for Veterinary Ethology UK meeting. Duncan and Petherick discuss cognition and animal welfare and mention 'cognitive processes of feeling, perceiving, being aware, remembering, learning, anticipating and recognising . . . the implications for welfare of the possession of these abilities'. Further, 'the extent to which animals are aware of their internal state while performing behaviour known to be indicative of the so-called "states of suffering" will determine how much they are actually suffering'. They then proceed to consider if animals think about the absence of real objects or events and the possible frustration that might ensue – could this be avoided by replacing such objects and events cognitively? – a sort of reverse scarecrow! Wiepkema takes a somewhat similar tack in discussing chronic stress, taken as occurring where there is a sudden decrease in the predictability of relevant events. The author writes 'Predictability implies expectancies demonstrated in anticipatory actions and/or when an environment change differs from its precedent. Under conditions of a changing predictability of relevant environmental events, emotional expressions occur'. Both Hurnik and Kendrick have touched upon such thoughts in different ways at this conference by giving much consideration to a 'neck up' approach – mercifully with some empathy.

Acknowledgements

The author would like to thank Dr Clive Phillips for help with some of the more agriculturally based references, Mr Alton Murphy of Alton Murphy Optometrists, North Wales, for the use of equipment in determining sheep refraction, and John Thomas and his staff at the farm of the University College of North Wales, Bangor, for supplying sheep and help in the collection of ocular data, and to Cindy Norton for help in the latter task.

Appendix

Retinoscopy

Retinoscopy is a standard clinical technique used in optometric and ophthalmic practice for determining objectively (i.e without the patient's report) refractive error and has been used as such for over a century.

When a moving beam of light is observed reflected from the rear inner surface of a subject's eye (i.e. the 'light reflex'), the direction and speed of movement of the light reflex relative to that of the light source depend upon the subject's refractive error present. Movement of the light source and light reflex in the same direction ('with' movement) indicates hyperopia ('long sight'), while movement of the light reflex in the opposite direction ('against' movement) indicates myopia ('short sight'). The relative speed of movement of light source and light reflex depends upon the *amount* of refractive error present and can be changed by placing appropriate lenses before the subject's eye. Under such conditions a lens power will eventually be reached where the apparent rate of movement of the light reflex becomes infinitely fast; additional lens power results in a subsequent reversal of the direction of movement of the light reflex. The lens power at this reversal point, minus the value of the testing distance (expressed in lens power terms, i.e. dioptres) will give a reliable value of the subject's refractive error.

References to retinoscopy abound in the optometric and ophthalmological literature. Emsley (1952) and Bennet and Francis (1962) present easily accessible detailed descriptions and theory of the technique. The above brief description of the technique has been liberally taken from the latter reference, and is also included in Piggins *et al.* (1983).

References

Anderson, P.K. (1979) Dugong behaviour: on being a marine mammalian grazer. *The Biologist* **61**, 113–114.

Backhaus, D. (1959) Visual acuity in ungulates. *Zeitschrift für Tierpsychologie* **16**, 454.

Baldwin, B.A. (1979) Operant studies on shape discrimination in goats. *Physiology and Behaviour* **23**, 455–459.

Baldwin, B.A. (1981) Shape discrimination in sheep and calves. *Animal Behaviour* **29**, 830–834.

Bazeley, D.R. and Ensor, C.V. (1989) Discrimination learning in sheep with cues varying in brightness and hue. *Applied Animal Behaviour Science* **23**, 293–299.

Bennet, A.G. and Francis, J.L. (1962) Retinoscopy and ophthalmoscopy. In: Davson, H (ed.), *The Eye*, vol. 4. Academic Press, London.

Brewster, D. (1860) *Memoirs of the Life, Writings, and Discoveries of Sir Isaac Newton*, Edinburgh. Edmonton & Douglas (cf.I. Newton 1680).

Carroll, L. (1871) *Through the Looking-glass, and what Alice Found There.* Macmillan, London.

Clarke, P.G.H. and Whitteridge, D. (1976) The cortical visual areas of the sheep. *Journal of Physiology (London)* **256**, 497–508.

Clarke, P.G.H., Donaldson, I.M.L. and Whitteridge, D. (1976) Binocular visual mechanisms in cortical areas I and II of the sheep. *Journal of Physiology (London)* **256**, 509–526.

Coile, D.C. and O'Keefe, L.P. (1988) Schematic eyes for domestic animals. *Ophthalmic and Physiological Optics* **8**, 215–220.

Dexler, H and Freund, L. (1906) Contributions to the physiology and biology of the dugong. *American Naturalist* **40**, 49–70.

Dunbar, R. (1991) Review of *How Monkeys See the World: Inside the mind of another species* by Cheney, D.L. and Seyforth, R.M. University of Chicago Press. *New Scientist*, 16 November, 48.

Duncan, L.J.H. and Petherick, J.C. (1989) Cognition: the implications for animal welfare. Abstract, Society for Veterinary Ethology Meeting, London, 1988. *Applied Animal Behaviour Science* **24**, 81.

Emsley, H.H. (1952) *Visual Optics*, vol. 1. Butterworths, London.

Evans, C.R. and Piggins, D.J. (1963) A comparison of the behaviour of geometrical shapes when viewed under conditions of steady fixation, and with apparatus for producing a stabilised retinal image. *British Journal of Physiological Optics* **20**, 261–274.

Geist, V. (1971) *Mountain Sheep: A study in behaviour and evolution.* University of Chicago Press, Chicago.

Gilbert, B.J. and Arave, C.W. (1986) Ability of cattle to distinguish among different wavelengths of light. *Journal of Dairy Science* **69**, 825–832.

Glickstein, M and Millodot, M. (1970) Retinoscopy and eye-size. *Science* **168**, 605–606.

Guilford, T. and Dawkins, M.S. (1991) Receiver psychology and the evolution of animal signals. *Animal Behaviour* **42**, 1–14.

Hartman, D.S. (1979) Ecology and behaviour of the manatee (*Trichechus manatus*). *American Society of Mammalogists Special Publication No. 5*, 153.

Hirschberg, J. (1882) Zur dioprik und ophthalmoskopie der Fisch und Amphibien augen. *Arch. Anat. (Physiol. Abth.)*, 493–526. Cited in O-J. Grüsser and U. Grüsser-Cornehls, Neurophysiology of the anuran visual system. Published in Llinas, R. and Precht, W. (eds.), *Frog Neurobiology: A Handbook.* Springer-Verlag, Berlin, 1976.

Hughes, A. (1977) The topography of vision in mammals of contrasting life style: comparative optics and retinal organisation. In: Crescitelli, F. (ed.), *Handbook of Sensory Physiology*, Vol. VII, 5. *The Visual System in vertebrates*. Springer-Verlag, Berlin.

Hughes A. and Whitteridge, D. (1973) The receptive fields and topographical organisation of goat retinal ganglion cells. *Vision Research* **13**, 1101–1114.

Hurnik, J.F., Piggins, D.J., Reinhart, B.S. and Summers, J.D. (1973) The effect of visual pattern complexity of feeders on food consumption of laying hens. *British Journal of Poultry Science* **15**, 97–105.

Kendrick, K.M. (1990) Through a sheep's eye. *New Scientist* 12 May, 62–65.

Kendrick, K.M. and Baldwin, B.A. (1987) Cells in temporal cortex of conscious sheep can respond preferentially to the sight of faces. *Science* **236**, 448–450.

Manda, M., Oku, Y., Ashidachi, A., Kubo, M., Kurohiji, I. (1989) Behavioural evidence for colour discrimination in cattle. *Japanese Journal of Zootechnical Science* **60**, 521–528 (Abstract).

Martin, G.R. (1988) Eye development and short sight. *Nature* **333**, 707.

Montgomery, G.G., Best, R.C. and Yamakoshi, M. (1981) A preliminary radiotelemetry study of the Amazonian manatee *trichechus inunguis* (Mammalia: Sirenia). *Biotropica* **13**(2), 81–85.

Muntz, W.R.A. (1978) A penetracáo de luz nas águas de rios amazônicos. *Acta Amazonica* **8**, 613–619.

Murie, J. (1872) On the form and structure of the manatee. *Zoological Society of London, Transactions* **8**, 127–202 (bound 1874).

Necker, L.A. (1832) Observations on some remarkable optical phenomena seen in Switzerland; and on an optical phenomenon which occurs on viewing a figure of a crystal or geometrical solid. *Philosophical Magazine* **1**, 329–337.

Nicolas, E. (1930) *Veterinary and Comparative Ophthalmology*. H. & W. Brown, London.

Penrose, L.S and Penrose, R. (1958) Impossible objects: a special type of illusion. *British Journal of Psychology* **49**. 31.

Petit, G. and Rochon-Duvigneaud, A. (1929) L'oeil et la vision de *L'Halicore dugong* ERXL. *Bulletin, Soc. Zool. de France* **54**, 129–138.

Piggins, D.J. (1970) Refraction of the harp seal *Pagophilus groenlandicus* (Erxleben 1777). *Nature* **227**, 78–79.

Piggins, D.J. (1978) Lilliputian eyes and vision. *Perception* **7**, 609–610.

Piggins, D.J. and Nichols, R.D. (1982) The pink room effect: tranquillization or methodological transgression? *Atti della Fondazione Giorgio Ronchi* **37**, 281–286.

Piggins, D.J. Muntz, W.R.A. and Best, R.C. (1983) The eye of the Amazonian manatee, *Trichechus inunguis:* ocular refraction and gross ocular anatomy, retinal histology and photopigment. *Marine and Behaviour Physiology* **9**, 111–130.

Prince, J.H. (1956) *Comparative Anatomy of the Eye*. Thomas, Springfield.

Pütter, A. (1903) Die Augen der Wassersäugetiere. *Zoologische Jahrbuch, Abteilung Anatomie* **17**, 99–402.

Reynolds, J.E. (1977) Aspects of the social behaviour and ecology of a semi-isolated colony of Florida manatees, *Trichechus manatus*. MSc Thesis, University of Miama, Coral Gables, 276 pp.

Riol, J.A., Sanchez, J.M., Eguren, V.G. and Gaudios, V.R. (1989) Colour perception in fighting cattle. *Applied Animal Behaviour Science* **23**, 199–206.

Rochon-Duvigneaud, A. (1943) Les yeux et la vision des vertebrés. Masson, Paris.

Sivak, J.G. and Allen, D.B. (1975) An evaluation of the 'ramp' retina of the horse eye *Vision Research* **15**, 1353–1356.

Soffie, M., Thines, G. and Falter, U. (1980) Colour discrimination in heifers. *Mammalia* **44**, 99–121 (Abstract).

Sorsby, A. *et al.* (1957) Medical Research Council (United Kingdom). *Special Research Report Series 293.*

Stratton, G.M. (1923) The colour red: and the anger of cattle. *Psychological Review* **30**, 321–325.
Swift, J. (1971) *Gulliver's Travels*. P. Turner, London. Oxford University Press (originally published 1726).
Tribe, D.E. and Gordon, J.G. (1949) The importance of colour vision to the grazing sheep. *Journal of Agricultural Science* **39**, 313–314.
Vosseler, J. (1924) Pflege und haltung der Seekühe (*Trichechus*) nebst beiträgen zu ihrer biologie. *Pallasia* **2**, 58–67, 113–133, 167–180, 213–230.
Walk, R.D. and Gibson, E.J. (1961) A comparative and analytic study of visual depth perception. *Psychological Monographs* **75**, 1–44.
Walls, G.L. (1942) *The Vertebrate Eye and its Adaptive Radiation*. Hafner, New York.
Wiepkema, P.R. (1989) Control and Cognition: essential elements of coping (Abstract). Society for Veterinary Ethology, London, 1988. *Applied Animal Behaviour Science* **24**, 81.
Woodyard, D. (1985) Research Programme in Information Processing and Social Behaviour in Captive West Indian Manatee. Fourth International Theriological Congress, Edmonton, Canada (Abstract).
Woodyard, D., Bowen, M., Suarez, R. and White, I. (1987) Review of Research on Problem-Solving Behaviour in Captive West Indian Manatees. 15th Annual International Marine Animal Trainers Association Conference, New Orleans, USA (Abstract).
Wright, W.D. (1975) *The Rays Are Not Coloured*. Hilger, London.

Further Reading

Bruce, V. and Green, P. (1985) *Visual Perception, Physiology, Psychology and Ecology*. Erlbaum, London.
Gregory, R.L. (1987) *Oxford Companion to the Mind*. Oxford University Press, Oxford.
Lythgoe, J.N. (1979) *The Ecology of Vision*. Clarendon, Oxford.
Wade, N.J. and Swanston, M. (1991) *An Introduction to Visual Perception*. Routledge, London.

9 Auditory Perception

H.E. HEFFNER and R.S. HEFFNER

Laboratory of Comparative Hearing, Department of Psychology, University of Toledo, Toledo, Ohio, USA

Summary

There have been three main sources of selective pressure on the ability of animals to perceive sound (Masterton and Diamond, 1973). The first has been the need to *detect* a sound, an ability that enables an animal to determine the presence of sound-producing objects in its environment, which in most cases are other animals. The second has been the need to *localize* the source of a sound so that an animal can either approach or avoid the sound source. Finally, an animal must be able to *identify* the biological meaning or relevance of the sound so that it may respond appropriately to the sound source.

Over the last decade our knowledge of the auditory abilities of animals in general and farm animals in particular has increased markedly. In the case of mammals, farm animals have been the object of study because it was necessary to determine the hearing abilities of large mammals in order to answer certain questions concerning the evolution of human hearing (e.g. R. Heffner and Heffner, 1983b). Because farm animals are relatively easy to handle, they made convenient subjects for experiments designed to answer questions which could not be addressed with standard laboratory animals.

The purpose of this chapter is to review what is known about the ability of farm animals to detect and localize sound in light of our knowledge concerning hearing in birds and mammals. In doing so, we have taken a broad definition of farm animals and, where information is available, have included pets, predators, and pests often encountered on farms. Also included in this chapter is a brief description of how to test the hearing ability of farm animals. Because little is currently known concerning the ability of animals to identify sounds, this topic is not covered.

Detection of Sound

Measuring absolute sensitivity

The basic test of hearing consists of determining the ability of an animal to hear pure tones at intervals throughout its hearing range. This is done by training the animal

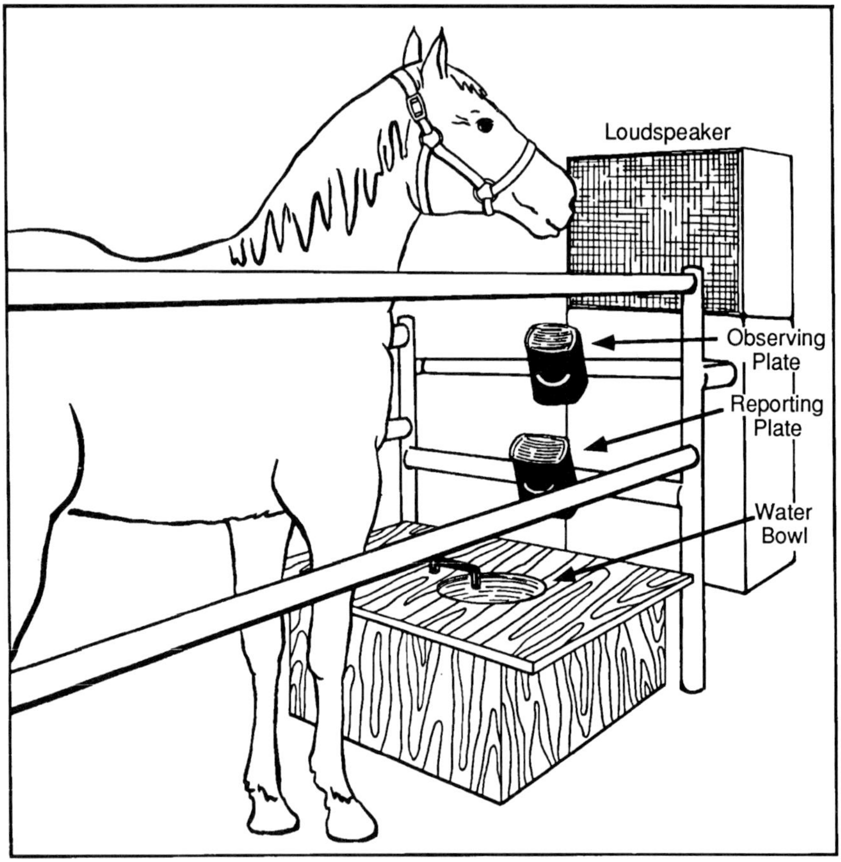

Fig. 9.1. Apparatus used to test hearing in horses and cattle. The animal was trained to place its nose on the observing plate and listen for a tone. Tones were presented at random intervals and the animal was given a water reward for touching the reporting plate within three seconds after onset of the tone (modified from H. Heffner and Heffner, 1983).

to respond to a tone and then reducing the intensity of the tone until the animal fails to respond. There are a number of behavioural procedures which can be used to train an animal to respond to sound. One procedure which has been used with farm animals is a 'go/no go' procedure in which an animal is trained to make a specific response whenever it hears a sound (e.g. R. Heffner and Heffner, 1983b).

In the go/no go procedure, a thirsty animal is led into a test stall and trained to place its nose on an 'observing' plate located in front of a loudspeaker (Fig. 9.1). Tones are then presented at random intervals and the animal is trained to break contact with the observing plate and touch a 'reporting' plate whenever it hears a tone. A response made in the presence of a tone is rewarded by delivering water into a bowl located below the reporting plate. However, if the animal responds when no tone is present, it is given a short wait or 'time out' of 5–15 s during which testing is halted. Testing is continued until the animal has received sufficient water which generally takes about 1 h each day.

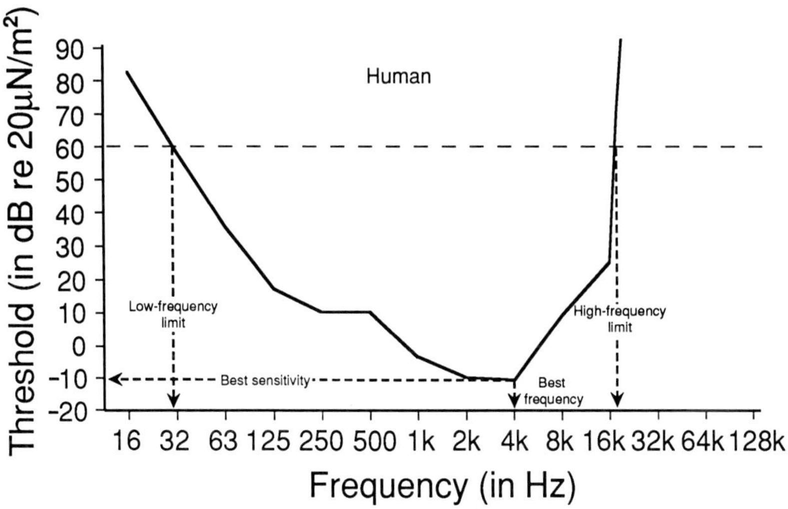

Fig. 9.2. Average audiogram for seven human subjects. To make the results comparable to those of animal studies, tones were presented from a loudspeaker placed in front of the subjects instead of using headphones (Heffner and Heffner, unpublished).

Once an animal has learnt to report reliably the presence of a tone, threshold testing is begun. This consists of gradually reducing the intensity of the tone until the animal no longer responds, implying that it can no longer hear the tone. The animal's threshold for that tone is then defined as the intensity at which it can detect the tone half of the time (i.e. 50% detection). Thresholds are then determined for tones of various frequencies taken in octave steps across the animal's hearing range. These thresholds are then used to construct an audiogram. It should be noted that obtaining and verifying a complete audiogram can be a lengthy procedure which takes eight to ten weeks of daily testing.

The audiogram

The pure-tone thresholds of an animal are used to construct an audiogram in which thresholds are plotted for tones of different frequency, an example of which is shown in Fig. 9.2. In this figure, each point represents the lowest intensity at which a particular frequency could be detected 50% of the time. Thus, large decibel (dB) values indicate that a tone had to be fairly intense before it could be detected while small values indicate that a less intense tone could be heard. Because sound, like temperature, is typically measured with a scale which has no absolute zero, negative values indicate only that the intensity of a sound is less than the 0 dB reference level.

The human audiogram shown in Fig. 9.2 demonstrates the characteristic shape of most normal audiograms. Beginning at the low frequencies, the audiogram shows that auditory sensitivity improves gradually as frequency is increased. Often, there is a range of best sensitivity in the middle of the audiogram followed by a gradual decrease in sensitivity which becomes a rapid decrease as the upper limit of hearing is approached.

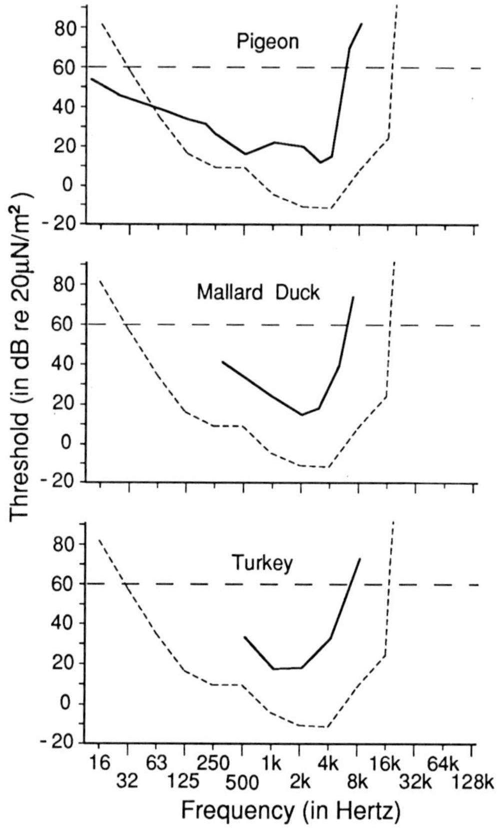

Fig. 9.3. Audiograms for the pigeon, mallard duck and turkey (solid lines). Human audiogram from Fig. 9.2 (dashed lines) is shown for comparison.

For comparative purposes, the audiogram can be described with four parameters (Masterton *et al.*, 1969). These are: (i) 'high-frequency limit' – arbitrarily defined as the highest frequency which an animal can hear at a sound pressure level (SPL) of 60 dB; (ii) 'low-frequency limit' – arbitrarily defined as the lowest frequency which can be heard at 60 dB; (iii) 'best frequency' – the frequency to which the animal is most sensitive; and (iv) 'best intensity' – the lowest intensity to which an animal can respond. As shown in Fig. 9.2, the corresponding values for humans are 17.6 kHz for our high-frequency limit and 31 Hz for our low-frequency limit with a best frequency of hearing around 4 kHz at which point our best sensitivity is −10 dB.

Hearing in birds

The audiograms of the mallard duck (*Anas platyrhynchos*), pigeon (*Columba livia*) and turkey (*Melegris gallopavo*) are shown in Fig. 9.3 (Trainer, 1947; Maiorana and Schleidt, 1972; Hienz *et al.*, 1977; Kreithen and Quine, 1979). Compared with humans (dashed line in each figure), it can be seen that the high-frequency hearing of birds is notably poor and, indeed, no bird is able to hear much above 10 kHz (cf. Fay, 1988). Although their best frequency of hearing is in the same range as that of humans, these three species are not as sensitive as humans. Less is known about the low-frequency limit of birds as few species have been tested at low frequencies. However, there is evidence that pigeons have excellent low-frequency hearing which they may use for navigation (Kreithen and Quine, 1979).

The audiograms of four additional species of birds which are commonly found around farms are provided for comparison – starling (*Sturnus vulgaris*), red-winged blackbird (*Agelaius Phoeniceus*), eastern sparrow hawk (*Falco sparverius*), and barn owl (*Tyto alba*) (Fig. 9.4) (Trainer, 1947; Konishi, 1973; Hienz *et al.*, 1977; Dooling *et al.*, 1986). The hearing abilities of these species are similar to those of the previous birds with one exception: the barn owl possesses greater sensitivity than the other birds, an adaptation which may be explained by the fact that it is a nocturnal predator which uses its hearing to locate prey.

In summary, the high-frequency sensitivity of birds is notable in that they do not hear significantly above 10 kHz. In most birds, both low-frequency sensitivity and best sensitivity are unremarkable with the exceptions of the pigeon, which appears to have exceptionally good low-frequency hearing, and the barn owl, which has excellent sensitivity at its best frequency of hearing.

Hearing in mammals

The hearing sensitivity of mammals differs from that of birds in several ways. First, most mammals possess the ability to hear frequencies above 10 kHz with the high-frequency limit of some mammals, such as bats and porpoises, extending beyond 100 kHz (cf. Fay, 1988). Second, whereas the auditory sensitivity of birds is relatively uniform, the hearing ability of mammals, particularly their ability to hear high and low frequencies, varies widely from one species to the next.

Hoofed mammals

The audiograms for horses (*Equus caballus*), sheep (*ovis aries*), goats (*Capra hircus*), pigs (*Sus scrofa*) and cattle (*Bos taurus*), are shown in Fig. 9.5 (Wollack, 1963; R. Heffner and Heffner, 1983b, 1990a). As illustrated in this figure, the high-frequency hearing of hoofed mammals not only surpasses that of birds, but it also exceeds the high-frequency hearing of humans (dashed lines in each figure). The upper frequency limit of these animals ranges from 33.5 kHz in horses to 42 kHz in sheep. These animals also have relatively good low-frequency hearing with low-frequency limits ranging from 23 Hz for cattle (surpassing human low-frequency hearing) to 125 Hz for sheep. The best frequency of hearing varies from 2 kHz for horses and goats to 10 kHz for sheep. Best sensitivity varies from −11 dB for cattle and goats, making them as sensitive as humans, to 9 dB for pigs. Overall, the hearing sensitivity of hoofed mammals is similar to that of humans with the most significant difference being the ability of these animals to hear frequencies above the human upper limit of hearing, i.e. ultrasonic sounds.

The audiograms shown in Fig. 9.5 were obtained with animals whose ears were free of any signs of damage or disease – pathological conditions, such as the presence

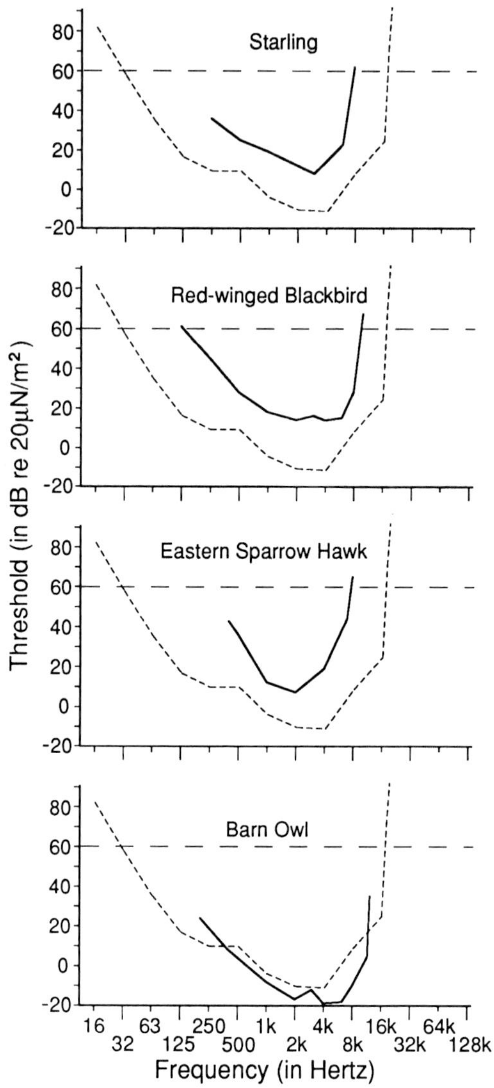

Fig. 9.4. Audiograms for the starling, red-winged blackbird, eastern sparrow hawk and barn owl (solid lines). Human audiogram is shown by dashed lines. Note the good sensitivity of the owl.

of ear mites, can substantially affect an animal's hearing. In particular, ear mites, which are often found in cattle and goats, have been shown to have a dramatic effect on an animal's ability to hear sound.

In the case of cattle, ear mites, *Raillietia auris*, are associated with redness, ulceration, and bleeding of the tissue lining the auditory canal as well as blockage of the canal by a thick plug of pus which may extend for 3 cm or more (R. Heffner and Heffner, 1983c). The pus, acting as an ear plug, attenuates middle frequencies

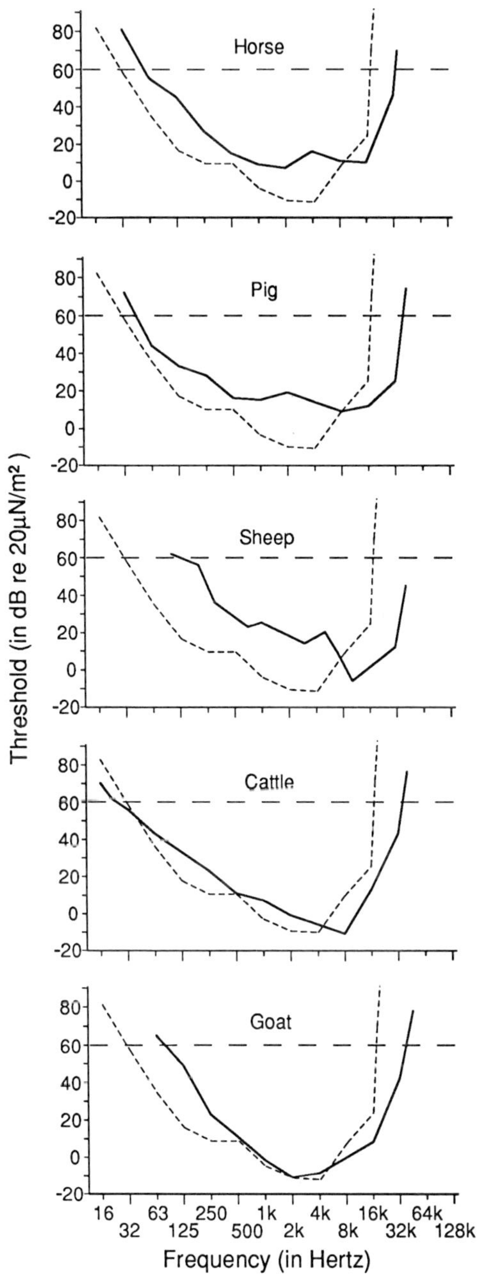

Fig. 9.5. Audiograms for the horse, pig, sheep, cattle and goat (solid lines). Human audiogram is shown by dashed lines.

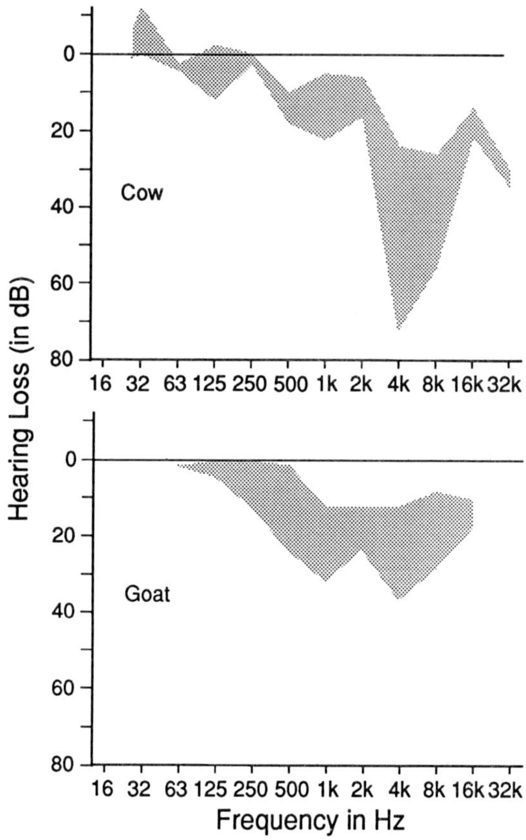

Fig. 9.6. Effect of ear mite infestation on the hearing of a cow (top) and a goat (bottom). Shaded areas indicate range of the hearing losses which varied markedly from day to day (R. Heffner and Heffner, 1983a and unpublished).

more effectively than low and high frequencies (R. Heffner and Heffner, 1983a). In cattle, this results in a hearing loss which is most severe around 4 and 8 kHz while frequencies below 500 Hz are unaffected (Fig. 9.6). This should not be considered an unusual condition as a survey of eight separate herds in southeastern Kansas found that 66% of the animals examined were infested with ear mites (R. Heffner and Heffner, 1983c). Ear mites are also found in goats and, although the effects are not as severe as in cattle, their presence does result in a definite hearing loss (Fig. 9.6; R. Heffner and Heffner, unpublished).

Carnivores

The audiograms of the domestic cat (*Felis domesticus*), dog (*Canis familiaris*) and ferret (*Mustela putorius*) show that these animals hear higher frequencies than hoofed mammals with their upper limits of hearing ranging from 42 kHz for the ferret to 79 kHz for the cat (Fig. 9.7) (Heffner, 1983; Kelly *et al.*, 1986; R. Heffner and Heffner,

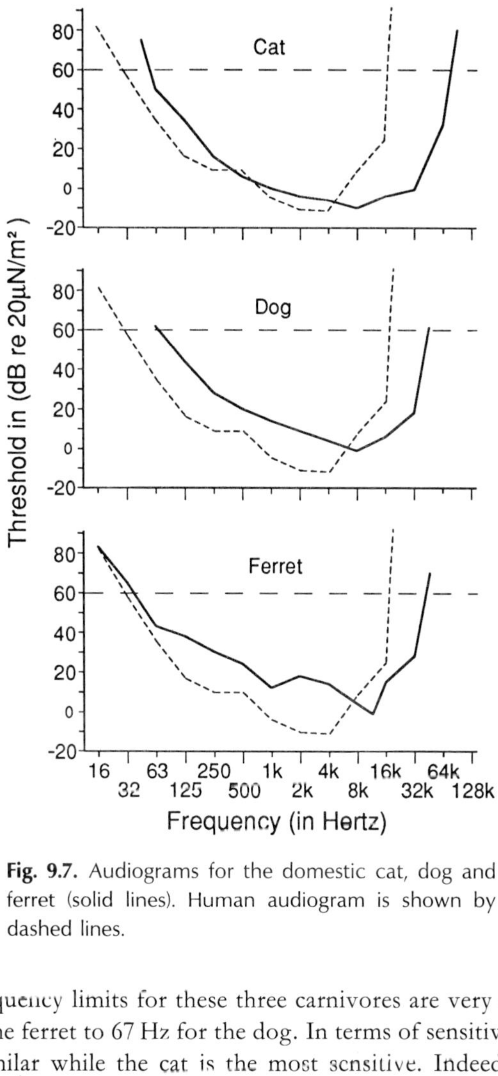

Fig. 9.7. Audiograms for the domestic cat, dog and ferret (solid lines). Human audiogram is shown by dashed lines.

1985b). Low frequency limits for these three carnivores are very similar and range from 36 Hz for the ferret to 67 Hz for the dog. In terms of sensitivity, the ferret and dog are very similar while the cat is the most sensitive. Indeed, cats share with humans, cattle, and goats the distinction of being the most sensitive of all mammals tested so far.

Glires

Glires is a cohort containing the orders Lagomorpha and Rodentia. Audiograms for the domestic rabbit (*Oryctolagus cuniculus*), domestic Norway rat (*Rattus norvegicus*), and wild house mouse (*Mus musculus*), shown in Fig. 9.8, indicate that, as a group, these animals possess better high-frequency hearing than any of the previous groups (Kelly and Masterton, 1977; H. Heffner and Masterton, 1980). The high-frequency hearing limits range from 49 kHz for the rabbit to 92 kHz for the wild house mouse. On the other hand, the low-frequency limits of these animals are not as low as the other mammals and range from 96 Hz for the rabbit to 2.3 kHz for the house

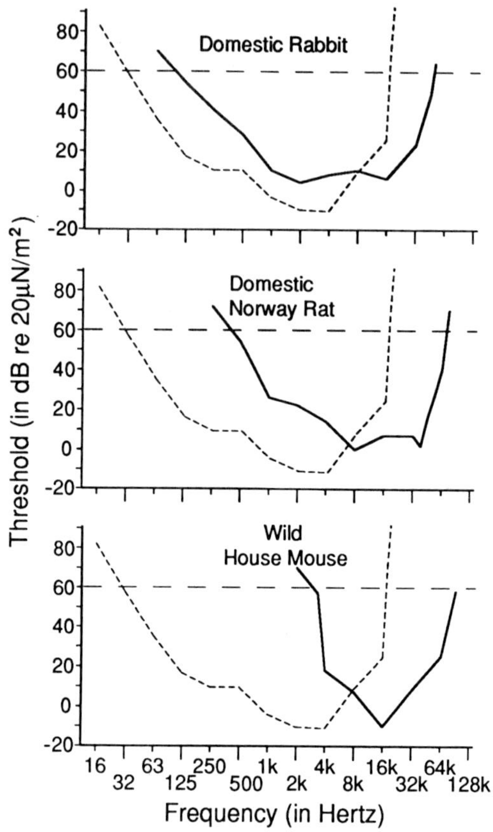

Fig. 9.8. Audiograms for the domestic rabbit, domestic Norway rat and wild house mouse (solid lines). Human audiogram is shown by dashed lines.

mouse. Indeed, the audiograms of the rat and mouse illustrate the phenomenon that animals with good high-frequency hearing often lack good low-frequency hearing.

Perception of ultrasound by mammals

Before continuing, it is of interest to note that there are currently a number of devices on the market which emit sounds above the human hearing range (i.e. ultrasound) for the purpose of repelling other animals. Claims have been made that high-frequency sound from 16 kHz to around 40 kHz can be used to repel deer, dogs, mice, rats, fleas, cockroaches and flies, and the list continues to grow despite the lack of evidence that these devices are effective.

Although such devices are usually advertised as being audible only to pests, all farm mammals are capable of hearing ultrasound (cf. Figs 9.5, 9.7 and 9.8) and one must consider whether ultrasonic repellers might stress these animals. In the case of ultrasonic flea collars, the question arises as to whether the ultrasound might cause a hearing loss in the animal wearing it. However, it may be that ultrasonic repellers cause no more stress in farm animals than in the animals they are supposed to repel.

Evolution of high-frequency hearing

The most striking difference between the hearing of birds and mammals is the ability of mammals to hear frequencies well beyond the range of birds (and other non-mammalian vertebrates as well; cf. Fay, 1988). Whereas birds are unable to hear much above 10 kHz, and amphibians and reptiles do not even hear this high, the average upper limit for mammals is 48 kHz with some species able to hear above 100 kHz. Such a significant difference in abilities leads to the question of *how* mammals are able to hear so high and *why* they do so.

How mammals hear high frequencies

It is generally accepted that one of the main factors accounting for the ability of mammals to hear high frequencies is the unique three-boned middle ear of mammals which transmits sound from the ear drum to the cochlea. Unlike non-mammalian vertebrates which possess a single middle ear bone, the mammalian middle ear contains an articulated chain of three auditory ossicles. Studies of the frequency response or transfer function of the middle ear of various vertebrates have demonstrated that the mammalian ear is far more efficient in transmitting high frequencies to the cochlea than are the middle ears of non-mammalian vertebrates (e.g. Saunders and Rosowski, 1979). Thus, mammals are able to hear higher frequencies than non-mammalian vertebrates because their middle ear permits them to transduce frequencies above 10 kHz.

Why mammals hear high frequencies

The second question is why mammals hear high frequencies, that is, what advantage does this ability confer upon them? Recent research has indicated that mammals use their high-frequency hearing primarily to localize the source of a sound. Specifically, high-frequency hearing enables mammals to use both pinna cues and binaural spectral-difference cues to localize sound.

In addition to the three-boned middle ear, mammals evolved another unique auditory structure: the external ear or pinna. One of the primary features of the pinna is that it acts as a directional filter, amplifying sounds originating from in front of the ear while attenuating sounds from behind. As such they enhance an animal's ability to pick out signals embedded in a noisy world. However, the possession of a pinna has another important consequence: by making the opening of the ear canal asymmetric, the pinnae enable an animal to determine whether a sound is coming from the front or back. In other words, the pinnae provide the primary, if not the only, means for avoiding front–back confusions. This is demonstrated by the fact that front–back localization is all but impossible when the pinnae are distorted or absent (Musicant and Butler, 1984; R. Heffner *et al.*, 1992).

In order to make front–back discriminations, however, it is necessary that the sound contain high frequencies. This is because the pinnae are not directional for low frequencies which can bend around the pinnae with little or no attenuation. Exactly how high the sound must be in order for the pinnae to play an effective role in sound localization depends on the size of the pinnae. In the Norway rat, front–back localization requires frequencies above 10 kHz and maximum performance is not attained unless frequencies up to at least 40 kHz are present (R. Heffner and Heffner, 1992a). Thus, the ability to hear high frequencies enables the pinnae to play an essential role in mammalian sound localization.

High-frequency hearing also plays a role in one of the two binaural locus cues in which the sound reaching the two ears is compared in order to determine the location of the sound source. Briefly, there are two primary binaural local cues: the difference in the time of arrival of a sound reaching the two ears and the difference in the frequency–intensity spectrum of the sound at the two ears. In both cases, the magnitude of the cues depends on the size of an animal's head. That is, the further apart the ears, the larger will be the difference in the time of arrival of the sound at the two ears. Similarly, the difference in the spectra of a sound reaching the two ears is greater for animals with wide-set ears. This is because the sound attenuation is greater over the longer distance between the ears and because animals with wide-set ears usually have large heads or large pinnae which shadow the high-frequency portion of a sound. As a result the sound reaching the ear nearest the sound source is of greater intensity and contains more high-frequency sound than the sound reaching the far ear.

Although the two binaural locus cues are readily available to animals with large heads, the effectiveness of either cue is diminished in animals with close-set ears. In the case of the time cue, the available difference may be so small that the auditory system can detect only gross changes in sound direction. Similarly, a small head is less effective than a large head in blocking sounds with the result that it produces a smaller spectral difference. However, an animal with a small head can increase the size of the spectral difference cue available to it if it is able to hear frequencies that are high enough to be effectively shadowed by its head and pinnae. Thus, the smaller an animal's head the higher it must hear in order to obtain a usable spectral-difference cue.

The correspondence between head size and high-frequency hearing in mammals is illustrated in Fig. 9.9. In this figure, head size is defined as the 'functional' distance between the two ears, where functional distance is defined as the time it takes for sound to travel from one ear to the other. As can be seen, mammals with small heads and close-set ears are better able to hear high-frequency sounds than species with large heads and wide-set ears. That the high-frequency hearing ability of mammals enhances the ability of small mammals to use the binaural spectral-difference cue is indicated by the fact that removing frequencies above 10 kHz degrades the ability of the Norway rat to make left–right judgements regarding the location of a sound source (R. Heffner and Heffner, 1992a).

It can also be seen in Fig. 9.9 that there are three known species which do not hear as high as would be expected from their head size. These are the pocket gopher (*Geomys bursarius*, R. Heffner and Heffner, 1990b), blind mole rat (*Spalax ehrenbergi*, Bronchti *et al.*, 1989), and naked mole rat (*Heterocephalus glaber*, R. Heffner and Heffner, 1991), rodents which spend almost their entire lives underground. However, not only have these animals lost the ability to hear high frequencies, they have also lost virtually all ability to localize sound (R. Heffner and Heffner, 1990b

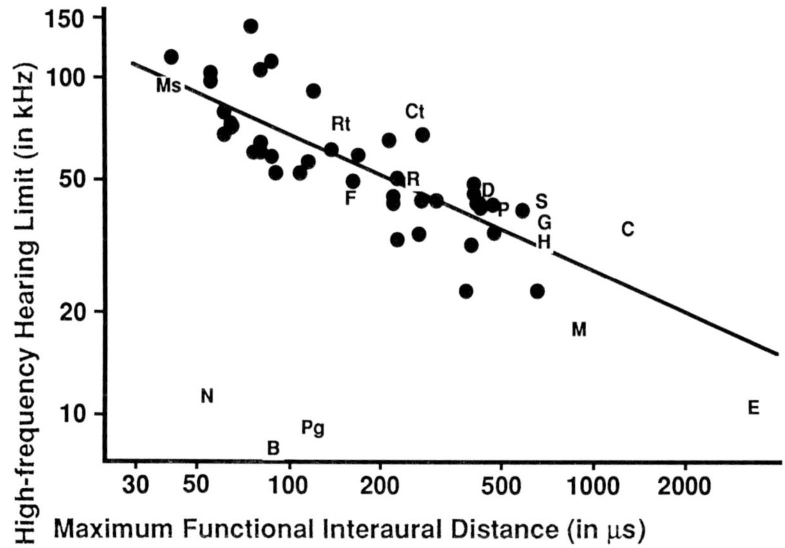

Fig. 9.9. Relationship between maximum functional interaural distance and the 60 dB high-frequency hearing limit in mammals. B, blind mole rat; C, cattle, Ct, cat; D, dog; E, elephant; F, ferret; G, goat, H, horse; M, man; Ms, house mouse; N, naked mole rat; P, pig; Pg, pocket gopher; R, rabbit; Rt, Norway rat; S, sheep. For species represented by filled circles, see R. Heffner and Heffner, 1985a.

and unpublished). Evidently, animals which have adapted to the one-dimensional world of an underground habitat have little use for sound localization. Thus, rather than attenuating the link between high-frequency hearing and sound localization in mammals, these animals make the point that mammals which lose their ability to localize sound also lose their high-frequency hearing.

Of the two roles of high-frequency hearing in sound localization, it appears that the use of high frequencies for pinna cues may be more fundamental than their use in the binaural spectral difference cue. The evidence for this conclusion comes from research on the ability of farm animals to localize sound. Specifically, it has been discovered that some large mammals, such as horses, pigs, and cattle, do not use the binaural spectral cue, but, instead, rely on the binaural time difference cue for making left–right locus decisions (R. Heffner and Heffner, 1986, 1989, 1992a). However, these animals do hear high frequencies and they do require their high-frequency hearing to use pinna cues to make front–back discriminations. Therefore, of the two functions, it appears that the necessity of high-frequency hearing in order to use pinnae cues is the more ubiquitous and may have played the greater role in the evolution of mammalian high-frequency hearing.

Finally, it should be noted that some small rodents and insectivores vocalize in the ultrasonic range thereby giving rise to the possibility that mammals evolved high-frequency hearing in order to use high frequencies for communication (e.g. Sales and Pye, 1974; Geyer and Barfield, 1979). Indeed, a number of species of mice and rats emit vocalizations in the frequency range above 30 kHz and vocalizations above 50 kHz are not uncommon. However, many if not most mammals which possess ultrasonic hearing do not appear to send or receive ultrasonic communications. In particular, dogs, pigs, sheep and goats, all of which hear above 40 kHz, as well as horses and cattle, which hear above 30 kHz, do not use ultrasonic sounds for communication (cf. Kiley, 1972; Fox and Cohen, 1977; Walser *et al.*, 1981; R. Heffner and Heffner, 1990a). Thus, the fact that high-frequency hearing is not closely linked with the use of ultrasonic communication suggests that those species which do vocalize in the ultrasonic range are making use of an ability which initially served another function.

Variation in low-frequency hearing

It has long been noted that low-frequency hearing varies widely among mammals (Masterton *et al.*, 1969). Whereas the elephant has a low-frequency hearing limit of 17 Hz, the little brown bat hears only as low as 10.3 kHz (see R. Heffner and Heffner, 1982). Thus, the low-frequency limits of mammals span a range of more than nine octaves. In comparison, high-frequency hearing limits vary over a range of only four octaves.

In searching for possible explanations of this variation, several factors have been examined: body weight, head size and high-frequency hearing (R. Heffner and Heffner, 1990a). The results of multiple regression analysis have indicated that, of these three factors, only high-frequency hearing is significantly correlated with low-frequency hearing ($r = 0.78$, $P < 0.01$). That is, there appears to be a trade-off such that animals with good high-frequency hearing generally have poor low-frequency hearing and vice versa.

However, there are exceptions to this relationship which suggest that adaptation to unusual environments may result in atypical abilities (R. Heffner and Heffner, 1990a). The first exception are marine mammals that hear under water, such as the porpoise and beluga whale, which possess good high-frequency and good low-frequency hearing. Because sound transmission in these animals can bypass the middle ear, this finding suggests that the middle ear may limit the frequency range of an animal. The second exception are underground mammals, such as pocket gophers and mole rats, which cannot hear well above 10 kHz, yet have unexceptional low-frequency hearing. These animals have given up high-frequency hearing, apparently because they do not use sound localization in their underground environment, and their hearing has not extended into the lower frequency range, apparently because there has been no selective pressure to hear low frequencies.

Given that there are important exceptions to the relationship betwen high- and low-frequency hearing, the question arises as to how farm animals fit the general pattern. Examining this relationship among the remaining mammals (i.e. disregarding underground and underwater animals) reveals that there are still seven species which deviate significantly from the regression line. The deviant species are indicated by open circles in Fig. 9.10 and their removal from the analysis raises the correlation coefficient from 0.78 to 0.95. However, as shown in Fig. 9.10, all of the hoofed mammals conform closely to this relationship – their good low-frequency hearing is accompanied by modest high-frequency hearing. Similarly, the mouse, rat and rabbit also fit the relationship.

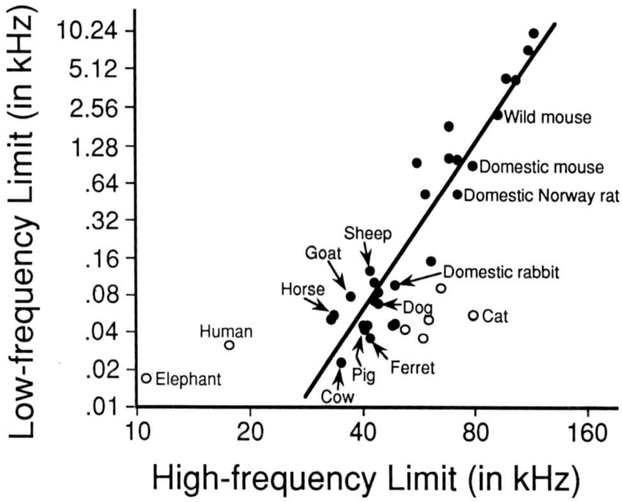

Fig. 9.10. Relationship between 60 db high-frequency and 60 dB low-frequency hearing limits. Open circles represent animals which deviate significantly from the relationship. For species represented by unlabelled circles, see R. Heffner and Heffner, 1985a.

Of the seven species which deviate significantly from this relationship, the domestic cat is one of five whose low-frequency hearing extends beyond that predicted by its high-frequency hearing (open circles to the right of the regression line in Fig. 9.10). This indicates that these animals have a broader range of hearing than most other mammals. Indeed, cats with their hearing range extending from 55 Hz to 79 kHz have the broadest range of hearing of any terrestrial surface dweller (cf. Fig. 10.7). Thus, cats, along with gerbils, least weasels, kangaroo rats and bushbabies, demonstrate that it is possible for an animal to achieve good high- and good low-frequency hearing.

Localization of Sound

Determining sound localization ability is more complex than determining absolute thresholds because localization ability varies depending on the relative location of the sound source and the type of sound being localized. First, animals are more accurate in localizing in the horizontal plane than in the vertical plane. Second, localization acuity is best for sound sources located directly in front with acuity steadily decreasing for locations off to the side (e.g. R. Heffner and Heffner, 1988). Finally, spectral composition is a factor in that broad-band noise with an abrupt onset is generally easier to localize than narrow-band noise, and pure tones are the most difficult of all. Indeed, this latter fact appears to be exploited by some animals whose tonal calls make it more difficult for predators to localize them (Marler, 1955; Brown and May, 1990).

Most studies have measured sound localization acuity in the horizontal plane for sound sources located in front of the animal. Generally, this has involved having an animal perform a left–right discrimination. Thus, unless otherwise noted, it can be assumed that an animal's sound localization acuity refers to its ability to perform a left–right discrimination using a brief complex sound, such as a noise burst or click.

Sound localization cues

As previously described, there are two sets of sound localization cues. The first set of cues are the binaural locus cues which consist of the difference in the time of arrival of a sound at the two ears and the difference in the frequency–intensity spectra of a sound at the two ears. These two cues are effective for localizing sound in the horizontal plane with their effectiveness decreasing as the location of the source moves from the front to the side.

The second set of cues are monaural cues which arise from the differences in the spectrum of a sound reaching a single ear from different locations. These differences are due to the fact that high frequency (short wavelength) sounds cannot readily bend around the head and pinna. As a result, the high frequency portion of a sound reaching an ear depends on the location of the sound source relative to that ear (e.g. Musicant and Butler, 1984). Because the primary source of these cues is the pinna, they are usually referred to as pinna cues. Pinna cues play an important role in horizontal localization by improving acuity for laterally placed sound sources (as opposed to sound sources located near midline) and by reducing front–back reversals. In addition, they provide the main cues for vertical localization.

Because birds lack an external ear and are unable to hear frequencies much above 10 kHz, it seems unlikely that they can use monaural spectral cues to localize sound. This conclusion is supported by the observation that, unlike mammals, birds are unable to localize sound in the horizontal plane with only one ear (Park and Dooling, 1991). Indeed, although some owls have evolved a facial ruff analogous to the mammalian pinna, they appear to use the binaural spectral cues which the ruff generates to perform vertical localization (Knudsen, 1980).

Although birds lack pinnae, they have another anatomical feature which mammals lack and which some birds may use for sound localization. Specifically, the middle ear cavities on each side are connected to each other by two air passages: an interaural canal lying ventral to the brain case; and eustachian tubes which fuse together before joining the nasopharynx (Knudsen, 1980). It has been proposed that this arrangement may enhance sound localization by allowing sound reaching one ear to travel through the head to partially cancel out the sound reaching the other ear thereby producing an intensity difference between the two ears (Lewis and Coles, 1980). However, it should be noted that the degree to which birds rely on these interaural passages for sound localization has not been established and the transmission of sound between the two ears in the barn owl appears to be too insignificant to play a role in sound localization (Moiseff and Konishi, 1981).

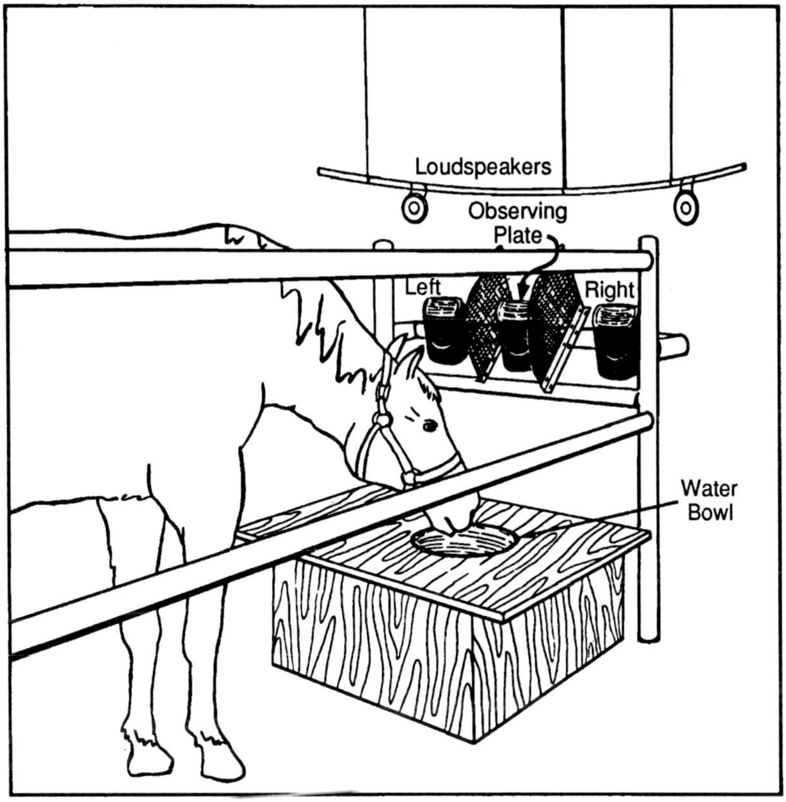

Fig. 9.11. Apparatus used to test sound localization in horses and cattle. An animal initiated a trial by touching the observing plate with its nose. It then touched either the left or right response plate to indicate whether a sound had come from a loudspeaker on its left or right side, respectively. Correct responses were rewarded by delivering water into the water bowl. The hardware–cloth panels separating the response plates served to prevent an animal from responding to the side plates too rapidly (modified from H. Heffner and Heffner, 1984).

Measuring sound localization acuity

There are two basic procedures for determining sound localization ability. The most common is to measure an animal's 'two-point sensitivity' by training it to discriminate between the same sound presented from one of two loudspeakers and then reducing the angular separation between the two speakers until the animal is no longer able to perform the discrimination. The second procedure exploits an animal's natural tendency to orient to the source of a sound and determines sound localization acuity by measuring the accuracy (i.e. the amount of error) with which an animal orients its head to a sound.

An animal's two-point locus acuity can be determined using a two-choice procedure. In this procedure a thirsty animal is led into a test stall and trained to place its nose on an observing plate located in front of an array of loudspeakers (Fig. 9.11).

A sound is presented from a loudspeaker located to the left or right of the animal and the animal is trained to touch one of two response plates located to its left or right. A response to the plate located on the same side as the sound source is rewarded by delivering water into a bowl located below the observing plate while an incorrect response is followed by a 5–15 s wait before the animal can try again. Thresholds are typically defined as the smallest angle which yields a performance of 75% correct (i.e. half way between perfect and random performance).

The orienting response is used to measure localization accuracy by measuring the accuracy of an animal's head orientation to a sound. Although this procedure has occasionally been used with mammals (e.g. Thompson and Masterton, 1978), its greatest success has been in measuring the sound localization ability in owls (Knudsen *et al.*, 1979). By using a search coil technique to measure head orientation, Knudsen and his colleagues have been able to measure the accuracy with which the barn owl (*Tyto alba*) localizes sound. Furthermore, by rewarding correct orientations with food, they have been able to maintain an animal's orientation response for long periods.

Birds

Although a number of studies have measured the ability of birds to localize sound in the horizontal plane, it is difficult to compare the results from birds because of the existence of important procedural differences between studies. In particular, various studies have used different localization stimuli, such as noise, tones, and bird calls, all with good reason, but with the result that the thresholds are not comparable. Comparisons are further complicated by the fact that several different definitions of thresholds have been used. However, one procedural difference that does not appear to affect thresholds is the use of different behavioural methods such as whether the animal is trained using positive reward or punishment, or whether it is required to locate the source of the sound or simply report a change in the location of the sound source (cf. R. Heffner and Heffner, 1988). However, the proficiency with which a test is administered can and does affect its outcome.

Most studies of bird sound localization have examined songbirds and the general procedure has been to train the animal using a two-choice left–right discrimination. The results of these studies have indicated that thresholds for songbirds for noise stimuli commonly range from 20° to 30° separation (for reviews, see Knudsen, 1980; Park and Dooling, 1991).

Turning to farm animals, sound localization behaviour has been studied in chickens (*Gallus domesticus*) by exploiting the natural behaviour of setting hens to search for chicks when they emit a distress call. By placing a chick behind one of two cardboard screens and observing the ability of a hen to locate it, Engelmann estimated the localization acuity of the hens to be about 4° (Engelmann, 1928). However, before concluding that hens are better able to localize than songbirds, it should be noted that the same study, using similar techniques, significantly overestimated the localization acuity of cats and dogs. Thus, this result needs to be confirmed with modern techniques before it can be accepted.

Sound localization thresholds have been determined for the pigeon (*Columba livia*) using a heart-rate conditioning procedure (Lewald, 1987). In this procedure, an animal is habituated to a sound presented from one location while the occasional presentation of the sound from a different location is paired with electric shock. This results in a change in the animal's heart rate whenever it detects a change in location. By measuring heart rate for different angles of separation, it has been shown that pigeons can detect angles as small as 4° (Lewald, 1987). However, it should be noted that threshold was defined as the smallest angle eliciting a statistically significant change in heart rate (i.e. $P = 0.025$, one-tailed distribution). In contrast, most other studies define threshold as the smallest angle at which an animal scores 75% correct which is usually well above the statistical threshold. Thus, while the pigeon probably localizes sound more accurately than songbirds, the 75% correct definition of threshold would place their threshold at a somewhat larger angle than 4°.

Finally, the sound localization ability of the barn owl (*Tyto alba*) has been carefully studied by measuring their orientation to sound (Knudsen *et al.*, 1979). These results have shown that the barn owl can orientate to a frontally placed sound source with an average error of about 2°, an accuracy which is comparable to a two-choice 75% threshold of 3.5°. Thus, the barn owl, a nocturnal predator which uses its hearing to locate prey, has the best sound localization acuity of any bird yet tested.

Mammals

The ability of mammals to localize sound has been determined for over 20 species representing nine different orders of mammals from marsupials and insectivores to proboscidea and artiodactyls (cf. R. Heffner and Heffner, 1992a). Furthermore, tests performed on several species by different laboratories using different methods have yielded comparable results (cf. R. Heffner and Heffner, 1988). Thus, not only does there exist information on the localization abilities of a variety of mammals, but there is reason for confidence in the validity of the data.

Because a knowledge of the exact location of a sound source would appear to be of obvious importance, it seems reasonable to assume that all animals are under strong selective pressure to localize sound as accurately as possible. Yet the results of studies to date have indicated that not all mammals do so. Indeed, the ability of mammals to localize sound has been shown to range from around 1° (e.g. elephants and humans) to more than 25° (e.g. gerbils and cattle) with some species (i.e. the pocket gopher) completely unable to localize brief sounds (R. Heffner and Heffner, 1992a, c).

Hoofed mammals

The sound localization thresholds of the four species tested so far span a wide range (Fig. 9.12). Three of the species, goats, horses and cattle, have thresholds of 18°, 25° and 30°, respectively, which place them well below the median acuity for mammals of 12°. On the other hand, pigs, with a threshold of 4.6°, are among the most accurate localizers.

Carnivores

Not surprisingly, carnivores tend to be among the more accurate localizers. As shown in Fig. 9.12, thresholds for the cat, dog and ferret are 5.7°, 8° and 8.5°, respectively.

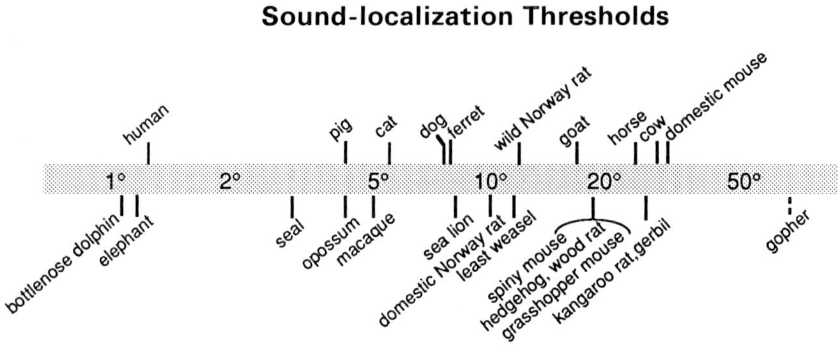

Fig. 9.12. Sound localization acuity in the horizontal plane for 25 mammals. With the exception of the gopher (dashed line) all animals were tested using brief broad-band stimuli. (Note log scale.)

Rodents

Although rodents tend to be among the less accurate mammals, the 12.8° threshold of the wild Norway rat places it close to the median for mammals (H. Heffner and Heffner, 1985). Although data are not available for the wild house mouse, a recent study shows domestic mice to have a threshold of 33° (R. Heffner, unpublished).

Variation in mammalian sound localization acuity

Until recently it was believed that all mammals were under strong selective pressure to localize sound as accurately as possible. The fact that some animals are unable to localize as accurately as others was explained in terms of the variation in the magnitude of the locus cues available to each animal. That is, because the magnitude of the binaural locus cues is largely determined by the size of the head and pinnae, it was expected that any variation in localization acuity was due to species differences in head and pinna size. Indeed, the observation that humans were more accurate than cats which, in turn, were more accurate than rats appeared to support this idea.

As more species were tested it became apparent that a large head did not automatically result in good localization acuity. Indeed, this theory became untenable when it was discovered that horses and cattle, despite their large heads, have poorer acuity than many small rodents (H. Heffner and Heffner, 1984; R. Heffner and Heffner, 1992b). Furthermore, although a small interaural distance may be a limiting factor in acuity, it was found that some small species possess relatively good acuity when compared to other species of equal or larger head sizes (cf. the least weasel and grasshopper mouse, Fig. 9.12).

In searching for the source of the variation in acuity, a number of factors have been considered. Among these are the possibility that various lifestyles, such as predator vs. prey, nocturnal vs. diurnal, or surface dwelling vs. underground, might be associated with variation in localization acuity. However, all of these possibilities have major exceptions and none has remained satisfactory when subjected to statistical analysis (R. Heffner and Heffner, 1992c).

There is, however, one factor which does seem to account for the variation in

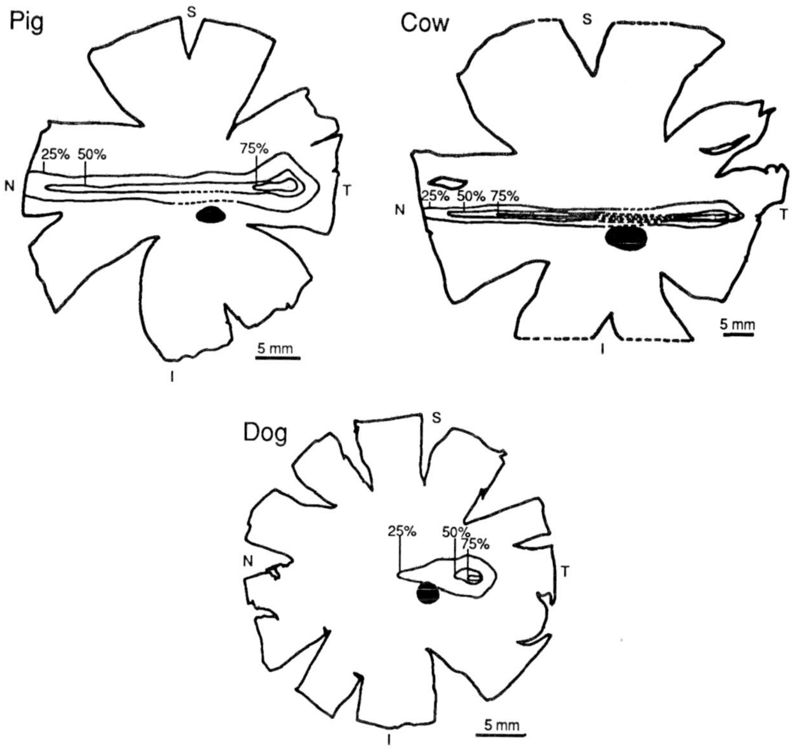

Fig. 9.13. Drawings of the retinae of a pig, cow and dog, illustrating the variation in the width of the field of best vision as estimated from the density of retinal ganglion cells. The contour lines encompass densities 25%, 50% and 75% of the maximum density for each species. The field of best vision is operationally defined as the part of the retina containing ganglion cells with packing densities equal to or greater than 75% of maximum. The dog has a narrow field of best vision whereas the cow, with its visual streak, has a very broad field. The pig has a visual streak combined with a small area of increased density located temporally. I, inferior; N, nasal; S, superior; T, temporal (data from R. Heffner and Heffner, 1992c).

locus acuity. In examining the utility of sound localization, it has been observed that one of the most consistent responses to an unexpected sound is the orienting reflex in which the head and eyes are turned towards a sound source (Pumphrey, 1950). When visually orienting to a sound, however, it is not simply the head or the eyes that are directed to the sound source, but the part of the visual field with the best visual acuity. In species such as humans, the area of best vision, i.e. the fovea, subtends an angle of only 1–2°. As a result, auditory localization has to be very precise in order to direct the fovea to the sound source. However, most mammals have a broader area of best vision which in some cases covers nearly the entire horizon of the eye and is referred to as a visual streak (Fig. 9.13; cf. Hughes, 1977). Species with progressively broader fields of best vision, therefore, should require correspondingly less precise auditory information in order to place a sound source within their field of best vision.

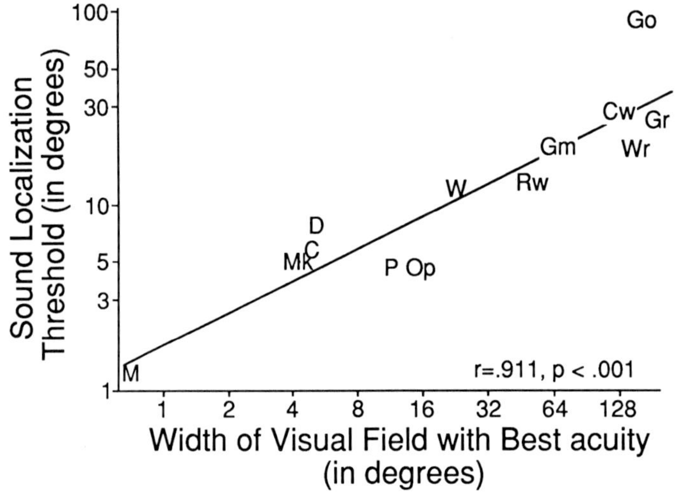

Fig. 9.14. Relationship between width of the field of best vision and sound localization threshold for 13 species of mammals (note log scales). Species with a narrow field of best vision have much smaller sound localization thresholds (greater acuity) than species with a broad field. C, cat; Cw, cow; D, dog; Gm, grasshopper mouse; Go, gopher; Gr, gerbil; M, man; Mk, macaque; Op, opossum; P, pig; Rw, wild Norway rat; W, least weasel; Wr, wood rat (modified from R. Heffner and Heffner, 1992c).

In order to determine the correlation between sound localization acuity and the size of the field of best vision, we have obtained anatomical measures of the visual acuity in animals for which sound localization thresholds are available. Because the density of ganglion cells in the retina is known to correspond to behavioural measures of visual acuity (e.g. Rolls and Cowey, 1970), measuring the density of ganglion cells in retinal wholemounts can provide an anatomical estimate of the relative acuity of the different parts of the retina. Given a picture of the variation in the density of ganglion cells, it is possible to arrive at a measure of the size of the field of best vision.

Using an arbitrary definition of field of best vision as the horizontal width (in degrees) of the region of the retina containing ganglion cell densities equal to or greater than 75% of the maximum density, we have determined the size of the field of best vision in 13 species (R. Heffner and Heffner, 1992c). These measurements have shown that the size of this field varies from approximately 1.5° in humans to around 180° in some species such as cattle (cf. Fig. 9.13). The relation between the width of the field of best vision and sound localization thresholds is shown in Fig. 9.14. As indicated in this figure, the width of the field of best vision is positively correlated with sound localization threshold ($r = 0.91$, $P < 0.001$). That is, animals with narrow fields of best vision, such as cats and humans, have lower thresholds (better sound localization acuity) than those with broader fields of best vision, such as cattle.

Although the sample is as yet small, it includes a broad range of species from five orders of mammals encompassing surface and underground dwellers, nocturnal and diurnal activity patterns, predators and prey, and body sizes ranging through more than five orders of magnitude. Furthermore, included in this sample are animals with good sound localization acuity as well as those with poor acuity and animals with large as well as those with small areas of best vision. Thus, although we are continuing to increase our sample, there is no reason to believe that the high correlation between sound localization acuity and width of the area centralis is the result of a restricted or unrepresentative sample.

Effect of Domestication on Hearing

When discussing the sensory abilities of domesticated animals, the question often arises as to whether their abilities have been reduced owing to protection from danger or selective breeding which characterize domestication. From the information we have available, it appears unlikely that the hearing abilities of farm animals have been reduced as a result of domestication (e.g. R. Heffner and Heffner, 1992b). With regard to auditory sensitivity, the high-frequency sensitivity of domesticated mammals is well within the range of that expected for animals of their size (cf. Fig. 9.9). Similarly, low-frequency sensitivity is not unique with the exception of the domestic cat which has unexpectedly good low-frequency hearing (cf. Fig. 9.10). Finally, in terms of best sensitivity, domestic mammals appear to have either average or exceptionally good hearing as cats, cattle and goats are among the most sensitive vertebrates.

Turning to sound localization, it might be speculated that the poor acuity of horses, goats and cattle may have arisen through domestication. However, there are several reasons why this is unlikely (for a detailed discussion, see R. Heffner and Heffner, 1992b). First, there are domestic species which possess good acuity. For example, domestic pigs are among the most accurate mammals and the domestic Norway rat is slightly more accurate than the wild Norway rat (Kavanagh and Kelly, 1986; R. Heffner and Heffner, 1989). Second, when mammals have lost the ability to localize sound (e.g. the pocket gopher), this loss has been accompanied by a general reduction in overall sensitivity including a loss of high frequency hearing (R. Heffner and Heffner, 1990b). However, the fact that domesticated animals have good high-frequency hearing and that cattle and goats have excellent sensitivity indicates that they do not fit the profile of an animal whose sound localization has degenerated. Finally, as was previously noted, the poor localization ability of horses, goats and cattle is predicted by the correspondence in mammals between the width of the field of best vision and localization acuity. In short, they appear to localize as accurately as necessary in order to direct their gaze to the source of a sound. Thus, there is no reason to believe that the hearing of domestic animals has undergone any reduction as the result of domestication.

Acknowledgement

Supported by NIH grant DC 00179.

References

Bronchti, G., Heil, P., Scheich, H. and Wollberg, Z. (1989) Auditory pathway and auditory activation of primary visual targets in the blind mole rat (*Spalax ehrenbergi*): I. 2-deoxyglucose study of subcortical centers. *Journal of Comparative Neurology* **284**, 253–274.

Brown, C.H. and May, B.J. (1990) Sound localization and binaural processes. In: Berkley, M.A. and Stebbins, W.C. (eds), *Comparative Perception*, vol. 1. Wiley, New York, pp. 247–284.

Dooling, R.J., Okanoya, K., Downing, J. and Hulse, S. (1986) Hearing in the starling (*Sturnus vulgaris*): Absolute thresholds and critical ratios. *Bulletin of the Psychonomic Society* **24**, 462–464.

Engelmann, W. (1928) Untersuchungen über die Schallokalisation bei Tieren. *Zeitschrift für Psychologie* **105**, 317–370.

Fay, R.R. (1988) *Hearing in Vertebrates: A psychophysics databook*. Heffernan Press, Worcester, MA.

Fox, M.W. and Cohen, J.A. (1977) Canid communication. In: Sebeok, T.A. (ed.), *How Animals Communicate*. Indiana University Press, Bloomington, IN, pp. 728–748.

Geyer, L.A. and Barfield, R.J. (1979) Introduction to the symposium: Ultrasonic communication in rodents. *American Zoologist* **19**, 411.

Heffner, H.E. (1983) Hearing in large and small dogs: Absolute thresholds and size of the tympanic membrane. *Behavioral Neuroscience* **97**, 310–318.

Heffner, H.E. and Heffner, R.S. (1983) The hearing ability of horses. *Equine Practice* **5**, 27–32.

Heffner, H.E. and Heffner, R.S. (1984) Sound localization in large mammals: Localization of complex sounds by horses. *Behavioural Neuroscience* **98**, 541–555.

Heffner, H.E. and Heffner, R.S. (1985) Sound localization in wild Norway rats (*Rattus norvegicus*). *Hearing Research* **19**, 151–155.

Heffner, H. and Masterton, B. (1980) Hearing in glires: domestic rabbit, cotton rat, feral house mouse, and kangaroo rat. *Journal of the Acoustical Society of America* **68**, 1584–1599.

Heffner, R.S. and Heffner, H.E. (1982) Hearing in the elephant: Absolute sensitivity, frequency discrimination, and sound localization. *Journal of Comparative and Physiological Psychology* **96**, 926–944.

Heffner, R.S. and Heffner, H.E. (1983a) Effect of cattle ear mite infestation on hearing in a cow. *Journal of the American Veterinary Medical Association* **182**, 612–614.

Heffner, R.S. and Heffner, H.E. (1983b) Hearing in large mammals: Horses (*Equus caballus*) and cattle (*Bos taurus*). *Behavioural Neuroscience* **97**, 299–309.

Heffner, R.S. and Heffner, H.E. (1983c) Occurrence of the cattle ear mite (*Raillietia auris*) in southeastern Kansas. *The Cornell Veterinarian* **73**, 193–199.

Heffner, R.S. and Heffner, H.E. (1985a) Hearing in mammals: The least weasel. *Journal of Mammalogy* **66**, 745–755.

Heffner, R.S. and Heffner, H.E. (1985b) Hearing range of the domestic cat. *Hearing Research* **19**, 85–88.

Heffner, R.S. and Heffner, H.E. (1986) Localization of tones by horses: Use of binaural cues and the superior olivary complex. *Behavioral Neuroscience* **100**, 93–103, 137.

Heffner, R.S. and Heffner, H.E. (1988) Sound localization acuity in the cat: Effect of azimuth, signal duration, and test procedure. *Hearing Research* **36**, 221–232.

Heffner, R.S. and Heffner, H.E. (1989) Sound localization, use of binaural cues and the superior olivary complex in pigs. *Brain, Behaviour and Evolution* **33**, 248–258.

Heffner, R.S. and Heffner, H.E. (1990a) Hearing in domestic pigs (*Sus scrofa*) and goats (*Capra hircus*). *Hearing Research* **48**, 231–240.

Heffner, R.S. and Heffner, H.E. (1990b) Vestigial hearing in a fossorial mammal, the pocket gopher (*Geomys bursarius*). *Hearing Research* **46**, 239–252.

Heffner, R.S. and Heffner, H.E. (1991) Hearing in subterranean mammals: naked mole rats (*Heterocephalus glaber*). *Abstracts of the Fourteenth Meeting of the Association for Research in Otolaryngology*, 24.

Heffner, R.S. and Heffner, H.E. (1992a) Evolution of sound localization in mammals. In: Popper, A.N., Fay, R.R. and Webster, D.B. (eds), *The Evolutionary Biology of Hearing*. Springer-Verlag, New York, pp. 691–716.

Heffner, R.S. and Heffner, H.E. (1992b) Hearing in large mammals: Sound-localization acuity in cattle (*Bos taurus*) and goats (*Capra hircus*). *Journal of Comparative Psychology* **106**, 107–113.

Heffner, R.S. and Heffner, H.E. (1992c) Visual factors in sound localization in mammals. *Journal of Comparative Neurology* **317**, 219–232.

Heffner, R.S., Koay, G. and Heffner, H.E. (1992) Importance of high frequencies and the contribution of the pinnae to sound localization in chinchillas. *Abstracts of the Fifteenth Meeting of the Association for Research in Otolaryngology* 50.

Hienz, R.D., Sinnott, J.M. and Sachs, M.B. (1977) Auditory sensitivity of the red-winged blackbird (*Agelaius phoeniceus*) and brown-headed cowbird (*Molothrus ater*). *Journal of Comparative and Physiological Psychology* **91**, 1365–1376.

Hughes, A. (1977) The topography of vision in mammals of contrasting life style: Comparative optics and retinal organisation. In: Crescitelli, F. (ed.), *Handbook of Sensory Physiology*, vol. VII/5. Springer-Verlag, New York, pp. 613–756.

Kavanagh, G.L. and Kelly, J.B. (1986) Midline and lateral field sound localization in the albino rat (*Rattus norvegicus*). *Behavioral Neuroscience* **100**, 200–205.

Kelly, J.B. and Masterton, B. (1977) Auditory sensitivity of the albino rat. *Journal of Comparative and Physiological Psychology* **91**, 930–936.

Kelly, J.B., Kavanagh, G.L. and Dalton, J.C.H. (1986) Hearing in the ferret (*Mustela putorius*): thresholds for pure tone detection. *Hearing Research* **24**, 269–275.

Kiley, M. (1972) The vocalizations of ungulates, their causation and function. *Zeitschrift für Tierpsychologie* **31**, 171–222.

Knudsen, E.I. (1980) Sound localization in birds. In: Popper, A.N. and Fay, R.R. (eds), *Comparative Studies of Hearing in Vertebrates*. Springer-Verlag, New York, pp. 289–322.

Knudsen, E.I., Blasdel, G.G. and Konishi, M. (1979) Sound localization by the barn owl (*Tyto alba*) measured with the search coil technique. *Journal of Comparative Physiology* **133**, 1–11.

Konishi, M. (1973) How the owl tracks its prey. *American Scientist* **61**, 414–424.

Kreithen, M.L. and Quine, D.B. (1979) Infrasound detection by the homing pigeon: A behavioral audiogram. *Journal of Comparative Physiology* **129**, 1–4.

Lewald, J. (1987) The acuity of sound localization in the pigeon (*Columba livia*). *Naturwissenschaften* **74**, 296–297.

Lewis, B. and Coles, R. (1980) Sound localization in birds. *Trends in Neurosciences* **3**, 102–105.

Maiorana, V.A. and Schleidt, W.M. (1972) The auditory sensitivity of the turkey. *Journal of Auditory Research* **12**, 203–207.

Marler, P. (1955) Characteristics of some animal calls. *Nature* **176**, 6–8.

Masterton, B. and Diamond, I.T. (1973) Hearing: Central neural mechanisms. In: Carterette, E.C. and Friedman, M.P. (eds), *Handbook of Perception*, vol. 3: *Biology of Perceptual Systems*. Academic Press, New York, pp. 407–448.

Masterton, B., Heffner, H. and Ravizza, R. (1969) The evolution of human hearing. *Journal of the Acoustical Society of America* **45**, 966–985.

Moiseff, A. and Konishi, M. (1981) The owl's interaural pathway is not involved in sound localization. *Journal of Comparative Physiology* **144**, 299–304.

Musicant, A.D. and Butler, R.A. (1984) The psychophysical basis of monaural localization. *Hearing Research* **14**, 185–190.

Park, T.J. and Dooling, R.J. (1991) Sound localization in small birds: Absolute localization in azimuth. *Journal of Comparative Psychology* **105**, 125–133.

Pumphrey, R.J. (1950) Hearing. *Symposia of the Society for Experimental Biology* **4**, 1–18.

Rolls, E.T. and Cowey, A. (1970) Topography of the retina and striate cortex and its relationship to visual acuity in rhesus monkeys and squirrel monkeys. *Experimental Brain Research* **10**, 298–310.

Sales, G. and Pye, D. (1974) *Ultrasonic Communication by Animals.* Chapman and Hall, London.

Saunders, J.C. and Rosowski, J.J. (1979) Assessment of hearing in animals. In: Rintelmann, W.F. (ed.), *Hearing Assessment.* University Park Press, Baltimore, MD, pp. 487–529.

Thompson, G.C. and Masterton, R.B. (1978) Brain stem auditory pathways involved in reflexive head orientation to sound. *Journal of Neurophysiology* **41**, 1183–1202.

Trainer, J.E. (1947) The auditory acuity of certain birds. In: Cunningham, G. W. (ed.), *Abstracts of Theses*, Cornell University, pp. 246–251.

Walser, E. S., Walters, E. and Hague, P. (1981) A statistical analysis of the structure of bleats from sheep of four different breeds. *Behaviour* **77**, 67–76.

Wollack, C.H. (1963) The auditory acuity of the sheep (*Ovis aries*). *Journal of Auditory Research* **3**, 121–132.

10 Olfaction and Taste

G.C. PERRY

Department of Animal Husbandry, University of Bristol, Langford House, Langford, Bristol, UK

Summary

Olfaction and taste assume great importance in species unable to indulge in extensive vocal communication. For social species it is important to recognize fellow members and while vision may have a role to play, olfaction offers an alternative, and in some cases, an extremely sensitive discrimination technique. The transition from casual observation to scientific investigation is described for olfaction. Although a relatively new science it has already provided evidence of powerful influences on reproductive efficiency. This chapter presents data from studies with laboratory animals and illustrates how this work has extended to domesticated species. The wider appreciation of the role of olfaction has already led to the modification of certain modern management practices.

The ability to discriminate between materials by taste is often claimed. Grazing animals are known to avoid certain plants but it is not clear whether such behaviour is based on taste *per se*. There is not much scientific evidence to support the role of taste in animals. There are taste receptors on the tongues of animals and there are also clear differences between species in taste responses.

The roles of olfaction and taste are reviewed together with generalized descriptions of their mode of operation.

Introduction

All our domesticated species demonstrate a degree of social living. This ability to spend time in some type of group formation implies an ability to communicate between individuals in order to maintain group cohesion and co-ordination of activities. As a higher primate, man has developed sophisticated technology in the form of media systems which satisfy the communicatory needs of society life. Other animals, however, do not use such obvious external aids but nevertheless are able to perceive or transfer information from and to their conspecifics. In a comparison between man and other animals, the end result may be similar but the biological sophistication of the animal-based system may be as 'impressive' as the technological principles developed by man.

Why olfaction? Information may be obtained through one or more of the sensory routes of sight, sound, touch, smell and taste. It could be argued that taste is a limited and specialized sensory input since it may be an evolved form of tactile input in which very specific information is sought. The other four senses can provide a much wider spectrum of information. Some of them, however, have limitations. Touch requires proximal contact between the animals or between the perceiver and the object in question. Sound may be more effective at a longer distance but carries the risk that it betrays the presence or location of the informer. Sight can also be effective over distances but its success may be influenced by terrain or the structure of the receiver's eyes. For example, if the eye is not suitably constructed then information may not be received during periods of darkness.

Smell offers many advantages as a communication system. It operates during light and dark. It can operate over considerable distances, subject to environmental factors such as wind direction. It can incorporate a temporal element based on the decay rates of the constituent materials and which may provide a basis for marking territories which, in turn, permits a degree of protection or security for social groups. Smell can also be used for individual recognition and can be used to establish a bond between mother and young.

Scientific Interest in Olfaction

All of the cited advantages are convenient hypotheses and, apart from anecdotal casual observation, had little or no scientific support up to the 1950s. Then evidence began to accumulate which identified odours as a vehicle for communication. The early data supported two main types of response. In the first type, exposure to a particular odour appeared to have an effect on reproductive efficiency and in the second type perception of an odour elicited a direct and specific change in animal behaviour. This differentiation led to the concept of primer and releaser agents.

Sherrington (1906) studied the nervous systems of animals and highlighted the role of the olfactory system in the search for food and a mate, but it was not until the 1950s that behavioural changes associated with odours were first reported.

Olfactory Influences on Reproduction

That first evidence came from laboratory mice. The oestrous cycle of the mouse is most frequently of 4–6 days' duration. Lee and Boot (1955, 1956) noted the occurrence of prolonged dioestrous intervals in mice housed in groups of four and suggested that these resulted from pseudopregnancies. The incidence of pseudopregnancy decreased if the olfactory bulbs were removed but not if body contact between animals was prevented. It was therefore suggested that an odour was responsible for the occurrence of pseudopregnancy.

Whitten (1959) confirmed the suppression of the cyclicity effect. Two initial

Table 10.1. Occurrence of oestrus in single or group caged mice (n = 30) (from Whitten, 1959).

Treatment	Mean frequency of oestrus between days 10 and 26 ± SE
Individually caged days 1–26	2.90 ± 0.14
Group caged days 27–52	0.57 ± 0.12
Group caged days 1–26	0.83 ± 0.15
Individually caged days 27–52	2.57 ± 0.24

Group vs. individual $P < 0.001$.

treatments, individual or group housed, were applied for 26 days and then they were reversed for a further 26 days. Vaginal smears were taken from each female for the last 16 days of each treatment phase (Table 10.1).

The prolonged cycles shown by the large group (n = 30) were induced within a relatively short time and suggested a powerful stimulus. The equally rapid reversal of the effect when housing conditions were changed indicated the ability to revert to cycles of normal duration and confirmed the potency of the 'off–on' stimulus (Table 10.2).

It was subsequently shown by Bronson and Whitten (1968) that the suppression of oestrous cyclicity was independent of auditory, visual or tactile cues and was due to chemical signals associated with female mouse urine (Albone, 1984).

The change in duration of the succeeding cycle following the introduction of a male questions the suggestion of Lee and Boot (1955, 1956) that the extended cycle length was due to pseudopregnancy, since pseudopregnancy duration is not altered by the presence of a male.

Whitten (1959) provided additional evidence to show that the suppressive effect of group housing was not due to overcrowding. He confined groups of female mice (n = 30) in cages and then introduced one male mouse, in a wire basket, into each cage. Vaginal smears were observed daily (Table 10.3).

Bruce (1959) presented another striking effect of a perceived odour. If a stud

Table 10.2. Duration of oestrous cycle in the presence and absence of males (from Whitten, 1958).

	Males present			Males absent		
Cycle number	1	2	3	4	5	6
Duration of cycle (days)	4.5	4.9	4.8	6.3	5.8	5.6
	Males absent			**Males present**		
Cycle number	1	2	3	4	5	6
Duration of cycle (days)	5.7	5.8	5.8	4.6	4.3	4.4

Table 10.3. Effect of introduction of a male on incidence of oestrus in grouped mice (from Whitten, 1959).

	Number of females exhibiting oestrus	
No. of females	on day male was introduced	within 4 days of introduction of male
30	2	24
30	1	26
30	1	23

male mouse was removed from a recently mated female and replaced with a male from a different strain, 70–80% of the newly mated females returned to oestrus within 7 days. In other words the consequences of the first mating were blocked. The effect was elicited even if body contact with the male was prevented (Bruce, 1960). Bruce and Parkes (1961) and Parkes and Bruce (1962) produced data which showed that a secretion deposited in the environment could also produce the blocking effect. They transferred newly mated female mice to cages which had previously been occupied by alien males. The blocking effect was similar to that which followed exposure to the males. Transfer to the alien male cage for a short period, however, produced a reduced effect. This suggested that a threshold concentration may be required. The females appear to be equally vulnerable during the first four days after mating but were less so on the fifth day when implantation occurs and the effect could not be induced on the sixth day.

The pregnancy block effect depends on the ability of the female to distinguish between a male of her own strain and an alien male. If the females are rendered anosmic the blocking effect does not occur, indicating once again that the effect is brought about by an odour which must be perceived by the recipient.

These examples demonstrate three effects:

1. A female–female oestrus cyclicity suppression.
2. A male effect resulting in apparent oestrus synchronization.
3. A male blocking effect of a genetically different male's mating attempts.

The apparent oestrus synchronization should perhaps be regarded as a male stimulatory effect since before introduction of the male all the females were in the same anoestrous state.

The original findings that odour cues could influence reproductive factors and success were obtained from studies with adult, sexually mature mice. Attention switched to pre-pubescent females and Vandenbergh (1973) provided evidence of a male pheromone which accelerated the onset of sexual maturity. He also suggested the existence of a female pheromone which inhibited the rate of puberty attainment. The acceleration effect was observed in both group and solitary housed females but was greatest for the individually housed mice (Table 10.4).

Again, the results indicated that there was a responsible factor associated with males and that the presence of the male is not necessary to elicit a response. It is clear, however, that males are able to exert their influence through their soiled bedding

Table 10.4. Male influences on the attainment of sexual maturity in mice (from Vandenbergh *et al.*, 1972).

	Mean age (days ± SE) at first oestrus	
	group caged females	singly caged females
Male present	39.6 ± 0.6	28.6 ± 0.6
Male soiled bedding	42.3 ± 1.0	33.0 ± 0.6
No Male	54.6 ± 2.9	35.9 ± 1.0

which suggests the deposition of a secretion on the material which retains its potency over a period of time.

The data in Table 10.4 also suggested that group caged juvenile mice may exert a delaying effect on the age at which puberty is attained. This was supported by unpublished data cited by Vandenbergh (1973) (Table 10.5).

The above data (cited by Vandenbergh, 1973) were obtained by Drickamer who published additional results (Drickamer and Hoover, 1979; Drickamer, 1982) which confirmed the suppressive effect of female group size on attainment of puberty and also showed that urine from singly caged oestrous females and pregnant lactating mice can accelerate puberty in young females and induce oestrus in adults.

Vandenbergh (1973) proposed a model of the interaction between adult male and female stimuli on pubertal development. With male and female stimuli present, a moderate rate of sexual maturation occurs but, in the absence of male stimuli, juvenile female stimulation reduces the rate of sexual maturation. Conversely, in the absence of stimuli from other females, male stimulation increases the rate of sexual maturation. He indicated that sexual maturation cannot be completely inhibited by female factors nor accelerated beyond a certain point by male factors.

Farm animal studies – sheep

Although these accounts were obtained from studies with laboratory mice, they heralded a period of intense study into pheromonal factors influencing reproductive success. Limited investigations into odour communication in farm livestock had taken place earlier but structured approaches integrating biological and chemical studies began around 1970. One of the earliest findings in farm animals came from Schinckel (1954a,b) who observed synchronization of lambing in Merino sheep and associated it with the shortening of oestrous cycles as the result of the introduction of adult rams into a flock of ewes previously isolated from male contact. While sheep

Table 10.5. Confirmatory evidence of Table 10.4 (from Vandenbergh, 1973).

Females/cage	Mean age (days ± SE) at first oestrus
1	38.0 ± 1.5
2	44.0 ± 1.5
4	54.3 ± 3.3
6	56.2 ± 1.3

may be regarded as seasonal breeders, this effect of synchronization is similar to the effect in segregated female mice exposed to a male (Whitten, 1959). Such similar responses removed doubts that laboratory mice were sufficiently different from domesticated farm species that the results of studies with them were of little or no significant value to the farmer.

It is now recognized that these odour effects do not have a direct influence on the behaviour of animals. Instead the stimulus elicits some physiological change hence these effects are said to be mediated by primer pheromones. The term pheromone was first proposed by Karlson and Luscher (1959) and was used to describe a substance secreted to the environment by an organism, which once perceived by another individual of the same species, elicited a specific reaction. In addition to the primer effect, the reaction could take the form of an immediate behavioural change. This type is now referred to as a releaser pheromone.

The awareness that a primer pheromonal effect existed in sheep awoke interest in chemical signalling in farm species. However, despite the known existence of the sheep primer pheromone, it was the identification of a releaser pheromone which provided the impetus for the next stages of olfactory research with farm animals.

Over the centuries man must have been aware of a temporal basis to mating activities in domesticated species. Unlike man and some of the other higher primates, adult members of the domesticated species do not indulge in mating attempts at intervals throughout the oestrous cycle. Mating attempts are concentrated into a short period. This suggests some form of communication which enables the male (or female) to recognize the appropriate physiological changes leading to sexual receptivity and seek out a partner from the social group. The interactions between two partners' courtship behaviour may then be structured to enable mating to take place at a time which maximizes the chances of fertilization.

Farm animal studies – pigs

A major breakthrough occurred when food scientists investigating a reported 'objectionable' odour in pig meat identified a compound isolated from the fat of sexually mature boars which was not detected in castrated or female pigs. The compound was an odoriferous C_{19}–Δ16 unsaturated androgen steroid ketone, 5α–androst–16–ene–3–one (Patterson, 1968a). A chance encounter between scientists in a laboratory led to the suggestion that the smell of the androgenic ketone was very similar to the odour generally present in the saliva of adult boars at mating time. The corresponding C_{19}–Δ16 steroid alcohol, 5α–androst–16–ene–3–ol was identified in much higher concentration in the boar submaxillary salivary gland (Patterson, 1968b). Reed *et al.* (1974) showed that the ketone and alcohol forms were equally effective in eliciting a standing behavioural oestrous response when applied in aerosol form to the heads of gilts in late pro-oestrus. These findings agreed with an androstenol theory of sex odours (Sink, 1967) and provided evidence that an androgenic steroid secreted in the saliva of sexually mature male pigs induced an immediate behavioural response in oestrous conspecifics. The extirpation of the submaxillary gland from 6-week-old entire male piglets renders the same boars incapable of eliciting the standing response from oestrous females when sexually mature (Perry *et al.*, 1972).

Signoret (1970) had previously recognized the role of olfaction in the reproductive behaviour of pigs when he observed that 60% of oestrous gilts which could not be induced to show a standing response, did so when placed in a boar's pen in the absence of any other visual auditory or tactile stimuli. Nosing and sniffing of the genitalia accounted for 47% of interactions between boars and oestrous gilts and a further 43% involved head-to-head contact. When anoestrous gilts and boars met, the equivalent figures were 53% and 27%. The increased level of interest shown by the boar to the head of the oestrous gilts seemed puzzling since it occurred at a time when one might have expected the male to focus his attention on the perineal region. The subsequent isolation of the androgenic steroids from the saliva of the boar, together with the evidence that it elicited an oestrous standing response (Perry *et al.*, 1972) offered an explanation for the boar's behaviour. The isolation, characterization and subsequent determination of the *in vivo* synthesis of the pheromone and its effect on the behaviour of a conspecific constitutes a full description of the first mammalian pheromone system to be identified.

The pro-oestrous gilt is seemingly aware of her changing physiological state and approaching sexual receptivity. When she encounters a boar she orientates towards him. Since she does not attempt to move away from his presence the boar then approaches her head and blows (breathes) into her face. The moist breath contains the androgen steroids and when the oestrous gilt recognizes this, she immediately adopts the standing posture. The elapsed time from initial head-to-head encounter to adoption of the standing posture may be as short as 5 seconds.

Cannulation of the boar's submaxillary salivary glands and the parotid salivary glands showed that whereas the flow of saliva from the parotid glands was stimulated by food and was then relatively high, the flow of saliva from the submaxillary gland was continuous and at a low rate. This suggested that the submaxillary gland acted as a reservoir for the releaser steroids. These observations support a convenient hypothesis since a mature boar must be prepared to test any female he meets and cannot wait for the steroids to be synthesized or transported from another site. The continuous secretion also offers explanation for the accumulation of the steroids in the fat tissues of the sexually mature male. In a subsequent report Perry *et al.* (1980) showed that extirpation of the submaxillary glands from young boars induced other behaviour changes in the boars when mature and appeared to reduce their libido.

Biochemical studies have shown that the C_{19}–Δ 16 steroids are formed in the pig testis. The 3α–hydroxy steroid is transported via the blood to the submaxillary salivary gland. Following sexual arousal, blood levels of this steroid and of testosterone increase synchronously leading to a long-term elevation of the ketone form in fatty tissues. Gennings *et al.* (1974) reported that the ketone is rapidly metabolized *in vitro* to the 3α– alcohol in 75% yield by sow epithelial tissue. This would explain the variations noted in the breath of aroused boars by Perry *et al.* (1980) and also account for the findings of Reed *et al.* (1974) that both the alcohol and ketone forms were equally effective in their ability to elicit the standing response, since behavioural adaptation to both odours by the oestrous female would be desirable in order to secure a mate.

Mechanisms for detection

Identification of a secretion and association with a rapid behavioural change in a conspecific suggests involvement of the central nervous sytem. Attention has focused on the olfactory epithelium and the vomeronasal organ (VNO). The VNO is well developed in most mammals including the human fetus (Gasser, 1975) and recently evidence has been produced which suggests that the human does possess a structure which resembles a VNO (Moran *et al.*, 1991).

Anatomical studies have shown that receptors in the VNO have access through the accessory olfactory bulb (AOB) to a system of central neural connectors which are not available to the olfactory neuroepithelium proper (Raisman, 1972; Broadwell, 1975; Heimer, 1975; Scalia and Winans, 1976). The accessory olfactory bulb is connected to the hypothalamus which is involved in the generation of the cyclic pre-ovulatory surge of gonadotrophins in rodents (Raisman, 1972).

Attempts to differentiate between the VNO and olfactory epithelium have not always provided clear-cut evidence since the techniques used to isolate or block one area, e.g. zinc sulphate infusion or surgical scraping of the cribiform plate or bulbectomy, do not necessarily leave the other area fully intact and functional. Nevertheless, there appears to be a mechanism whereby molecules can be recognized at receptor sites and which is associated with the higher centres of the nervous system.

The existence of such a system may explain why most mammals have evolved a behavioural response known as the Flehmen. Most often this involves an upper lip curl and neck extension and an inhalation. Samples of inspired air would then reach the VNO located in the upper air passages. The technique may be compared to that of a wine taster who takes a sample of wine into the mouth and draws air over it. The resultant effect is a much wider spectrum of perceived flavours when compared with smell of material in the glass or a taste. In reality, the wine taster is behaving like a biological gas chromatograph and the prolonged passage of air over the material in the mouth is removing molecules according to their molecular weight and/or volatility. It is convenient to assume that molecules liberated into the air possess volatile characteristics but polar compounds may be in aqueous solution when inspired and could be conveyed to receptor sites.

The Flehmen response is often seen when animals investigate each other during courtship. However, Stevens *et al.* (1982) concluded that the Flehmen was not a reliable indicator of an oestrous signal since rams showed this behaviour in response to urination by non-oestrous ewes. They suggested that the Flehmen behaviour is a general olfactory investigation mechanism. Their results did suggest that rams spent significantly more time investigating oestrous ewes than metaoestrous or ovariectomized ewes which suggested they were aware of the changed physiological status of the oestrous ewes. The earlier studies with mouse primer pheromones had clearly implicated urine as the carrier medium for information but the numbers of urinations by oestrous ewes were significantly lower than that for non-oestrous controls, implying that urine might not be the vehicle for chemical information. Stevens (1983) studied the effects of urine, vaginal mucus and skin material from oestrous and non-oestrous sheep but was unable to elicit similar levels of male interest previously demonstrated to tethered oestrous females accessible to olfactory investigation. Washing or covering the perineal region of oestrous ewes did, however, reduce ram interest and she concluded that skin secretions from the posterior vagina and/or perineal skin surface were possible sites for the source of secretion which indicates the stage of the oestrous cycle.

Farm animal studies – cattle

These observations supported data from field studies with cattle which showed that the bull monitors all the cows within a herd by olfactory investigation and apparently identifies impending oestrus by demonstrating a significant increase in olfactory investigation time as early as four days before standing oestrus. Very little mounting behaviour took place before the day of standing oestrus and Stevens suggested that the cow might release two types of signal: the first alerted the bull to changing physiological state and impending oestrus and the second acting as a mount releaser.

Following the suggestion of Stevens (1983) that the likely source(s) of the signals might be the posterior vagina or perineal skin region, Blazquez *et al.* (1987a,b) undertook a histological and histochemical survey of the reproductive tract caudal to the cervix but failed to identify any sites thought capable of acting as specialized secretory glands. Attention switched to the vulval skin area (Blazquez *et al.*, 1987c) where differences were identified in the sweat and sebaceous glands when compared with similar glands in the neck region or other mucocutaneous sites such as the lips or eyelids. They showed that the vulval sweat and sebaceous glands were significantly greater in volume than those from the other regions and that adult cows possessed glands of greater volumes than sexually immature heifers or steers. Their results showed that the adult morphology was present in 15- and 18-month-old but not in 6- and 9-month-old heifer calves.

Blazquez *et al.* (1987d) gave sexually immature female calves implants of oestradiol–17β and observed that sebaceous gland volume in the perineal region was significantly greater than that of control animals, whereas similar glands in the neck were significantly smaller. Oestrogen administration is recognized to reduce sebaceous gland size (Strauss *et al.*, 1983), therefore the observed increase of gland volumes in the perineal region suggested a specialized type of sebaceous gland. Mykytowycz (1970) had previously described specialized skin glands and their role in chemical communication.

In controlled behavioural tests, Blazquez *et al.* (1988) were able to demonstrate that induction of sweat gland discharge in mid-cycle cows through the local administration of adrenaline significantly increased the incidence of bull olfactory behaviour.

The involvement of true sweat glands in oestrous signalling would be expected to lead to an increase in bull olfactory investigation in high temperature countries. Detailed field observations (French *et al.*, 1989), however, produced no evidence of such an effect. In an attempt to investigate temperature and sweat gland relationships, Blazquez *et al.* (in preparation) measured cutaneous moisture evaporation (CME) rates from the lumbar dorsal and perineal regions of dioestrous adult cows and bulls at two different ambient temperatures of 10°C and 36°C dry bulb. In the thermoneutral environment there was no difference between the CME rates of the two regions in females. Following exposure to the high temperatures, the lumbodorsal CME rates in males and females and the scrotal region in males showed a significant rise but there was only a slight rise in the CME rate from the perineal region of the cows, thus confirming the field observation that high ambient temperatures did not result in an increased secretion containing oestrous signals. In a second experiment perineal and lumbo-dorsal CME rates of two cows were measured from day −7 to day +1 of the oestrous cycle. The perineal CME reached a maximum on the day of oestrus in both cows but there was no effect of stage of cycle on lumbo-dorsal CME rate.

These results confirm the earlier suggestions that the perineal sweat and sebaceous glands are specialized and support the hypothesis that they are the source of the secretion which indicates the cow's physiological stage of oestrus.

The development of intensive management systems for farm livestock has taken place without recognition that primer pheromonal effects may influence production traits. Segregation of replacement female breeding stock is often practised in pig production and may delay the onset of puberty. It has been shown by Kirkwood and Hughes (1979) that the introduction of a mature boar can hasten the onset of puberty and synchronize pubertal oestrus as long as the gilts are between 160 and 170 days of age when the boar stimulus is introduced.

The management of large herds of adult cows in the absence of a bull makes the detection of oestrus difficult for the stockperson. Williamson *et al.* (1972) found that 21% of cows in a herd of 52 animals did not exhibit standing behaviour even though there was positive evidence from a heatmount detector and a positive ovarian follicle. These cows were under continuous observation. A review by Esslemont and Bryant (1974) produced a list of eight publications in which the incidences of silent heats as a proportion of all heats observed ranged from 4 to 43.4%. This may be evidence for the suppression of behavioural oestrus in all female groups, although there are no data to indicate whether such an effect is mediated by odour or by some other social factor.

One possible reason for the relatively poor oestrus detection rate in dairy herds was identified by French *et al.* (1989). They conducted detailed observations of a large dairy herd running with a bull. The herd consisted of 113 cows of which 73 were known to be pregnant. The remaining 40 non-pregnant cows included those which had been inseminated but had failed to conceive, cows that had not been seen in oestrus since calving and cycling cows which were to be sold at the end of the lactation. Approximately 150 behavioural activities were recorded and from the subsequent principal axis factor analysis, groups of activities were identified. By retrospectively plotting these activities against the day of standing behaviour, the behaviour most often used by the farmers to identify oestrus, it appeared that the bull was able to identify pro-oestrous cows four days before standing behaviour occurred. In contrast, cows did not appear to increase their attention in pro-oestrous cows until 2 days before oestrus. This suggests a sexual dimorphism in the ability to detect oestrous odours which may be dependent upon the concentration of the secretion. In practical terms this reduces the period over which cows will demonstrate interest in the target animal.

Mother–young recognition and acceptance is a well-known example of the role of odours in some farm species. The difficulties experienced in fostering lambs are well recognized by shepherds who often resort to removing the skin from a dead lamb and putting it over an orphan lamb in order to persuade the bereaved mother

Table 10.6. A tentative scheme for the composition of flavours (from Solms, 1971).

Type of material	Perceived by
Low boiling point compounds and high boiling point compounds with odour effects	Odour
Non-volatile compounds with taste and tactile effects	Taste

to accept it. In parts of South America this practice is exploited for wool production. Newborn male vicunas are covered with the skin of dead alpacas and fostered on to a female alpaca to which the vicuna becomes sexually imprinted. When it matures it mates with a female alpaca and the resulting hybrid produces vicuna wool in alpaca quantities (Hodge, 1946).

Possible Role for Taste

The data of French *et al.* (1989) also confirmed observations made by Stevens (1983) that as the time of standing oestrus approached, the bull also began to lick the perineal region of the cow. The combination of olfactory and gustatory behaviours might suggest the presence of more than one secretion. The first material may be volatile and attract the attention of the bull and then the spectrum of compounds changes so that later secretions contain non-volatiles which are detected through taste or tactile effects. Solms (1971) suggested a tentative flavour scheme which could develop as the more volatile compounds were replaced (Table 10.6).

Taste perception

By definition, tastes are limited to oral sensations of sweet, sour, salt and bitter (Moncrieff, 1967). These sensations are direct taste effects. In man it has been shown that there are different thresholds for the detection of each of the sensations. For example, bitter sensations are detected at lower molecular concentrations than sweet ones (von Skramlik, 1963). For example 10^{15} molecules of strychnine on the tongue are sufficient to elicit a bitter sensation compared with 10^{20} molecules of sugar which elicit a sweet sensation. In man, the bitter taste is often associated with harmful compounds, exemplified by alkaloids, and since some animals are known to reject poisonous plants it could be assumed that a similar order of ranking exists in threshold detection levels.

Both olfaction and taste rely on the recognition of specific molecules and this process must operate quickly, which suggests the involvement of the higher regions of the central nervous system.

The mammalian taste buds are located on the tongue. Receptor cells are grouped around a pore and the whole structure constitutes a taste bud. The taste cells have a very short life of a few days (Beidler, 1963) and are innervated by approximately

50 nerves entering the base of the bud and splitting to provide 200 or more branches. Very few cells are not in anatomical contact with the nerve fibres (Beidler, 1970). An individual taste bud will respond to one of the four taste sensations and specific groups of taste buds for each of the taste sensations are localized in particular regions of the tongue. The mean number of taste buds varies from 24 in the chicken (Lindenmaier and Kare, 1958) to 25 000 in the calf (Weber *et al.*, 1961). However, the actual number of buds does not appear to correlate well with the ability to discriminate since the cow does not respond behaviourally to chemicals which elicit strong rejection in chickens (Kare and Ficken, 1963).

It has been shown that taste buds develop *in utero* in the sheep and by 14 weeks of gestation are morphologically similar to adult taste buds (Bradley and Mistretta, 1975). This finding raises the question of the function of taste in animals, and in particular the need for mature taste buds in a neonate that relies on maternal milk for its nutrients at a time when the ability to discriminate between sources of food is not required.

Taste buds could help to regulate food intake on a quantitative basis by monitoring the molecular concentration. However, it is recognized that animals can regulate their caloric intake over a wide range of dietary dilutions. It has also been shown that despite direct manipulation of the taste of food through the addition of dextrose (Teitelbaum, 1955) or quinine hydrochloride (Teitelbaum and Epstein, 1962), rats maintained their caloric intake although the sweet and bitter thresholds have been much altered. Additionally, caloric intake regulation continues in the absence of taste and smell (Bellows and van Waggenen, 1939).

The role of taste, therefore, is unclear. It is known that there are clear differences between species in taste responses (see review by Kare and Ficken, 1963). Within species, wide variations in taste thresholds have been noted and most of the variation is of the continuous type, although there are specific instances in which one animal preferred and another rejected a particular material (Ficken and Kare, 1961).

It is possible that apparent responses based on taste may be due to some other feature of the material. For example, some grazing species will reject particular grasses that are hairy and coarse.

Physiological factors within the animal may also influence taste thresholds or preferences. Bell (1963) described studies with goats in which the sodium taste threshold was lowered by increasing the sodium content of the plasma.

Evidence of taste preferences and thresholds in farm animal species supports the earlier statement that bitter materials generally have the lowest thresholds concentration for acceptance. Data from studies with cattle and calves, sheep and goats, pigs and chickens have been published by Hafez (1962) and describe species preferences and dislikes but they do not mention a phenomenon sometimes reported by farmers in which an animal maintained on a diet of constant composition will voluntarily reduce its intake or reject the food. A subsequent change in dietary composition sometimes leads to a restoration to previous intake levels. This suggests a satiation or habituation effect associated with taste and highlights the need for more detailed studies.

References

Albone, E.S. (1984) *Mammalian Semiochemistry*, Wiley & Sons, Chichester and New York.

Beidler, L.M. (1963) Dynamics of taste cells. In: Zotterman, Y. (ed.), *Olfaction and Taste, Proceeding of the 1st International Symposium*, Stockholm, 1962, pp. 133–145.

Beidler, L.M. (1970) Physiological properties of mammalian taste receptors. In: Wolstenholme, G.E.W. and Knight, J. (eds), *Taste and Smell in Vertebrates*. A Ciba Foundation Symposium, London, 1969, pp. 51–67.

Bell, F.R. (1963) The variation in taste thresholds of ruminants associated with sodium depletion. In: Zotterman, Y. (ed.), *Olfaction and Taste. Proceedings of the 1st International Symposium*, Stockholm, 1962, pp. 299–307.

Bellows, R.T. and van Waggenen, W.P. (1939) The effect of resection of the olfactory, gustatory and trigeminal nerves on water drinking in dogs without and with diabetes insipidus. *American Journal of Physiology*, **126**, 13–19.

Blazquez, N.B., Batten, E.H., Long, S.E. and Perry, G.C. (1987a) Histology and histochemistry of the bovine reproductive tract caudal to the cervix. Part I. The vestibule and associated glands. *British Veterinary Journal* **143**, 328–337.

Blazquez, N.B., Batten, E.H., Long, S.E. and Perry, G.C. (1987b) Histology and histochemistry of the bovine reproductive tract caudal to the cervix. Part II. The vagina and associated structures. *British Veterinary Journal* **143**, 337–343.

Blazquez, N.B., Batten, E.H., Long, S.E. and Perry, G.C. (1987c) A quantitative morphological examination of bovine vulval skin glands. *Journal of Anatomy* **155**, 153–163.

Blazquez, N.B., Long, S.E., Perry, G.C. and Watson, E.D. (1987d) Effect of oestradiol–17β on perineal and neck skin glands in heifer calves. *Journal of Endocrinology* **115**, 43–46.

Blazquez, N.B., French, J.M., Long, S.E. and Perry, G.C. (1988) A pheromonal function for the perineal skin glands in the cow. *Veterinary Record* **123**, 49–50.

Blazquez, N.B., Long, S.E., Mayhew, T.M., Perry, G.C., Prescott, N.J. and Wathes, C.M. Perineal skin glands in the cow: sweat gland discharge and morphometry (submitted for publication).

Bradley, R.M. and Mistretta, C.M. (1975) The developing sense of taste. In: Denton, D.A. and Coghlan, J.P. (eds), *Olfaction and Taste V. Proceedings of the 5th International Symposium*, Melbourne, 1974, pp. 91–98.

Broadwell, R.D. (1975) Olfactory relationships of the telencephalon and diencephalon in the rabbit. I. An autoradiographic study of the efferent connections of the main and accessory bulbs. *Journal of Comparative Neurology* **163**, 329–345.

Bronson, F.H. and Whitten, W.K. (1968) Oestrus-accelerating pheromone of mice; assay, androgen-dependency and presence in bladder urine. *Journal of Reproduction and Fertility* **15**, 131–134.

Bruce, H.M. (1959) An exteroceptive block to pregnancy in the mouse. *Nature (London)* **184**, 105.

Bruce, H.M. (1960) A block to pregnancy in mice caused by the proximity of strange males. *Journal of Reproduction and Fertility* **1**, 96–103.

Bruce, H.M. and Parkes, A.S. (1961) An olfactory block to implantation in mice. *Journal of Reproduction and Fertility* **2**, 195–196.

Drickamer, L.C. (1982) Delay and acceleration of puberty in female mice by urinary clonosignals from other females. *Developmental Psychobiology* **15**, 433–445.

Drickamer, L.C. and Hoover, J.E. (1979) Effects of urine from pregnant and lactating female house mice on sexual maturation of juvenile females. *Developmental Psychobiology* **12**, 545–551.

Esslemont, R.J. and Bryant, M.J. (1974) Economic and husbandry aspects of the manifestation and detection of oestrus in cows. Part II. The manifestation of oestrus in cattle *Agricultural Development and Advisory Service Quarterly Review* **13**, 19–26.

Ficken, M.S. and Kare, M.R. (1961) Individual variation in the ability to taste. *Poultry Science* **40**, 1402.

French, J.M., Moore, G.F., Perry, G.C. and Long, S.E. (1989) Behavioural predictors of oestrus in domestic cattle, *Bos taurus*. *Animal Behaviour* **38**, 913–919.

Gasser, R.F. (1975) *Atlas of Human Embryos*. Harper and Row, New York.

Gennings, J.N., Gower, D.B. and Bannister, L.H. (1974) Studies on the metabolism of the odoriferous ketones, 5–androst–16–ene–3–one and 4,16–androstadien–3–one by the nasal epithelium of the mature and immature sow. *Biochimica and Biophysica Acta* **369**, 294–303.

Hafez, E.S.E. (ed.) (1962) *The Behaviour of Domestic Animals*. Baillière, Tindall and Cox, London.

Heimer, L. (1975) Olfactory projections to the diencephalon. In: Stump, F.W.E. and Grant, L.D. (eds), *Anatomical Neuroendocrinology*. Kargan, Basel, pp. 30–39.

Hodge, W.H. (1946) Camels of the clouds. *National Geographic Magazine* **89**, 641–656. Cited by Stoddard, D.M. (1980) *The Ecology of Vertebrate Olfaction*, Chapman & Hall, London and New York, p. 203.

Kare, M.R. and Ficken, M.S. (1963) Comparative studies on the sense of taste. In: Zotterman, Y. (ed.), *Olfaction and Taste. Proceedings of the 1st International Symposium*, Stockholm, 1962, pp. 285–297.

Karlson, P. and Luscher, M. (1959) 'Pheromones': a new term for a class of biologically active substances. *Nature (London)* **183**, 155–156.

Kirkwood, R.N. and Hughes, P.E. (1979) The influence of age at first boar contact on puberty attainment in the gilt. *Animal Production* **29**, 231–238.

Lee, S. van der and Boot, L.M. (1955) Spontaneous pseudopregnancy in mice. *Acta Physiologie Pharmacologie Neer* **4**, 442–443.

Lee, S. van der and Boot, L.M. (1956) Spontaneous pseudopregnancy in mice. II. *Acta Physiologie Pharmacologie Neer* **5**, 213–215.

Lindenmaier, P. and Kare, M.R. (1958) The taste end-organs of the chicken. *Poultry Science* **38**, 545–550.

Moncrieff, R.W. (1967) *The Chemical Senses*. Leonard Hill, London.

Moran, D.T., Rowley, J.C., Jafek, B.W., Stensaas, L.J. and Berliner, D.L. (1991) Frequency of occurrence and ultrastructure of the vomeronasal (Jacobson's) organ in man. *Proceedings of Chemical Signals in Vertebrates VI*, University of Pennsylvania, Philadelphia, Penn., USA, p. 82.

Mykytowycz, R. (1970) The role of skin glands in mammalian communication. In: Johnson, J.W.Jr, Moulton, D.G. and Turk, A. (eds), *Advances in Chemoreception*, vol. 1. Appleton–Century–Crofts, New York, pp. 327–360.

Parkes, A.S. and Bruce, H.M. (1962) Pregnancy block in female mice placed in boxes soiled by males. *Journal of Reproduction and Fertility* **4**, 303–308.

Patterson, R.L.S. (1968a) 5–androst–16–ene–3–one; compound responsible for taint in boar fat. *Journal of Science in Food Agriculture* **19**, 31–38.

Patterson, R.L.S. (1968b) Identification of 3–hydroxy–5–androst–16–ene as the musk odour component of boar submaxillary gland and its relationship to the sex odour taint in pork meat. *Journal of Science in Food Agriculture* **19**, 434–438.

Perry, G.C., Patterson, R.L.S. and Stinson, C.G. (1972) Submaxillary salivary gland involvement in porcine mating behaviour. *7th International Congress of Animal Reproduction and Artificial Insemination*, Munich, pp. 395–399.

Perry, G.C., Patterson, R.L.S., Macfie, H.J.H. and Stinson, C.G. (1980) Pig courtship behaviour: pheromonal property of androstene steroids in male submaxillary secretion. *Animal Production* **31**, 191–199.

Raisman, G. (1972) An experimental study of the projection of the amygdala to the accessory olfactory bulb and its relationship to the concept of a dual olfactory system. *Experimental Brain Research* **14**, 395–408.

Reed, A.C.B., Melrose, D.R. and Patterson, R.L.S. (1974) Androgen steroids as an aid to the detection of oestrus in pig artificial insemination. *British Veterinary Journal* **130**, 61–67.

Scalia, F. and Winans, S.S. (1976) New perspectives on the morphology of the olfactory system: olfactory and vomeronasal pathways in mammals. In: Doty, R.L. (ed.), *Mammalian Olfaction, Reproductive Processes and Behaviour*. Academic Press, New York, pp. 7–28.

Schinckel, P.G. (1954a) The effect of the ram on the incidence and occurrence of oestrus in ewes. *Australian Veterinary Journal* **30**, 189–195.

Schinckel, P.G. (1954b) The effect of the presence of the ram on the ovarian activity of the ewe. *Australian Journal of Agricultural research* **5**, 465–469.

Sherrington, C.S. (1906) *The Integrative Action of the Nervous System*. Yale University Press.

Signoret, J.P. (1970) Reproductive behaviour of pigs. *Journal of Reproduction and Fertility* (Suppl. 11), 105–117.

Sink, J.D. (1967) Theoretical aspects of sex odor in swine. *Journal of Theoretical Biology* **17**, 174–180.

Skramlik, E. von (1963) The fundamental substrates of taste. In: Zotterman, Y. (ed.), *Olfaction and Taste, Proceedings of the 1st International Symposium*, Stockholm, 1962, pp. 125–132.

Solms, J. (1971) Non-volatile compounds and the flavor of foods. In: Ohloff, G. and Thomas. A.F. (eds), *Gustation and Olfaction*. Academic Press, New York and London, pp. 92–109.

Stevens, K. (1983) The role of olfaction during oestrus detection in sheep and cattle. Unpublished PhD Thesis, University of Bristol.

Stevens, K., Perry, G.C. and Long, S.E. (1982) Effect of ewe urine and vaginal secretions on ram investigative behaviour. *Journal of Chemical Ecology* **8**, 23–29.

Strauss, J.S., Downing, D.T. and Ebling, F.J. (1983) Sebaceous glands. In: Goldsmith, L.A. (ed.), *Biochemistry and Physiology of the Skin*, Oxford University Press, Oxford, pp. 569–595.

Teitelbaum, P. (1955) Sensory control of hypothalamic hyperphagia. *Journal of Comparative and Physiological Psychology* **51**, 135–141.

Teitelbaum, P. and Epstein, A.N. (1962) The lateral hypothalamic syndrome recovery of feeding and drinking after lateral hypothalamic lesions. *Psychological Review* **69**, 74–90.

Vandenbergh, J.G. (1973) Acceleration and inhibition of puberty in female mice by pheromones. *Journal of Reproduction and Fertility* (Suppl. 19), 411–419.

Vandenbergh, J.G., Drickamer, L.C. and Colby, D.R. (1972) Social and dietary factors in the sexual maturation of female mice. *Journal of Reproduction and Fertility* **28**, 397–405.

Weber, W., Davies, R.O. and Kare, M.R. (1961) Unpublished data cited in Kare, M.R. and Ficken, M.S. (1963) *Proceedings of the 1st International Symposium*, Stockholm, 1962, p. 295.

Whitten, W.K. (1958) Modification of the oestrous cycle of the mouse by external stimuli associated with the male. Changes in the oestrous cycle determined by vaginal smears. *Journal of Endocrinology* **17**, 307–313.

Whitten, W.K. (1959) Occurrence of anoestrus in mice caged in groups. *Journal of Endocrinology* **18**, 102–107.

Williamson, N.B., Morris, R.S., Blood, D.C., Cannon, C.M. and Wright, P.J. (1972) A study of oestrous behaviour and oestrus detection methods in a large commercial dairy herd. *Veterinary Record* **91**, 58–62.

11 Tactile and Pain Perception

A. LIVINGSTON, S. LEY and A. WATERMAN

Departments of Pharmacology and Veterinary Surgery, University of Bristol, UK

Summary

The relationship between animals and tactile and pain stimuli is very complex and in some ways confused by human experience. We know from electrophysiological studies that mammals have the same spectrum of peripheral receptors in terms of touch, heat, pressure, chemical and polymodal receptors as humans and we know that the wiring of the peripheral, spinal and brain pathways for these stimuli is similar in all species, but at this point we come across the first problem. That is, that pain is a perception, and it does not exist as such until the higher centres of the brain perceive it as pain. This can move us into the realms of philosophy, with questions such as: Do you perceive pain in the same way as I do? It is best to leave philosophy at this point, but we must always bear these concepts in mind when we consider animal pain. Since we are dealing with an experience, a good general rule is that if a stimulus is painful to man then it is reasonable to believe that it is painful to animals. At first glance this may seem obvious, but if we look further at the situation it is not completely so.

Since the lack of communication makes it impossible to relate or quantify the pain as we can with fellow humans, we are forced to rely on the animal's response to the painful stimulus. Here we have our first problems, since different species respond quite differently to pain. The best example of this is the degree of difference in response between herd ruminants and carnivores, which is probably related to survival and selection pressure in the ruminant. A second problem associated with assessment of pain in animals is that responses to acute and chronic pain may be quite different, and there may be a neurological basis for this. Finally, we are faced with the problems of dissociating the responses to stress from those to pain, since it is possible to have a stress response without pain, but it is debatable whether one can have pain without stress. Despite these difficulties, progress has been made in the study of the concept of pain in animals and also its alleviation, and some of these findings will be described.

Perceptions

The relationship between *tactile* and *pain perception* would appear to be a very close one, and in some cases would appear to be one of degree. For instance the perception of warmth can develop into the painful stimulus of burning or the perception of touch can develop into the painful stimulus of pinching or squeezing. The *receptors* at the periphery of the nervous system which provide the initial detection of the stimulus are of different kinds and can respond to different stimuli, however, and although the progression from warmth to burning would seem to be a gradual development, in fact more than one type of receptor may be involved. This can be seen by applying a gradually increasing temperature to the skin which will suddenly become painful at a particular temperature. This is the *threshold for pain* and represents the point at which *nociceptors* (the term applied to pain receptors) become activated. This, of course, only represents the input side of the system and the nervous system can integrate and modify this input in a variety of ways. The activation of the pain receptor results in a series of impulses being passed up a sensory nerve fibre to the central nervous system. These sensory nerve fibres associated with pain are usually very fine fibres either of the myelinated *Aδ type* or a slower conducting unmyelinated fibre known as a *C fibre*. For most of the body these fibres pass through the dorsal roots of the spinal cord and into the spinal cord itself. In many instances the information is then passed through the spinal cord to the motor neurones which activate skeletal muscles in the region associated with the pain stimulus to produce what is known as a *reflex arc* which produces the reflex withdrawal of the body from the pain producing stimulus. That is, if a foot is damaged the foot is rapidly removed from the stimulus without *conscious thought* and without involvement of the higher centres of the brain or any perception of the pain as such, hence the damage can be limited almost before the pain is felt. However, as we know, the pain is eventually felt and what has happened in the spinal cord is that in addition to the messages sent to the skeletal muscles, messages are also sent up the spinal cord to the brain (Fig. 11.1). The first point in the spinal cord where this splitting of the messages takes place is a region known as the lamina II (or *substantia gelatinosa*) of the *dorsal horn* of the grey matter in the *spinal cord* and this represents the first processing of the *pain information*. As the message passes up through the brain there are many other regions such as the *medial thalamus*, the *periaqueductal grey* and the *locus coeruleus* where further processing of the information takes place. These regions pass information to other areas of the brain which produce the ancillary responses to the pain stimulus such as effects on emotions, blood pressure and hormones. Eventually, the information reaches the cortex and it is at this level that the actual perception of pain takes place. All the regions of the nervous system below this level are merely concerned with the passing of information associated with the *noxious stimulus* and pain as such is only experienced when the messages reach the *cortex*. This is not the end of the story because the various regions of the brain are capable of modifying the noxious stimulus information locally and also by sending nerve fibres which descend back down to the spinal cord which can modulate the information passing up from the injured area. Knowledge of the chemical transmitter agents involved in the passage of

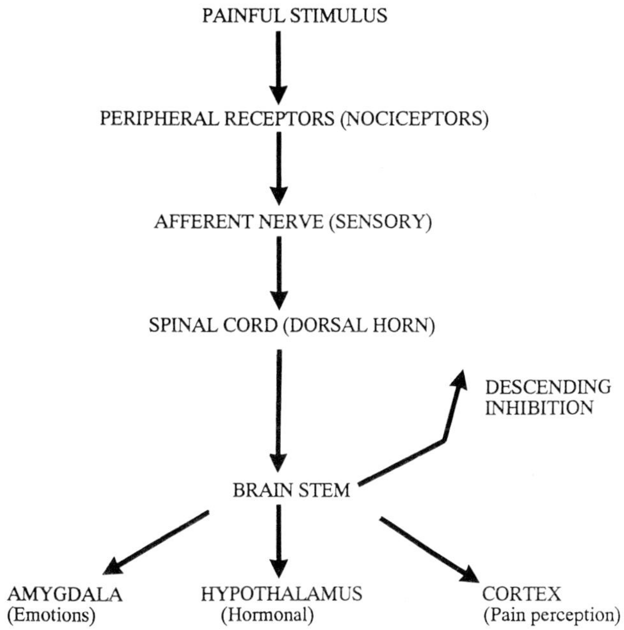

Fig. 11.1. The pathway followed by stimuli from the source to the brain.

information in these modulating systems allows us to use drugs which will themselves cause modification of the pain input and hence help us control pain as a perception.

For the majority of animal species, certainly for man and the farm animals, the arrangement and structure of the nervous system as described is highly conserved, that is the '*wiring*' of the *nervous system* appears almost identical, apart from slight species variations in the relative importance of the various chemical transmitters of the nerves involved in the system. This allows us to make a general, but very important, assumption with regard to the processing of pain information in farm animals. That is, if an experience causes pain in man, it is reasonable to expect it to cause a similar pain experience in the animal. Of course, it is possible to dispute such a statement on several grounds:

1. The fact that most animals have relatively less cortical tissue in their brains than man, but there is no evidence that the regions associated with pain perception are under-represented in these areas.
2. Many responses seen, in the absence of verbal corroboration, may merely represent a *spinal reflex* and not involve the higher centres; however we know the pathways to the higher centres exist in the same way as in man and, therefore, the assumption that they are there to serve a function can be made. Furthermore, animals rapidly learn to avoid potentially painful stimuli, therefore they must have perceived the stimulus as unpleasant or harmful.
3. The proposition that animals do not show the same overt response to pain as man suggests that they do not perceive it as unpleasant. For instance man will often show *vocalization*. This proposition will be considered in more detail later.

One final aspect in the perception of pain in both animals and man which must be considered is that of the *time* modality. The concepts of *acute* and *chronic pain* have received a great deal of attention over the last few years since there seem to be some fundamental differences in the processing by the nervous system of these two parameters. Our experiences with human sufferers from chronic pain have shown that several of the concepts which apply to acute pain are not directly applicable in the chronic condition. For instance, it is not always possible to demonstrate clear lesions and hence discrete peripheral stimulation of receptors in chronic pain states, nor is it always possible to alleviate the pain by equivalent doses of pain killing drugs which we know stop acute pain. There is accumulating evidence that the presence of chronic pain can lead to changes in the central and peripheral nervous systems which may represent the body's attempts to adapt to the continual presence of the stimulus. Studies in laboratory animals (Woolf, 1983), farm animals (Ley *et al.*, 1989) and man (McQuay *et al.*, 1988) have all indicated that a degree of plasticity is present in the nervous system which may lead to changes in perception in the chronic pain state.

Responses

In terms of the human *appreciation of pain*, the actual response to the pain is often, perhaps misguidedly, used as the indicator of the *severity of pain*; even so it is generally accepted that considerable variation in response can occur, either with individuals who tolerate pain undemonstratively such as boxers, or that expectation of the pain stimulus can diminish the overt response to it. However, what is often completely misunderstood is the enormous variation in behavioural response which can take place between different species. If one looks at man, then the response could be classed as demonstrative, with vocalization and overt behavioural actions. The same applies to dogs, cats and pigs all of whom show vocalization and clear behavioural responses (either aversive or aggressive) in response to an acute stimulus and also clear behaviours in the presence of chronic pain. If this pattern is then

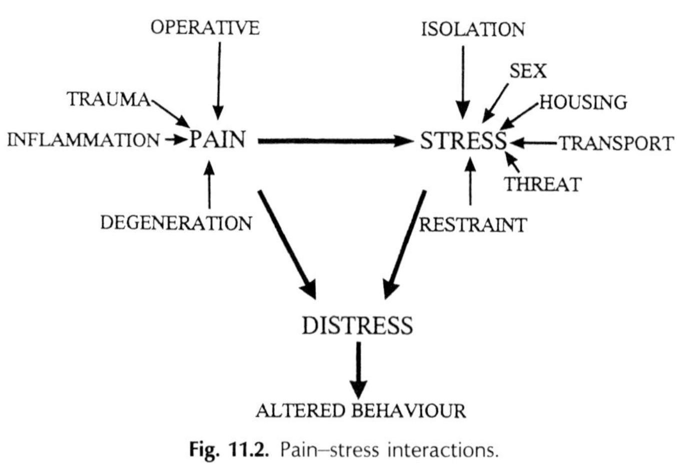

Fig. 11.2. Pain–stress interactions.

compared to sheep, goats and cattle then a clear division is seen: vocalization in response to acute pain is not often demonstrated by ruminants in pain and very often the behavioural responses such as aggression are muted; horses and donkeys tend to fall between the two groups in their activity. The other major division between these groups of animals is that the 'non-demonstrators' tend to be the herbivorous herd animals while the 'demonstrators' tend to be either the carnivorous predators or the relatively unchallenged omnivores. Thus it is possible that the overt demonstration of pain is a luxury afforded only by those species which are not subjected to predation. The herbivorous herd animals could find it a distinct advantage in terms of survival not to demonstrate unusual behaviour patterns within the herd which would attract the attention of predators roaming the periphery of the herd whose attentions usually fall on an individual behaving in some unusual manner. However, this theory does not answer the question, do they not show the behaviour because they have been selected not to, or do they have some intrinsic mechanism, perhaps involving endogenous opioids, which means that they do actually feel less pain? With regard to the latter suggestion all we can offer at the moment is the observation that in terms of threshold to mechanical pressure the threshold is very similar in man, sheep and horses, at about five Newtons of force (Ley *et al.*, 1989, Chambers *et al.*, 1990) which would support the proposition that the farm animals have been selected not to demonstrate pain and that they actually do perceive it in a similar manner.

This leads on to another aspect of pain perception and that is the interaction with what is generally known as stress (see Chapter 15). While it is clearly possible to have stress in the absence of pain in farm animals, under conditions such as transport, isolation and batch mixing, it is not clear if one can predict the occurrence of pain without concurrent stress (Fig. 11.2). While acute pain would be expected to cause stress, it is not clear if chronic pain will invariably be accompanied by stress. One of the major problems with this situation is how to try to evaluate these situations in general and stress in particular. Using models such as isolation, it is possible to measure various hormonal and physiological parameters whose changes can be associated reliably with the onset of stress. However, it is important in these studies to ensure that the observation and sampling of the animal do not themselves cause stress. When we transfer these studies to acute pain investigations we have to be particularly careful that the measurement techniques themselves do not cause stress rather than the presence of pain stimulus. When these studies are extended to look at chronic pain conditions the picture becomes more complicated since we have already suggested that it is conceivable that chronic pain and stress are not so closely linked. One example of the difficulty in this area is the concentrations of *plasma cortisol* found in sheep. Several authors (Jephcott *et al.*, 1987, 1988) have shown that following procedures which should cause acute pain or stress in sheep, there is a clear rise in plasma cortisol concentration which is in keeping with the results of other species. However, when plasma cortisol concentrations were measured in sheep suffering chronic pain there was a significant depression in the concentration (Ley *et al.*, 1991b). Thus dependence on the concentration of this hormone cannot simply be used to measure the presence of pain.

Studies on the behaviour patterns of animals experiencing stress have provided some very useful information, and it is unfortunate that the same degree of information is not available for farm animals experiencing pain, although some excellent observations have been made with respect to laboratory animals. The most useful document available at present with regard to farm animals is the small booklet issued by the Association of Veterinary Teachers and Research Workers and the Universities Federation for Animal Welfare, entitled *Guidelines for the Recognition and Assessment of Pain in Animals* which provides a series of observations which can be made on animals and can be used to assess the likelihood of the existence of pain. This includes such observations as behaviour patterns, appearance, posture, gait, response to handling, appetite and weight, and emphasizes that these signs must be taken as a whole and that the significance of individual signs can vary with the nature of the pain and the species of animal.

Control

As indicated earlier, it is difficult to evaluate pain in animals because of the variations in individual and species response, because of the different types of pain that occur and because of the complicating stress responses that may accompany the pain state. One other approach to evaluating pain in farm animals is the use of *analgesic* or pain-killing drugs. This is a form of circular argument, but the rationale is this. If we know a drug alleviates pain in man, and since the neuronal arrangements associated with pain perceptors are similar in man and animals, then if that drug alters the behaviours and responses in an animal back to normality we can predict the behaviours and responses were due to pain in the animal. Again though, if only life were that simple: many drugs like the opioids not only act on pain pathways but also alter mood and can produce *sedation* and *euphoria* or *dysphoria*, so it is difficult to be certain that what is seen is purely a result of pain control. Other drugs such as the α_2 adrenergic agonists produce only sedation and analgesia, particularly in farm animals, but have wide ranging effects on hormone levels and physiological parameters. The simplest system of drug use, with the least complications, involves the utilization of local anaesthetics as field or nerve blocks, but these are only applicable for surface or peripheral sources of pain, and will also block motor output to the muscles, and hence modify some responses that way (Fig. 11.3).

Another complication has been shown to occur in states of chronic pain. Studies on sheep (Ley *et al.*, 1990, 1991a) have shown that if they suffer chronic pain from foot rot the analgesic effects of opioids and α_2 adrenergics are significantly reduced. That means that an animal suffering chronic pain is more sensitive to acute pain. It also means that the conventional dose regimes used to counteract acute pain are probably inadequate when used to treat chronic pain. One other interesting series of observations in man also has implications in this area. During general anaesthesia for surgery it is normal to keep the anaesthetic level as low as possible while monitoring an adequate level for the surgeon to operate. At these levels the information from the receptors is passing to the higher centres of the brain, but is blocked at this level by the general anaesthetic. It appears that it is possible that this barrage of impulses can cause changes in the central nervous system which render the person more susceptible to pain following recovery from the anaesthetic. McQuay *et al.* (1988) found that following orthopaedic surgery people who were given analgesic drugs such as opioids before and during surgery, required less post-operative analgesia

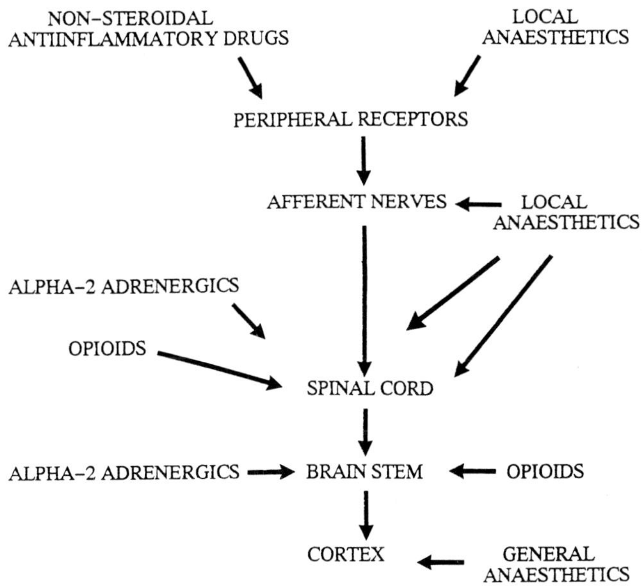

Fig. 11.3. The control of pain using drugs.

than those who had the general anaesthetic alone, and a similar study following amputations using a local anaesthetic block in conjunction with general anaesthesia showed a similar effect (Bach *et al.*, 1988). Another observation in sheep (Ley *et al.*, 1989) showed that animals apparently cured of foot rot showed an increased sensitivity to pain for several weeks after the initial lesions had gone and the animals were no longer lame.

Thus, although analgesic studies have limitations in the diagnosis of pain in farm animals, they have provided us with some information on the nature of pain, particularly of the chronic form, which may help us in our understanding of what happens in farm animals.

Summary

There is a great deal which we do not understand about pain in farm animals, but there are several points which we can make.

If a lesion or procedure causes pain in man it is reasonable to believe it will cause pain in animals. The use of behavioural responses as an indicator of pain must take into account species differences, and accept that man is not a good example for comparison with most farm animals. The use of hormonal and physiological changes as parameters of pain can be complicated by the concurrent stress responses. Parameters associated with acute pain cannot necessarily be used as indicators in cases of chronic pain. The use of analgesic drugs, while subject to variation in different pain states, should always be considered where pain is expected to occur, and may have a significant role in conjunction with surgery under general anaesthesia. Finally, to re-emphasize, if we would expect pain to occur in man we should undertake adequate analgesic treatment to control that expected pain in farm animals.

References

Bach, S., Noreng, M.F. and Tjellden, N.U. (1988) Phantom limb pain in amputees during the first 12 months following limb amputation, after pre-operative lumbar epidural blockade. *Pain* **33**, 297–301.

Chambers, J.P., Livingston, A. and Waterman, A.D. (1990) A device for testing nociceptive thresholds in horses. *Journal of the Association of Veterinary Anaesthetists* **17**, 42–44.

Jephcott, E.H., McMillen, I.C., Rushen, J.P. and Thorburn, G. (1987) A comparison of the effects of electro-immobilisation and, or, shearing procedures on ovine plasma concentrations of β-endorphin/β-lipotrophin and cortisol. *Research in Veterinary Science* **43**, 97–100.

Jephcott, E.H., McMillen, I.C., Congdon, P. and Thorburn, G.D. (1988) Electroimmobilisation and ovine plasma cortisol concentrations: effect of current intensity, current duration and diazepam. *Research in Veterinary Science* **44**, 21–24.

Ley, S.J., Livingston, A. and Waterman, A.E. (1989) The effect of chronic clinical pain on thermal and mechanical thresholds in sheep. *Pain* **39**, 353–357.

Ley, S., Waterman, A. and Livingston, A. (1990) The analgesic action of fentanyl on the mechanical threshold response in sheep in chronic pain with lameness caused by foot rot. In: Van Ree, V.M., Mulder, A.H., Wiegant, V.M. and Van Wimersma Greidanus, T.B. (eds), *New Leads in Opioid Research*. pp. 52–54.

Ley, S., Waterman, A. and Livingston, A. (1991a) The influence of chronic pain on the analgesic effects of the α_2 adrenoceptor agonist, xylazine, in sheep. *Journal of Veterinary Pharmacology and Therapeutics* **14**, 141–144.

Ley, S.J., Livingston, A. and Waterman, A.E. (1991b) Effects of chronic lameness on the concentrations of cortisol, prolactin and vasopressin in the plasma of sheep. *Veterinary Record* **129**, 45–47.

McQuay, H.J., Carroll, D. and Moore, R.A. (1988) Post-operative orthopaedic pain, the effect of opiate premedication and local anaesthetic blocks. *Pain* **33**, 291–295.

Woolf, C.J. (1983) Evidence for a central component of post-injury pain hypersensitivity. *Nature* **306**, 686–688.

12 Cognition

K.M. Kendrick

Agricultural and Food Research Council, Institute of Animal Physiology and Genetics Research, Cambridge, UK

Summary

The study of how animals perceive their environment is central not only to our appreciation of their welfare needs but also to our understanding of the way the brain processes such essential information. While much can be gained from the use of behavioural observation and simple choice tests, a more direct approach to an understanding of animal cognition is through the study of how the brain actually processes the sensory information it receives. If we can establish how this information is accurately used to trigger appropriate emotional and behavioural responses, and how learning and the environment influence this process, it will be a major advance in our understanding of how the brain functions. We will also further our appreciation of animal cognition and the ability of animals to adapt to the various environments we create for them.

During the past five years we have employed the method of recording the electrical output of single nerve cells in the sheep brain to investigate how it carries out the complex and essential function of recognizing animals, humans and food. By establishing whether cells of specific brain regions change their activity in response to salient images, such as food or faces, we can directly determine the animals' use of sensory information for recognition. We can also investigate the effects of individual preferences (if one food or individual is preferred to another), changes in internal state (whether the animal is hungry or not) and learning on such recognition.

Work to date has provided striking evidence of a sophisticated organization and accuracy of visual recognition of objects (food) and individuals (human and animal faces) by sheep which is similar to that described in monkeys and humans. One important principle to emerge is that even though these visual recognition cells respond faster than it is possible for the animal to make some form of emotional or behavioural response, their activity is clearly influenced by the emotional or behavioural significance of what is seen. For example, the sight of preferred foods is a more potent stimulus than that of less preferred ones and, if the animal is not hungry, the sight of food has no effect at all. Similarly, faces of humans and dogs are categorized by different cells to those responding to faces of sheep. This strategy of organizing recognition on the basis of the emotional or behavioural significance, rather than on visual similarities *per se*, optimizes the speed with which the animal can respond to the appearance of objects or individuals. This speed advantage may, of course, be offset by a corresponding reduction in the accuracy of identification.

Learning also has an important influence on visual recognition. Cells responding to the sight of food generally only require a single trial for the animal to establish that a novel food is palatable before they develop the ability to respond to the sight of that food. More indirect indications of learning are that some cells are influenced by the size of horns on an animal's head, the clothing worn by humans, and whether a sheep is socially familiar. It is also possible to show that visual recognition cells can change their coding priorities in response to changes in an animal's motivational priorities.

Work is also described which has looked at the influence of learning and motivational state on the way the brain processes information relating to olfactory recognition, particularly in the context of maternal recognition of offspring. There is a potential to use an analysis of sensory coding strategies employed by farm animals for understanding whether neural constraints on cognitive behaviour occur as a result of being reared under difficult conditions.

Introduction

The term 'cognition' refers to the mental facility of 'knowing'. In the context of human and animal behaviour this knowledge is initially gained through the processes whereby an individual senses and interacts with important objects and individuals in its environment, experiences their value and learns something about their physical properties in order to be able to recognize, and respond appropriately to, them on future occasions. It also includes the processes by which the human or animal gains knowledge about itself. At a higher level it covers the way objects and individuals are categorized, how they can become associated with one another and how context can affect such associations. At the highest level this knowledge includes not only an animal or human's individual perceptions of their environment but also their capacity to see it from the point of view of another individual. This ability to see things from another animal's point of view has only been clearly demonstrated in primates although we should not rule out the possibility that it may also be shown by sub-primate species as well. The study of cognitive behaviour, therefore, covers all aspects of an individual's acquisition and use of knowledge about self and environment that allows it to gain the most value from its interaction with the environment.

Why should we study cognitive behaviour in farm animals? The simple answer to this is that it gives us important insights into how such animals perceive and interact with the various environments we create for them. If we can do this we can begin to assess how changing an animal's environment influences its mental as well as its physical well-being. The much quoted idiom 'healthy body, healthy mind' accurately expresses the common assumption that physical health must lead to mental health, whereas even a cursory consideration of mentally disturbed humans or animals shows that many such individuals are in perfect physical shape. It is important to monitor an animal's physical and mental health separately. Whereas it is relatively easy to measure an animal's physical health the same is clearly not true of its mental health. The animal can only indicate the state of its mental health through changes in its behaviour and whereas drastic reactions, such as the performance of stereotypic behaviours, are obvious, milder forms of mental disturbance might not produce such noticeable behavioural changes. How do we measure mental health in farm animals? The most important first step towards answering this question is to gain some understanding of their cognitive behaviour.

In spite of the fact that it has been recognized for some time that it is important to understand how animals interact with their environment we really know very little about how they perceive, value and gain knowledge about their environment. The main reason for this is that merely observing an animal's behaviour, using an ethological approach, cannot reliably provide objective indices of these mental processes. One approach to an understanding of the way farm animals perceive their environment is to establish how they learn to recognize and categorize important objects and individuals in their environment and whether there are constraints on these faculties imposed by genetic factors, quality of maternal care, rearing environment, etc. We are currently attempting to do this in the sheep by combining the approach of studying how this animal's brain processes important sensory information and how differences in rearing environment, learning and motivational state influence this. We hope that this combined neuroethological approach of studying brain function with studies observing the behaviour of the animals, can provide objective indices of cognitive behaviour in these, and potentially other species as well.

Very little is known about cognitive behaviour in farm animals and the influence that early rearing conditions and the environment might have on it. In this chapter I will illustrate some of our findings in sheep using a neuroethological approach and, where appropriate, review some of the findings from other workers relevant to our understanding of cognitive behaviour in sheep as well as other farm animal species. Finally, I will outline some potential areas where further research could be beneficial.

How Is Knowledge about the Characteristics and Palatability of Food Acquired?

How is food recognized?

It is clearly necessary for all animals to learn rapidly to locate and recognize palatable foods in their environment, and also to avoid toxic or otherwise unpalatable ones. How do they gain this knowledge and what factors influence this process? Most animals use sight, smell and taste to recognize food, although they can also easily learn to associate auditory cues with imminent arrival of food (i.e. the sound of a food trolley or a tone, for example). How is this sensory information used to optimize accurate recognition of known palatable foods and how is knowledge of new foods acquired? A secondary set of questions to be asked concerns whether the sensory recognition of food relates to an animal's individual preferences for specific foods, its motivation to eat them (i.e. is the animal hungry or not?) and the rewarding properties that ingestion of food has. A further related question is whether the recognition of food is linked in some ways to similar recognition of water and minerals such as salt.

In sheep, evidence suggests that the visual sense is the most important for recognition of food, since anosmic individuals do not have any major problem in discriminating between foods (Baldwin *et al.*, 1977; Milne *et al.*, 1982). This contrasts with animals such as pigs, which may have to root for food, where olfaction plays a more important role in recognizing food since visual cues are clearly not available under such circumstances (Hafez and Signoret, 1969). Sheep, as is the case with most farm animals, have a good visual sense. They can see almost completely around themselves (290°), although they only have good acuity in the 40° of their frontal eye field where the two eyes overlap. Sheep (Baldwin, 1981), chickens (Sutherland, 1961), goats (Baldwin, 1979) and calves (Baldwin, 1981) have all been shown to be able to perform quite complex shape discrimination tasks and are in a position to use the visual sense to discriminate between foods.

Our own work on visual recognition of food has concentrated on the sheep. Sheep are known to exhibit food preferences both in terms of preferential grazing of specific clovers and types of grass, and in the order in which they ingest palatable foods (such as hay, oats and various concentrates) when housed inside. The visual sense appears to be important in this discrimination between foods (Bazely, 1988). We have investigated whether areas of the brain known to be important for the control of food intake also play a role in the control of visual recognition of food. To do this we have recorded the electrical activity of single neurones in these regions in conscious sheep while they view, and subsequently ingest, different foods. The technique for making such recordings from the brain has been described in detail elsewhere (Kendrick and Baldwin, 1991). We have found that there are cells within the lateral hypothalamus and zona incerta of the sheep brain which respond to the sight, but not smell, of food as opposed to other objects (Kendrick and Baldwin, 1986a,b). The majority of the cells only respond when food is moved towards the mouth, which probably reflects the fact that sheep normally have to lower their heads to eat, giving them the perception of apparent movement. A small number of cells do, however, respond to the static visual presentation of food. We have been able to show that the activity of such cells is influenced by the animal's individual food preferences and its state of satiety. It is even possible to demonstrate the influence of learning on them, either in the context of learning to associate non-food objects with food, or learning to recognize novel palatable foods. It is the findings on the effects of learning that are the most revealing in terms of increasing our understanding of the cognitive recognition capabilities of these animals. Figure 12.1 illustrates that cells in the zona incerta which respond to the sight of food can be conditioned to respond to the sight of a coloured bottle associated with a food reward. There are individual differences between animals for the number of trials required to learn this association, at the cellular level, but generally they learn quite slowly requiring between 7 and 13 trials. The cells also only appear to retain this information for 1 h or so. These data imply that the abilities of sheep to learn and retain information about whether something is associated with food is poor. Behaviourally we know that this is not the case. Figure 12.1 also shows a vastly improved speed of learning and retention if, instead of using a totally artificial learning paradigm, we use novel foods. In this instance generally only one or two trials are required to learn to recognize a novel palatable food and the cells will continue to respond to the sight of this food even one or two months afterwards without having had access to this food again. These data give a very different picture of a sheep's ability to learn to

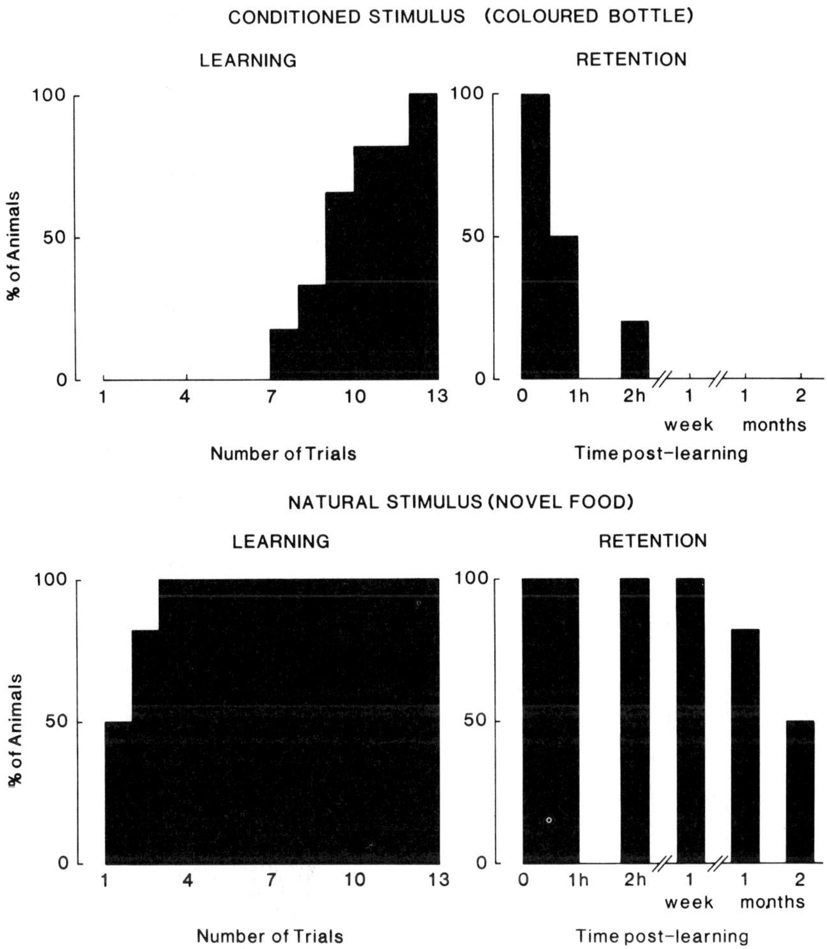

Fig. 12.1. Histograms show how cells responding to the sight of food in six different sheep require more trials to associate the appearance of a conditioned stimulus (a coloured bottle), and retain this information for a shorter time than the same type of cells, in the same animals, do for a natural stimulus (a novel food). Each animal had a mean figure calculated for it based on the activity profiles of two to four different cells for each learning condition.

recognize new foods and show quite quite clearly that this species, and most other non-primate species, should not really have its 'intelligence' judged on the basis of its ability to form artificial associations between objects and rewards. Similar support for the assertion that farm animals perform better in learning tasks using natural rather than artificial stimuli come from work on imprinting in chickens (visual learning – Horn and McCabe, 1984; Johnson *et al.*, 1985; Johnson and Horn, 1988; auditory learning – Van Kampen and Bolhuis, 1991). Indeed, with relatively few exceptions (goats performing triple choice delayed response tasks – Soltysik and Baldwin, 1972) farm animals do not perform well on complex learning tasks involving the making of artificial associations and yet they clearly have no trouble in learning complex information about the environment they live in.

We have not carried out any specific studies looking at the way food aversions are learned in sheep although there is a large body of evidence showing that they can easily learn to avoid food associated with toxic substances such as lithium chloride (Burritt and Provenza, 1990). The cells that we record from in the brain which respond to the sight of foods only respond to the sight of those foods that the animal will eat. We have never, therefore, found any instances of cells responding to foods which the animal does not find palatable and consequently will not eat. The probable reason for this is that the areas of the brain containing these food responsive cells are also sites associated with rewarding aspects of stimuli (Rolls *et al.*, 1980) but not their negative aspects. It seems probable, therefore, that other brain regions may be involved in the process of recognizing and avoiding toxic foods. The one piece of evidence we do have concerns the effects of ingestion of rolled oats which are a highly palatable food for most sheep but which can also cause stomach ache when too many of them are eaten. We noticed that in two sheep where this happened oats subsequently became a relatively ineffective stimulus in eliciting responses from cells responding to the sight of food, whereas previously they had been the second most effective. Clearly, therefore, palatable foods that have caused some degree of physical discomfort to the animal become less attractive to the animal to ingest and this is reflected in its ability to activate food recognition circuits within the brain.

How is recognition of food related to similar recognition of salt and water?

Not only do animals have to recognize food they also have to recognize minerals and water as well. Are there separate mechanisms in the brain to assist in recognition of these other important substances? Our evidence suggests that there may be considerable interactions between the neural mechanisms controlling the recognition of food, water and salt. Indeed we have direct evidence that cells responding specifically to the sight of food, when the animal's main motivational priority is to ingest food, can switch their priorities and respond specifically to the sight of salt when the animal is sodium deficient (Kendrick and Baldwin, 1989a). Figure 12.2 illustrates this phenomenon using both electrophysiological recording data and data from an analysis of the *in vivo* release of the inhibitory amino acid transmitter GABA which is related to the activity changes of the cells responding to the sight of food and salt (Kendrick *et al.*, 1991a). At this level of processing the same cells can potentially code for the recognition of a variety of substances dependent upon the animal's current needs.

The importance of food taste and texture

Apart from having different appearances or smells, all foods that an animal eats have distinctive tastes and textures. How do animals use this information? Obviously one answer to this is that information on the taste and texture of food provides additional information for correctly identifying palatable substances. Indeed, for example, sheep can learn to ingest preferentially hay treated with palatable chemicals

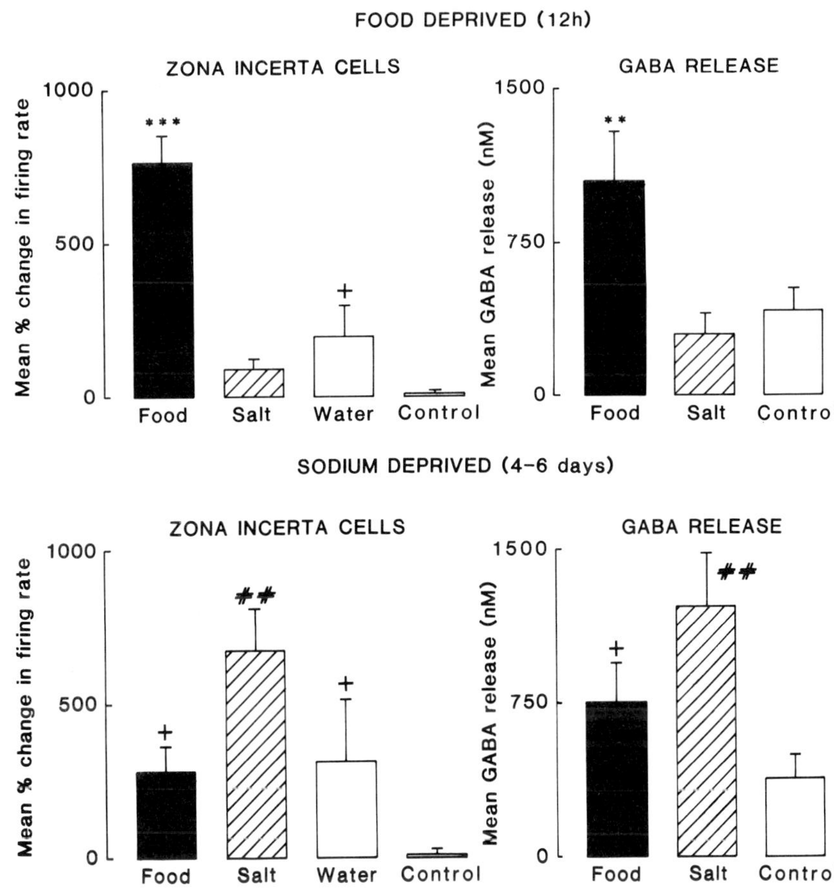

Fig. 12.2. Histograms show (top left) mean % changes in firing rate of cells in the zona incerta of five Dalesbred sheep, in response to the sight of food, water and salt when the animals were food deprived for 12 h. The response to the sight of food is significantly greater than that to water and salt, as is the mean release of GABA in this region (measured by microdialysis – top right). When the animals are sodium deprived for 4–6 days (via a parotid fistula) the responses of the cells and the GABA release are greater to the sight of salt. $^{***}P < 0.01$, $^{**}P < 0.02$ compared to salt, water and control (empty bowl); ##$P < 0.02$ compared to food, water and control; $+P < 0.05$ compared to control.

(Gherardi and Black, 1991; Gherardi *et al.*, 1991). As such one might expect some form of overlap between neural mechanisms subserving the visual recognition of food, described above, and those involved in processing its taste and texture. This is indeed the case since at least some cells which respond to the sight of food also respond to its ingestion (Kendrick and Baldwin, 1986b). Further, since cells responding to the sight of food cease to do so if the animal is not allowed to eat it, the sight and taste of food must always be linked together with its rewarding properties. The role of the texture of food in providing specific identification cues is also supported by our data which show that cells which respond to the ingestion of foods as solids do not do so when the same foods are converted into liquid form (the taste should obviously be similar in the two instances although there is a large difference in texture – Kendrick and Baldwin, 1986b).

What conclusions can be drawn from the principles of food recognition which are relevant to our understanding of cognitive behaviour?

1. Specific neural circuits exist in the brain to facilitate the accurate recognition of food, salt and water, and they are strongly influenced by an animal's preferences and motivational state.
2. The effects of learning on the activity of these cells illustrate an important consideration when attempting to assess an animal's intelligence. Learning and its retention are much better when natural as opposed to artificial associations are required.
3. Changes in motivational priorities to ingest different classes of substances (i.e. food, salt and water) may, in part, be subserved by alterations in the coding priorities of the same neural circuits. This type of system where the same neural circuits can be activated by different objects, dependent upon an animal's current needs, would help reduce competition between motivational priorities occurring and avoid counterproductive behavioural conflict.
4. Evidence suggests that these neural systems, specialized for the recognition of ingestible substances, are also an integral part of reward mechanisms in the brain. Probably, therefore, when the animal sees a palatable substance it simultaneously experiences the satisfaction that subsequent taste, ingestion and digestion of the substance will bring, before it actually eats it. If this is the case then by studying the activity of these neural systems we are actually investigating the animal's conscious, cognitive appreciation of the significance of food.

How Do Animals Recognize One Another and Other Animals and Humans?

Visual recognition of individuals through perception of faces and body shape

Although, as I have already discussed, most domestic animals have good eyesight very little attention has been paid to the possibility that they might use this to recognize other animals and humans. Instead, much more attention has been paid to the use of olfaction to carry out this important task and I will discuss this in the next section. Once again, our own work in this context has looked exclusively at sheep (Kendrick and Baldwin, 1987, 1989a; Kendrick, 1990a, 1991). It has long been established that in horned ungulates, the presence of horns and their size are important indicators of an individual animal's dominance and gender (Collias, 1956; Geist, 1968; Lincoln, 1972) and this feature must be perceived using vision. The most direct behavioural evidence for recognition of individuals through their faces comes from work done by Alexander and Shillito-Walser in the late 1970s. They found that postpartum ewes found it difficult to recognize their lambs if the appearance of the lamb's head was altered by colouring it black (Alexander and Shillito-Walser, 1977). The head has also been shown to be an important stimulus for imprinting in chickens (Johnson, 1988). Visual cues have also been found to be important for lambs to recognize their mothers from about 3 weeks of age (Shillito and Alexander, 1975; Alexander and Shillito-Walser, 1978) and my own observations on hand-reared lambs suggest that the head (i.e. human head in this instance) is important for this recognition since covering it slows the lamb's approach to its human caretaker. Indeed, physiological evidence has shown that the visual system of the newborn lamb is virtually fully mature at birth and it is therefore capable of using this sense to learn about the environment from day 1 of life (Ramachandran *et al.*, 1977).

In monkeys, and probably humans, there are specialized neural circuits within the temporal lobe of the brain for assisting in the visual recognition of faces and other body regions (Gross *et al.*, 1972; Perrett *et al.*, 1982). We have found that sheep also have specialized neural circuits in this region for responding to faces (Kendrick and Baldwin, 1987; Kendrick, 1990a, 1991). These 'face' cells in the temporal cortex respond extremely quickly following the appearance of the stimulus (90–180 ms) and would appear to be involved in a system specialized for extremely rapid identification of individuals. As with the food recognition cells, these cells responding to faces also often respond better to moving images than to static ones. The responses of the cells in this region to different faces give a great deal of information about cognitive aspects of individual recognition. The first important finding from these studies is that cells do not code for faces *per se* but for specific groups of faces or facial features. The basis for this form of coding seems to be to co-process facial images which have a similar emotional/behavioural significance for the animal rather than using visual similarity. Thus, for example, one sub-population of cells codes for faces with horns and how large the horns are, a second group codes only for faces of sheep of the same breed (see Fig. 12.3), and a third group codes for faces of humans and dogs. This last category is the best example of facial images being co-processed by the same cell on the basis of emotional significance rather than visual similarity, since human and dog faces are not physically similar but their emotional significance for the sheep is. Behaviourally, the animals also have specific cognitive requirements for identifying faces with horns (perception of dominance and gender) and animals of the same breed (most sheep breeds prefer to stay with members of their own breed – Winfield and Mullaney, 1973; Shillito Walser *et al.*, 1981). The cells coding for recognition of faces of sheep of the same breed also respond preferentially to socially familiar animals and this is the most specific form of coding relating to identity of individual animals that we can find. Indeed, it appears that the cells are mainly involved in the categorization of different faces in terms of their behavioural/emotional relevance to the animal rather than for identifying specific individuals. The ability to perform such a categorization must be learned. One implication from our studies is that learning must play an important role in shaping the categorization strategies used by these cells. Further, since the temporal lobe has pathways linking it with other brain regions involved in the control of memory and emotional behaviour, it is in an ideal position to interact with these centres to evolve the optimal specialization for rapid identification of important categories of individuals. The other important implication arising from these observations is that individuals are not first recognized and then subsequently their significance felt; the two processes are interrelated and probably simultaneous. Indeed I have recently concluded from these experiments

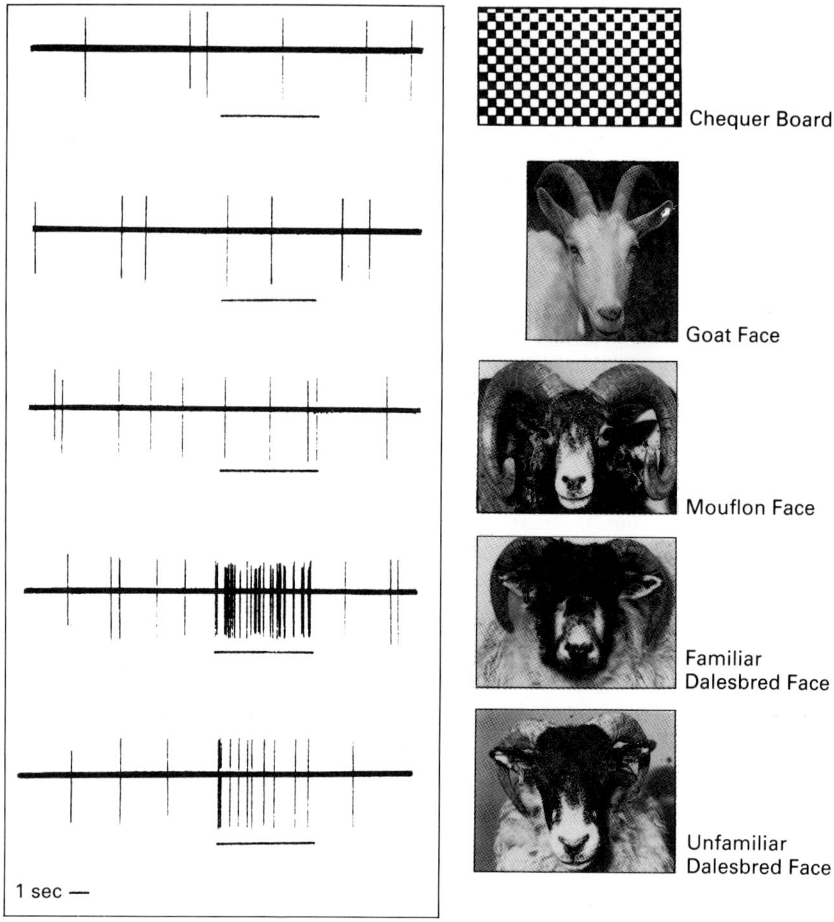

Fig. 12.3. The selective response of a single neurone in the temporal cortex of the brain of a Dalesbred sheep to faces of other Dalesbred sheep, and particularly familiar individuals. The vertical deflections on the traces on the left of the picture are the output of the single cell. The horizontal bars under the traces indicate the period of time (5 s) when the picture of the face (shown on the right) was presented.

that for sheep, and probably other animals that '. . . their perceptions, like ours, are not merely coloured by their emotions – they are intimately linked with them' (Kendrick, 1990b).

We have yet to demonstrate directly the influence of learning on these face responsive temporal cortex cells, and indeed there may be some innate predisposition for using certain characteristics to identify faces. For example, our evidence suggests that the eyes and a correctly orientated outline of the face are the most essential features for identifying a human face (Kendrick, 1990a, 1991). Direct eye contact between a human and a sheep, or a dog and a sheep, is a well-known ploy used to control the behaviour of sheep. Indeed, one of the essential prerequisites for a good sheep-dog is often referred to as being able to 'give the eye' to the sheep. An innate predisposition for preference of facial stimuli, particularly the eyes, has been proposed both for chickens and humans (see Johnson, 1988). However, overall, it would seem most likely that there may be an innate predisposition to attend to face-like stimuli rather than an ability to identify specific classes of individual. The fact that face responsive cells in the sheep temporal cortex do not respond to the sight of upside-down faces (Kendrick and Baldwin, 1987), whereas in arboreal primates they do (Perrett *et al.*, 1982), suggests that these cells are greatly influenced by an animal's habitat and behavioural requirements (i.e. a sheep would not have had to learn to recognize upside-down faces whereas an arboreal primate would) and are not simply part of an innate recognition system.

The identification of categories of individual through perception of their faces is not possible for sheep, or indeed for other animals and humans, over long distances, and yet in the wild they have been reported to identify accurately predators visually at distances in excess of a kilometre (Geist, 1971). How do they do this? The answer to this is probably through learning physical characteristics of body shape and posture. We have found that sheep (Kendrick and Baldwin, 1989a), like monkeys (Perrett *et al.*, 1985), also have cells in the temporal cortex that are responsive to the sight of body shape and posture. At this stage, in sheep, we have only managed to demonstrate this type of coding for the sight of a human shape, although we would anticipate that similar cells responding to animal body shapes exist as well. As with the facial recognition cells described above, the body shape recognition cells are not identifying specific individuals but humans as a category. Thus, the cells respond equally to any individual, irrespective of what sex they are, their clothes, build, whether their head is visible or whether their front or back view is presented. To do this the animals must have learned something specific about the human body shape which allows accurate identification to be made despite the many various guises in which a human might appear. The result of this learning is incorporated in the coding strategy employed by these cells since if a human does something unusual to alter his or her appearance, by crawling instead of walking, or walking sideways towards the animal for example, then the cells do not respond (see Fig. 12.4). We know from behavioural evidence that humans can approach sheep much closer by crawling towards them and it would appear that the sheep's normal concept of a human is characterized by a bipedal posture.

For the cells responding to the sight of faces and those responding to body shape there is a strong influence of context. Figure 12.5(b) shows, for example, that the cells only respond to horns if they are seen in the context of a face. Similarly, we have found that the majority of cells coding for horned faces only respond to a frontal view and less to a profile view or a view of the back of the head (Kendrick, 1990a,b, 1991). The horns are visible in each case and yet is only when the horns are seen in the context of direct eye contact that the cell responds. Indeed we now also have some preliminary evidence that if a frontal view of the head with horns is shown, but with the eyes obscured, then the cells do not respond. Obviously some kind of social or aggressive encounter is more likely to occur when direct eye contact occurs between individuals. Cells responding to the sight of the human body shape also demonstrate the importance of context since the majority of them only code for an approaching human shape rather than a stationary or retreating one (Kendrick and Baldwin, 1989a). Clearly, the human shape is being perceived in the context of what

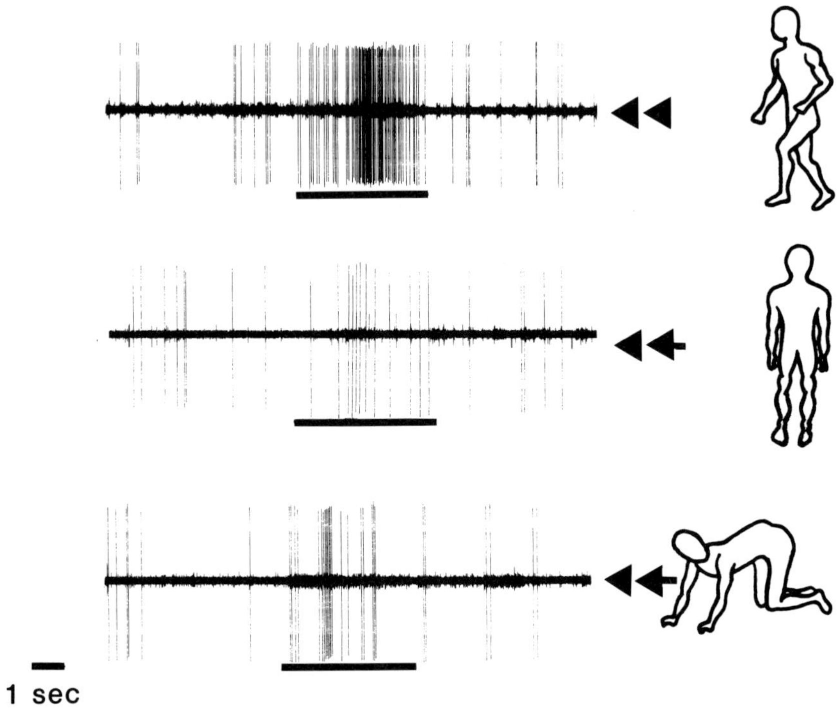

Fig. 12.4. Traces show the strong response (vertical deflections on the traces) of a temporal cortex cell in a Dalesbred sheep to a human approaching showing a frontal view (top trace) but not with only a side view (middle trace) or when crawling (bottom trace).

it is doing, and an approaching human is a much more important stimulus for a sheep to attend to than a stationary or retreating one.

The best example we have of the effects of context on neural coding in the temporal cortex of the sheep brain is from the activity profiles of a very small population of cells that respond not to the sight of humans *per se*, but to what they are wearing. Figure 12.5(a) gives an example of a temporal cortex cell that responds equivalently to the sight of three different humans only when they are wearing a white laboratory coat. The coat by itself, or a coat which is held, but not worn, by the human have no effect at all. The coat must be seen in the context of being worn by the human in order to influence the activity of the cell. This is not simply a fear response since the cell does not respond to a frightening stimulus such as a loud noise. Clearly the animal's perception of a human must change dramatically when the human puts on a white coat. Since white coats are rarely worn in the context of the experiments that the sheep are subjected to, it is unlikely that this is a learned negative reaction to humans wearing white coats. A possible explanation might be in terms of the human wearing a white coat looking more like a sheep, although we still need more data to support this hypothesis.

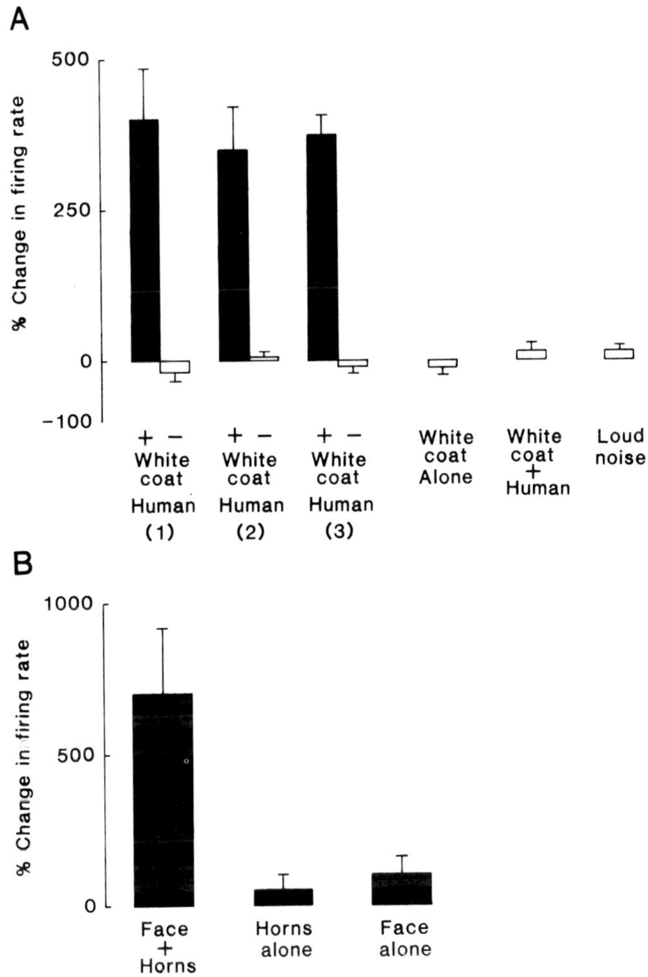

Fig. 12.5. (a) Histograms show the changes in firing rate of a single cell in the temporal cortex of a Dalesbred sheep caused by the approach of any human wearing a white coat. No effect of the white coat alone or the white coat held by a human, or a loud noise is seen. (b) Mean percentage changes in the firing rate of temporal cortex cells (data from five Dalesbred sheep) responding to the sight of faces with horns. Only when the horns are seen in the context of a face do the cells alter their response.

Olfactory recognition of individuals

There is a large amount of evidence showing that farm animals can distinguish between individuals using the sense of smell (pigs – Meese *et al.*, 1975; goats and calves – Baldwin, 1977; sheep – Baldwin and Meese, 1977). Very little is known about how olfactory information is processed by the brain in farm animals and still less about the influence of learning. We have made an intensive investigation of how the selective olfactory recognition bond between a ewe and its lambs is made. This selective bond is formed within 2 h postpartum in maternally experienced animals although first-time mothers usually take longer than this (Poindron and Lévy, 1990). The formation of the olfactory recognition is triggered by signals from the vagina and cervix, during labour contractions (Kendrick *et al.*, 1991c), provoking profound changes in the neurochemistry of the olfactory bulb (Kendrick *et al.*, 1991b) and probably other brain regions as well. Such changes only occur at parturition when the animals have had experience of rearing lambs before and so there is an important learning element involved. Parturition also causes changes in the electrophysiological responses of olfactory bulb neurones to lamb odours (Fig. 12.6) and in the neurochemical release (noradrenaline and acetylcholine) that occurs in response to them (Kendrick *et al.*, 1991b). Thus we can see that the olfactory processing system in the brain alters its coding priorities dependent upon the animal's current behavioural requirements, in exactly the same way that the cells in the zona incerta which respond to the sight of food do when the sheep's motivational priorities change.

Auditory recognition of individuals

Although farm animals exhibit a range of vocalizations, and are therefore in a position to use the auditory sense to recognize one another, we know very little about how the brain utilizes this information. We have found cells in the temporal lobe of the sheep which respond specifically to different frequencies of sound, but so far have found only a few which respond better to complex vocalizations as opposed to pure tones. There is, however, much better evidence for specific coding of natural sounds by cells in the auditory neostriatum of the avian brain (Scheich *et al.*, 1979). All animals, and humans, are exposed to sounds *in utero* and this may even assist in an offspring's rapid recognition of its mother's voice postpartum (Armitage *et al.*, 1980). However, for the most part, farm animals do not under normal circumstances constantly vocalize to one another, and vocalizations may be primarily used as a mode of expression and communication rather than as an aid to recognition. Work on auditory recognition of lambs by ewes has also reached this conclusion that hearing is not a primary recognition sense (Shillito-Walser, 1978). Clearly, however, animals can indicate their emotional state and communicate important information such as 'danger' to other animals through their auditory sense. It is important for us to establish how farm animals learn to use this information and how it comes to be employed as a means of expression as well as of communication.

Combination of information from the different senses to assist in individual recognition

While we have good evidence for neural circuits combining information on the sight of foods with that of their taste and texture, we have not been able to show a similar integration of multisensory information in the sheep brain relating to recognition of individuals. Such multisensory integration must occur, and the senses are clearly important for cross-checking one another. The way that multisensory information

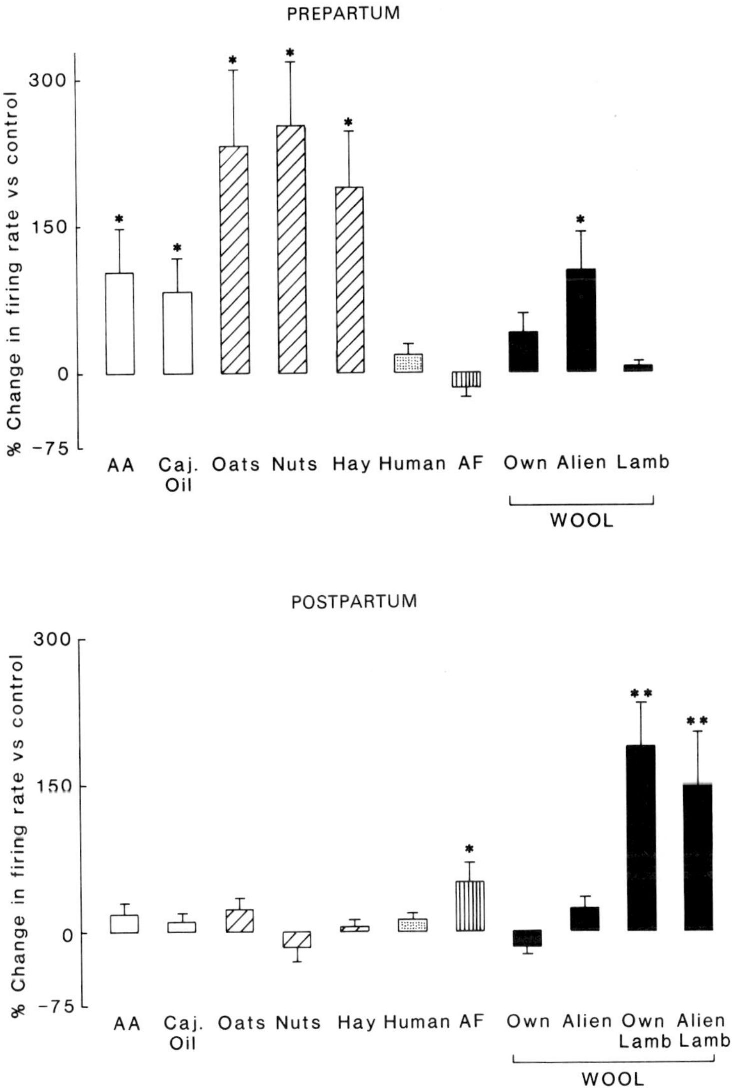

Fig. 12.6. Histograms show the mean percentage change in firing rate vs. control for olfactory bulb neurones (mitral and glomerular cells) responding primarily to biological odours in four Dalesbred sheep. Prior to parturition, food odours are the most effective odours (hatched bars) and no cells respond to amniotic fluid (vertical bars) or lamb wool odours. After parturition, lamb odours (whether own lamb or alien lamb) are the most effective stimuli and there is also an increase in the ability of amniotic fluid to evoke a response. It would appear that, following parturition, the attentional priorities of the sheep have switched from food odours to lamb odours. AA = amyl acetate; Caj oil = cajput oil; AF = amniotic fluid.

is integrated by the brain, and the ultimate influence that this has on an animal's cognitive representation of the world, is an important area for further research.

What conclusions can be drawn from the principles of sensory recognition of individuals which are relevant to our understanding of cognitive behaviour?

1. In farm animals the visual and olfactory systems seem to be primarily involved in the process of individual recognition and visual information is coded more on the basis of similar emotional/behavioural significance to the animal than on the basis of physical similarity.
2. The visual system in sheep, and probably in other farm animals, can be used to identify categories of individuals either by their faces or through body shape. Like primates, they have specialized neural circuits in the temporal lobe to assist with this. Such sophisticated systems for recognizing categories of individuals clearly underline the importance that the interactions with other animals and humans have for farm animals.
3. The importance of moving stimuli both in the context of face and body shape recognition suggests that sheep are better at discerning moving images than static ones. Indeed, one of the areas of the sheep brain which is specialized for the analysis of visual movement (the superior colliculus) is extremely large compared to other species.
4. Learning, context and behavioural/emotional significance all influence the processing of sensory information relating to individual recognition.
5. The sensory systems involved in identifying specific categories of individual are linked to centres of the brain controlling emotional behaviour and memory. Indeed the evidence suggests that it is impossible for an animal to identify an individual independently from experiencing the emotional/behavioural significance that it has. As such we should not regard sensation and emotion as separate systems but as an integrated continuum.

Learning about other Aspects of the Environment

Apart from being able to locate and recognize food and other animals and humans, animals also learn about other important aspects of their environment such as shade, boundary fences, where to shelter from the rain, etc. In all these cases the knowledge that has been acquired takes the form of a flexible concept which can easily be used to generalize across situations (i.e. similar to the concept of a human shape discussed above). One of the most fascinating examples of this type of concept formation in sheep is that of a boundary fence. Sheep are capable of discerning that there is a physical barrier between them and a human no matter what form that barrier may take (wire fence, chain fence, gate, etc.). Even with a single-stranded electric fence the sheep appears to know that what keeps it in will also keep humans out. Thus, as with case of an approaching human shape being different from a receding one, the behavioural significance of a potentially threatening stimulus, such as a human, can be radically altered by a simple thing like a fence between the sheep and that human. As with most accurate generalized concepts of this kind one would anticipate that they are learned over a period of time and must be constantly updated as a result of experience. This type of complex concept learning may be more sensitive to the influence of placing animals in a stressful or restricted environment over long periods. As far as I am aware there have been no experiments carried out which directly test this hypothesis but they would not be difficult to do.

The use of robot milking and wool shearing machines also raises interesting questions regarding animal perceptions of these new developments in their environment. The interactions between animals and man-made machines is obviously an area for further research if we are to try to design such machines to make them acceptable to the animals. It is possible, for example, that for farm animals, learning about machines may be akin to learning the kind of artificial associations which we classically use in our human intelligence tests and which, as I have already mentioned, are clearly much harder tasks for these animals than similar learning about associations between natural stimuli.

Learning by Example

Learning about the environment for both animals and humans is not always achieved by direct experience but often through observing, or listening to, the experiences of others. Animals, just like humans, are exposed throughout their lives to the influence of other individuals, some of whom are clearly more important than others. For all animals and humans the effect of parental attitudes has enormous, and potentially irreversible, influence on the cognitive behaviour of their young. In farm animals the best documented effects of maternal influence on the behaviour of offspring are in terms of diet selection (Matthews and Kilgour, 1980; Kovalcik and Kovalcik, 1986; Lynch and Bell, 1987; Squibb *et al.*, 1990). While it might be argued that this is simply an example of a 'mindless' imitation of mother's behaviour the fact that lambs retain a dietary preference taught to them by their mothers for up to three years, without having eaten the food in the intervening period, argues strongly against this (Green *et al.*, 1984). Instead, it would seem more likely that the knowledge the lamb has of the palatability of the food is integrated with some emotional property associated with the lamb's past relationship with its mother.

From my own research, I have observed a similar influence of maternal behaviour on lamb cognitive behaviour in terms of attitudes towards humans. In three cases we have cross-fostered one member of a set of twin lambs on to another ewe. In each case one of the pairs of mothers for these genetically similar animals had a marked wariness of humans and would always avoid rather than approach one. The other mother would approach a human rather than avoid it. The lambs brought up by their prospective mothers adopted the maternal attitude towards humans within 2 days (before this time the lambs did not normally avoid humans). After weaning the lambs were separated from their respective mothers and reunited. The lamb from the pair that had learned to avoid humans continued to do so for up to 3 months even though its twin showed the opposite behaviour. It would seem once again, therefore, that when an animal learns something from its mother the acquired knowledge is linked to a memory of mother, and her emotional significance, and is extremely difficult to reverse.

Although mother may be the most important role model for most animals, clearly the experiences and behaviours of other animals can continue to influence an animal throughout its life. The dietary preferences of lambs can, for example, be influenced by other lambs as well as by their mothers (Provenza and Burritt, 1991). Another good example of the influence that animals have on each other's behaviour is the establishment of dominance hierarchies. An individual can decide, for example, its relationship with one particular animal as a result of a conflict between that animal and another one. It does not actually have to fight the dominant animal itself to gain this knowledge. Indeed, next to mother, the dominant social animals in a group probably exert the most profound influence on the behaviour of their subordinates. In my own experience, for example, a sheep that I had reared by hand became the dominant animal in a flock of ewes and juveniles and rapidly influenced the other animals to approach me in a field whereas previously they had avoided me.

Environmental and Developmental Constraints on Cognitive Behaviour

Stress, housing conditions and novelty

There is little doubt that stress, inappropriate housing conditions and an impoverished environment can all reduce an animal's quality of life and lead to the expression of abnormal behaviour. The effect of these factors on an animal's cognitive behaviour is largely unknown although one would obviously predict that they would have a major impact. Humans or animals under constant stress do not learn well simply because they are unable to attend selectively to the changes in their environment that the learning task requires. Similarly, if either animals or humans live in an environment where they do not encounter 'novelty' they also subsequently find it hard to acquire new knowledge and can react negatively to environmental change. All this implies that for the normal development and expression of cognitive behaviour an animal's environment should not be stressful and should also incorporate some periodic element of change or novelty.

'Critical periods' in development

Not only is an animal's interaction with the environment profoundly influenced by the behaviour of its mother during early development, but there is also a large amount of evidence that any experience in early life has greater effects than similar experiences later on. This concept that there are 'critical periods' during an animal's, or human's, life where experience will have a more profound and long-lasting effect than at any other time of life has received widespread support. The most important question in terms of the influence of such critical periods for the development of cognitive behaviour is what happens if the animal is denied the relevant experiences during them. For example, I have already discussed the importance of the maternal influence on an offspring's subsequent cognitive behaviour: what happens if there is an absence of maternal care during this important developmental period? Original studies on monkeys (Mason, 1960; Harlow, 1969) showed that, if they were reared in isolation, their subsequent social and sexual behaviour was drastically influenced. Similar findings have been reported in cattle (Broom and Leaver, 1978; Broom, 1981), and social behaviour in lambs is also disturbed by rearing in isolation (Zito *et al.*, 1977; Moberg and Wood, 1982). Sexual behaviour can also be influenced by rearing in isolation in both pigs (Hemsworth *et al.*, 1978) and sheep (Zenchak and Anderson, 1980).

The implications of the above work are that a combination of social interaction with other young animals, through play for example, and maternal care exert important influences on the development of specific behaviour patterns. The effects of isolation rearing on the development of these behaviour patterns can be permanent. It would seem likely that this critical period of life is an important time for brain development. Part of this development will be in terms of learning specific motor patterns, although it is difficult to see how the effects of isolation could be mediated purely through an inability to learn new motor skills. The potentially irreversible damage caused by being reared in isolation might instead be the failure to learn to form appropriate associations between the performance of specific behaviours and their emotional significance. As I have already pointed out, an analysis of the effects of learning on the brain shows that neural circuits seem to code simultaneously for recognition, context and emotional/behavioural significance. One of the important factors in early brain development may be the organization of the brain to do this and the result of not being able to do so might lead to the animal not being able to associate the performance of a particular behaviour with its emotional significance. This is, of course, one of the problems with autistic children, and the result of rearing an animal in isolation might produce something akin to an autistic animal. If this is the case then the animal will also have profound learning and communication problems.

Future Areas of Research

I have already mentioned some future areas of research that would be beneficial to pursue in the context of understanding farm animal cognition and here I shall outline some additional ones. As I mentioned in the introduction to this chapter, we know very little about cognitive behaviour in farm animals and yet such knowledge is essential if we are to appreciate fully their welfare needs. The approach we have adopted of investigating the way the brain processes important information is novel and has allowed us to reveal the presence of highly specialized, complex and sophisticated neural systems for analysing sensory information from the environment. The methods of coding information and the remarkable influence of context, learning and motivation on such sensory analysis systems proves beyond doubt that we are witnessing the basis of an animal's conscious interaction with its environment and not just the workings of simple, innate, dispassionate sensing mechanisms. The neural mechanisms subserving recognition of both food and individuals are inextricably tied into mechanisms governing their emotional and behavioural significance, and ultimately the feelings of pleasure, pain or fear that their recognition evokes in the animal. Obviously we are constrained to investigate these systems in a direct stimulus–response paradigm where the object we want the animal to recognize is actually presented to it. However, since animals readily seek out specific objects and individuals which are not currently in view it follows that these neural recognition systems might also be active when an animal is searching for specific objects/individuals. In this respect, it is interesting to note that temporal cortex cells responding to food or faces sometimes continue to show changed activity even in the absence of these stimuli. Furthermore, contextual cues from the environment must aid the search for important objects and individuals, and one might anticipate that neural systems subserving their sensory recognition might also simultaneously code for the environmental contexts, and spatial location, that such things are normally found in. Information directly relating to these questions will throw light not only upon the way that animals organize information from the environment but will also give us insights into their 'thoughts' as they interact with it.

The recently developed techniques of tracing the functional neuroanatomy of the brain by mapping changes in the expression of proto-oncogenes (c-*fos* and c-*jun* for example) present a powerful additional tool to the electrophysiological approach I have described (see review by Doucet *et al.*, 1990, and a demonstration of the effects of learning on the brain using this technique in chickens – Anokhin and Rose, 1991). It should be possible to identify how different parts of the brain interact with one another, and are modified by experience using these combined approaches. Additionally, much information about farm animal cognitive behaviour can still be gleaned through the use of well-designed and controlled behavioural experiments which allow the animal to tell us what it recognizes and prefers about objects and individuals in the environment. A combination of all these techniques should also allow us to examine directly the pathological alterations in brain/behaviour resulting from stress, boredom, absence of maternal care and disease.

There is obviously an urgent need for more research on the influence of developmental critical periods, maternal care, stress and environmental/housing constraints that animals are exposed to on the development and adaptability of cognitive behaviours. The mainly indirect evidence available to date clearly indicates that all these factors are important and one would anticipate that irreversible changes in the organization of the brain may result from them. The coding strategies I have described for sensory recognition circuits in the sheep brain might prove to be extremely sensitive indicators of the influence that adverse rearing or environmental conditions might have on an animal's cognitive behaviour.

References

Alexander, G. and Shillito-Walser, E. (1977) Importance of visual cues from various body regions in maternal recognition of the young in Merino sheep (*Ovis aries*). *Applied Animal Ethology* **3**, 137–143.

Alexander, G. and Shillito-Walser, E. (1978) Visual discrimination between ewes by lambs. *Applied Animal Ethology* **4**, 81–85.

Anokhin, K.V. and Rose, S.P.R. (1991) Learning-induced increase in immediate early gene messenger RNA in the chick forebrain. *European Journal of Neuroscience* **3**, 162–167.

Armitage, S.E., Baldwin, B.A. and Vince, M.A. (1980) The fetal sound environment of sheep. *Science* **208**, 1173–1174.

Baldwin, B.A. (1977) Ability of goats and calves to distinguish between conspecific urine samples using olfaction. *Applied Animal Ethology* **3**, 145–150.

Baldwin, B.A. (1979) Operant studies on shape discrimination in goats. *Physiology and Behaviour* **23**, 455–459.

Baldwin, B.A. (1981) Shape discrimination in sheep and calves. *Animal Behaviour* **29**, 830–834.

Baldwin, B.A. and Meese, G.B. (1977) The ability of sheep to distinguish between conspecifics by means of olfaction. *Physiology and Behavior* **18**, 803–808.

Baldwin, B.A., McLaughlin, C.L. and Baile, C.A. (1977) The effect of ablation of the olfactory bulbs on feeding behaviour in sheep. *Applied Animal Ethology* **3**, 151–161.

Bazely, D.R. (1988) Doctoral Thesis, University of Oxford.

Broom, D.M. (1981) *Biology of Behaviour*. Cambridge University Press, Cambridge, 257 pp.

Broom, D.M. and Leaver, J.D. (1978) The effects of group housing or partial isolation on later social behaviour of calves. *Animal Behaviour* **26**, 1255–1263.

Burritt, E.A. and Provenza, F.D. (1990) Food aversion learning in sheep: persistence of conditioned taste aversions to palatable shrubs (*Cercocarpus montanus* and *Amelanchier alnifolia*). *Journal of Animal Science* **68**, 1003–1007.

Collias, N.E. (1956) The analysis of socialization in sheep and goats. *Ecology* **37**, 228–239.

Doucet, J.P., Squinto, S.P. and Bazan, N.G. (1990) Fos-Jun and the primary genomic response in the nervous system. Possible physiological role and pathophysiological significance. *Molecular Neurobiology* **4**, 27–54.

Geist, V. (1968) On the inter-relation of external appearances, social behaviour and social structures of mountain sheep. *Zeitschrift für Tierpsychologie* **25**, 199–215.

Geist, V. (1971) *Mountain Sheep: A study in behaviour and evolution*. University of Chicago Press, Chicago.

Gherardi, S.G. and Black, J.L. (1991) Effect of palatability on voluntary feed intake by sheep. I. Identification of chemicals that alter the palatability of a forage. *Australian Journal of Agricultural Research* **42**, 571–584.

Gherardi, S.G., Black, J.L. and Colebrook, W.F. (1991) Effect of palatability on voluntary feed intake by sheep. II. The effect of altering the palatability of a wheaten hay on long-term intake and preference. *Australian Journal of Agricultural Research* **42**, 585–598.

Green, G.C., Elwin, R.L., Mottershead, B.E., Keogh, R.G. and Lynch, J.J. (1984) Long term effects of early experience to supplementary feeding in sheep. *Proceedings of the Australian Society of Animal Production* **15**, 373–380.

Gross, C.G., Rocha-Miranda, C.E. and Bender, D.B. (1972) Visual properties of neurons in inferotemporal cortex of the macaque. *Journal of Neurophysiology* **35**, 96–111.

Hafez, E.S.E. and Signoret, J.P. (1969) The behaviour of swine. In: Hafez, E.S.E. (ed.), *The Behaviour of Domestic Animals*. Ballière, Tindall and Cassell, London, pp. 341–390.

Harlow, H.F. (1969) Age-mate or peer affectional system. *Advances in the Study of Behaviour* **2**, 333–383.

Hemsworth, P.H., Findlay, J.K. and Beilharz, R.G. (1978) The importance of physical contact with other pigs during rearing on the sexual behaviour of the male domestic pig. *Animal Production* **27**, 201–207.

Horn, G. and McCabe, B.J. (1984) Predispositions and preferences: effects on imprinting of lesions to the chick brain. *Animal Behaviour* **32**, 288–292.

Johnson, M. (1988) Memories of mother. *New Scientist*, 18 February, 60–62.

Johnson, M.H. and Horn, G. (1988) Development of filial preferences in dark-reared chicks. *Animal Behaviour* **36**, 675–683.

Johnson, M.H., Bolhuis, J. and Horn, G. (1985) Interaction between acquired preferences and developing predispositions in an imprinting situation. *Animal Behaviour* **33**, 1000–1006.

Kendrick, K.M. (1990a) Neural processing of visual recognition of individuals in sheep. In: Zayan, R. and Dantzer, R. (eds), *Social Stress in Domestic Animals.* Kluwer Academic Publishers, The Netherlands, pp. 144–156.

Kendrick, K.M. (1990b) Through a sheep's eye. *New Scientist*, 12 May, 62–65.

Kendrick, K.M. (1991) How the sheep's brain controls the visual recognition of animals and humans. *Journal of Animal Science* **69**, 5008–5016.

Kendrick, K.M. and Baldwin, B.A. (1986a) The activity of neurones in the lateral hypothalamus and zona incerta of sheep responding to the sight or approach of food is modified by learning and satiety and reflects food preference. *Brain Research* **375**, 320–328.

Kendrick, K.M. and Baldwin, B.A. (1986b) Characterization of neuronal responses in the zona incerta of the subthalamic region of the sheep during ingestion of food and liquid. *Neuroscience Letters* **63**, 237–242.

Kendrick, K.M. and Baldwin, B.A. (1987) Cells in temporal cortex of conscious sheep can respond preferentially to the sight of faces. *Science* **236**, 448–450.

Kendrick, K.M. and Baldwin, B.A. (1989a) The effects of sodium appetite on the responses of cells in the zona incerta to the sight or ingestion of food, salt and water in sheep. *Brain Research* **492**, 211–218.

Kendrick, K.M. and Baldwin, B.A. (1989b) Visual responses of sheep temporal cortex cells to moving and stationary human images. *Neuroscience Letters* **100**, 193–197.

Kendrick, K.M. and Baldwin, B.A. (1991) Single-unit recording in conscious sheep. In: Conn, P.M. (ed.), *Methods in Neuroscience*, vol. 4. Academic Press, San Diego, pp. 3–15.

Kendrick, K.M., Hinton, M.R. and Baldwin, B.A. (1991a) GABA release in the zona incerta of the sheep in response to the sight and ingestion of food and salt. *Brain Research* **550**, 165–168.

Kendrick, K.M., Lévy, F. and Keverne, E.B. (1991b) Neurochemical changes underlying the formation of olfactory memory in sheep. *Current Separations* **10**, 96–97.

Kendrick, K.M., Lévy, F. and Keverne, E.B. (1991c) Importance of vaginocervical stimulation for the formation of maternal bonding in primiparous and multiparous parturient ewes. *Physiology and Behavior* **50**, 595–600.

Kovalcik, K. and Kovalcik, M. (1986) Learning ability and memory testing in cattle of different ages. *Applied Animal Science* **15**, 27–35.

Lincoln, G.A. (1972) The role of antlers in the behaviour of the Red Deer. *Journal of Experimental Zoology* **182**, 233–250.

Lynch, J.J. and Bell, A.K. (1987) The transmission from generation to generation in sheep of the learned behaviour for eating grain supplements. *Australian Veterinary Journal* **64**, 291–292.

Mason, W.A. (1960) The effects of social restriction on the behavior of rhesus monkeys: I. Free social behavior. *Journal of Comparative and Physiological Psychology* **53**, 582–589.

Matthews, L.R. and Kilgour, R. (1980) Learning and associated factors in ruminant feeding behavior. In: Ruckebusch, Y. and Thivend, P. (eds), *Digestive Physiology and Metabolism in Ruminants.* AVI Publishing Co., Westport, USA, pp. 123–144.

Meese, G.B., Conner, D.J. and Baldwin, B.A. (1975) Ability of the pig to distinguish between conspecific urine samples using olfaction. *Physiology and Behaviour* **15**, 121–125.

Milne, J.A., Hodgson, J., Spence, A.M. and Baldwin, B.A. (1982) Effect of ablation of the olfactory bulbs on the diet ingested by grazing sheep. *Applied Animal Ethology* **8**, 261–268.

Moberg, G.P. and Wood, V.A. (1982) Effect of differential rearing on the behavioral and adrenocortical response of lambs to a novel environment. *Applied Animal Ethology* **8**, 269–279.

Perrett, D.I., Rolls, E.T. and Caan, W. (1982) Visual neurones responsive to faces in the monkey temporal cortex. *Experimental Brain Research* **47**, 329–342.

Perrett, D.I., Smith, P.A.J., Mistlin, A.J., Chitty, A.J., Head, A.S., Potter, D.D., Broennimann, R., Milner, A.D. and Jeeves, M.A. (1985) Visual analysis of body movements by neurones in the temporal cortex of the macaque monkey: a preliminary report. *Behavioural Brain Research* **16**, 153–170.

Poindron, P. and Lévy, F. (1990) Physiological, sensory, and experiential determinants of maternal behaviour in sheep. In: Krasnegor, N.A. and Bridges, R.S. (eds), *Mammalian Parenting*. Oxford University Press, Oxford, pp. 133–156.

Provenza, F.D. and Burritt, E.A. (1991) Socially induced diet preference ameliorates conditioned food aversion in lambs. *Applied Animal Behaviour Science* **31**, 229–236.

Ramachandran, V.S., Clarke, P.G.H. and Whitteridge, D. (1977) Cells selective to binocular disparity in the cortex of newborn lambs. *Nature* **268**, 333–335.

Rolls, E.T., Burton, M.J. and Mora, F. (1980) Neurophysiological analysis of brain stimulation reward in the monkey. *Brain Research* **194**, 339–357.

Scheich, H., Langner, B. and Bonke, D. (1979) Responsiveness of units in the auditory neostriatum of the guinea-fowl (*Numida meleagris*) to species-specific calls and synthetic stimuli. II. Discrimination of iambus-like calls. *Journal of Comparative Physiology* **132**, 257–276.

Shillito-Walser, E.E. (1978) A comparison of the role of vision and hearing in ewes finding their own lambs. *Applied Animal Ethology* **4**, 71–79.

Shillito, E. and Alexander, G. (1975) Mutual recognition amongst ewes and lambs of four breeds of sheep (*Ovis aries*). *Applied Animal Ethology* **1**, 151–165.

Shillito-Walser, E., Willadsen, S. and Hague, P. (1981) Pair association between lambs of different breeds born to Jacob and Dalesbred ewes after embryo transplantation. *Applied Animal Ethology* **7**, 351–358.

Soltysik, S.S. and Baldwin, B.A. (1972) The performance of goats in triple choice delayed response tasks. *Acta Neurobiologica Experimenta* **32**, 73–86.

Squibb, R.C., Provenza, F.D. and Balph, D.F. (1990) Effect of age of exposure on consumption of a shrub by sheep. *Journal of Animal Science* **68**, 987–997.

Sutherland, N.S. (1961) The methods and findings of experiments on the visual discrimination of shape by animals. *Experimental Psychology Society Monographs* No. 1.

Van Kampen, H.S. and Bolhuis, J.J. (1991) Auditory learning and filial imprinting in the chick. *Behaviour* **117**, 303–319.

Winfield, C.G. and Mullaney, P.G. (1973) A note on the social behaviour of a flock of Merino and Wiltshire horn sheep. *Animal Production* **17**, 93–95.

Zenchak, J.J. and Anderson, G.C. (1980) Sexual performance levels of rams (*Ovis aries*) as affected by social experiences during rearing. *Journal of Animal Science* **50**, 167–174.

Zito, C.A., Wilson, L.L. and Graves, H.B. (1977) Some effects of social deprivation on behavioral development of lambs. *Applied Animal Ethology* **3**, 367–377.

Section III

Animal Responses to the Environment

Recently, intensive animal production systems have been developed to enable large numbers of animals to be housed in a relatively small area and thereby reduce the capital and labour requirements for livestock husbandry. Consideration of the animals' responses to the environment was initially largely confined to productivity, in an anthropocentric attempt to maximize economic return. More recently zoocentric concern by the general public has forced detailed examinations of the welfare of livestock in intensive systems. Initially a great deal was discussed with little scientific basis but eventually it has proved possible to quantify animals' environmental responses sufficiently for objective decisions to be made on whether systems or parts of systems are morally acceptable or not. However, many experiments, especially preference and operant conditioning tests, have raised further questions, particularly the influences of genotype and previous experience. Future research will no doubt expand our knowledge of these influences to reduce further the subjectivity of welfare decisions and the excessively anthropomorphic assessment of husbandry systems.

13 Behaviour

J.F. Hurnik

Department of Animal and Poultry Science, University of Guelph, Canada

Summary

The acceptability of management systems for farm animals has traditionally been dominated by economic considerations. Strongly emerging ethical concerns regarding the quality of life of farm animals demand more serious attention to the animals' psychological properties and social tendencies. Morally acceptable housing systems are expected to assure adequate harmony between the genetic predispositions of farm animals and their overall environment. Familiarity with animals' cognitive abilities, learning capacities and motor activities is a necessary prerequisite for the successful design of animal housing. Behavioural response can be defined as an observable interaction between the organism's motivational state and the perceived attributes (stimuli) of its environment. This chapter classifies the interaction between motivational and environmental factors into four distinct categories: attraction; aversion; deprivation; and placation. Behavioural signs characteristic of each of the four classes permit a reasonable interpretation of animal desires. However, since desire may or may not be identical to need, such interpretation is reliable only when animal desires and animal needs concord fully with each other. Otherwise, interpretation of animal responses may be unreliable for long-term assessment of the quality of their environment. Consequently, in addition to the interpretation of behaviour responses, dependable evaluation of newly designed animal production systems should also include a comprehensive analysis of the physical and psychological health of the animals involved, including the incidence of mortality.

Introduction

The intention to detect the rules which guide animal responsiveness to their surroundings has attracted a large number of biologists from many scientific disciplines. Pavlov, Skinner, Tinbergen, Lorenz, just to name some well-known pioneers in behavioural sciences, have helped us to understand many of the psychophysiological phenomena which govern animal behaviour.

The design of animal facilities and management systems has traditionally been dominated by short-term economic considerations. In the past, scientists in animal agriculture did not utilize very systematically the potential benefits of ethological information to improve the productivity of farm operations and better the well-being of farm animals. Strongly emerging ethical concerns for the quality of life of farm animals demand serious attention of the designers of contemporary production systems to behavioural requirements of farm animals, including their psychological propensities and social tendencies.

The Concept of Harmony between Animals and their Surroundings

In order to minimize misunderstanding, let me first briefly qualify that in this chapter behavioural response is defined as *any observable action of the organism that results from stimulation* (Hurnik *et al.*, 1985), regardless if such stimulation has its origin within or outside the organism and regardless of whether or not the observed response is influenced or uninfluenced by conditioning processes. An observed behaviour can be defined as a manifested interaction between the genotype of the organism and its perceived environment. Genotype and environment are the fundamental determinants of all organic forms of life; all of its functions and actions. Each individual animal is a unique combination of its genotype and environmental experiences. Psychophysiological states of animals form the motivational base of manifested behaviours and are governed by the same biological principle.

Optimal biological functioning of an organism occurs only when it lives in the most appropriate surroundings. Let us call such an ideal blend between the organism and its surroundings a *state of harmony* and assume that, under such conditions and only under such conditions, the best overall biological functioning of the organism is assured and the maximum quality of life is reached. Optimal biological functioning does not necessarily mean highest performance, such as the fastest speed, domination of the largest territory, the most powerful response, or the strongest reproduction. Highest performance may lead to lower quality of life by weakening the organism through exhaustion and greater susceptibility to infections or injuries or early mortality. In relation to animal quality of life, i.e. its well-being or suffering, the concept of biological functioning must encompass the organism's overall capacity to interact with its surrounding and adjust to changes as they occur during its life.

The concept of harmony, as used in this discussion, rests on an optimal convergence – the best possible blend – between the organism and its environment, achievable only through full adaptation of the organism to its surroundings or full adjustment of the surroundings to the genetic and ontogenic predispositions of the organism. The term *harmony* should thus be understood as a theoretical framework, an abstract notion of optimal well-being. Different surroundings may emphasize different biological functions of the organism. An animal can have good quality of life without having all the capabilities of its wild predecessors. For example, an animal living in an area totally free of predators will never need to use its ability to avoid or escape a predator as long as it stays in that area. In this case, the relevance

MOTIVATION (Animal intention)		STIMULUS (Environmental factor): Present	STIMULUS (Environmental factor): Absent
	To Acquire	ATTRACTION	DEPRIVATION
	To Avoid	AVERSION	PLACATION

Fig. 13.1. Psychological states resulting from different combinations of stimulus presence or absence and animal motivation.

of a specific function to the quality of animal life can be derived only from correlation to other functions that are important in this environment. Farm animals, as do all other animals, possess a range of behavioural expectations regarding their surroundings in order to live as physically and psychologically healthy individuals. If the opportunity to exercise these expectations is restricted, harmony cannot be achieved and the quality of their lives may be significantly reduced. In the context of agricultural ethics, discussion of such a concept does not, in itself, fully define our obligations towards farm animals. Moral acceptability of a given production system depends on joint consideration of animal quality of life, achieved productivity and ecological consequences (Hurnik, 1990).

Classification of Interaction between Motivational and Environmental Factors

An animal's surroundings consist of numerous components or factors which can generally be defined as stimuli. According to genetic predisposition or ontogenic experience, an animal might be motivated to acquire or avoid certain stimuli and such stimuli might be either present or absent in the animal's environment. A combination of the organism's intention to acquire or avoid a given stimulus and the presence or absence of such a stimulus in its surroundings can be classified into four basic psychological categories: attraction; aversion; deprivation; or placation (Fig. 13.1).

Attraction

Attraction results from the organism's motivation to experience a given stimulus. Behavioural indicators of attraction are:

- consummatory responses
- satiation responses.

The utilization of required and available stimuli is facilitated through stimulus-specific consummatory actions such as consumption, investigation, playing or copulation. Consummatory activity, if completed, leads to satiation, cessation of a particular response and refocusing on other forms of stimulation. For example, a hungry animal attracted to feed may engage in eating and after reaching satiation, redirect its attention to a source of water. When transferred to a new pen, an animal will closely investigate its surroundings before turning to other activities such as resting or eating.

Aversion

Aversion occurs when an organism is motivated to avoid a stimulus present in its surroundings. Behavioural indicators of aversion include a broad spectrum of actions, which can be classified as:

- neutralization responses
- resignation responses.

Neutralization responses tend to minimize the effects of the aversive stimulus and they may include avoidance, escape, threat signals and aggression or attempts to ignore or appease the aversive stimulus. For example, given the opportunity, an animal will avoid zones of thermal discomfort; a subordinate male in a sexually heterogeneous flock or herd may not attempt to copulate in the presence of a dominant, aggresive male, but instead display signs of appeasement or subordination.

Resignation responses are characterized by cessation of active resistance to the aversive stimulation. Resignation can be manifested as helplessness, characterized by immobility and lack of defence, even if the subject is able to respond, or lethargy, characterized by the organism's inability to respond due to physical exhaustion. Chickens and pigs subjected to repeated cannibalistic events often cease vocalizing or running away when pecked or bitten (perhaps to avoid attracting additional cannibalistic peers); in a lethargic stage these animals are immobile, too weak to ambulate or even to approach feeders or waterers. Powerful, extended or repeated aversive stimulation often leads to neurotic or maladaptive behavioural disorders.

Deprivation

Deprivation results from the inability to experience a desired stimulus. Typical behavioural indicators of deprivation are:

- searching responses
- compensatory responses.

Searching for a deprived stimulus is characterized by actions that increase the probability the organism will experience this stimulus. Spatial search, vocalization, disregard or challenge of obstacles, and willingness to accept punishment in order to reach the desired goal are the most common signs of the early stage of deprivation. Experienced animal farmers can readily identify vocalizations of thirsty or hungry animals.

Compensatory responses occur if the organism does not succeed in obtaining the deprived stimulus. These include willingness to accept suboptimal stimuli which possess some attributes common to the deprived stimulus (*stimulus compensation*) or operational similarity in the action patterns by which the missing stimulus is consummated (*action compensation*). The distinction between stimulus compensation and action compensation accounts for potential differences between biological mechanisms controlling the organism's stimulus substance satiation and mechanisms controlling the organism's action satiation to attain this stimulus. In modern production systems, animals are fed balanced, readily available diets of high nutritional density. The efficiency of feed intake is substantially higher and the eating time to satisfy metabolic requirements of these animals is much shorter than in traditional, extensive systems. Consumption of readily available, high density diet requires 50–80% less eating time to satisfy animals' metabolic requirements. As a result, these animals are nutritionally satiated but partially deprived of eating activity, thus engaging in feather pecking, playing with the feed, sham-chewing or other forms of compensatory actions. The reverse, i.e. action satiation but stimulus deprivation, may also occur if consummation efficiency is very low and action satiation occurs before an adequate amount of the stimulus substance is attained. Such an animal would be satiated in eating activity, but deprived in feed intake.

Stimulus compensation can occur in two ways. *Stimulus approximation* involves refocusing on functionally or perceptually similar stimuli, whereas *stimulus displacement* involves refocusing on completely different, seemingly irrelevant, stimuli. Stimulus approximation occurs when a rooster suddenly deprived of females becomes sexually attracted to male peers. An animal deprived of food and showing aggression toward its groupmates illustrates stimulus displacement.

The spectrum of action compensations appears to be generally broader than stimulus compensation (Table 13.1). An animal may perform actions resembling those needed to consummate the deprived stimulus. A hen deprived of feed may peck on the empty feeder or the walls of the pen using motions very similar or identical to those needed for feed intake. These behaviours can be called *redirected activities*. In another situation, an animal may perform actions that are only rudimentary or incomplete when compared to those needed to attain the deprived stimulus. These actions, usually incomplete compared to consummation of the deprived stimulus, are often repeated over an extended period of time. A hen housed in the restrictive environment of a cage or a horse housed in a small pen with very limited opportunity to walk may pace, using repeated one or two-step movements. These behaviours can be categorized as *stereotypic activities*. An animal may also perform actions necessary to consummate the deprived stimulus without the presence of such a stimulus in its surroundings. A prepartum sow housed in a farrowing crate may perform head, leg and body movements identical to those used to build or readjust a litter nest without any nest-building material being available. These behaviours can be called *vacuum activities*. In some cases, an animal deprived of a desired stimulus may engage in actions that appear completely irrelevant to those needed to consummate the deprived stimulus. These actions can be classified as *displacement activities*. A hen in a battery cage which has no access to a nest, may display an increased tendency for pacing or feather pecking prior to egg laying.

Complete or extended deprivation of a substance essential to animal health or life, will lead to morbidity, progressive debility or mortality. After unsuccessfully searching for water or feed and a period of compensatory responses, deprived hens

Table 13.1. Psychological states and their characteristic behavioural indicators.

Psychological state of the organism	Behavioural indicators
Attraction	• consummatory responses • satiation responses
Aversion	• neutralization responses • resignation responses – helpnessness – lethargy
Deprivation	• search for responses • compensatory responses – stimulus compensation stimulus approximation stimulus displacement – action compensation redirected activities stereotypic activities vacuum activities displacement activities
Placation	• restoratory (contentment) responses

will show signs of reduced mobility, increased recumbency, loss of vigour and progressive bodily deterioration. Debility, even if corrected, usually has a long-lasting efect on the organism's physical and psychological capabilities.

Placation

Placation is a psychological state that develops when a stimulus, which the organism wants to avoid, is neutralized. Behavioural indicators of placation are:

- contentment–restoration responses.

Birds in battery cages or young pigs in a pen attempt to escape if an unfamiliar person approaches close to the cage. When the person leaves, they cease their escape attempts, calm down and turn to other activities.

These four psychological states and corresponding behavioural indicators relate to elementary units of animal surroundings – a single stimulus. In reality, the surroundings are composed of (or lacking) an array of stimuli that might have different psychological implications at various times. The arousal potential of stimuli which coincide in time might be different or similar, generating motivational complexities or conflicts. A thirsty animal may be attracted to water but, at the same time, experience aversion to an aggressive peer standing close to the water. Another factor contributing to the complexity of an animal's psychological state is its past experience. If an animal experiences some form of reinforcement generated through its own action, conditioning can occur, which will influence future responses to the same or similar situations. Animals housed in close confinement of battery cages, gestation or farrowing crates often engage in cage pecking, sterotypic champing, vacuum nest building or pacing. If such activities become rewarding, e.g. by reducing the animal's awareness of deprivation, or incidentally reinforced by some positive stimulation, e.g. feed delivery, these responses may become conditioned and even generalized. Animals may then display these behaviours in response to situations that tend to generate fear, frustration and conflicts, or excitement and anticipation of reward. The laying hen in a battery cage may then display stereotypic pacing prior to egg laying, when exposed to novel objects, or at the beginning of feed delivery. Sows in gestation crates very often show an increased level of sterotypic champing and redirected rooting prior to feeding. Attention to ontogenic experiences is of particular significance for an interpretation of animal actions. Behavioural analyses which ignore the role of acquired psychological association may lead to misidentification of the prime causes of animal actions.

Attraction and placation are usually considered well-being-positive psychological states, whereas deprivation and aversion are well-being-negative. Deeper analysis of the above states, however, reveals a more complex relation to animal well-being or suffering. To deal with this topic it is first necessary to address the concepts of animal needs and desires.

Animal Needs and Desires

The concept of quality of animal life is evaluative (Rollin, 1981, 1989). This does not mean that the evaluation is necessarily subjective and unreliable. Judgements about the quality of human life, well-being or suffering, are also evaluative, yet they are based on objective criteria. Whether an animal is or is not in acceptable harmony with its surroundings is a function of subjective conditions. For an animal to live and be healthy, certain conditions must be satisfied. Nevertheless, an animal might be healthy, but uncomfortable. To avoid discomfort, further conditions must be satisfied. Conditions which must be satisfied for the animal to live, stay healthy and be comfortable may be called *needs*. The quality of life of animals housed in certain systems might be considered to be good if animal's needs are well satisfied.

It might be suggested by some that the quality of an animal's life is a function satisfying its desires rather than its needs. The term *desire* refers to the *organism's motivation to acquire, control or experience features of its environment* (Hurnik *et al.*, 1985). Desires are interpreted operationally from observed behaviour and may or may not indicate underlying needs. It appears safer to regard an animal's quality of life to be a function of satisfaction or dissatisfaction of needs rather than desires (Hurnik and Lehman, 1988). Animals may occasionally want to have substances, or to do things that harm them, as exemplified in Table 13.2. In such cases, they may even experience serious suffering if their desires are satisfied. Furthermore, suppose that some animals have desires and that, sometimes, satisfying these desires is not in itself harmful. In such situations, it is still possible that satisfying these desires would occur at the expense of satisfying some serious needs. If this were to happen, one would not be able to agree that the animal was better off because its desires had been satisfied. Of course, in many cases, satisfaction of a desire would benefit the animal or, at least,

Table 13.2. Examples of beneficial or harmful consequences of psychological states.

Psychological state	Stimulus context	Consequence for the animal
Attraction	Nutritionally balanced diet	Beneficial
	Nutritionally detrimental diet	Harmful
Aversion	Poisonous substances	Beneficial
	Medication	Harmful
Deprivation	Contact with infected peer	Beneficial
	Space for muscular exercise	Harmful
Placation	Safe resting site	Beneficial
	Unsafe resting site	Harmful

not harm nor interfere with its needs. Only in such cases would satisfaction of the desire be good for the animal.

Evaluation of animal well-being or suffering requires estimating the extent to which their life-, health- and comfort-sustaining needs are satisfied (Hurnik and Lehman, 1988). To assess accurately the quality of animal life in different production systems, we must take into account not only all of their needs but also the degree to which each need is satisfied. This raises complex questions. It may be difficult for all of the animals to satisfy all their needs completely. For example, newly born animals need an environment that has sufficient warmth and fresh air. Achieving adequate warmth may conflict with achieving sufficient fresh air. If we have two animals in such circumstances, one has to breathe a certain proportion of ammonia, but can easily maintain sufficient body temperature whereas the other has fresher air but is shivering from cold. Which of the two is suffering more? It is reasonable to assume that the better an individual's needs are satisfied, the better are its prospects for a relatively long life. If two groups of animals of a similar genotype are compared, and in one group the needs are satisfied more completely than the other, longevity in the former group would probably be greater. This suggests that longevity, subject to certain restrictions, might be taken as a reasonably reliable indicator of animal well-being (Hurnik and Lehman, 1988).

It is possible that some might object to the use of longevity as an indicator of animal quality of life on the ground that animals in captivity may live longer than their non-captive counterparts. It may also be argued, however, that all agricultural practices require some kind of captivity and the termination of life of farm animals occurs before their natural death. The comparisons are not made between captive and non-captive animals, but among various production systems in which some spatial restriction, weather and predator protection, dietary control and medical assistance to animals will always be practised. Possibly, certain restrictions or qualifications should be taken into account in applying longevity as an indicator of an individual animal's well-being or suffering. Nevertheless, there are sufficient reasons to support its use in our time. Longevity assessment has several important advantages. It is a broadly based variable that provides relevant weight to satisfaction of important and less important biological requirements and avoids potentially serious inaccuracies in the human interpretation of animal needs and animal desires. It combines the cumulative influences of harmful and beneficial stimulation experienced by the animal during its life. Longevity is an objectively measurable parameter which offers immediately applicable comparisons of different production systems. Longevity studies also provide an opportunity to monitor behavioural responses indicating inadequate complexity of surroundings (such as neutralization, resignation or compensation responses), to measure animal productivity and to assess ecological impact, thus broadening the objective base for comparative assessment of various production systems prior to their commercialization. Longevity studies would be time consuming and expensive. Nevertheless, such effort appears to be unavoidable in our search for advanced systems of animal agriculture.

Conclusion

Determining which animal production system is morally acceptable is not a simple matter. Review of animal welfare literature (e.g. Harrison, 1964; Singer, 1975; Dawkins, 1980; Narveson, 1980; Rollin, 1989; Wood-Gush and Vestergaard, 1989) indicates that the concept of animal well-being is not sufficiently clear, despite many experiments conducted in the last two decades (e.g. Sorensen, 1981; Wegner, 1985). Motivational studies to identify animal desires, using operant conditioning or choice preferences, often account for interesting experiments that may facilitate an understanding of animal responses to specific situations. However, generalization of the results of such studies may be seriously handicapped, because these responses change substantially with genotype, environmental experiences, age, reproduction stage, social circumstances, health states and many other variables.

This chapter classifies the interaction between motivational and environmental factors into four distinct categories: attraction; aversion; deprivation; and placation. Behavioural symptoms, characteristic of each of the four classes, permit a relatively accurate interpretation of animal desires. Nevertheless, for an assessment of animal needs, such interpretation may be appropriate only in a situation when animal desires and animal needs fully concord with each other. In opposite cases, analyses of animal responses would be unreliable for long-term assessment of the quality of animal surroundings. As a consequence of such weakness in the interpretation of behaviour responses, more reliable assessment of animal quality of life in newly designed animal production systems also requires a comprehensive analysis of the physical and psychological health of animals involved, including the incidence of mortality.

References

Dawkins, M.S. (1980) *Animal Suffering: The science of animal welfare.* Chapman and Hall, London.

Harrison, R. (1964) *Animal Machines: The new factory farming industry.* Vincent Stuart, London.

Hurnik, J.F. (1990) Animal welfare: Ethical aspects and practical considerations. *Poultry Science* **69**, 1827–1834.

Hurnik, J.F. and Lehman, H. (1988) Ethics and farm animal welfare. *Journal of Agricultural Ethics* **1**, 305–318.

Hurnik, J.F., Webster, A.B. and Siegel, P.B. (1985) *Dictionary of Farm Animal Behaviour*. University of Guelph, Guelph, Canada.
Narveson, J. (1980) Animal rights revisited. *Animal Regulation Studies* **2**, 223–236.
Regan, T. (1984) *The Case for Animal Rights*. University of California Press, Berkeley and Los Angeles.
Rollin, B.E. (1981) *Animal Rights and Human Morality*. Prometheus Books, New York.
Rollin, B.E. (1989) *The Unheeded Cry and Animal Consciousness: Animal pain and science*. Oxford University Press, Oxford and New York.
Singer, P. (1975) *Animal Liberation: A new ethics for our treatment of animals*. Avon Books, New York.
Sorensen, L.Y. (ed.) (1981) First European Symposium on Poultry Welfare, 1981, Koge. World Poultry Science Association – Danish Branch.
Wegner, R. (ed.) (1985) Second European Symposium on Poultry Welfare, 1985, Celle. World Poultry Science Association – German Branch.
Wood-Gush, D.G.M. and Vestergaard, K. (1989) Exploratory behaviour and the welfare of intensively kept animals. *Journal of Agricultural Ethics* **2**, 161–170.

14 Animal Welfare: its Scientific Measurement and Current Relevance to Animal Husbandry in Europe

D.M. BROOM

Department of Clinical Veterinary Medicine, University of Cambridge, Cambridge, UK

Summary

The welfare of an individual is its state as regards its attempts to cope with its environment (Broom, 1986). Poor welfare occurs in situations in which the effects on the animal are adverse in that there is reduced fitness or clear indications that fitness will be reduced, i.e. the animal is stressed, and in other situations where the individual has difficulty in coping with its environment. Welfare is improved by increased control over the environment. Welfare should not be defined solely in terms of subjective experiences. There is a wide variety of measurements of welfare. Behaviour measurements are of particular importance but should be combined with physiological, immunological, injury, disease, mortality risk, growth and reproduction measures in order to obtain complete information.

The Concept of Welfare

The difficult conditions which animals often encounter during their lives can have effects on the individual which are adverse in that they ultimately reduce its fitness. In order to try to avoid such adversity, animals use a wide variety of coping methods. Coping responses help the animal to control its interaction with its environment and to maintain mental and bodily stability. Both coping responses and direct effects of the environment affect the state of the animal. *The welfare of an individual is its state as regards its attempts to cope with its environment* (Broom, 1986).

Failure to cope means that fitness is reduced. An effect on an individual reduces fitness if it lowers life expectancy, reduces the number of offspring produced per breeding period or increases the interval between breeding periods. If the control systems of an animal are over-taxed to such an extent that there is an actual or potential reduction in fitness then that animal is stressed (Broom, 1988a; Broom and Johnson, in preparation). The welfare of an individual is poor in these cirumstances and also when the individual has difficulty in coping with the conditions encountered. If coping with the environment is easy, however, the welfare is good. Indeed the state of an individual in respect of coping with its environment may be positively affected by pleasurable sensations.

Welfare varies on a continuum from very poor to very good and will fluctuate during life. Fluctuations in welfare are indicated by a wide range of measures. The idea that welfare varies over a range and can be measured using a variety of indicators has also been emphasized by Curtis (1986) and Duncan (1987). The measurement of welfare must be objective and quite separate from any ethical considerations. Once the description of the state of the individual, based on the various measures of welfare, has been made, decisions about whether or not the welfare is ethically tolerable can be taken.

An important concept in relation to understanding animal welfare is the control which an individual has over its environment. Welfare is poorer when the individual lacks control, or feels pain or fear, or is affected by the consequences of lack of control. Conversely, a characteristic of situations in which welfare is good is that in them the individual has effective control over its interactions with its environment. The methods which animals use to try to maintain control and to cope with difficulties include a range of physiological and behavioural responses such as changes in heart rate, adrenal action, movements which reduce pain and actions which result in avoidance of excessive cooling. If there is some degree of loss of control, the consequence may be injury, or behavioural pathologies such as self-mutilation, or impaired immune system function. The amount that individuals are having to do to try to cope with difficulties and the consequences of failure to cope are measures of how poor their welfare is. If it is apparent from observed physiology and behaviour that the individual is effectively controlling its interactions with its environment then welfare can be assessed as being at the good end of the range. Individuals can then engage in low priority behaviour and there may be some indications from behaviour or brain functioning of pleasurable sensation. In general, however, it is more difficult to quantify the extent to which welfare is good.

Welfare in Relation to Suffering and Other Subjective Feelings

The subjective feelings of an individual are an important aspect of its welfare. Pleasurable or unpleasant feelings are a part of the state of an individual as regards its attempts to cope with its environment. Dawkins (1980) has discussed suffering in some detail and (Dawkins, 1990) has stated that 'suffering occurs when unpleasant subjective feelings are acute or continue for a long time because the animal is unable to carry out those actions that would normally reduce risks to life and reproduction in those circumstances'. This description of the circumstances in which suffering occurs has clear links with the definition of welfare explained earlier in this chapter. Unpleasant subjective feelings will affect the welfare of the individual. Suffering and poor welfare will often occur together but welfare should not be defined solely in terms of subjective experiences (Broom, 1990, 1991a, b). It is possible for the state of an individual to be affected without suffering occurring and most of the measurements of welfare do not give clear information about subjective experiences. Examples of situations in which welfare is poor but there is no suffering include those in which there is an injury or severe disease condition but the individual is asleep or narcotized. The injury or disease consequence is one aspect of the state of the individual, which means that welfare is poor. Any associated pain or other unpleasant feeling is another aspect of the state which will also have a substantial effect on welfare. Welfare is poor if there is an injury and worse if there is also suffering.

Measures of Animal Welfare

A few examples of measures of animal welfare are presented here and discussed in relation to what information they give us about the effects of the environment on the animal. Since there are many different methods of coping, and many signs of failure to cope, any one of these measures can indicate poor welfare. Good growth does not necessarily mean good welfare. For more extensive reviews of the subject see Fraser and Broom (1990, Chapter 29) and Broom and Johnson (in preparation, Chapters 5–7).

Physiological measures

Heart rate varies according to activity level but it also changes when animals are preparing for emergency action. The well-camouflaged hen willow grouse (*Lagopus lagopus*) shows bradycardia (lowered heart rate) when sitting on a nest and approached by a predator (Gabrielsen *et al.*, 1977). This must have the effect of reducing the likelihood that a movement will betray the presence of the bird to the predator. A sheep confronted with a man and a dog, however, will show tachycardia (increased heart rate) superimposed on any heart rate which is a consequence of recent action (Baldock and Sibly, 1990). The increased heart rate is preparation for rapid flight should this be necessary. In both situations, measurement of heart rate gives information about how much the individual is having to try to cope with the situation. There may be no long-term adverse consequences of the situation but the welfare of an individual which shows a very substantial heart rate response is less good than that of an individual which shows a small response. It may be that the subjective experience parallels the heart rate response but we do not know this. Fluctuations in heart rate do occur in people or other species which are anaesthetized and do not have subjective experiences.

Adrenal cortex activity is greater in severe operations on sheep than in minor operations (Shutt *et al.*, 1987) and greater after rough handling of hens being taken to slaughter than after gentle handling (Broom *et al.*, 1986: Knowles and Broom, 1990). There are many other examples of such adrenal cortex responses (Dantzer and Mormède, 1979) and in comparative studies of the effects of treatments a greater response indicates poorer welfare than a smaller response. As with heart rate measurement, suffering must often be greater when the increase in adrenal cortex activity is greater. Indeed Rushen (1986b) has shown that the sort of situations which result in larger adrenal cortex responses cause more subsequent avoidance than do those which elicit smaller responses. However, adrenal cortex responses can occur when a cut is made in tissue during a farm operation carried out under anaesthesia (D. Mellor, personal communication). The animal could not be said to be suffering but a cut is in itself an indication of poorer welfare, and the adrenal response in such conditions reinforces this. Some adrenal responses occur during courtship and mating when no assumption of poor welfare would be made so care must be taken to interpret these measures taking due account of the context in which they occur.

Behavioural measures

One response to long-term confinement in a small pen is reduced activity and unresponsiveness, perhaps associated with self-narcotization using endogenous opiates (Broom, 1987, 1988a). The process leading to such an extreme response is very likely to involve suffering but if there really is ultimate self-narcotization, it may be that no suffering is occurring at that time. Nevertheless the welfare of an individual which can only cope with its environment by self-narcotizing must be considered to be very poor. Similarly, the welfare of a human drug addict is poor when under the influence of a narcotic drug but there is no suffering at that time, even if there is a great suffering at other times. It is the welfare of the individual throughout these periods which concerns us, not just the period when there is suffering.

Stereotypies are also important indicators of poor welfare (Fraser and Broom, 1990) and their significance in relation to suffering is discussed by Mason (1991b). Whatever the causation of stereotypies, they are shown in situations which are difficult and it is likely that at least some of them are useless and energetically costly signs of brain function pathology (Dantzer, 1986; Mason, 1991a). People with psychiatric disorders, very severely deprived animals and individuals subjected to extreme frustration show stereotypies. They are interpreted as an indication of some temporary or long-lasting psychological problem in man and in other species. Hence they are clearly indicators of poor welfare but as Mason (1991b) explains, while they probably indicate suffering in the past, we have much less confidence in saying that the individual suffers when it is showing them.

Immunological measures, injury and disease

When animals are exposed to difficult conditions, a consequence which may occur frequently is immunosuppression. The efficiency of antibody responses and of cell-mediated immunity can be reduced by a variety of factors including extreme temperature conditions and situations where there is much aggressive behaviour (Kelly, 1980; Siegel, 1985; Broom, 1988b). Any housing or management condition which results in impaired immune system function is clearly not a good one for the animal and welfare is poorer if there is immunosuppression than if there is none. This is still true if no disease challenge occurs and while disease usually results in suffering there need not be any suffering associated with immunosuppression.

The point that injury and disease indicate poor welfare, irrespective of the occurrence of suffering, has been made already. The idea that clinical signs of disease or injury can be indicators of suffering or potential suffering is widely accepted (e.g.

Dawkins, 1988). Information about the severity of pain or discomfort can be obtained from behavioural and physiological measures (Morton and Griffiths, 1985; Fraser and Broom, 1990). The extent of body tissue damage and the degree of disturbance of physiological and behavioural processes give information about how poor is the welfare of the individual. We do not know the actual feelings of the animal but it is reasonable to deduce something about suffering from such measurements. However, it is more precise to speak of effects on welfare, for we know that the state of the individual is altered, than to speak of suffering and, once again, we cannot equate poor welfare with suffering.

Fitness measures

If two animal husbandry systems are compared and the mortality rate is significantly higher in the first than in the second then we can say that the welfare is less good in the first (Hurnik and Lehman, 1988; Broom, 1986, 1988a, 1991a). Similarly, welfare is less good in an animal, perhaps one in a zoo, which cannot reproduce even when a suitable partner is provided, than in an animal of the same species kept in conditions which do allow it to reproduce. In both of these comparisons, the animal whose welfare is poorer may have suffered but we often do not know whether this is so and while we should consider the welfare problem to be exacerbated by suffering, there is still a problem in the absence of suffering.

Preference assessment

It is clear that suffering is one of the most important aspects of poor welfare but that welfare encompasses some effects on animals in addition to subjective experiences. Much valuable information relevant to animal welfare can be obtained from behavioural studies in which the preferences of individuals are assessed. For example, the extent to which a situation is avoided on a second encounter can provide information about how aversive the first encounter was (Rushen, 1986a). Also, as emphasized by Dawkins (1983, 1990) and Duncan and Kite (1987) sophisticated preference tests can give information about the value of a resource to an individual. A strong positive preference indicates that the welfare of the individual is likely to be improved by acquisition of that resource, although care is needed in the interpretation of preference tests (Duncan, 1978; Dawkins, 1990; Broom, 1990, 1991a, in press). The subjective feelings of the individual during the test situation are being assessed in such sophisticated preference tests. Once this information has been obtained, new conditions for the animal can be designed and compared with existing conditions using a range of welfare indicators.

Current Relevance to Animal Husbandry in Europe

The extent of concern about animal welfare in Europe is shown by the fact that Members of Parliament in the United Kingdom and Members of the European Parliament in the European Community receive more letters about animal welfare than about any other topic. Many of these letters concern farm animals and the ways in which they are kept, transported and slaughtered. Some European countries have a government committee on animal welfare, in some cases a specific committee on farm animal welfare. The committees, the Ministries of Agriculture and the European Commission take much note of scientific evidence when discussing animal husbandry, and formulating new laws or Codes of Practice for use on farms. Scientists studying animal welfare play a considerable part in influencing public opinion and legislation. Indeed it is very important that new laws and recommendations should be based on scientific evidence about animal welfare rather than on emotional responses.

Following the Recommendations of the Standing Committee of the European Convention for the Protection of Animals Kept for Farming Purposes, most European countries have recently brought into force laws whose aim is to improve the welfare of farm animals. The Convention was formulated by the Council of Europe which now includes almost all European countries. Some stricter laws, directives and regulations have been brought in by the European Community and by various individual countries. The laws in Switzerland, Sweden, Germany, the United Kingdom, the Netherlands, Denmark and Norway are among the most progressive.

Examples of recent developments include the requirements for the stunning of animals prior to slaughter with details of stunning currents, etc., limitations on journey duration and stocking density during transport, prevention of certain operations on farm animals, requirements for the use of anaesthetic or for limitations as to age of animal for other operations, and many requirements concerning the housing and management of animals. Following recent evidence concerning the welfare of hens, calves and pigs certain housing systems are being prohibited. The use of small crates for rearing veal calves is now illegal in Switzerland, Sweden and the United Kingdom and is likely to be banned in the whole of the European Community soon. Stalls and tethers for pregnant sows are banned in Switzerland, Sweden and the United Kingdom and the use of tethers is being phased out in the whole European Community. The small battery cage is banned in Switzerland and will shortly be banned in Sweden and probably in The Netherlands.

At the same time that these new laws are being brought into force, laws to restrict imports of animal produce unless there has been acceptable slaughter, transport and housing are under consideration. These laws are already starting to affect non-European countries who export to Europe. For example, laws concerned with improving animal welfare, as well as those concerned with meat hygiene and quality are affecting the procedures used in slaughterhouses in non-European countries. Another effect on countries outside Europe is a moral one. People in Japan, North America, Australia, New Zealand and many other countries are not happy to be thought of as barbarians because of the way in which they treat animals. Neither are they pleased to be told by Europeans that in the area of world progress which relates to the humane treatment of animals, they are backward countries. The costs of improving animal welfare are not large so the change in attitudes to animals and consequential legislation are sweeping the world rapidly.

Conclusions

Welfare is a characteristic of an animal which varies from very poor to very good and which can be measured.

The term stress is best used for poor welfare in which the effects on the animal are adverse in that there is reduced fitness or clear indications that fitness will be reduced.

Welfare is also poor when the individual has difficulty in coping with its environment. It is good when an individual has effective control over the factors which affect its life. The welfare of an individual is its state as regards its attempts to cope with its environment.

Behavioural, physiological, immunological, injury, disease, mortality risk, growth and reproduction measures are all of value in assessing welfare. Behaviour measurements are of particular importance but a wide range of measures should be used.

Suffering is an important aspect of poor welfare but welfare can be poor in the absence of suffering, for example when injured or diseased but asleep, anaesthetized or treated with analgesics, or when the immune system is not working properly.

Sophisticated studies of animal preferences are very valuable guides to how to improve animal welfare. They need to be followed up by designing new conditions based on the preferences and comparing these with existing conditions.

Public pressure concerning farm animal welfare is very great in many European countries. Legislators are influenced by this but depend upon scientific evidence about animal welfare when formulating new legislation. Laws about housing, transport and slaughter are now stricter in all European countries following Council of Europe Recommendations. Recent developments in housing and management include the banning in some countries of veal-calf crates, sow stalls and tethers and battery cages. Other countries are affected by these developments because of trade restrictions and the spread of moral concerns about animal welfare.

References

Baldock, N.M. and Sibly, R.M. (1990) Effects of handling and transportation on heart rate and behaviour in sheep. *Applied Animal Behaviour Science* **28**, 15–39.

Broom, D.M. (1986) Indicators of poor welfare. *British Veterinary Journal* **142**, 524–526.

Broom, D.M. (1987) Applications of neurobiological studies to farm animal welfare. In: Wiepkema, P.R. and van Adrichem, P.W.M. (eds), *Biology of Stress in Farm Animals: An integrated approach. Current Topics in Veterinary Medicine and Animal Science* **42**, 101–110. Martinus Nijhoff, Dordrecht.

Broom, D.M. (1988a) The scientific assessment of animal welfare. *Applied Animal Behaviour Science* **20**, 5–19.

Broom, D.M. (1988b) The relationship between welfare and disease susceptibility in farm animals. In: Gibson, T.E. (ed.), *Animal Disease – a Welfare Problem*. British Veterinary Association Animal Welfare Foundation, London, pp. 22–29.

Broom, D.M. (1990) The importance of measures of poor welfare. *Behavioral and Brain Sciences* **13**, 14.

Broom, D.M. (1991a) Animal welfare: concepts and measurement. *Journal of Animal Science* **69**, 4167–4175.

Broom, D.M. (1991b) Assessing welfare and suffering. *Behavioural Processes* **25**, 117–123.

Broom, D.M. (in press) A usable definition of animal welfare. *Journal of Agricultural Ethics.*

Broom, D.M. and Johnson, K.G. (in preparation) *Stress and Animal Welfare.* Chapman and Hall, London; Routledge, Chapman and Hall, New York.

Broom, D.M., Knight, P.G. and Stansfield, S.C. (1986) Hen behaviour and hypothalamic–pituitary–adrenal responses to handling and transport. *Applied Animal Behaviour Science* **16**, 98.

Curtis, S.E. (1986) The case for intensive farming of food animals. In Fox, M.W. and Mickley, L.D. (eds), *Advances in Animal Welfare Science 1986/87.* The Humane Society of the United States, Washington DC, pp. 245–255.

Dantzer, R. (1986) Behavioral, physiological and functional aspects of stereotyped behavior: a review and a reinterpretation. *Journal of Animal Science* **62**, 1776–1786.

Dantzer, R. and Mormède, P. (1979) *Le stress en élevage intensif.* Masson, Paris, 118pp.

Dawkins, M. (1980) *Animal Suffering.* Chapman and Hall, London and New York, 149pp.

Dawkins, M. (1983) Battery hens name their price: consumer demand theory and the measurement of animal needs. *Animal Behaviour* **31**, 1195–1205.

Dawkins, M. (1988) Behavioural deprivation: a central problem in animal welfare. *Applied Animal Behaviour Science* **20**, 200–225.

Dawkins, M. (1990) From an animal's point of view: motivation, fitness and animal welfare. *Behavioral and Brain Sciences* **13**, 1–9.

Duncan, I.J.H. (1978) The interpretation of preference tests in animal behaviour. *Applied Animal Ethology* **4**, 197–200.

Duncan, I.J.H. (1987) The welfare of farm animals: an ethological approach. *Science Progress* **71**, 317–326.

Duncan, I.J.H. and Kite, V.G. (1987) Some investigations into motivation in the domestic fowl. *Applied Animal Behaviour Science* **18**, 387–388.

Fraser, A.F. and Broom, D.M. (1990) *Farm Animal Behaviour and Welfare.* Baillière Tindall, London; Saunders, New York, 437pp.

Gabrielsen, G.W., Kanwisher, J.W. and Steen, J.B. (1977) Emotional bradycardia: a telemetry study on incubating willow grouse, *Lagopus lagopus. Acta Physiologica Scandinavica* **100**, 255–257.

Hurnik, J.F. and Lehman, H. (1988) Ethics and farm animal welfare. *Journal of Agricultural Ethics* **1**, 305–318.

Kelley, K.W. (1980) Stress and immune function. A bibliographic review. *Annales Recherches Vétérinaire* **11**, 445–478.

Knowles, T.G. and Broom, D.M. (1990) The handling and transport of broilers and spent hens. *Applied Animal Behaviour Science* **28**, 75–91.

Mason, G.J. (1991a) Stereotypies: a critical review. *Animal Behaviour* **41**, 1015–1037.

Mason, G.J. (1991b) Stereotypies and suffering. *Behavioural Processes* **25**, 103–115.

Morton, D.B. and Griffiths, P.H.M. (1985). Guidelines on the recognition of pain, distress and discomfort in experimental animals and an hypothesis for assessment. *Veterinary Record* **116**, 431–436.

Rushen, J. (1986a) The validity of behavioural measures of aversion: a review. *Applied Animal Behaviour Science* **16**, 309–323.

Rushen, J. (1986b) Aversion of sheep for handling treatments: paired choice experiments. *Applied Animal Behaviour Science* **16**, 363–370.

Shutt, D.A., Fell, L.R., Cornell, R., Bell, A.K., Wallace, C.A. and Smith, A.I. (1987) Stress-induced changes in plasma concentrations of immunoreactive β endorphin and cortisol in response to routine surgical procedures in lambs. *Australian Journal of Biological Science* **40**, 97–103.

Siegel, H.S. (1987) Effects of behavioural and physical stressors on immune responses. In: Wiepkema, P.R. and van Adrichem, P.W.M. (eds.), *Biology of Stress in Farm Animals, Current Topics in Veterinary Medicine and Animal Science* **42**, 39–54.

15 Stress: a General Overview

R. EWBANK

Universities Federation for Animal Welfare (UFAW), Potters Bar, Herts, UK

Summary

Animals, in the living world, are continually having to make adjustments in their bodily states and lifestyles according to the varying demands made upon them by changes in their environment. Unless they adapt, they will not maintain internal homeostasis, bodily integrity and the ability to grow and reproduce, i.e. they will not survive or thrive. The adaptations take place through a variety of interrelating anatomical, physiological, biochemical, immunological, behavioural and psychological mechanisms. Some of the responses of these biological mechanisms are specific to the particular challenge, some are more general in nature. But the response, whatever its form, is at a biological cost – whether it be in energy, time or allocation of otherwise wanted resources. If the challenge is of low intensity, the necessary response will be small and well within the capacity of the animal; if the needed response is substantial then the adaptation may have been made at a cost detrimental to the animal's other biological enterprises.

The tendency nowadays, in stress theory, is to think of a quick response (the 'fight or flight' adrenaline reaction) followed by a slower developing but much more prolonged corticosteroid response as making up the main but not sole parts of the stress mechanisms. A chronic stressor may induce the corticosteroid response without there being an obvious fight/flight reaction.

This pituitary/corticosteroid pathway is central to stress research and any serious investigation nowadays has to take note, among other things, of the levels of blood corticosteroids and/or the ability of the anterior pituitary to respond, via the CNS, to standard stimuli and/or the ability of the adrenal cortex to respond to measured doses of ACTH.

In using the word 'stress' I conform to the ideas of those who hold that its use should be limited to substantial responses to noxious or potentially noxious stimuli (stressors), i.e. an animal which is being stressed is having to carry a biological cost for adapting (coping) or attempting to adapt to damage or threat of damage from the environment.

The term 'distress' is used to indicate a severe stress response in which there is some indication that the animal is conscious of what is going on, and finds it unpleasant. There should be behavioural signs of discomfort, pain, emotional distress, i.e. signs of suffering.

Confusions over the Meaning of the Term Stress

The word stress has come to be associated in several ways with varying parts of the adaptation mechanisms. Unfortunately there is much confusion as to the precise way in which the term should be used.

To some, stress is the thing or things in the environment which are disturbing/damaging the animal, to others it is the physiological response to the disturbing/damaging challenge and to yet others – admittedly mainly working in the human field – it is the psychological problem of being unable to cope with the demanding environmental situation. Many agricultural/biomedical workers nowadays tend to use the term only in its 'response to challenge' sense.

It will perhaps help our understanding if we first look at the way the word was originally used in the biomedical sciences, then go on to see how its use has widened and finally to consider stress definitions and suggest how the term could be most usefully employed in the future.

Original Use of the Term Stress

The work of Hans Selye in the 1930s, 1940s and 1950s (Selye, 1956) – the first user of the term in an adaptive mechanism sense – has been central to the development of the stress concept. He was impressed by the non-specific bodily responses made by animals (in the initial work, rats) to a whole variety of noxious or potentially noxious environmental stimuli (stressors). These non-specific responses – the triad of pathological changes (enlargement of adrenals, gastric ulceration, shrinkage of lymphatic tissue) and the three timephases (the initial alarm reaction, the stage of resistance and, if the stressor was intense and/or prolonged in its action, the stage of exhaustion) came to be known as the General Adaption Syndrome. Much work over the years has shown that there is an underlying biological pathway:

STIMULI (stressors) → receptor organs → CNS → anterior pituitary → ACTH → adrenal cortex → corticosteroids in blood → target/tissue organs → RESPONSE

Stressors stimulate the receptor organs and there can be a resulting release of corticosteroids into the blood. One of the main functions of these corticosteroids is to help mobilize the body's defence mechanisms in preparation for exertion and tissue repair. Continued high levels of blood corticosteroids, however, may be damaging to the animal – for example, by interference with its ability to produce antibodies (see Dantzer and Kelley, 1989, for further discussion).

This pituitary/corticosteroid pathway has remained central to stress research and any serious investigation nowadays has to take note, among other things, of the levels of the blood corticosteroids and/or the ability of the anterior pituitary to respond, via the CNS, to standard stimuli and/or the ability of the adrenal cortex to respond to measured doses of ACTH.

The earlier concepts of Walter Cannon and his co-workers (Cannon and de la Paz, 1911; Cannon, 1929) on the emergency reactions of animals to sudden environmental stimuli have been incorporated into the stress theory. An animal will be startled, its sympathetic nervous system will be activated above its resting level and the adrenal medulla and, at times, other organs/tissues will secrete adrenaline/noradrenaline into the blood. The function of the emergency reaction is the quick mobilization of the body's resources for the 'fight or flight' response: the heart rate is increased, blood sugar levels are raised, blood is redistributed from skin and viscera to muscles and brain, etc.

The tendency nowadays, in stress theory, is to think of a quick response (the 'fight or flight' adrenaline reaction) followed by a slower developing but much more prolonged corticosteroid response as making up one of the main parts of the stress mechanism. A chronic stressor may induce the corticosteroid response without there being an obvious fight/flight reaction (interestingly, Walter Cannon talked about such 'adjustment responses' in his lectures – see Cannon, 1935).

Figure 15.1 gives an indication of how Selye (1956) saw the overall physiological interactions in 1956. For a recent full account of the stress responses of the neuroendocrine system see the review of Becker (1986).

A recent complication of the stress story is the possible involvement of the β endorphins (endogenous morphine-like substances). It is now known that many animals can produce these hormonal/neurotransmitter substances and that they may, among other functions, have a pain relieving effect (natural opiates). Conditioned fear stimuli, for example, can result in endorphins being released from the anterior pituitary alongside ACTH and, it has been suggested, giving the animal an increased pain threshold, i.e. having an analgesic effect. It has also been proposed, on somewhat thin evidence, that the persistence of some stereotyped behaviours in animals may partly be a result of them becoming 'addicted' to endorphins during a prolonged period of intense adaptation (distress) and then finding that they can produce the endorphins needed to satisfy their 'addiction' by performing stereotyped movements.

Some authorities, e.g. Mason, 1971 and Burchfield, 1979, have held that there are cognitive and emotional elements involved in animals' stress responses. Levine (1990) has written a review of the relationships between cognition, psychological well-being and stress in laboratory animals.

No matter how these various biological mechanisms work together it does seem that animals respond to stimuli by specific and non-specific adaptive responses. It is these non-specific adaptive responses which have classically been labelled as the stress response.

In many cases the animal's response to environmental challenge is successful; the source of potential danger is removed and/or the resulting effects are handled by the stress and other adaptive mechanisms; the biological cost is well within the spare capacity of the animal; the animal survives and thrives. If, however, the potential source of damage to the animal is not readily removed or otherwise overcome by adaptive processes, the damage from the primary source is likely to continue; the stress mechanism will remain activated and this, in itself, can have deleterious effects. Figure 15.2 gives an overview of adaption and the stress response.

Prolonged unsuccessful stress responses and to some extent successful adaptation, but at a high biological cost, can produce long-term effects such as:

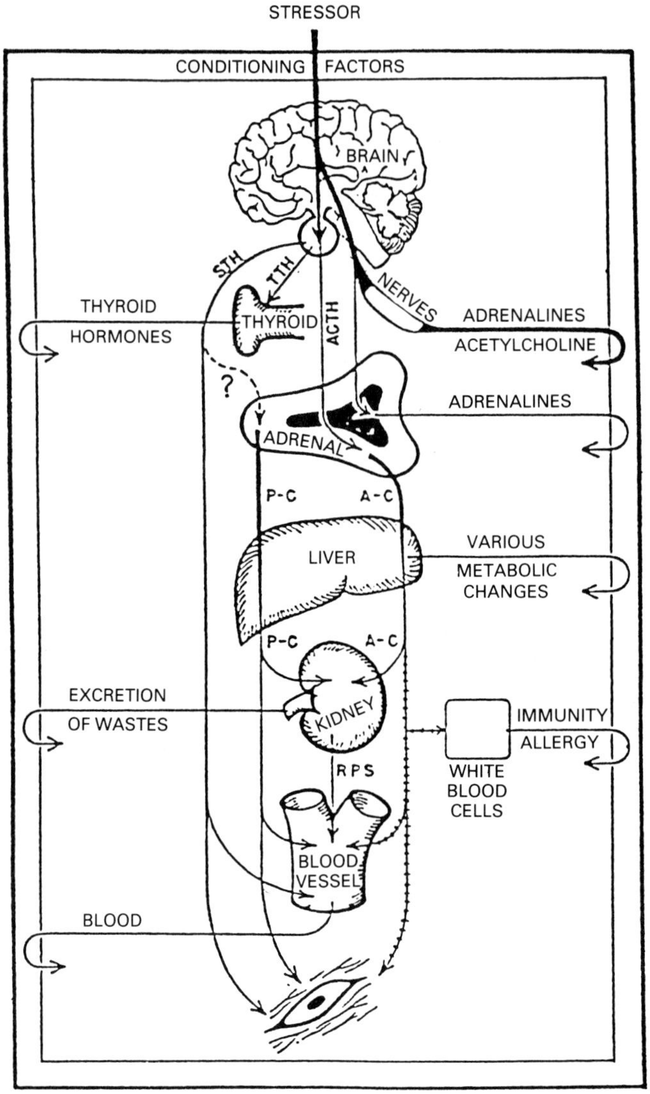

Fig. 15.1. Stress – the overall picture (Selye, 1956, *The Stress of Life*; reproduced by permission of McGraw-Hill Inc., New York).

1. structural damage to the animal, e.g. gastric ulceration;
2. interference with immune function;
3. decrease of growth;
4. impairment of reproduction;
5. changes in behaviour – signs of suffering?

Some of these long-term effects together with the relatively earlier developing short-term changes in blood adrenaline/corticosteroid levels and the behaviours

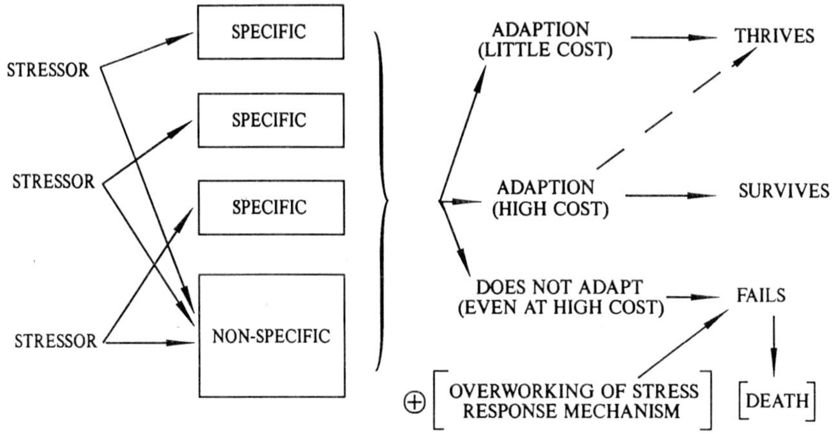

Fig. 15.2. An overview of adaption and the stress response.

associated with the fight/flight reaction, have been observed and measured in the attempt to quantify the stress response.

One of the main difficulties in the practical use of the term stress is a consequence of the conflicting ideas held by those who maintain that the term should be restricted to responses which are damaging in themselves and/or which are indicative of potentially harmful threats to the animal and those who believe that the term should cover even the mildest (non-harmful, very low biological cost) forms of adaptation. Selye was of the latter group and held that 'complete freedom from stress is death' (Selye, 1973). He also maintained that extremely pleasant experiences could be stressful to animals as could be the absence of environmental stimulation.

Table 15.1 indicates how some authors have variously used stress terms in relation to the level of animals' responses to stressors.

Definitions of Stress

There have been many attempts to define stress in general terms. A useful one would seem to be:

> Stress is the non-specific response of the body to any demand made upon it. (Selye, 1974)

Perhaps an even better attempt has been made by Archer (1979) in his somewhat dated but still very useful book *Animals under Stress*:

> Stress is the prolonged inability to remove a source of potential danger, leading to activation of systems for coping with danger beyond their range of maximum efficiency.

I find that in my current usage of the term, I tend to conform to the ideas of those who hold that its use should be limited to substantial responses to noxious or potentially noxious stimuli (stressors); i.e. an animal which is being stressed is having

Table 15.1. Use of stress terms in relation to the levels of stress response.

	Level of stress response		
	Relatively low level	Medium level	Relatively high level
Source	Harmless to animal, fully adaptive	Some damage to animal, probably adaptive	Damage to animal, possibly adaptive, unpleasant – causes suffering?
Selye (1956)	←	Stress	→
Ewbank (1973)	← Stress →	← Overstress	→
Selye (1974)	← Stress	→	← Distress →
Fraser et al. (1975)		← Stress	→
Ewbank (1985)	← ? Physiological Stress →	← Overstress →	← Distress →
Ewbank (this chapter)		← Stress →	← Distress →

Source: Adapted from Ewbank, 1985.

to carry a biological cost of adapting (coping), or attempting to adapt, to damage or threat of damage from the environment.

I like to try and indicate the type of stressor, e.g. heat, social, disease challenge, noise, and its probable intensity and possible duration, i.e. low or high strength, long or short acting, and then to suggest the severity of the stress response, i.e. mild or severe. In a mild response, as I see it, the animal is readily adapting and at a low biological cost. In a severe stress response the animal is coping (just?) but at a high biological price.

I try to restrict my use of the term distress to a severe stress response in which there is some indication that the animal is conscious of what is going on, and finds it unpleasant. There should be behavioural signs of discomfort, pain, emotional distress, i.e. signs of suffering. These signs should be noticeable by the stockperson, the person in charge of the animals or the inspecting veterinary surgeon. This usage of the term largely fits in with the UK legal terminology where, in the various animal protection and cruelty Acts, there is the general offence of causing unnecessary pain, suffering or distress. In many ways distress is seen as emotional suffering. We should not think that emotional suffering is a new concept which should be treated warily – it has been recognized in UK law for many years. The 1911 Protection of Animals Act makes it an offence, for instance, to 'infuriate, or terrify any animal'.

Recommendations

It is important that we should clear up in our own minds our understanding of the stress concept. Much scientific discussion becomes confused and much practical advice is disregarded when uncertain terminologies are employed. There is a real need to come to some general agreement as to how to define and use such terms as stressor, stress and distress.

Might I recommend that, in future, we:

1. restrict usage of the term *stress* to the largely adaptive response of an animal to a stressor;
2. define a *stressor* as a damaging or potentially damaging stimulus which is likely to induce a stress response;
3. always specify, if possible, the *type of stressor* (e.g. heat, social, noise);
4. always try and indicate the *intensity* (high, medium, low) and *duration of action* (long or short acting) of the stressor;
5. if possible suggest the *severity of the stress* response (e.g. mild, severe);
6. limit the employment of the term *distress* to a severe stress response accompanied by behavioural signs suggesting that the animal finds it unpleasant.

It has been put forward by some that the term stress has, over the years, been so widely used or misused that it can now have no place in the study of any such subject as 'farm animals and the environment'. This I dispute. If we can employ it with accuracy and care, it still has a most useful role to play.

References

Archer, J. (1979) *Animals under Stress.* Edward Arnold, London.

Becker, B.A. (1987) The phenomenon of stress: concepts and mechanisms associated with stress-induced responses of the neuro-endocrine system. *Veterinary Research Communications* **11**, 443–456.

Burchfield, S.R. (1979) The stress response: a new perspectve. *Psychosomatic Medicine* **41**, 661–672.

Cannon, W.B. (1929) *Bodily Changes in Pain, Hunger, Fear and Rage*, 2nd edition. Appleton, New York.

Cannon, W.B. (1935) Stresses and strains of homeostasis, *American Journal of Medical Sciences* **189**(1), 1–14.

Cannon, W.B. and de la Paz, D. (1911) Emotional stimulation of adrenal secretion. *American Journal of Physiology* **27**, 64–70.

Dantzer, R. and Kelly, K.W. (1989) Stress and immunity: an integrated view of relationships between the nervous and immune systems. *Life Sciences* **44**, 1995–2008.

Ewbank, R. (1973) The use and abuse of the term stress in husbandry and welfare. *Veterinary Record* **92**, 709–710.

Ewbank, R. (1985) Behavioral responses to stress in farm animals. In: Moberg, G.P. (ed.), *Animal Stress.* American Physiological Society, Bethesda, Maryland, pp. 71–79.

Fraser, D., Ritchie, J.S.D. and Fraser, A.F. (1975) The term 'stress' in a veterinary context. *British Veterinary Journal* **131**, 653–662.

Levine, S. (1990) Stress and cognition in laboratory animals. In Rollin, B.E. and Kesely, M.L. (eds), *The Experimental Animal in Biomedical Research*, vol. I. *A Survey of Scientific and Ethical Issues for Investigators.* CRC Press, Boca Raton, USA.

Mason, J.W. (1971) A re-evaluation of the concept of non specificity in stress theory. *Journal of Psychiatric Research* **8**, 323–333.

Selye, H. (1956) *The Stress of Life.* McGraw-Hill, New York.

Selye, H. (1973) The evolution of the stess concept. *American Scientist* **61**, 692–699.

Selye, H. (1974) *Stress without Distress.* Lippincott, Philadelphia.

Further Reading

Dantzer, R. (1989) Neuroendocrine correlates of control and coping. In: Steptoe, A. and Appels, A. (eds), *Stress, Personal Control and Health.* John Wiley, Chichester (UK), pp. 277–294.

Friend, T.H. (1991) Behavioral aspects of stress. *Journal of Dairy Science* **74**, 292–303.

Gray, J.A. (1987) *The Psychology of Fear and Stress*, 2nd edition. Cambridge University Press, Cambridge.

Lefcourt, A.M. (1986) Usage of the term 'stress' as it applies to cattle. *Vlaams Diergeneeskundig Tijdschrift* **55**, 258–265.

Mason, J.W. (1975) A historical view of the stress field. Part I and Part II. *Journal of Human Stress* **1**, (1) 6–12 and (2) 22–36.

Mench, J.A. (1991) Stress in birds. *Acta XX Congressus Internationalis Ornithologica* **3**, 1905–1914

Moberg, G.P. (ed.) (1985) *Animal Stress.* American Physiological Society, Bethesda, Maryland.

Puglisi-Allegra, S. and Oliverio, A. (eds) (1990) *Psychobiology of Stress.* Kluwer Academic Publishers, Dordrecht (Netherlands).

Selye, H. (1975) Confusion and controversy in the stress field. *Journal of Human Stress* **1**, (2) 37–44.

Smith, J.A. and Boyd, K.M. (eds) (1991) *Lives in the Balance. The ethics of using animals in biomedical reseach.* Oxford University Press, Oxford, pp. 45–77.

Stephens, D.B. (1980) Stress and its measurement in domestic animals: a review of behavioural and physiological studies under field and laboratory situations. *Advances in Veterinary Science and Comparative Medicine* **24**, 179–210.

Warburton, D.M. (1991) Stress and distress in response to change. In: Box, H.O. (ed.), *Primate Responses to Environmental Change.* Chapman and Hall, London, pp. 337–356.

Wiepkema, P.R. and van Adrichen, P.W.M. (eds) (1987) *Biology of Stress in Farm Animals: an integrative approach.* Martinus Nijhoff, Dordrecht (Netherlands).

16 Environment-dependent Disease

J.Y. Blom

Danish Institute for Fisheries Technology and Aquaculture, Hirtshals, Denmark

Summary

It is widely acknowledged that disease in farm animals is a major constraint in farming, leading to low productivity and economic losses. As a disease is the ultimate consequence of animal–environment maladaptation, it also forms a crude indicator of animal welfare.

With increasing costs of production (expensive manpower and buildings) there has been a dramatic change during recent decades towards highly intensive production systems. The inevitable outcome has been adverse environmental conditions, resulting in disease problems and subsequent economic losses.

The environment in a broad sense includes the combination of conditions in which farm animals live, including external factors like housing, penning and tying-up, ventilation and bedding, and the internal environment (infection and disease).

Research conducted during the last decade, following the development of sophisticated epidemiological techniques, has focused on the qualitative and quantitative importance of several external and internal environment factors, influencing animal health.

This chapter draws examples from epidemiological studies, focusing on major disease entities, clearly environment-dependent. Examples are presented on bovine mastitis and leg disorders, respiratory disease in housed calves and pigs and reproductive disorders, stressing the importance of utilizing epidemiological evidence in the design and management of farm animal production systems.

Introduction

The evolution of husbandry – man's exploitation and improvement of natural herds – is the root of health problems in farm animals. Natural herds have for centuries reached a complete biological equilibrium and balance with their environment and have adapted evolutionary to the prevailing conditions in their specific ecosystems. Natural herds are often intermingled with other species, and thus geographically spread, and in these conditions droplet infections play an insignificant role.

Table 16.1. Structural changes in Danish dairy and pig production.

	% of dairy herds		% of national herd	
Herd size of cows	1970	1985	1970	1985
1–19	85	51	63	18
20–49	14	38	32	49
> 49	1	12	5	33
No. of herds ('1000)	96 325	39 014	1152	951
	% of pig herds		**% of national herd**	
Herd size of pigs	1970	1985	1970	1985
1–99	81	56	44	10
100–499	19	33	52	38
> 499	0	12	4	5
No. of herds ('1000)	118 263	44 222	8290	9089

Source: After Anon, 1988.

Natural herds show low rates of reproduction and production, and therefore man very early domesticated herds to secure a regular, safe and convenient supply of food. Gradually domestication moved from aggregated flocks of animals to a pastural ecosystem, as man began selecting and culling animals to improve profitability. The aggregation of animals changed the equilibrium of the ecosystem, and the spread of infectious diseases – a prevailing challenge to the veterinary profession in modern intensive husbandry systems.

The evolution of husbandry systems includes a third step, the agricultural phase, where man began growing feeds for the animals. They were housed for part of the year for their protection and for the farmer's convenience. Today, in most affluent countries animal husbandry has changed from 'farming' to become an 'industrial' undertaking. The animals are often permanently housed, food is brought to them, either home grown or purchased from elsewhere. The production systems, although advantageous from a management, marketing and profitability point of view, are vulnerable ecosystems, and even minute shortfalls in the layout of housing systems, the feeding or management can easily result in suffering, infectious and environment-dependent diseases.

The structural changes in animal production have been dramatic in the last 10–15 years. During this period average herd sizes have grown significantly in Denmark as in other countries (Table 16.1), and the traditional mixed herds are replaced by specialized monospecies herds. Danish statistics show that the productivity in animal production has grown by 26%, and the productivity per man year by 126% in 15 years (Table 16.2). This latter development has been achieved by specialization, rationalization and mechanization of virtually all procedures in today's farms.

The intensification of animal production has been a continuous process through centuries, but the pace has increased dramatically within the last 30 years. The evolutionary process before the 1950s, and the relatively lower productivity per

Table 16.2. Index of development in farm animal and labour productivity in Denmark.

	1970–1974	1985
Crop production	100	197
Animal production	100	126
Productivity per man year	100	226

Source: After Anon, 1988.

animal were able to cope with erroneous developments in environmental as well as husbandry methods. 'Trial and error' evolution took place with limited adverse effects on animal welfare and health. But as the pace increased in the 1960s, the technical evolution soon overtook the knowledge of the biological impact of engineering improvements, and with often severe effects on animal welfare and health, before faulty design and husbandry systems were traced and corrected.

Most disease problems associated with intensive husbandry systems are of a multifactorial nature. The combined effect of adverse environmental and management factors alters the equilibrium of the ecosystem – the threshold of adaptation to the environment is exceeded, and disease results. Figure 16.1 outlines important environment factors affecting farm animal health.

Experience shows that new production systems and environments give rise to unexpected and often dramatic disease events, and the numerous farm building engineering and veterinary studies reflect the lack of basic epidemiologic knowledge of the effect of environmental and management changes on animal behaviour, welfare, productivity and health of livestock.

This chapter does not attempt to cover to the full extent all reported relationships between environment and disease, as the number of possible interrelationships

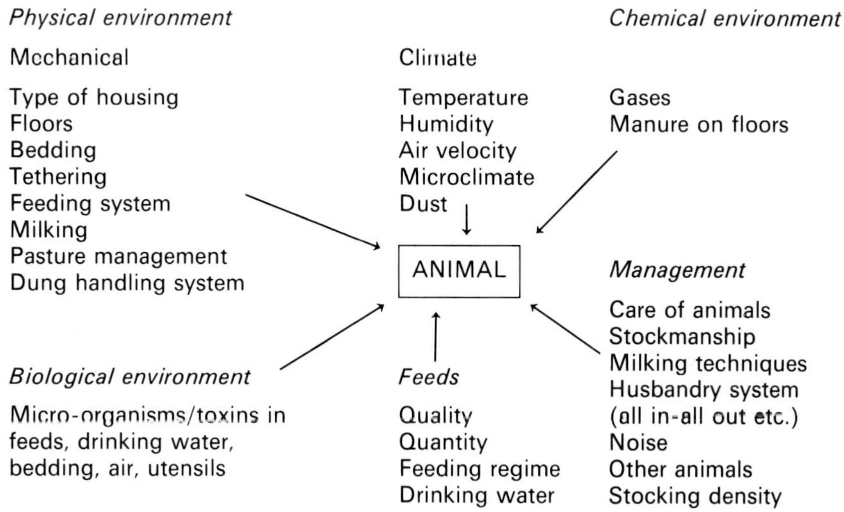

Fig. 16.1. Essential environment factors affecting animal health.

are many, but will highlight some central disease entities clearly associated with farm building engineering and management in dairy cattle and pigs.

Traumatic Injuries in Dairy Cows

Until the late 1950s only minor changes in dairy cow housing occurred. The traditional tie stall barn was improved, as stalls were appropriately sized for larger breeds, thus reducing traumatic injury. New tethering devices, designed to facilitate easy handling of the growing number of cows in the barns, often resulted in cow discomfort and trauma (Mortensen, 1978). Gutter cleaners were introduced to mechanize manure handling, but in North Europe liquid manure handling was soon introduced, whereby new tie stall barns were designed with short stalls with gratings.

The high capital investment in tie stall barns and equipment to attain labour efficiency caused farmers to look for alternatives. Loose housing with deep straw bedding and a separate milking parlour worked well under good management and adequate use of straw, but was soon to be replaced by free stall housing, as straw became scarce in many dairy herds, and free stall houses were more easy to manage.

Comparative studies on the effect of different housing systems on the health conditions of dairy cows were soon to emerge following the intensification of production and the diversification in dairy housing designs. Ekesbo (1966) made a thorough survey of Swedish dairy farms, and the evolution of veterinary epidemiology set off numerous reports into the effects of various housing systems, fixtures and fittings and management on animal health. Many reports focus on traumatic injuries, as they may be crude indicators of unfavourable conditions in the physical environment.

Traumatic injuries in dairy cows may be classified into two major groups: systematic and accidental, according to type, development and course (Table 16.3).

Systematic injuries affect to a varying degree the majority of cows in the herd, and they are caused primarily by the design and construction materials in the cows' resting area. Every housing system exhibits characteristic problems, but there is still a wide variation in type and severity of lesions within housing systems due to differences in design and layout, flooring, bedding, management, etc.

The effect of housing system *per se* on the occurrence of traumatic injuries was studied by Blom (1981, 1982). The cumulative number of traumatic injuries to limbs and body was found to be higher in tied herds (Fig. 16.2). Gjestang and Gravås (1976) observed a higher incidence of foot and leg disorders in herds that were housed all year ('zero grazing'). A survey of the literature (Krohn, 1986) concludes that the long-term effect of zero grazing is considerable, as concerns disease and reproduction. Furthermore, a recent Danish study concludes that summer grazing is economically feasible, if housing conditions are not optimal (Kristensen *et al.*, 1986).

Floor design is critical to cow comfort in many ways, but no single floor material is able to fulfil all the cow's requirements (Irps, 1983). Concrete and similar materials are most frequently used for stall floors. Traditionally, straw or sawdust is used to provide a warm and protective bedding, but unfortunately bedding has been abolished in many housing systems, due either to the labour requirement involved in straw handling or to its assumed incompatibility with liquid manure handling. The

Table 16.3. Injury-predisposing environmental factors in dairy cows.

Type of injury	Predisposing factors
Systematic injuries	
Rubbing injuries and pressure injuries to knee and hock joint areas	Abrasive surfaces in lying area Sharp bedding material Incorrectly adjusted neck bar Restrictive tie systems Short stalls with manure gratings Feeding trough and neck rests of cubicles too high and sharp-edged
Pressure injuries to dorsal part of neck	Neck bar placed too high Feeding trough too low
Injuries to costal arch, hip and ischial areas	Incorreclty adjusted partitions Sharp-edged partitions
Teat injuries	Short and narrow stalls Wet and slippery flooring Restrictive tie systems No bedding
Accidental injuries	
Limb fractures	Falls on slippery floors Restrictive tie systems
Fracture of vertebral column	Inadequate partitions in front of cubicles
Suffocation	Incorrectly placed partitions in cubicles Restrictive tie systems
Muscle injuries (Downer cows)	Short stalls Dung channels too deep

Source: After Blom, 1981.

no-litter system is most frequently used in tied herds with short stalls and manure gratings. In these barns cows are forced to stand with their hind legs on the gratings. Consequently this is associated with an increased number of injuries to the hind feet and legs, and teats (Ekesbo, 1966; Klein-Hessling, 1966; Groth and Eichler-Steinhauff, 1978). Other studies show that longer stalls and straw are fully compatible with liquid manure handling, and that the frequency of traumatic injuries may be reduced by these means (Geer and Grommers, 1975; Blom, 1982; Schmidt *et al.*, 1985). A Danish study revealed that the incidence of teat injuries in free stall housing was decreased significantly by use of long stalls (Anon, 1976).

To avoid the shortcomings of the no-litter system mats of rubber and other materials have been used in many barns. Walberg (1978), however, found no differences in limb lesions in a study of different floor materials. Nygaard (1979) observed significantly fewer limb injuries on floors with a smooth texture than on floors with rough texture. Furthermore, the number of lesions was 50% higher on cold floors than on warm floors. He suggested that the cows had tried to reduce the heat loss on cold floors by changing lying position in a way that led to higher

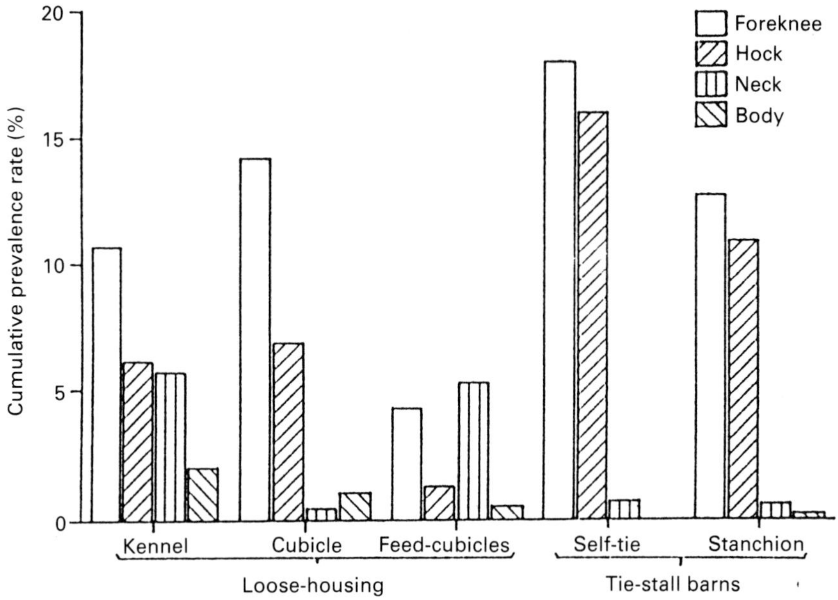

Fig. 16.2. Cumulative prevalence rate of severe traumatic injuries in different housing systems (after Blom, 1981).

pressure on the limbs. He also noted that cold floors are generally more wet than warm floors, and thus more injury-predisposing, as also observed by Blom (1981).

It is generally found that rubber mats without the use of straw bedding are not to be recommended as they tend to get wet and slippery despite their good friction characteristics. The use of bedding, preferably long or chopped straw, is thus imperative for the prevention of traumatic injuries (Blom, 1982; Schmidt *et al.*, 1985).

The design of tethering devices in tie-stall housing affects the incidence of traumatic injuries considerably. When cows are tethered to a degree that allows little freedom of movement, they tend to lie down for longer periods, and the ties impede the process of getting up and lying down, thus increasing the susceptibility to traumatic injuries (Mortensen, 1978; Geer and Grommers, 1975; Nygaard, 1979). This is especially observed in the so-called self-tying stanchions. The ties in this system are often of the yoke type, and studies have shown that the incidence of traumatic injuries is significantly increased (Blom, 1981).

The design of the feeding trough and barrier has been shown to be of major importance to cow health and comfort (Ekesbo, 1966; Blom, 1981). Blom (1982) observed a higher prevalence of neck injuries when neck bars were incorrectly adjusted, and Matzke (1975) and Groth and Eichler-Steinhauff (1978) found swellings and wounds on the costal arch, hip and ischial areas of cows when stall partitions were incorrectly adjusted.

Accidental injuries, as shown in Table 16.3, tend to affect only a few animals in the herd, but as the course of events is often fatal to the animal, these injuries are of utmost importance.

Traumatic injuries may often be regarded as of minor importance to animal

Table 16.4. Frequency of environment-dependent diseases in different housing systems.

Disease	Frequency (number of treatments/100 cows)				
	Tie-stall barns			Free-stall barns	
	Zero grazing	Dung channels	Manure gratings	Solid floors	Slatted floors
Mastitis	94	63	64	60	40
Teat injuries	20	9	12	5	4
Limb disorders	13	6	10	7	2
Foot disorders	12	2	14	31	31
Endometritis	8	4	5	11	5

Source: After Jørgensen, 1975.

health. They do, however, indicate conditions of discomfort and reduced animal welfare, and there is ample evidence that these injuries may form a portal of entry to diseases. Jørgensen (1975) demonstrated the effect of housing systems on environment dependent diseases (Table 16.4).

In a Danish study (Blom, 1982) it was found that only a small proportion of knee injuries needed veterinary assistance. Nevertheless, all injuries are signs of a suboptimal environment and should be prevented.

Tail Tip Necrosis in Young Fattening Bulls

Fattening bulls were traditionally raised in stalls in North Europe. As housing systems for dairy cows changed during the last two decades from solid manure handling to liquid manure handling, the housing of fattening bulls changed. Today the majority of fattening bulls are group housed on slatted floors in many countries.

As husbandry systems change, so do disease patterns. Tail tip necrosis evolved as a disease problem in Denmark as in other European countries (e.g. Kunz and Vogel, 1978; Martig and Leuenberger, 1978: Madsen and Nielsen, 1985). The numerous reports all conclude that tail tip necrosis due to traumatic injury to the tail is associated with close confinement.

Behavioural studies show that densely stocked bulls on slatted floors were more frequently disturbed in their rest by penmates, and were forced to move from a lying position more often than bulls in deep litter (Larsson *et al.*, 1984). There were also indications of more frequent tail tramping in close confinement than in pens with more liberal space allowance.

A Danish study (Madsen and Nielsen, 1985) revealed that tail tip necrosis is a multifactorial disease, and that salient factors are slatted concrete floors, close confinement, hot climate and body weights beyond 200 kg. Further analysis of health data showed that other disease problems were significantly higher in problem herds (Table 16.5).

In some countries the disease problem is prevented by amputation of the tails of the young bulls (Lenk, 1981). This procedure is, however, ethically problematical, and is not allowed in Denmark. The data from Madsen and Nielson (1985) also indicate that the cosmetic prevention of tail tip necrosis does not alleviate the underlying problem and other disease problems in affected herds.

Based on the preliminary Danish study, a controlled study to reveal the relationship between space allowance and the incidence of tail tip necrosis was conducted

Table 16.5. Disease in calf fattening herds as related to occurrence of tail tip necrosis.

	Incidence rates, average in control herds	Herds with tail tip necrosis		Control herds		
		Number of herds		Number of herds		
	%	< average	> average	< average	> average	p*
Pneumonia	22.8	4	16	15	8	< .01
Enteritis	3.0	10	10	18	5	ns
Foot rot	3.8	6	14	15	8	< .05
Mortality	4.8	2	18	16	7	< .0001

Source: After Madsen and Nielsen, 1985.
*Level of significance. Differences between case and control herds (χ^2 test).

Table 16.6. Productivity, tail amputations and economic results in calf fattening from 250 kg in different stocking densities.

	Stocking density (m^2 per calf)		
	1.9 m^2	1.5 m^2	1.3 m^2
Daily gain (g)	1161	1116	1010
Total gain (kg)	174	167	152
Tail amputations	0.08	0.10	0.13
Net return per calf (DKK)	387	370	214
Yearly net return per pen (DKK)	3703	4292	2899

Source: Madsen *et al.*, 1987.

(Madsen *et al.*, 1987). The study showed that tail tip necrosis may be reduced in quantity and severity by giving a larger space allowance to fattening bulls. The study also demonstrated that these simple means of disease prevention had a positive effect on productivity and economy, as the daily weight gain and net return per calf are higher in moderately stocked pens (Table 16.6).

Traumatic Injuries and Diseases in Neonatal Pigs

In spite of a vast number of studies into the aetiology of preweaning mortality in pigs, a preweaning mortality of 20% is still reported in many countries (Nielsen *et al.*, 1974; Backstrom *et al.*, 1982; English and Morrison, 1984).

Death losses caused by crushing of the piglet by the dam, and other fatal traumatic injuries account for the largest mortality group among live born pigs in

Table 16.7. Distribution of preweaning mortality in pigs.

Cause of death	Percentage of total deaths
Dead antepartum	14
Dead intrapartum	16
Traumatized	22
Weak, undersized	11
Splayleg	12
Starvation	4
Infectious diseases	15
Malformations, miscellaneous	6

Source: After Svendsen and Bengtsson, 1982.

most herds. Mortality due to traumatic injuries varies greatly from herd to herd, and it is apparent that these losses are related to housing system and management routines (Bille *et al.*, 1974).

Svendsen and Bengtsson (1982) reported on causes of mortality in more than 700 litters and revealed that traumatic injury accounts for 22% of the deaths (Table 16.7). Most of these pigs had been crushed by the sow, while a few had been stepped on, bitten, crushed between the sow and pen fittings, or traumatized by pen fittings. The percentage of pigs that died with trauma secondary to some obviously predisposing factor, or as a secondary diagnosis, was 2.2%. These pigs were primarily weakborn, underweight, splaylegged, starved, ill due to diarrhoea, joint inflammation, or generalized infections and were crushed or overlain by the sow as a result of these difficulties.

Ethological studies of housing systems show that the neonatal pigs are at the highest risk within the first 24–36 hours postpartum, and while the sow is moving. Apart from direct trauma caused by the sow, a major part of perinatal deaths is associated with thermal conditions of the piglet. Cold exposure of piglets early in life reduces serum levels of passively acquired colostral immunoglobulins (Blecha and Kelley, 1981). It is suggested that cold exposure reduces nursing vigour, thus resulting in reduced intake of immunoglobulins and energy, leaving the piglets weak and exposed to infections (Kelley, 1982) (Fig. 16.3).

Based on the findings of behaviour and production studies, Svendsen *et al.* (1986) summarized some basic environment factors important in the prevention of perinatal mortality in pigs.

1. *Heating* is essential for the microenvironment of the piglet. The fewest losses were observed in farrowing units with a room temperature of 22°C, and the sow maintained fixed in a farrowing crate on totally perforated floors without straw bedding. Movable heat lamps were placed over the piglets parallel to the sow's udder. Svendsen *et al.* (1986) note that the provision of insulated and draught-free creep areas on each side of the sow enable the pigs to locate the heated area quickly after birth, and the piglets are less prone to become chilled, thereby reducing the danger of trauma. The establishment of warm areas, however, necessitates fixation of the sow.
2. *Fixation* of the sow is advised for the first few days after farrowing. The design

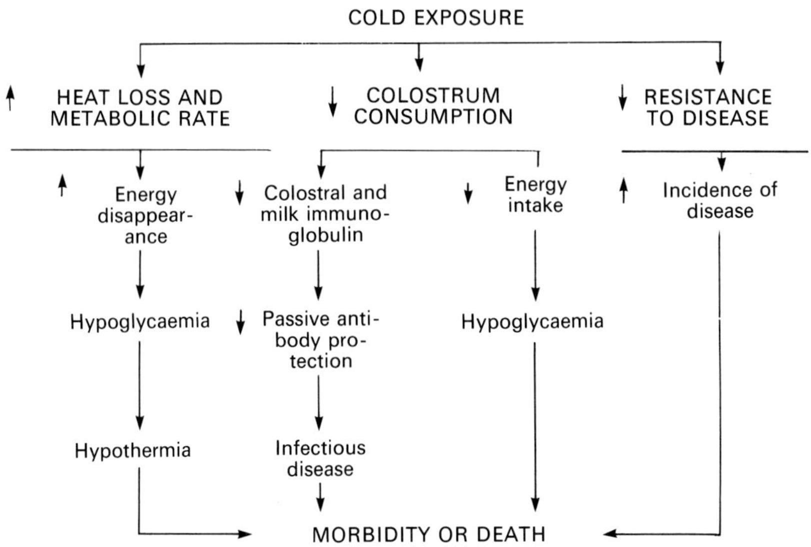

Fig. 16.3. Effects of cold exposure on newborn domestic mammals (after Kelley, 1982).

of the fittings is aimed at directing the movements of the sow when she lies down in order to reduce the risk of traumatic injuries to the piglets. On the other hand, fixation of the sow has been disputed in some countries. Based on a survey of Swedish sow herds, Bäckstrom (1973) reported a higher incidence rate of traumatic injuries and disease in sows that were confined in crates at farrowing. Since then fixation systems have been considerably improved, based on very thorough studies of behavioural needs of the sow (Baxter and Schwaller, 1983).

3. *Flooring.* The use of flooring that permits the sow to control the speed of her descent as she lies down, and provides a well-insulated, clean and non-slippery surface for the piglets has been shown to reduce the occurrence of traumatic injuries. The relationship between flooring type, preference and injury was reviewed by Webb and Nilsson (1983).

Health Problems Associated with Housing of Sows

Sow housing today reflects the industrialization of pig production, and the trends in sow housing were for some 20 years set by technical and economic considerations.

Thus confinement systems for dry sows and the lactation period were introduced. The most common system for dry sows is a confinement system with stalls and tethers, generally without straw. Lactating sows are confined in farrowing crates designed to minimize crushing of the piglets.

Confinement systems for sows have caused concern because the animals are severely restricted in their movement. They are unable to exercise and perform normal behaviour patterns. It is apparent that subsequent stereotypic behaviour patterns arise in close confinement (Jensen, 1980), especially in the absence of straw.

Sows in pens with slatted floors and in tethered systems show a higher incidence

Table 16.8. Incidence of foot lesions in sows during the dry period and lactation as related to flooring.

	Flooring		
	Without slats	Partly slatted	Statistical difference
Number of sows	3520	588	
Foot lesions (%)	3.3	6.3	***

Source: After Bäckström, 1973.
$^{***}P < 0.001$.

of foot lesions than sows in pens without slats (Bäckstrom, 1973) (Table 16.8) Muirhead (1981) reported 14% of a sow herd being lame due to faulty slats. When the slats were replaced the figure fell to 1%. Loose housing or outdoor exercise has been shown to be healthier as compared to confinement (Bäckstrom, 1973) (Table 16.9).

Table 16.9. Disease and farrowing performance as related to sow confinement during gestation and farrowing.

	Sow free in pen	Sow confined in crate	Statistical difference
Number of sows	654	1283	
% Mastitis–metritis–agalactia	6.7	11.2	**
% Farrowing time < 8 h	2.3	5.4	**
% Total morbidity	12.8	24.1	***
% Medically treated	10.4	16.8	***

Source: After Bäckström, 1973.
$^{**}P < 0.01$; $^{***}P < 0.001$.

No doubt cleanliness of the sow is also an important factor in health problems, both in the development of foot disorders as well as in reproductive performance. As confined sows are often forced to lie in their own dung and urine, this can lead to urogenital tract infections and subsequent losses in reproductive performance (Muirhead, 1983).

Foot Disorders in Dairy Cows and Beef Calves

Lameness due to foot disorders causes major losses in many dairy herds, but nevertheless it is one of the most neglected conditions in dairy husbandry. In a survey of British dairy herds the yearly incidence rate of lameness was 5.5% (Russel and Weaver, 1978), but the true incidence rate is most likely much higher. A Danish study conducted in 30 dairy herds revealed a high prevalence rate of foot disorders (Thysen *et al.*, 1981a) (Table 16.10).

The prevalence and incidence of foot disorders is influenced by inheritance, feeding, housing system, infectious agents and management (Amstutz, 1985), but the environment is of utmost importance in the development of clinical foot disorders.

In tied herds lameness is often associated with wet pasture conditions and stony

Table 16.10. Site and severity of foot disorders in dairy cows (n = 2637).

		Affected cows (%)			
		Forefeet		Hind feet	
Disorder	Severity	Medial claw	Lateral claw	Medial claw	Lateral claw
Foot rot	1	32.9		32.1	
	2	24.0		25.1	
	3	1.4		5.9	
	4	0.0		0.9	
Contusion of the sole	1	2.5	0.3	0.8	0.8
	2	11.2	2.1	5.9	17.5
	3	9.0	1.9	2.9	22.9
	4	2.9	0.2	1.0	9.6
	5	1.4	0.3	0.5	11.5
Chronic laminitis	1	1.0	0.6	0.2	0.4
	2	0.0	0.1	0.1	0.2
Foreign body in sole		4.7	4.5	1.0	4.8

Source: After Thysen *et al.*, 1981a.

driveways (Ekesbo, 1970; Russel and Weaver, 1978). Thysen *et al.* (1981b) found the prevalence rate of foot rot and contusion of the sole to be higher in free stall housing than in tied herds. In a study of 5111 cows in 36 dairy herds, the number of treatments for foot disorders was observed to be higher in tied herds than in free stall housing (Blom, 1982) (Table 16.11). It is assumed that the major predisposing environmental factors are moisture and dung in walking areas and stalls. The Danish investigations show a significant influence of housing systems on foot disorders. However, large variations within housing systems indicate that management factors also greatly influence the prevalence and incidence of foot disorders in dairy cows.

Several reports have shown that intensive carbohydrate feeding and low structural content of the feeding ration may precipitate contusion of the sole (Peterse,

Table 16.11. Incidence rate of clinical foot disorders in different housing systems.

	Incidence (number per 100 cow years)					
	Free-stall housing				Tie-stall housing	
Disorder	Kennel	Uninsulated	Isolated	Feed cubicle	Self-tie	Tie-barns
Panaritium	7.1[b]	13.7[a]	11.3[a]	18.2[a]	6.4[bc]	3.6[c]
Contusion of the sole	0.4[c]	5.5[a]	2.6[b]	0	0	1.2[c]
Other disorders	0.4	0.7	0.6	0.8	0	0.4
Total	7.9[c]	19.9[a]	14.5[b]	19.0[a]	6.4[c]	5.2[c]

Source: After Blom, 1982.
Values with different alphabetical codes differ significantly ($P < 0.01$)

Table 16.12. Incidence of injury and infection to the feet of beef cattle on concrete slatted floors or on straw.

	Incidence (%)	
Injury or infection	Slatted floors	Straw
Conditions observed after slaughter		
Overgrowth of heel horn	22–35	1–6
Abrasion of abaxial wall	57	46
Necrotic pododermatitis	77	64
Traumatic pododermatitis	8–12	3–4
Septic traumatic pododermatitis	20	6
Clinical lameness	4.8	2.4

Source: After Murphy, 1978.

1985). This condition may be alleviated by proper feeding of the cow in early lactation, and a strict claw trimming programme (Enevoldsen, 1986).

Solid floors in free stall housing are generally regarded as more disease-predisposing than slatted floors, but a controlled study in Denmark did not confirm this hypothesis (Thysen *et al.*, 1981b). The effect of floor material on the incidence of injuries and infections to the feet of beef cattle was surveyed by Murphy (1978). As can be seen from Table 16.12, all conditions were more frequent on slatted floors.

Mastitis in Dairy Cows

Despite intense research efforts clinical mastitis continues to be the predominant clinical entity in intensive dairy production. The annual losses are mainly caused by loss of milk, reduced milk yield and culling of unproductive animals.

Mastitis is caused by introduction of micro-organisms from the environment (milking machine or stall floors) through the teat canal. Numerous reports have focused on physical damage to the teat canal and shown it to be the major aetiological factor involved (Agger, 1981). Depending on the cow's immunological status and the micro-organism, subclinical or clinical mastitis will develop.

The introduction of sulphonamides and penicillin some 50 years ago dramatically changed the attitude towards mastitis prevention. For a couple of decades the method of therapy and prophylaxis of mastitis was the use of antibiotics, and the microbiology of mastitis was highlighted. Admittedly, mastitis due to specific micro-organisms was successfully combated, but mastitis due to opportunistic bacteria remained a challenge to dairymen. Epidemiological studies of bovine mastitis have since then focused on the effect of various environmental factors making up the web of causation of mastitis.

The effect of housing system has been studied by several authors, and some results are compiled in Table 16.13. As can be seen from the table the incidence and prevalence of teat injuries and the incidence of clinical mastitis are generally higher in tied herds than in free stall housing. It must be emphasized, however, that

Table 16.13. The effect of housing type on annual incidence of mastitis and teat injuries.

Disease	Tie-stall barn	Free-stall housing	Source
Incidence of teat lesions (%)	23.6[1] 22.2[2]	3.3	Ekesbo, 1966
Incidence of mastitis (%)	36.3[1] 38.1[2]	23.3	
Incidence of teat lesions (%)	8.7[1]	0.9	Grommers, 1967
Prevalence of teat lesions (%)	3.40[1] 5.53[2]	1.09[3]	Maton and De Moor, 1975
Teat lesions per 100 cows	10.0	3.0	Jørgensen and Nielsen, 1977
Mastitis treatments per 100 cows	60.0	45.0	
Incidence of teat lesions (%)	6.4	3.8	Bakken, 1981
Incidence of mastitis (%)	31.7	12.1	
Incidence of udder diseases (%)	14.8	12.1	Schubert *et al.*, 1982

[1]Straw bedding; [2]no straw and metal gratings; [3]deep litter.

differences in recording criteria presumably vary between the cited studies, as did environmental factors such as housing design, use of bedding, grazing, etc.

Very few reports describe the influence of housing system on subclinical mastitis. Scandinavian reports all show that subclinical mastitis is more frequent in tied herds (Bakken, 1981; Blom *et al.*, 1985). The effects of individual stall design and management factors have been studied, especially in Scandinavia.

Stall design studies have shown that short stalls are precipitating mastitis (Anon, 1976; Bakken and Gudding, 1982) and that stall width and the use of partitions between cows significantly influence udder health (Schmidt Madsen, 1978).

The tethering devices in tied herds are of major importance for the health condition of dairy cows (Mortensen, 1978; Blom, 1981; Bakken, 1981), especially if tethering is restrictive, thus impeding the natural behaviour of the cow when rising and lying down. Abnormal behaviour patterns enhance the risk of teat injuries.

Floor type and bedding materials have been shown to be important for udder health. Ekesbo (1970) found the incidence of teat injuries and subsequent mastitis significantly higher in stalls without bedding, irrespective of housing system (Table 16.14). A recent Danish study (Schmidt *et al.*, 1985) concludes that sawdust used as bedding in concrete stalls, or limestone on rubber mats resulted in a higher frequency of subclinical and clinical mastitis, as compared to stalls with straw bedding. They also showed that sawdust bedding was heavily contaminated with opportunistic bacteria.

Frequent application of straw bedding, thus preventing soiling of stalls and

Table 16.14. Incidence of teat injuries and mastitis as related to housing system and use of bedding.

	Incidence (%)			
	Loose housing		Tie-stall barns	
Injury	Bedding	No bedding	Bedding	No bedding
Teat injuries	1.4	10.1	3.9	10.1
Mastitis	6.8	17.3	8.6	16.8

Source: After Ekesbo, 1970.

cows, was found to reduce the incidence of clinical mastitis (Blom *et al.*, 1985; Schmidt *et al.*, 1985) (Fig. 16.4).

The possible influence of the milking machine on the occurrence of mastitis has been the objective of several studies, but Schmidt Madsen and Klastrup (1980) found that malfunction of the milking machine accounted for only 6% of the variability of prevalence of mastitis in problem herds.

In summary, only 20–25% of the variability in udder health between housing types may be attributed to environment factors (Schmidt Madsen, 1978; Bakken, 1981). It is generally assumed that many other factors contribute to the occurrence of bovine mastitis, such as management routines, milking, genetic improvement and breed.

Calf Respiratory Disease

Respiratory disease continues to be the major problem for calf rearers in most countries in intensive husbandry systems, and especially in specialized fattening herds. Calf mortality is recorded at 15% per year in Denmark, including a perinatal mortality of approximately 5%. The estimated incidence rate of respiratory disease

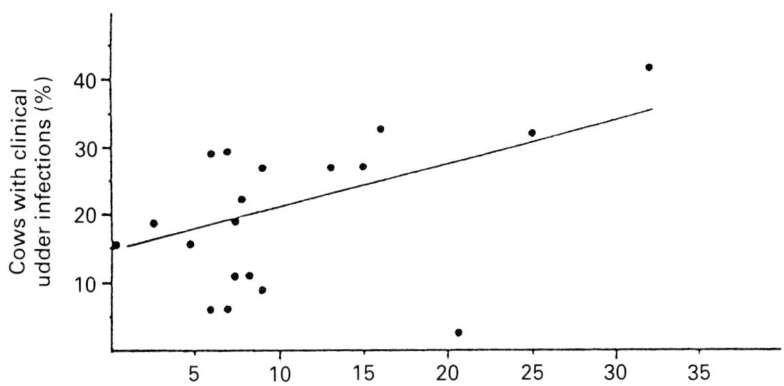

Fig. 16.4. The association between the incidence of clinical mastitis and cleanliness of cows (after Blom *et al.*, 1985).

in calves is approximately 30% in dairy herds, and often much higher in specialized fattening herds. Serious financial losses occur as a result of deaths, culling, veterinary treatments and depressed growth rates (Østergaard *et al.*, 1986).

Research into calf respiratory disease was for many years focused on the isolation and identification of micro-organisms which might have an aetiological role, and the elucidation of the pathological and immunological processes involved. The primary objective of such work has been the development of effective vaccines. Bacteria, viruses, mycoplasmas and other infectious agents have been associated with calf respiratory disease (Bitsch *et al.*, 1976; Allan, 1978; Stott *et al.*, 1978) acting either alone or together, and their effects modified by climatic, husbandry and immune factors.

Under field conditions calves are exposed to a variety of micro-organisms, which can be isolated from the respiratory tract of both sick and healthy animals. Yet, attempts to reproduce the clinical syndrome under laboratory conditions have generally failed. The many attempts to prevent calf respiratory disease by the use of vaccines have shown no promise for alleviating this syndrome. Therefore, for the last 15 years there has been much effort to elucidate the role of environmental and husbandry factors involved in the development of calf respiratory disease, as this syndrome is a typical example of a multifactorial disease.

In intensive husbandry systems in temperate climates calves are traditionally housed in insulated and mechanically ventilated houses. There has been a steady trend towards the construction of large houses with controlled climate systems, thus following the trend-setting engineering practice of poultry and pig housing.

Major research efforts have been directed towards the influence of temperature and humidity on calf respiratory disease (Maund and Turner, 1979; Webster, 1979). Research in Denmark showed a tendency towards an increased incidence of respiratory disease at a relative humidity level above 85% and/or at temperatures below 10°C in calf houses (Blom and Thysen, 1980). But it is generally accepted that the temperature itself is of no major importance, provided that dry bedding is applied, and the calves are well protected against draught.

Further Danish investigations into the role of relative humidity were conducted in several calf houses. The investigations showed that in winter periods it was difficult to maintain relative humidities below 70% at air flows of 18–25 m^3 per calf per hour, and heating costs were high (Blom *et al.*, 1984b). Based on these findings the ventilation rates were recommended to be at least 30 m^3 per calf per hour in insulated calf houses. The investigations showed no differences in the incidence of calf respiratory disease among humidity levels (Table 16.15).

In parallel studies of calf respiratory disease in uninsulated, naturally ventilated calf houses and calf hutches the incidence rates of respiratory disease were considerably lower. These findings and measurements of the number of airborne bacteria in different housing systems (Blom *et al.*, 1984a) (Table 16.16) have focused on the effect of airborne bacteria on respiratory disease. Pritchard *et al.* (1981) showed the area of lung consolidation of veal calves at slaughter to be related directly to the level of airborne bacteria, and thus their challenge to the lower respiratory tract. Jones *et al.* (1982) showed that relative humidity, reflecting the ventilation rate in the calf house, has a significant modulatory effect on the number of airborne bacteria. They also showed the importance of stocking density on the number of airborne bacteria (Fig. 16.5).

Table 16.15. Incidence rate of disease in dairy calves at different humidity levels.

House	Max. humidity (% RH)	Respiratory disease (%)	Enteritis (%)
1A	70	51	16
1B	85	53	16
2A	70	34	15
2B	85	25	17

Source: After Blom *et al.*, 1984b.

In dairy herds, calf houses are traditionally stocked continuously. Experience has shown that a build-up of infections during months with a high stocking rate may lead to unsatisfactory health conditions (Roy, 1978; Miller *et al.*, 1978; Blom *et al.*, 1984b). Controlled studies of the effect of batch stocking (all in–all out) were carried out in Denmark (Østergaard *et al.*, 1986). The investigations showed that batch stocking resulted in a significantly lower incidence of calf respiratory disease, enteritis and other diseases and a significantly higher live weight gain than at continuous stocking (Table 16.17).

Respiratory Disease in Growing Pigs

Pig production has changed profoundly in the last decade in many countries, and in many herds thousands of slaughter pigs are housed under one roof. The major disease problem in growing pigs is respiratory disease, and respiratory disease accounts for major losses in productivity. Approximately 70% of disease records in

Table 16.16. Mean values of aerial bacteria and fungi in three calf housing systems.

	Insulated, heated and mechanically ventilated	Insulated and naturally ventilated	Uninsulated, naturally ventilated
Air temperature (°C)	16.7 (± 3.8)	14.9 (± 5.7)	10.5 (± 6.7)
RH (%)	72 (± 3)	75 (± 7)	79 (± 8)
Airborne bacteria colony forming particles ($\times 10^3$ m^{-3})			
Total	77.8 (± 27.5)[b]	101.6 (± 41.1)[a]	67.6 (± 36.4)[b]
Below 4.7 μm	30.9 (± 13.0)	42.1 (± 19.5)	33.4 (± 21.4)
Proportion below 4.7 μm (%)	39	40	43
Airborne fungi	119.3 (± 99.5)[a]	127.1 (± 98.5)[a]	40.5 (± 41.5)[b]

Source: After Blom *et al.*, 1984a.
Values with different alphabetical codes differ significantly ($P < 0.05$).

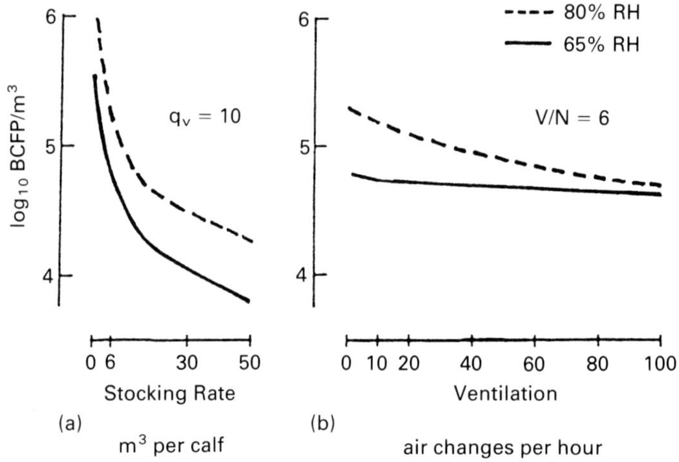

Fig. 16.5. Predicted effects on concentrations of airborne bacteria colony-forming particles (BCFP) ($< 4.7 \mu m$) in calf houses of: (a) varying stocking rate at a ventilation rate of $10\,m^3 h^{-1}$ and RH of 65% or 80%; (b) varying ventilation rate at a fixed stocking rate of $6\,m^2$/calf (after Jones *et al.*, 1982).

Danish abattoirs are related to respiratory disease, and 50% of the slaughter pigs have records of pneumonic lesions (Anon, 1991).

As the majority of the clinical entities in the conglomerate of pig respiratory diseases are clearly associated with a specific pathogen, a number of these infections have been successfully controlled, either by use of specific vaccines or medication programmes, or by eliminating the pathogens from the herd (Specific Pathogen Free herds). Nevertheless, attempts to control respiratory disease by vaccination and medication have not solved all disease problems in pig production, as the environ-

Table 16.17. Health and growth of dairy calves (1–16 weeks) as related to management system.

	Continuous stocking	Batch stocking	Statistical significance
Herd 1 (SDM), no. of calves	205	204	
Calves treated for:			
Respiratory disease (%)	65	29	$P < 0.001$
Enteritis (%)	23	8	$P < 0.001$
Mortality (%)	6	3	$P < 0.245$
Daily weight gain (g)	557	613	$P < 0.001$
Herd 2 (Jersey), no. of calves	72	91	
Calves treated for:			
Respiratory diseases (%)	88	39	$P < 0.001$
Enteritis (%)	49[1]	52[1]	ns
Mortality (%)	4	9	$P < 0.243$
Daily weight gain (g)	354	428	$P < 0.001$

Source: After Østergaard *et al.*, 1986.

[1]Primary cause: *Salmonella dublin* infection, which is easily transmitted from section to section.

Table 16.18. Air quality in herds with and without clinical atrophic rhinitis (AR).

AR status	− AR	+ AR
No. of herds	29	15
Aerial bacteria, total (× 10 CFU[1]/m^3)	14.1	36.5
Gram negative bacteria (CFU[1]/m^3)	517	777
Dust, total (mg/m^3)	1.86	2.58
Dust, respirable (mg/m^3)	0.43	0.59
Endotoxins (× 10 EU[2]/m^3)	33	48
Ammonia (ppm)	8.36	9.27

Source: After Bækbo, 1989.
[1] CFU = colony forming units; EU[2] = endotoxin units. Sampler used: Cyclon sampler.

mental conditions and management systems used can have a profound effect on the pigs' health.

Pig health is a balance between the challenges of infectious agents on the one hand and their resistance on the other. Stressors such as heat, cold, overcrowding, weaning, noise and restraint have been shown to alter the immune system of animals (Kelley, 1980), but the relative importance of these factors has yet to be fully understood.

Growing pigs adapt easily to temperatures within the thermal neutral zone (15–30°C), but the effective environmental temperature is adversely influenced by insufficient insulation, and draughts along floors and walls, that come through doors and windows (Bäckstrom and Hoefling, 1982). Kosztolich (1966) observed that chills due to draughts lowered disease resistance. On the other hand, modulation of air temperature and air movements may be used to influence the pigs' choice of position in pens (Ingram, 1981). The pig is tolerant to a wide range of humidities (60–90%).

The major effect of air temperature and humidity, however, is their influence on survival of airborne bacteria and lung clearance of pigs (Curtis *et al.*, 1976; Muller *et al.*, 1988). Kovacs *et al.* (1967) recorded the presence of 110 to 430 dust particles per millilitre of air with between 4 and 11 bacteria attached to each particle. In a recent study Bækbo (1989) observed dust and airborne bacteria negatively correlated to outdoor temperature, thus reflecting the ventilation rate. He also noted that high stocking density, the use of dry feeds and the use of straw bedding were associated with high amounts of airborne dust, and an increase in the number of airborne bacteria. When fattening units with and without clinical atrophic rhinitis were compared, a distinct tendency to poorer air quality in units with atrophic rhinitis was revealed, and a significant correlation was observed between the occurrence of airborne toxigenic *Pasteurella multocida* and clinical atrophic rhinitis (Table 16.18).

The design of pens has been studied by Flesja *et al.* (1982), who found that solid-sided pens were associated with a reduction in the incidence of pneumonia. The partitions provide a barrier to the spread of airborne infection.

Flesja *et al.* (1982) also observed that the risk of pneumonia and pleurisy seemed to be increased if pens were stocked with more than 12 pigs. Bäckstrom and Bremer (1978) observed a higher frequency of respiratory disease in houses with less than

Table 16.19. Respiratory disease in growing pigs in two trial periods as related to management and husbandry system.

	Trial period	No. of herds	No. of pigs slaughtered	Incidence of pneumonia (%)	
				trial period (1)	trial period (2)
All in–all out	1	7	2 118	3.2	
	2	6	898		8.0
Continuous	1	29	7 614	11.1	
throughput	2	28	5 433		21.9
Solid manure	1	19	3 873	7.6	
	2	19	3 301		15.8
Liquid manure	1	37	10 305	10.6	
	2	32	7 641		22.4

Source: After Larsson and Bäckström, 1971.

0.5 m^2 per pig lying area and less than 3 m^3 per pig air space. These observations were supported by Tielen (1978). In a large survey in Sweden, Lindquist (1974) observed a lower incidence of respiratory disease in houses with less than 500 growing pigs, and when more than 3 m^3 air space and 0.7 m^2 floor per pig was provided.

The importance of the management and husbandry system was studied by Larsson and Bäckstrom (1971) (Table 16.19). The survey indicated that the incidence of pneumonia was significantly lower where strictly 'all in–all out' procedures were used. They also found the incidence of pneumonia to be significantly lower in systems with solid manure handling than in houses with liquid manure handling. These results may be due to draughty slatted floors or the failure to remove noxious slurry gases. Since then, liquid manure systems and the use of slatted floors for growing pigs has increased but the design of the systems has been improved.

Concluding Remarks

The last decades have shown remarkable developments in food animal productivity and labour productivity, through major developments in agricultural engineering, physiology, genetics, biological engineering and animal husbandry management systems.

These results have certainly been to the benefit of the consumer, as prices of meat and milk and other agricultural products have been stable throughout this period. On the other hand, consumers and society have severely questioned the ethics of these achievements – are they made at the expense of animal welfare, suffering and disease?

The data presented in this chapter clearly show that certain husbandry systems are putting a lot of strain on the animals, and the growing volume of reports on animal behaviour and welfare show that somewhere in the evolutionary process, farmers and agricultural engineers put too little emphasis on the basic needs of animals in their considerations of new housing and husbandry systems (Signoret, 1981; Sybesma, 1981; Baxter *et al.*, 1983).

Therefore, there is a need to look further into the possibilities of providing farm animals with adequate space and stimuli, and at the same time secure for the farmer good working environments and a reasonable profitability. The literature cited above shows that this is in fact possible. It is absolutely necessary that for the next decade agricultural engineers, farmers, ethologists and veterinarians work together to ensure that profitable farming meets the basic needs of farm animals.

References

Agger, J.F.G. (1981) Teat lesions and foot disorders in dairy cows. PhD Thesis. Royal Veterinary and Agricultural University, Copenhagen.

Allan, E.M. (1978) Pulmonary bacterial flora of pneumonic and non-pneumonic calves. In: Martin, W.B. (ed.), *Respiratory Diseases in Cattle*. Martinus Nijhoff, The Hague, pp. 305–316.

Amstutz, H.E. (1985) Prevention and control of lameness in dairy cattle. *Veterinary Clinics of North America: Food Animal Practice* **1**, pp. 25–37.

Anon (1976) Nyere løsdriftstalde med sengebåse. Erfaringer fra praksis. Beretning nr. 18. De landbrugstekniske undersøgelser, Ørritslevgård.

Anon (1988) Report concerning intensive production systems in agriculture. Report 1126, Ministry of Justice, Copenhagen.

Anon (1991) Danish Bacon and Meat Council. Personal communication.

Bäckström, L. (1973) Environment and animal health in piglet production. A field study of incidence and correlations. *Acta Veterinaria Scandinavica* (Suppl. 41).

Bäckström, L. and Bremer (1978) The relationship between disease incidences of fatteners registered at slaughter and environmental factors in the herds. *Nordisk Veterinærmedicin* **30**, 526–533.

Bäckström, L., and Höefling, D.C. (1982) *Veterinary Clinics of North America*, 259–276.

Bäckström, L., Connors, J., Price, W. and Morkoc, A. (1982) Mastitis–Metritis–Agalactia (MMA) in the sow: A field survey of MMA and other farrowing disorders under different gestation and farrowing housing conditions. *Proceedings of the 7th International Pig Veterinary Society Congress*, Mexico, p. 175.

Bækbo, P. (1989) Air quality and health in pig houses. PhD Thesis, Royal Veterinary and Agricultural University, Copenhagen.

Bakken, G. (1981) Subclinical mastitis in Norwegian diary cows. *Acta Agriculturae Scandinavica* **31**, 273–286.

Bakken, G. and Gudding, R. (1982) Environmental effects in udder health in Norwegian dairy herds. *Proceedings of the 2nd International Livestock Environment Symposium*, Ames, Iowa, pp. 540–544.

Baxter, M.R. and Schwaller, C.E. (1983) Space requirements for sows in confinement. In: Baxter, S.H., Baxter, M.R. and MacCormack, J.A.C. (eds), *Farm Animal Housing and Welfare*. Martinus Nijhoff, The Hague, pp. 181–195.

Baxter, S.H., Baxter, M.R. and MacCormack, J.A.C. (eds) (1983) *Farm Animal Housing and Welfare*. Martinus Nijhoff, The Hague.

Bille, N., Nielsen, N.C., Larsen, J.L. and Svendsen, J. (1974) Preweaning mortality in pigs. 3. Traumatic injuries. *Nordisk Veterinærmedicin* **26**, 617–625.

Bitsch, V., Friis, N.F. and Krogh, H.V. (1976) A microbiological study of pneumonic calf lungs. *Acta Veterinaria Scandinavica* **17**, 32–42.

Blecha, F. and Kelley, K.W. (1981) Effects of cold and weaning stressors on the antibody-mediated immune response of pigs. *Journal of Animal Science* **53**, 439–447.

Blom, J.Y. (1981) Traumatic injuries to dairy cows as related to housing systems. In: Østergaard, V. and Hindhede, J. (eds), *Report 515 from the National Institute of Animal Science*, Copenhagen, pp. 73–104.

Blom, J.Y. (1982) The incidence of clinical foot disorders in dairy cows. In: Østergaard, V. and Hindhede, J. (eds), *Report 532 from the National Institute of Animal Science*, Copenhagen, pp. 128–140.

Blom, J.Y. and Thysen, I. (1980) Klimaets indflydelse på kalves sundhed. In: Østergaard, V. and Hindhede, J. (eds), *Report 502 from the National Institute of Animal Science*, Copenhagen, pp. 138–153.

Blom, J.Y., Madsen, E.B., Krogh, H.V. and Wolstrup, J. (1984a) Numbers of airborne bacteria and fungi in calf houses. *Nordisk Veterinærmedicin* **36**, 215–220.

Blom, J.Y., Thysen, I., Østergaard, V. and Møller, F. (1984b) Calf health and weight gain as related to climate, iron and immune status and disease treatment. *Report 570 from the National Institute of Animal Science*, Copenhagen.

Blom, J.Y., Thysen, I., Madsen, P.S. and Petersen, E.E. (1985) Udder health in dairy cows in relation to housing system. In: Østergaard, V. (ed.), *Report 588 from the National Institute of Animal Production*, Copenhagen, pp. 103–124.

Curtis, S.E., Kingdon, D.A., Simon, J. and Drummond, J.G. (1976) Effects of age and cold on pulmonal bacterial clearance in the young pig. *American Journal of Veterinary Medicine* **37**, 299–301.

Ekesbo, I. (1966) Disease incidence in tied and loose housed dairy cattle. *Acta Agricultura Scandinavica* (Suppl. 15).

Ekesbo, I. (1970) Traditional and modern barn environment as related to animal health and welfare. *Proceedings 11th Nordiske Veterinarkongres*, Bergen pp. 56–63.

Enevoldsen, C. (1986) Claw disorders in dairy cows, prevention and control. In: Østergaard, V. and Hindhede, J. (eds), *Report 615 from the National Institute of Animal Production*, Copenhagen, pp. 132–141.

English, P.R. and Morrison, V. (1984) Causes and prevention of piglet mortality. *Pig News and Information* **5**, 369–397.

Flesja, K.I., Forus, I.B. and Solberg, I. (1982) Pathological lesions in swine at slaughter. *Acta Veterinaria Scandinavica* **23**, 169–183.

Geer, D. and Grommers, F.J. (1975) Preventie van teepelbeschadigingen bij melkoeien door stalverbeetering in groupstallen. *Tijdschrift voor Diergeneeskundig* **100**, 533–538.

Gjestang, K.E. and Gravås, L. (1976) Coatings of floors for dairy cows. *Report 87. Department of Agricultural Structures*, Agricultural University of Norway.

Grommers, J.F. (1967) Dairy cattle housing and health. PhD Thesis. University of Utrecht, The Netherlands.

Groth, W. and Eichler-Steinhauff, H.J. (1978) Haltungsbedingte Schaden beim Milchvieh. *Fortschritte der Veterinarmedizin* **28**, 34–43.

Ingram, D.L. (1981) Physiology and behaviour of young pigs in relation to the environment. In: Sybesma, W. (ed.), *The Welfare of Pigs*. Martinus Nijhoff, The Hague, pp. 33–42.

Irps, H. (1983) Results of research projects into flooring preference of cattle. In: Baxter, S.H., Baxter, M.R. and MacCormack, J.A.C. (eds), *Farm Animal Housing and Welfare*. Martinus Nijhoff, The Hague, pp. 200–215.

Jensen, P. (1980) The effect of confinement on the behaviour of dry sows: an ethological study. Report 2 Institutionen for husdyrshygien med hovslagarskolan. Swedish Agriculture University, Uppsala.

Jones, C.R., Wathes, C.M. and Webster, A.J.F. (1982) Release and clearance rates of airborne bacteria within a controlled climate calf house. *Proceedings of the Second International Livestock Environment Symposium*, Ames, Iowa, pp. 529–533.

Jørgensen, M. (1975) Disease–environment recording. *Report 431 from the National Institute of Animal Science*, Copenhagen, pp. 105–112.

Jørgensen, M. and Nielsen, S.M. (1977) Annual Meeting, National Institute of Animal Science, Copenhagen (unpublished).

Kelley, K.W. (1980) Stress and the immune system. *Annales de Recherches Vétérinaires* **11**, 445–478.

Kelley, K.W. (1982) Immunobiology of domestic animals as affected by hot and cold weather. *Proceedings of the 2nd International Livestock Environment Symposium*, Ames, Iowa, pp. 470–482.

Klein-Hessling, P. (1966) Gesundheitsschaden bei der einstreulosen Gitterrostaufstallung. *Der Tierzuchter* **18**, 79–81.

Kosztolich, O. (1966) Ambient climate and enzootic pneumonia with special reference to the determination of climate. *Wiener Tierärztliche Monatsschrift* **53**, 687–693.

Kovacs, F., Nagy, A. and Sallaz, J. (1967) The effect of certain environmental factors on the health and production of pigs. II. Data on the dust and living germ content as well as on the chemical contamination of the air in pig houses of closed systems. *Magyar Allatorvosok Lapja* **22**, 496–502.

Kristensen, E.S., Henneberg, U. and Hindhede, J. (1986) Summer feeding systems: Effect on performance and economics in dairy herds. In: Østergaard, V. and Hindhede, J. (eds), *Report 615 from the National Institute of Animal Science*, Copenhagen, pp. 26–51.

Krohn, C.C. (1986) Exercise in cows. *Ugeskrift for Jordbrug*, 1364–1366.

Kunz, W. and Vogel, O. (1978) Tail tip inflammation: a new problem in fattening bull production. *Tierärztliche Umschau* **33**, 344–353.

Larsson, J., Konggard, S.P., Bisgaard Madsen, E. and Nielsen, K. (1984) Tail tramping in young bulls. II. Behaviour in relation to stocking rate and type of production unit. *Report 569 from the National Institute of Animal Production*, Copenhagen.

Larsson, K. and Bäckström, L. (1971) Veterinary inspection statistics and environmental factors in conjunction with fattening pig production. *Svensk Veterinartidning* **23**, 496–502.

Lenk, T. (1981) On wound infections in fattening bull units with bulls in pens, with special reference to prophylactic tail amputation. *Monatshefte der Veterinärmedizin* **36**, 699–702.

Lindquist, J.O. (1974) Animal health and environment in the production of fattening pigs. A study of disease incidence in relation to certain environmental factors, daily weight gain and carcass classification. *Acta Veterinaria Scandinavica* (Suppl. 51).

Madsen, E.B.M. and Nielsen, K. (1985) A study of tail tip necrosis in young fattening bulls on slatted floors. *Nordisk Veterinærmedicin* **37**, 349–357.

Madsen, E.B.M., Thysen, I., Ingvartsen, K.L. and Østergaard, V. (1987) Health and productivity of young bulls on slatted floors at different stocking densities. In: Østergaard, V. and Hindhede, J. (eds), *Report 628 from the National Institute of Animal Science*, Copenhagen, pp. 209–218.

Martig, J. and Leuenberger, W. (1978) Tail tip necrosis in a Swiss fattening bull unit. *Schweizer Archiv für Tierheilkunde* **120**, 447–453.

Maton, A. and De Moor, A. (1975) A study of the relations between the housing conditions and behaviour patterns and injuries in dairy cattle. *Vlaams Diergeneeskundig Tijdschrift* **44**, 1–18.

Matzke, P. (1975) Haltungsschaden bei Milchvieh in modernen Aufstallungsformen. *Tierärztliche Praxis* **3**, 15–26.

Maund, B.A. and Turner, P.J. (1979) Effect of ventilation rate on calf performance. *Experimental Husbandry* **35**, 85–94.

Miller, W.M., Harkness, J.W., Richards, M.S. and Pritchard, D.G. (1978) Studies of calf respiratory diseases in a large commercial veal unit. In: Martin, W.B. (ed.), *Respiratory Diseases in Cattle*. Martinus Nijhoff, The Hague, pp. 181–194.

Mortensen, B. (1978) Stanchion barn systems for dairy cows: An ethological evaluation. *1st World Congress on Ethology Applied to Zootechnics*, Madrid, pp. 181–193.

Muirhead, M.R. (1981) A comparison of different systems of sow housing and management. *Pig Veterinary Society Proceedings* **8**, pp. 18–28.

Muirhead, M.R. (1983) Pig housing and environment. *Veterinary Record* **113**, 587–593.

Müller, W., Dossow, A. and Dinter, P.S. (1988) Die Tenazitet von Bakterien im luftgetragenen Zustand. *Tierärztliche Umschau* **43**, 258–262.

Murphy, P. (1978) Foot and limb disorders of beef cattle. Clinical studies. *Symposium, Injuries to Animals due to Floor Surfaces*, Fulmer Grange, Slough, UK.

Nielsen, N.C., Christensen, K., Bille, N. and Larsen, J.L. (1974) Preweaning mortality in pigs. 1. Herd investigations. *Nordisk Veterinærmedicin* **26**, 137–150.

Nygaard, A. (1979) Environmental studies in housing for milk production. *Report 67, Department of Agricultural Structures*, Agricultural University of Norway.

Østergaard, V., Blom, J.Y., Thysen, I., Hansen, K. and Birkkjær, K.O. (1986) The effect of sectioning calf houses on calf health, weight gain and economy. *Report 610 from the National Institute of Animal Science*, Copenhagen.

Peterse, D.J. (1985) Laminitis and interdigital dermatitis and heel horn erosion. *Veterinary Clinics of North America: Food Animal Practice* **1**, pp. 83–91.

Pritchard, D., Carpenter, C.A., Mozaria, S.P., Harkness, J.W., Richards, M.S. and Brewer, J.I. (1981) Effect of air filtration on respiratory disease in intensively housed veal calves. *Veterinary Record* **109**, 5–9.

Roy, J.H.B. (1978) Management and nutrition practice of dairy calves. 29th Annual Meeting of the European Association of Animal Production, Stockholm, 7pp.

Russel, A.M. and Weaver, A.D. (1978) The Compton survey of bovine lameness 1977. *Report of the second symposium on bovine digital diseases*, Skara, Sweden, 109–112.

Schmidt, M., Jørgensen, M., Møller-Madsen, Aa, Jensen, H., Horvath, Z., Keller, P. and Konggaard, S.P. (1985) Straw as bedding for dairy cows. *Report 593 from the National Institute of Animal Science*, Copenhagen.

Schmidt Madsen, P. (1978) Environment data in Danish herds under mastitis control. *Proceedings of the 13th Nordic Veterinary Congress*, Turku, pp. 208–211.

Schmidt Madsen, P. and Klastrup, O. (1980) Some milking machine factors and their relationship to the infection level in mastitis-affected herds. *Proceedings of an International Workshop on Machine Milking and Mastitis*, Morepark, Ireland, pp. 137–141.

Schubert, U., Claus, J. and Ernst, E. (1982) Konstitution, Fruchtbarkeit und Leistung bei Milchkuhen in modernen Haltungssystemen. *Zuchtungskunde* **54**, 16–24.

Signoret, J.P. (ed.) (1981) *Welfare and Husbandry of Calves*. Martinus Nijhoff, The Hague.

Stott, E.J., Thomas, L.H., Collins, A.P., Hamilton, S., Jebbett, J. and Luther, P.D. (1978). In: Martin, W.B. (ed.), *Respiratory Diseases in Cattle*. Martinus Nijhoff, The Hague, pp. 230–240.

Svendsen, J. and Bengtsson, A.Ch. (1982) Production systems for rearing piglets: Methods for reducing the number of pigs traumatized by the sow. Research Leaders Meeting. Konsulentavdelingen, Reporter, Allmänt 39, Swedish University of Agricultural Science, Uppsala.

Svendsen, J., Bengtsson, A.Ch. and Svendsen, L.S. (1986) Occurrence and causes of traumatic injuries in neonatal pigs. *Pig News and Information* **7**, 159–170.

Sybesma, W. (ed.) (1981) *The Welfare of Pigs*. Martinus Nijhoff, The Hague.

Thysen, I., Buchwald, E., Smedegaard, H.H. and Blom, J.Y. (1981a) Foot disorders in dairy herds in relation to parity and stage of lactation. 32nd Annual Meeting of the European Association of Animal Production, Zagreb.

Thysen, I. Buchwald, E. and Smedegaard, H.H. (1981b) Injuries and disease in the claws of dairy cows. In: Østergaard, V. and Hindhede, J. (eds.) *Report 515 from the National Institute of Animal Science*, Copenhagen, pp. 49–72.

Tielen, M.J.M. (1978) Buildings, environmental conditions and disease. In: Proceedings of the 29th Annual Meeting of the European Association of Animal Production, Stockholm, p. 47.

Walberg, K. (1978) Floors in animal buildings. Part I, Report 2. Swedish University of Agricultural Science, Lund.

Webb, N.G. and Nilsson, C. (1983) Flooring and injury – an overview. In: Baxter, S.H., Baxter, M.R. and MacCormack, J.A.C. (eds), *Farm Animal Housing and Welfare*. Martinus Nijhoff, The Hague, pp. 226–259.

Webster, A.J.F. (1979) Housing and husbandry of the veal calf. In: Grunsell, C.S.G. and Hill, F.W.G. (eds), *Veterinary Annual*, Vol. 19, Scientechnica, Bristol, pp. 49–53.

17 Genotype–Environment Interaction

J.B. Owen

School of Agricultural and Forest Sciences, University of Wales, Bangor, UK

Summary

Farm animals have originated from wild progenitors who were adapted to often very different environments from those in current animal production systems. Varying degrees of amelioration of their genome have occurred since domestication in response to selection forces. This chapter considers the evidence for genotype–environment interaction, i.e. differential performance of breeds/strains within species when exposed to varying environments or, conversely, the differential effects of environmental factors when assessed in relation to different animal genotypes. Environmental aspects considered include physical and climatic aspects (temperature, daylength, etc.) as well as degree of confinement and group size.

It is concluded that important interactions do occur which involve susceptibility to disease, temperament and amenability to close interaction with man and his machines (e.g. milking machines).

Looking to the future it is considered that major decisions need to be made of how much man-made intervention in animal environments is feasible and prudent. The long-term answer may be to achieve the optimum compromise between changes in the environment and changes in the genome of the animals themselves.

Introduction

Over a long period within the environment they occupy, animals have evolved a complex blueprint for their development – the genome. In order to maximize reproductive fitness in the long term, the genome has to include provision for two complications. One is that animals exist within social groups interacting with their group members and, in turn, with the other animal and plant life of the environment; the success of the individual is therefore bound within a social context. The other complication is that the environment is not constant: it has seasonal variations within the year, between year fluctuations and in the longer term may show systematic or sometimes catastrophic change. These two complications necessitate the development of genes controlling social interaction and also genes that favour the maintenance of some degree of adaptability to environmental change.

The net result of evolutionary development was that pre-domestication, animals had evolved a complex genome well suited to their environmental niche but with obvious social/group adaptation and with a certain in-built flexibility in the face of environmental variation. Domestication appears to have been a traumatic experience, although in certain respects phased over a period of time. Archaeological evidence points to post-domestication 'stress' resulting in slower development, lowered size and other signs of depressed function. There is also evidence that 'disruptive' selection can operate on domestication leading to the expression of unusual phenotypic traits, e.g. in coat colour patterns (Belyaev and Trut, 1986; Zeuner, 1963).

The domestication of farm animals, starting with sheep (about 12 000 years ago), was accompanied and followed by a modification of their genome in several respects, particularly in relation to docility and several major production traits. In relation to the pre-domestication state, the subsequent changes are probably relatively small in many relevant respects and modern farm animal breeds still retain a large proportion of their original genetic make-up and its complex phenotypic expression.

With the modification of and variation in the environment for modern farm animals, it would be reasonable to expect genotype–environment interaction, whereby genotypes show differential expression of various traits in differing environments.

Measurement of Genotype–Environment Interaction

Genotype–environment interaction implies a relative change in the phenotypic expression of genotypes when observed in different environments, with respect to a defined trait.

The environmental aspects of interest in this chapter include climate, photoperiod, feed choice, feeding system, disease challenge, confinement and method of milk evacuation. In order to understand some of the problems of modern animal management some comparisons need to be drawn between the presumed environment of the wild progenitor and that of the present-day animal. Imposed nutritional constraints are not considered in this chapter although aspects of feeding behaviour are.

The genotypes of interest are mainly comparisons within species, of breeds or of identified individuals, e.g. sires within herds.

The main traits considered are (i) behavioural (ii) metabolic, physiological and structural and (iii) production traits including growth and body composition, reproduction and lactation, and resistance to disease.

Genotype–environment interaction can vary from a major change of ranking of genotypes with respect to the phenotype for a trait to a lesser effect of changed relativity within the rankings.

Genotype–environment interactions usually involve high sampling variance, their estimation thus usually necessitating large numbers for their precise evaluation.

Horst (1985) has described the various statistical approaches to the estimation of genotype–environment interaction, e.g. rank correlation, regression of the phenotypic values of genotypes on the different environment levels, estimation of the interaction term in factorial designs, genetic correlation of the performance of the same genotypes observed in two environments and specifically the correlation of the breeding values (BVs) of a genotype estimated in two environments.

The correlation of BVs estimated in two environments is often a convenient and practically useful method which has wide application, for instance in dairy cattle breeding.

Climate

Cold

Animals adapt in many ways – behaviourally, metabolically and structurally – to cold conditions. The extreme level of adaptation involves a period of hibernation although such an extreme manifestation does not concern us in this discussion. Interesting observations have been made on the wild house mouse (*Mus musculus*) in this context. Mice kept at low ambient temperatures, *circa* 0°C, over several generations (Eskimo mice) showed many adaptations, some of them genetic changes, as compared to mice kept at 20–23°C (Barnett and Dickson, 1989). In the control, warm environment mice changed little over 10 generations but in the cold the mice became progressively heavier and fatter, more early maturing and with lower early mortality. Part of the improvement related to a more concentrated milk by the Eskimo dams and improved nesting and mothering care of the young. When compared in the warm environment after 10 generations the Eskimo mice were more fertile, had a similar litter size and heavier young than the control mice. Such findings can only be interpreted very cautiously from the mouse model to much larger sized farm animals. They do, however, lend support to the unquantified observations that genetic changes do occur in many respects when farm animals are bred in cold environments, if only in superficial aspects such as coat length and thickness. These genetic divergencies are likely to contribute to the genotype–environment interaction observed when animals of different origin are compared under cold and under temperate conditions. Cold conditions can be interpreted in a wider sense than ambient temperature when the chill factors of air movement and moisture are taken into account, particularly when animals are out-wintered. Both cattle and sheep breeds bred for such chilling conditions perform relatively better in them than breeds developed in warmer conditions, because of their metabolic, structural and behavioural adaptation to such conditions. The problem in farm animal breeds, not encountered in some of the mouse model experiments, is that chilling conditions under farm conditions are normally confounded with a reduction in quantity and quality of food supply.

One interesting example of work on the application of genotype–environment interaction in this respect is the work of Slee and Stott (1986) on the body-temperature reaction of young lambs to a cold chill treatment. This test has been

Table 17.1. Ranking of predicted BV for 17 bulls from AI studs with progeny in the USA, Mexico and Colombia.

Sire	Mexico rank	Colombia rank	USA rank
1	1	6	1
2	2	7	8
3	3	2	10
4	4	8	11
5	5	5	3
6	6	17	6
7	7	16	13
8	8	3	2
9	9	10	9
10	10	9	4
11	11	14	17
12	12	15	15
13	13	1	12
14	14	4	16
15	15	11	14
16	16	13	5
17	17	12	7

Source: After Abubakar *et al.*, 1987.

shown to distinguish genotypes with respect to resistance to cold and has been applied to the selection for cold resistance in hill sheep.

Heat

In dairy cattle many studies of genotype–environment interaction have been made on more general milk production traits, particularly using the estimation of the correlation of the BV of dairy sires in more than one environment. These tend to show relatively little interaction for milk yield, even in quite divergent environments (Syrstad, 1990). In some cases, particularly for traits associated with fertility, there is evidence of interaction from comparisons involving extremes of environment, e.g. temperate with hot tropical conditions (Abubaker *et al.*, 1987) as shown in Table 17.1.

It is often difficult to disentangle the biological basis of the interactions since the environmental differences often involve the confounding of nutrition, heat stress and disease challenge. Mechanisms as diverse as urea recycling, grazing behaviour and tick resistance may be involved.

For beef cattle there are several reports of major genotype–environment interaction. Butts *et al.* (1971) compared two lines of Hereford cattle originating either in Montana or in Florida, USA, at the same two locations. Table 17.2 shows that important genotype–environment interactions were observed for weaning and yearling weights ($P < 0.01$).

Table 17.2. Least-square means for weight traits of cattle from two origins compared at two locations.

			Origin of lines	
Sex	Weight at:	Location	Montana	Florida
Male	Weaning (kg)	Montana	190	181
		Florida	189	210
Female	Weaning (kg)	Montana	183	166
		Florida	178	191
Female	Yearling (kg)	Montana	352	316
		Florida	285	302

Source: After Butts *et al.*, 1971.

Similarly for 5-year-old cows a genotype–environment interaction was observed ($P < 0.05$), arising from the fact that the cows of Montana origin located in Florida had a lower percentage pregnant and calf weaned than the other three groups. These results show that these genotypes excel when on 'home' as compared to 'away' ground.

Later analyses of these and additional data have confirmed the earlier results. For example Koger *et al.* (1979), in the first of a series of papers on the final results of this long-term experiment, have emphasized the genotype–environment interaction in relation to reproductive traits (Table 17.3).

In this report genotype–environment interaction is expressed in terms of the mean superiority of local over introduced lines and demonstrates the effect in a striking and practical manner. Like some other reports it is not demonstrable that heat is the main factor involved in the contrasts of environments.

Frisch (1981), however, does provide evidence that heat tolerance can be an

Table 17.3. Genotype–environment interaction for reproductive traits in Hereford cattle in Montana and Florida, USA.

Origin	No. of matings	Pregnancy rate (% ± SE)	Weaning rate (% ± SE)
Phase 1			
Local (L)	1412	84.1 ± 1.3	76.7 ± 1.3
Introduced (I)	1533	77.4 ± 1.2	70.6 ± 1.3
L − I		6.7 ± 1.8	6.1 ± 1.9
Significance level		$P < 0.01$	$P < 0.01$
Phase 2			
Local (L)	768	79.6 ± 1.9	71.5 ± 2.1
Introduced (I)	191	68.0 ± 3.5	59.0 ± 3.7
L − I		11.6 ± 4.0	12.5 ± 4.2
Significance level		$P < 0.01$	$P < 0.01$

Source: After Koger *et al.*, 1979.

Table 17.4. Least-square means of two traits for cattle selection lines under indoor (unstressed) conditions and outdoor heat stress.

Trait	Level of stress	Line: Selected	Line: Control	
Gain/day (kg)	Low (indoor)	0.79	0.85	($P < 0.10$)
	High (grazing)	0.25	0.15	($P < 0.05$)
Rectal temperature (°C)	Low (indoor)	39.83	40.03	($P < 0.05$)
	High (grazing)	39.47	39.98	($P < 0.01$)

Source: After Frisch, 1981.

important potential cause of genotype–environment interaction in beef cattle (Table 17.4).

This shows that under pen conditions of low heat stress the control line showed, if anything, higher gain than the line selected for 18-month weight in the hot tropical environment. Under hot tropical grazing conditions, however, the selected line showed higher gain than the control and there was a more marked restriction of rectal temperature in the selected than in the control line.

In sheep genotype–environment interaction is a well-known phenomenon because of the wide range of environments in which sheep are kept. Information on such interaction comes from the comparison of native and exotic genotypes under stressful environments. Olsson and Beyene (1990), for example, compared the Arsi sheep, native to the highland of Ethiopia, with crossbred sheep sired by introduced rams of the Bleu de Maine, Hampshire and Corriedale breeds (Table 17.5).

Many sheep breeds that have evolved in subtropical semi-arid areas of the world, have a fat tail which may have a significance analogous to the hump of the camel. Its function seems to be to act as an energy store to buffer seasonal fluctuation in feed supply. It may also be a means to maximize energy storage capacity without

Table 17.5. Performance of native Arsi sheep and its various crosses with exotic rams, under Ethiopian conditions.

Genotype	Barrenness (%)	Total weaning wt/ewe joined (kg)
Group 1 (years 1974–1979)		
Arsi	9.6 ± 4.8	13.7
25% Exotic	8.9 ± 5.7	16.2
58% Exotic	15.5 ± 5.5	17.5
Group 2 (years 1978–1987)		
Arsi	3.8 ± 2.6	14.3
50% Exotic	8.8 ± 2.4	17.5
75% Exotic	18.9 ± 2.5	15.5
87.5% Exotic	36.4 ± 3.7	12.6

Source: After Olsson and Beyene, 1990.

Table 17.6. Comparison of naked neck (NaNa) and normal feathered ($na^+ na^+$) dwarf lines of poultry at two temperatures, expressed relative to the full-feathered line.

	31°C		25°C	
Trait	NaNa	$na^+ na^+$	NaNa	$na^+ na^+$
Mean egg wt	107	100	102	100
Egg mass/28 d	113	100	100	100
g food/g eggs	88	100	102	100

Source: After Hartmann, 1986.

impeding heat loss capacity; however, observation of animals where the tail has been docked do not always lend support to this hypothesis (Marai *et al.*, 1987).

Poultry are the species of animal most studied in relation to genotype–environment interaction because of their high reproductive potential and ease of testing under standardized conditions. Horst (1985) has shown that substantial genotype–environment interaction has been reported, particularly for body weight, egg production and egg weight in various locations, housing systems, temperature and lighting regimes. Under hot conditions lighter bodied birds perform relatively better than their heavier counterparts. A major gene for dwarfism has also been shown to confer similar benefits of low body size in hot conditions. The naked neck major gene which reduces the feathering of the bird has also been shown to confer relatively greater advantages in high temperatures (Hartmann, 1986) as shown in Table 17.6.

Photoperiod

Many aspects of animal function are distinctly seasonal, including fibre growth and shedding, occurrence of oestrus in sheep and growth and reproduction in poultry. In many cases these functions have been shown to be related to the photoperiod and the role of melatonin, secreted from the pineal gland, in this respect, is now well documented.

An interesting study of genotype–environment interaction in sheep was carried out by Hulet *et al.* (1974) where ewes originating from Texas and from Idaho were observed at the two locations. Allowing for the fact that ewes take about six months to adjust to their changed breeding pattern at a new location, Fig. 17.1 shows the occurrence of oestrus in ewes in this experiment one year following transfer, expressed as the percentage of ewes in oestrus.

The results show that in Idaho the breeding pattern of the two ewe types is indistinguishable. In Texas, however, the breeding season pattern for both ewe types is displaced to earlier in the season. There is, however, a difference between genotypes, with the north-western ewes coming into season about one month earlier and staying in season about one month longer ($P < 0.01$). Although the biological interpretation of these results is difficult, they do demonstrate the need to assess

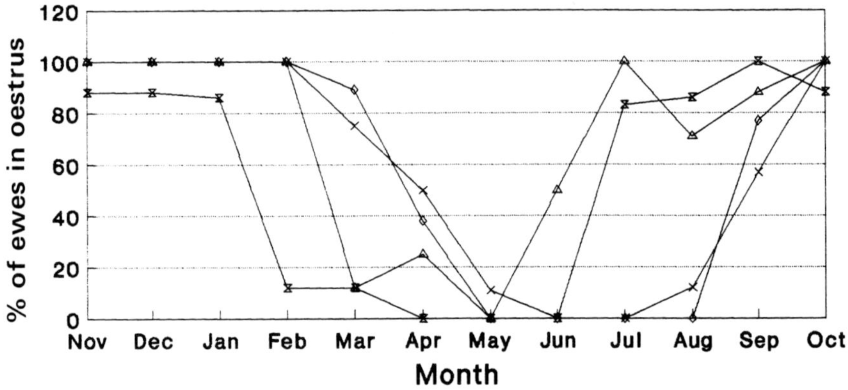

Fig. 17.1. Percentage of ewes in oestrus, 1970–71. —x— Texas ewes in Idaho; —◇— NW ewes in Idaho; —△— Texas ewes in Texas; —⧗— NW ewes in Texas.

breeding season performance of sheep being newly introduced to different locations, particularly where latitude differences are involved.

Aboul-Ela and Aboul-Naga (1987), in surveying the reproduction of subtropical fat-tailed sheep, have shown that although seasonality is much reduced at low latitudes, the small photoperiod changes are an important factor in the changes in oestrous activity of subtropical sheep. It can be concluded that a search for adaptable ewes for frequent breeding systems could well be concentrated on local sheep breeds or sheep breeds from other subtropical areas, rather than risk the complications of genotype–environment interaction when introducing sheep from northern (or higher) latitudes.

Since photoperiod changes may have a more pervasive effect on behaviour and production in animals than at present realized (Phillips and Schofield, 1989) the interactions noted for sheep breeding activity may have more widespread significance.

Photoperiod patterns are also known to affect poultry in several ways, particularly in relation to reproductive development, including egg laying traits. Timmermans and Sallevelt (1983) studied layers selected for egg production in two housing system environments. One environment was a battery cage system with a constant light regime and the other a floor deep litter system with natural lighting. Although the environments varied in several respects the authors considered the photoperiod contrast to be the main feature. The two genotypes, selected for total egg production, were White Leghorn (WL) and Medium Heavy (MH) in each of the two environments for eight generations. In the final test year, carried out under the constant daylength battery system, it was shown that indirect selection of the WL in natural daylength resulted in broadly superior performance to direct selection in constant daylength. For for MH genotype, in contrast, direct selection was superior as shown in Table 17.7.

The results for genotypes tested under natural environments are not given so that the genotype–environment interaction cannot be directly assessed. However, it is an indication that the selection environment is not always a clear guide to performance in other environments.

Table 17.7. Long-term selection followed by a constant daylength battery test.

Genotype	Selection environment	Egg production (g)	Egg number	Egg weight (g)
WL	Constant	1227	214	58.5
WL	Natural	1237	223	59.0
MH	Constant	1154	211	59.7
MH	Natural	1100	211	57.8

Source: After Timmermans and Sallevelt, 1983.

Feed Choice

The feeding behaviour of modern farm animals is dominated by their wild progenitor's genome with some modification that has occurred in the era since domestication (*circa* 12 000 years ago). The environment of present-day nomadic and settled range grazing herds/flocks probably still allows some rein to the genotype in terms of feeding behaviour and there is clearly a retention of many aspects such as territorial grazing (Provenza and Balph, 1988). There is little documented evidence to assess to what extent animals bred for generations under confinement in fenced fields, have lost the ranging behaviour of their wild pastoral relatives.

With regard to feed choice, however, there is evidence that animals subjected to new choices show a differential development of the innate wisdom associated with their wild environment. Pregnant ewes given free choice to feed materials had variable success and many chose very badly (Gordon and Tribe, 1951). Tribe (1955) went on to demonstrate differential wisdom by rats exposed to a choice of purified feed materials as compared to their normal wisdom when given the choice of familiar items. Another example is the differential response of rat genotypes to 'cafeteria feeding' where some genotypes exposed to this unfamiliar modern choice become obese while others do not (Rothwell and Stock, 1980).

Feeding System

There are several interesting examples of genotype–environment interaction related to feeding system. In pigs the comparison of restricted feeding with *ad libitum* feeding is important while in cattle the comparison of high forage with high grain feeding is relevant.

In relation to pig testing procedure it has been evident for some time that the feeding system on test is an important factor in the resultant genetic change (Owen and Morton, 1969) and that selection on an *ad libitum* system involves changes in different underlying components of the traits selected for than when selection is on the restricted system. Others like Steane *et al.* (1984) have subsequently confirmed that genotype × feeding system interactions occur for several pig traits, notably for growth rate, feed conversion efficiency, and lean content of the carcass.

The evaluation of dairy cattle sires has been examined in relation to the grain

Table 17.8. Hen housed production (% 18–57 weeks) in relation to strain and housing environment.

Dominance selected strain	Floor		Cages	
	Crowded	Uncrowded	Crowded	Uncrowded
High	41.0	36.8	37.8	51.0
Low	53.2	59.1	49.5	51.7
High − Low	− 12.2	− 22.3	− 11.7	− 0.7

Source: After Biswas and Craig, 1970.

feeding system under which their daughters have been kept. Many of these studies, such as that of Wiggans and Van Vleck (1978), have not detected important genotype–environment interactions between high forage and high grain feeding systems. However, Mao and Burnside (1969) did observe a major sire–herd environment interaction in relation to different levels of grain feeding in the summer. As with many other analyses of observational data in relation to genotype–environment interaction, the interpretation of significant interactions is difficult. In this case the difficulty lies in determining how far summer grain feeding levels are confounded with some other component of the herd environment which could be the primary cause of the interaction.

Group Size/Housing System

There is genotype–environment interaction in relation to many traits, for example fertility, when domesticated and wild animals are compared under free ranging conditions and confined man-controlled environments. The fertility of wild animals under confined conditions (as in zoos), even those bred in captivity, is generally low but also variable, some breeding freely others hardly at all. Most modern farm animals have been under intense selection for fertility under confined conditions and exhibit fertility levels similar to those of equivalent animals in the wild.

Housing system/genotype interaction in poultry

As with other animals there is considerable variation in the reported interaction between poultry genotypes and housing systems. Biswas and Craig (1970) carried out a comprehensive study of the differential effect of two types of housing environment on the performance of two strains of layers. The two strains had been previously selected (for five generations) for high (high) and low social dominance ability (low). The environments consisted of floor pens vs. cages under crowded or uncrowded conditions; the extreme of privacy being one bird per cage. The results showed large interaction of housing system and strain of hen for the egg production traits – total hen-housed egg production, rate of lay and age at first egg. There were also clear three-way interactions of strain of hen with housing system and stocking density in relation to hen-housed production and rate of lay (Table 17.8). It would appear that high dominance ability is a disadvantage under crowded conditions, presumably because of the scope for aggressive incidents.

Evidence for interaction with housing/milking system in cattle is more difficult to quantify although there is much variation in the performance of dairy cows on cubicle systems. Some cows persistently refuse to lie in cubicles as intended but so far there is little documented evidence that the difference between users and non-users is due to genotype rather than more ephemeral factors.

Evidence from more closely studied species indicates that there can be important genotype–environment interactions in relation to behaviour across a range of environments, presenting different levels of stress. For example, Gerlai and Csanyi (1990) have demonstrated differential behavioural responses to familiar and unfamiliar situations in strains and crosses of Paradise fish (*Macropodus opercularis*).

Disease Challenge

The existence of genetic resistance to (or tolerance of) disease challenge implies a genotype–environment interaction, since resistant genotypes could be expected to perform relatively better in a high challenge than in a low challenge environment. A striking example is the tolerance of tsetse-fly transmitted trypanosomiasis shown by some African breeds, notably the N'Dama cattle breed. These cattle are able to withstand the ill-effects of the causal parasite, notably anaemia, and to continue to thrive and show high efficiency of growth and lactation in tsetse-fly areas where non-tolerant animals suffer illness and death. Murray *et al.* (1991) describe an experiment carried out at the International Laboratory for Animal Diseases, Kenya. N'Dama cattle were transported from the Gambia as embryos and reared like the Boran comparison genotype from Boran (surrogate) mothers in an area of Kenya free from trypanosomes. Eight cattle from each genotype were successively challenged with various trypanosome infections from one year of age. The cattle were well fed and to prevent death were treated with a trypanocidal drug when a severe level of infection was reached (as judged by a blood packed red cell volume of 15% or less). The results were that no N'Dama cattle required treatment during the course of four different challenges while all the infected Boran cattle did require treatment. Infected Boran cattle showed severe weight loss and suppression of normal oestrous activity in the female as compared to uninfected controls. The infected N'Dama on the other hand showed normal weight gain and oestrous activity as compared to the uninfected animals.

Morris (1991), reviewing the position in Australia and New Zealand, showed that the heritability of resistance to many economically important animal diseases compared favourably with that for many production traits. He also reported the differential response of animals to differing levels of challenge.

A good example is that reported by Frisch (1981) from the beef cattle selection experiment previously discussed. This was carried out under dry tropical conditions at Rockhampton, Queensland. Differential performance of the line selected for 18-month weight and the control, unselected group, was observed according to the test environment. In the natural environment with high heat stress and challenge by helminth worms, the selected line grew faster than the control line. More detailed

Table 17.9. Comparison of performance of selection line cattle in stress and stress-free environment.

		Line	
Variable	Level of stress	Selected	Control
Gain (kg)	Low (drenched)	59	62
	High (undrenched)	71	42

Line × stress interaction $P < 0.05$.

Source: After Frisch, 1981.

examination revealed that selection line cattle outperformed the controls when either of the stresses was operative but not when the lines were compared in a stress-free environment (Tables 17.4 and 17.9).

Milk Extraction Method

Another area of genotype–environment interaction relates to the milking/suckling process in dairy animals. Non-dairy breeds of farm animals depend to a large extent on the suckling stimulus of their young to activate the reflex release of milk from the alveoli and ducts of the mammary gland to the main ducts. Highly developed dairy animals on the other hand have been bred for generations for the ability to respond to artificial milking by hand or by machine. In between we have dairy animals, particularly sheep, where both suckling and milking are practised concurrently. It is, therefore, to be expected that the relative lactation merits of animals can interact with the evacuation technique as it varies from suckling only through the intermediate practice to the extreme dairy system where the young are taken from the dam shortly after birth and subsequent milking is all artificial. Some undeveloped cattle breeds are known to release more milk for hand milking in the presence of the calf than without. In sheep, dairy breeds tend to give similar total yields of milk whether the evacuation is by suckling only or by a range of combinations of milking and suckling. This was demonstrated in observations of Syrian Awassi ewes subjected to five 'time of weaning' systems of management during a 150-day lactation (Fadel *et al.*, 1989) The systems ranged from suckling only (system 5) to a system of daily milking combined with suckling for the first 60 days (system 1) (Table 17.10).

Ewes from non-dairy breeds yield much less milk when subjected to milking than when subjected to suckling only.

Current developments in cow milking technique, towards the goal of fully automatic milking machines, may well reveal further genotype–environment interaction whereby some genotypes will be more amenable to such a fully automated system, which allows much freer expression of innate behaviour in this respect.

Table 17.10. Milk yield (kg) of Awassi ewes during a 150-day lactation.

	Management system				
Year	1	2	3	4	5
1984					
'milked' yield	78	68	40	13	0
Total yield	132	131	132	116	132
1985					
'milked' yield	95	85	52	22	0
Total yield	152	152	150	164	162

Source: After Fadel *et al.*, 1989.

Discussion

The review indicates that genotype–environment interaction is relatively unimportant over a range of commercial conditions within a limited region, in relation to the environments and traits studied in farm animals. Where the environmental differences are very marked, however, as in the contrast between temperate and tropical conditions, interactions are often important particularly for reproductive traits. It is not easy to design formal tests for genotype–environment interactions that may be relatively small and yet exert important subtle effects on animal welfare and production efficiency.

A major problem of animal husbandry nowadays is the increasing ease with which animal genotypes, particularly in the form of semen or embryos, can be transported from their region of origin. This means that present-day animal genotypes have to contend with one of a range of environments, not only in terms of location but in terms of disease challenge and nutrition. In particular, genotypes developed in high input husbandry systems in temperate areas are being exposed to tropical and subtropical conditions and high disease challenge systems, often compounded with seasonal nutritional constraints.

One of the important problems in interpreting much of the evidence presented is that in addition to the difficulty of establishing genotype–environment interaction in the first place, the precise biological components of the interaction are often difficult to disentangle. Under farm conditions environmental factors such as climate, seasonal availability and quality of feed, pathogen challenge and stockmanship are all confounded in a complex way.

Genotype–environment interaction seems to arise from the occurrence of a change of environment from that in which the genotype has developed over several generations. If the changes are relatively small and phased over generations then most genotypes are sufficiently flexible to cope; in any case most environments in practice vary both within and between seasons such that adapted genotypes evolve to cater for this variability. With a change of environment animals seem to adapt relatively quickly both physiologically and behaviourally and by genetic change. The experiments reported on Eskimo mice confirm this possibility in a striking way. Such genetic adaptation suggests that the imbalance or maladaptation implied by genotype–

environment interaction, can be relatively quickly removed, through continued selection in the new environment. Undoubtedly since domestication several such genetic changes have occurred in farm animals.

Studies of the unconstrained natural behaviour of modern farm animal genotypes, as in the work of Jensen and of Wood-Gush on the domestic pig, reveal patterns of behaviour that have little to do with the modern piggery environment. Indeed Jensen and Redbo (1987) indicate that the perceived good practice of stockmen, who move the sow out of the farrowing quarters during the first day after farrowing, may well conflict with the natural inclination of the sow and result in increased piglet mortality. In such instances it would seem to be far more efficient to adapt stockmanship practice within the natural behaviour pattern of a species than to entail a longer period of genetic adaptation to discordant modern practices.

On a wider front there seem to be many ways in which husbandry practices could be adapted to minimize the necessary genetic changes in farm animals.

Long term aspects of selection

One constraint to the effective adaptation of farm animal genotypes to their modern environments stems from a technicality of breeding scheme operation. Since the generation length is the major denominator of rate of progress in a breeding scheme, most selection emphasis is placed on traits measured early in the animal's life. In the long run this inevitably weakens the selection for adaptive traits that ensure healthy longevity. The modern dairy cow tends to be bred for high performance as a young cow, thus inevitably decreasing the relative weight placed on traits associated with sustained performance. It is no wonder then that the modern cow has problems of mastitis, lameness and infertility that are generally not prominently displayed by the heifer in her first lactation. The longevity of New Zealand dairy cows has been reduced by 0.5 lactations in the last 30 years.

Use of modern health care

Another constraint to the adaptation of animals to modern environments is that imbalances between the animal's genome and these environments, as evidenced by ill-health in its widest sense, instigate a more intensive artificial intervention in the environment. Modern drugs and vaccines distort natural ecological adaptation and create an artificial environment to which the animal genome responds. Unfortunately these changes reduce innate resistance to pathogenic challenge and create a drug-dependent genotype. Since pathogens can and do adapt to drug attack there is a problem of continued imbalance which has long-term dangers. These include the possible development of more dangerous pathogens which are difficult to control artificially.

Selection strategy

There is a widespread conception, supported by some evidence (e.g. Frisch, 1981), that traits involved in adaptation to stressful environment, brought about by long-term selection, show a negative genetic correlation with production traits in farm animals. This can be explained, to some extent, by the fact that higher production, for example in the form of twinning or a high lactation drive, might be counter-productive in the longer term, leading to females that produce well in their first parity and then suffer the consequences in terms of lowered fertility, etc., arising out of over-depletion of body reserves. However, there is also evidence that where nutritional constraints are removed, there appears to be quick genetic adaptation to the constraints of the stressful environment (as in the Eskimo mouse experiment) and evidence that there is no adverse genetic correlation between adaptation and productive traits (Seifert, 1984). The N'Dama cattle, discussed in relation to trypanotolerance in Africa, once thought to be very poor in terms of productive traits, have more recently been shown to have good performance relative to live weight, both in terms of body gain and of lactation (Murray *et al.*, 1991).

An interesting general possibility suggested by the reviewed data, is that in cold environments large body size and a fat body composition seem to be part of the adaptation (e.g. the Eskimo mice) while in hot conditions the reverse is true with animals within species tending to be smaller in size and of lean body composition (apart from concentrated depots such as the tail and rump in sheep and the hump in camels).

Both the adaptation of farm animal genotypes and the amelioration of their environment, as imposed in man-made housing and management systems, are important elements in securing humane and efficient husbandry. Much more knowledge of the behavioural and other characteristics of the farm animal species under free ranging 'natural' conditions would allow a more fruitful balance to be struck between the animal's complex genome and the imposed management practices.

It is by no means clear that breeding stock should be selected in artificially 'stress-free' environments, characteristic of pedigree herds/flocks, since such environments are neither economical nor practicable in many situations where genotypes of world importance will need to perform.

References

Aboul-Ela, M.B. and Aboul-Naga, A.M. (1987) Improving fecundity in sub-tropical fat-tailed sheep. In: Marai, I.F.M. and Owen, J.B. (eds), *New Techniques in Sheep Production.* Butterworths, London.

Abubakar, B.Y., McDowell, R.E. and Van Vleck, L.D. (1987) Interaction genotype and environment for breeding efficiency and milk production of Holsteins in Mexico and Colombia. *Tropical Agriculture (Trinidad)* **64**, 17–22.

Barnett S.A. and Dickson, R.G. (1989) Wild mice in the cold: some findings on adaptation. *Biological Reviews* **64**, 317–340.

Belyaev, D.K. and Trut, L.N. (1986) The genetics and evolution of animal domestication: a revised interpretation. Paper presented by L.N. Trut at the 37th Meeting of the European Association of Animal Production, Budapest, 1986.

Biswas, D.K. and Craig, J.V. (1970) Genotype–environment interaction in chickens selected for high and low social dominance. *Poultry Science* **49**, 681–692.

Butts, W.T., Koger, M., Pahnish, O.F., Burns, W.C. and Warwick, E.J. (1971) Performance of two lines of Hereford cattle in two environments. *Journal of Animal Science* **33**, 923–932.

Fadel, I., Owen, J.B. and Kassem, R. (1989) Economic assessment of Awassi ewe production systems under semi-arid conditions. *Research and Development in Agriculture* **6**, 47–52.

Frisch, J.E. (1981) Changes occurring in cattle as a consequence of selection for growth rate in a stressful environment. *Journal of Agricultural Science* **96**, 23–38.

Gerlai, R. and Csanyi, V. (1990) Genotype–environment interaction and the correlation structure of behavioural elements in Paradise fish (*Macropodus opercularis*). *Physiology and Behavior* **47**, 343–356.

Gordon, J.G. and Tribe, D.E. (1951) The self-selection of diet by pregnant ewes. *Journal of Agricultural Science* **41**, 187–190.

Hartmann, W. (1986) Recent research on the breeding of egg production stock. *Proceedings of the 7th European Poultry Conference, Paris, 1986*, vol. 1, pp. 49–59.

Horst, P. (1985) Effects of genotype–environment interactions on efficiency of improvement of egg production. In: Hill, W.G., Manson, J.M. and Hewitt, D. (eds), *Poultry Genetics and Breeding*. British Poultry Science, Longman Group, Harlow, pp. 147–156.

Horst, P. and Mathur, P.K. (1990) Genetic aspects of adaptation to heat stress. In: Hill, W.G., Thompson, R. and Wooliams, J.A. (eds), *Proceedings of the 4th World Congress on Genetics applied to Livestock Production. XIV Dairy cattle genetics and breeding, adaptation, conservation*, pp. 286–297.

Hulet, C.V., Shelton, M., Gallagher, J.R. and Price, D.A. (1974) Effects of origin and environment on reproductive phenomena in Rambouillet ewes. 1. Breeding season and ovulation. *Journal of Animal Science* **38**, 1210–1217.

Jensen, P. and Redbo, I. (1987) Behaviour during nest leaving in free ranging domestic pigs. *Applied Animal Behaviour Science* **18**, 355–362.

Koger, M., Burns, W.C., Pahnisch, O.F. and Butts, W.T. (1979) Genotype by environment interactions in Hereford cattle: 1. Reproductive traits. *Journal of Animal Science* **49**, 396–402.

Mao. I.L. and Burnside, E.B. (1969) Sire by herd environment interaction for milk production. *Journal of Dairy Science* **52**, 1055–1062.

Marai, I.F.M., Nowar, M.S., Bahgat, L.B. and Owen, J.B. (1987) Effect of docking and shearing on growth and carcass traits of fat-tailed Ossimi sheep. *Journal of Agricultural Science* **109**, 513–518.

Morris, C.A. (1991) Screening and selection for disease resistance – repercussions for genetic improvement. In: Owen, J.B. and Axford, R.F.E. (eds), *Breeding for Disease Resistance in Farm Animals*. CAB International, Wallingford, pp. 123–135.

Murray, M., Stear, M.J., Trail, J.C.M., d'Ieteren, G.D.M., Agyemang, K. and Dwinger, R.H. (1991) Trypanosomiasis in cattle: prospects for control. In: Owen, J.B. and Axford, R.F.E. (eds), *Breeding for Disease Resistance in Farm Animals*. CAB International, Wallingford, pp. 203–223.

Olsson, A. and Beyene, T. (1990) Performance of indigenous and crossbred sheep in the highland of Ethiopia. In: Hill, W.G., Thompson, R, and Wooliams, J.A. (eds), *Proceedings of the 4th World Congress on Genetics applied to Livestock Production. XIV Dairy cattle genetics and breeding, adaptation, conservation*, pp. 406–409.

Owen, J.B. and Morton, J.R. (1969) The association of food conversion ratio, age at slaughter and carcass quality in pigs fed *ad libitum*. *Animal Production* **11**, 317–324.

Phillips, C.J.C. and Schofield, S.A. (1989) The effect of supplementary light on the production and behaviour of dairy cows. *Animal Production* **48**, 293–303.

Provenza, F.D. and Balph, D.F. (1988) Development of dietary choice in livestock on rangelands and its implications for management. *Journal of Animal Science* **66**, 2356–2368.

Rothwell, N.J. and Stock, M.J. (1980) Intrastrain differences in the response to overfeeding in the rat. *Proceedings of the Nutrition Society* **39**, 20A.

Seifert, G.W. (1984) Selecting cattle for beef production in humid and sub-humid tropical environments. In: *Proceedings of the 2nd World Congress on Sheep and Beef Cattle Breeding*, Pretoria, S. Africa, April 1984, pp. 160–164.

Slee, J. and Stott, A.W. (1986) Genetic selection for cold resistance in Scottish Blackface lambs. *Animal Production* **43**, 397–404.

Steane, D.E., Evans, D.G. and Kempster, A.J. (1984) A practical estimation of G/E in the United Kingdom. Paper presented at the 35th annual meeting of the European Association of Animal Production 1984.

Syrstad, O. (1990) Dairy cattle crossbreeding in the tropics: the importance of genotype–environment interaction. *Livestock Production Science* **24**, 109–118.

Timmermans, M.P.F.C.A. and Sallevelt, J.H.M.T. (1983) Genotype–environment interactions in a long-term selection experiment with layers. *Archiv für Geflugelkunde* **47**, 159–165.

Tribe, D.E. (1955) Choice of diets by rats 4. The choice of purified constituents during growth, pregnancy and lactation. *British Journal of Nutrition* **9**, 103–109.

Wiggans, G.R. and Van Vleck, L.D. (1978) Evaluation of sires in herds feeding differing proportions of concentrates and roughages. *Journal of Dairy Science* **61**, 246–249.

Zeuner, F.G. (1963) *A History of Domesticated Animals*. Hutchinson, London.

18 Effects of Early Environment on Behaviour

P. Le Neindre, I. Veissier, A. Boissy and X. Boivin

Institut National de la Recherche Agronomique, Laboratoire de la Lactation et de l'Elevage des Ruminants, Theix, France

Summary

This chapter reviews the effects of early experience of various environments on further behaviours. Behaviours related to relations between peers, between a dam and its offspring and between animals and humans are discussed as well as maintenance behaviours such as feeding, fear reactions and learning.

It is shown that, in beef cattle, the social environment of a young animal determines, at least in part, its subsequent social behaviour but not its maternal behaviour. Calves isolated from peers tend to be at the bottom of the social hierarchy. Moreover, the early social environment seems also to have an effect on sexual activities. Rams reared without females do not mount them correctly afterwards.

Reactions to humans are reduced by accustoming animals to human contacts. In mothered calves, these contacts are efficient especially when they are experienced soon after weaning. In a similar way, food items are more readily ingested when animals have experienced them during infancy. Again the dam is of major importance since the young usually eats similar items of food.

The infantile environment affects also general trends of behaviour by acting on fear reactions and learning abilities. Most of the available information concerns rodents, in which stimulations during the neonatal period reduce later fear reactions and enhance learning flexibility. Rats reared in social isolation and in poor environments persevere more in strategies that are no longer rewarding. In sheep and cattle, animals that have been mothered are more reactive to new stimuli; few effects on learning are observed.

It is concluded that early experience has long lasting effects on behaviours and that the social environment plays a major role.

Introduction

This chapter focuses on the effects of the neonatal and juvenile environments on the behaviour of cattle but also when necessary on other farm or laboratory animals.

Experience is one of the major factors regulating an animal's behaviour and in particular events occurring during the first periods of its life can have major consequences. The mammal's first environment is the uterus: fluctuations in the chemical composition of the maternal bloodstream and a wider range of physical events, which are only partly attenuated by the amniotic fluid, can have lasting effects on behaviour of the offspring (*prenatal effects*). Most of the domestic species, and in particular cattle, are social and these animals are very sensitive to any change in their social environment. After birth, the mother herself is a major component of this social environment and the stimulation she gives to her offspring is of crucial importance (*postnatal effects*). Then, as the young gains independence, the socio-physical features of the environment become more important, especially during the juvenile period (*juvenile effects*).

Husbandry of domestic animals can imply very important constraints on the animals, limiting their environment, and this is especially true for their social environment. These types of constraints are common in cattle husbandry. Under dairy management calves are in most cases isolated from their mothers from soon after birth then, and for some weeks, they are often isolated from other calves to limit the spread of diseases. After this they are, when not reared in crates like veal calves, grouped with young of the same age and gender.

In other husbandry systems young cattle are reared by their mothers at pasture in herds until the age of six to nine months. They have thus a lot of contact not only with their mothers but also with other young and adult cattle.

Neonatal and juvenile experiences can modulate the activities of animals when they are young, but they can also have long-term effects on different behaviours and especially on feeding and social behaviours. Studies of the latter effects are emphasized in this chapter, including not only the relations between peers but also relations with sexual partners, newborn animals and humans. However, it is often difficult to understand if a feature of the environment modifies some specific behaviours or if it affects more general characteristics of the animal such as reactivity and learning which will also be considered.

Social Behaviour

Development of the social behaviour of the young with its dam

Before discussing the effect of experience on social behaviour, it is necessary to describe the normal development of social behaviour in young cattle when they are with their mothers in a group with enough space. Findings of Le Neindre (1984) on cattle at pasture are reported here.

During the first two days after birth the calf and its dam are together and they spend most of the time isolated from the other animals. Then during the following week the calf remains alone for about 10% of the total time while its mother is in the herd. During this hiding period the calf has contact mainly with its mother which very quickly integrates the group of cows (Fig. 18.1a). Later on the calf very seldom is alone and most of the time its nearest neighbour is a calf (Fig. 18.1b), particularly when they are lying down. It lives in a 'crèche' with the other calves of the herd. The

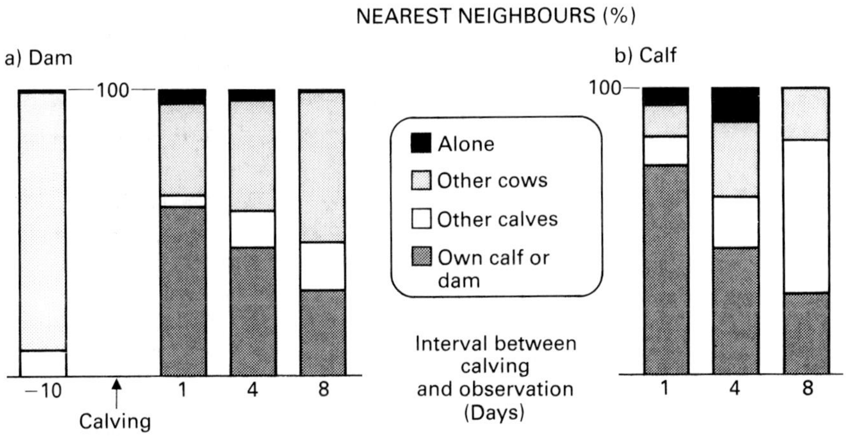

Fig. 18.1. Evolution of nearest neighbours with the interval between calving and observation. % of scans (one scan every thirty minutes) (after Le Neindre, 1984).

crèche has also been described by Sato *et al.* (1984). However, even if it is close to other calves, most of the interactions of the calf are with its dam during the first two months after birth. Every day until more than five months of age the calf suckles its dam in three to six bouts for a total of 30 min and it is licked by its dam for about the same amount of time.

The specific relationship between the young and its mother remains for a very long time and is still significant after the birth of the calf from the following gestation (Veissier *et al.*, 1990). However as they get older, calves, and especially the males, interact more with the other members of the herd. There is then a shift from a specific relation of the calf with its mother to relations with all the members of the herd. Like Bouissou (1985), we observed no relation between the rank in the dominance hierarchy of the dam and its calf. This is in contrast to conclusions on deer (de Vos *et al.*, 1987) and on primates (Imanishi, 1960).

Calves, therefore, develop specific affinities not only for their dams but also for other animals from the herd. Bouissou and Andrieu (1977) studied the affinities of Friesian heifers reared together from birth, six months or nine months of age. They observed specific relationships between animals from the same groups with an increased time spent close together and fewer aggressive interactions. The greatest effects were within calves reared together from birth.

In sheep, the development of social behaviour is slightly different. The lamb follows its mother from a few hours after birth and begins to interact with other lambs later than the calf. Four to six months after lambing the ewe actively rejects its young. However, the main features are the same: the mother is a very strong attachment object and is used as a base by the young to integrate into the social group.

Effects of the environment on the social behaviour development

Most dairy calves are isolated from their mothers soon after birth and reared in groups of males or females. Compared to calves reared by their mothers in a herd,

Table 18.1. Influence of breed and rearing conditions on hierarchical position (the higher the data the higher the hierarchical position).

Herds	FN	FM	SN		SM
SN/SM			4.7	**	12.0
FN/FM	8.7 NS	6.5			
SM/FM		6.1	**		12.7
SN/FN	3.0	**	9.9		

Source: After Le Neindre, 1989b.
Mann–Whitney test: NS not significant, ** significant difference, $P < 0.01$.
F, Friesian; S, Salers; N, non-mothered; M, mothered.

heifer calves living in these conditions had many more interactions with other calves (agonistic interactions: 16.6 vs. 1.0; non-agonistic interactions and licks: 6.6 vs. 0.3). This increased social activity can have detrimental effects on behaviour and production criteria for some calves, especially when space and food are limited (Stephens, 1974).

When mixing animals that were either reared in isolation or group reared, Warnick *et al.* (1977) and Broom and Leaver (1977) observed that those reared in isolation performed badly compared to the others and that their hierarchical ranks were lower. However, Bouissou (1985) and Le Neindre (1989a) found no effect of isolation when mixing Friesian animals that had been isolated or group-reared. The discrepancy between these two sets of data is probably due to the experimental procedures. In the first set of data isolated animals were mixed directly with grouped animals and they were dominated. In the second set isolated animals were mixed with calves from the same treatment for quite a long time before meeting the other calves. In the same experiment Friesians and Salers reared in the two systems were compared. If isolation has no effect on subsequent social behaviour of Friesians, it has a significant effect on Salers (Table 18.1).

Waterhouse (1978) observed that the social ability of calves individually penned was impaired much more than those without a visual isolation. It is possible that the lower rank of these animals is due not only to a lack of previous social contact but also to the generally poor environment in which they have lived.

Maternal Behaviour

Le Neindre (1989b) found no important effects of isolation on the subsequent maternal behaviour of primiparous cows. In the four-hour period after calving, the only significant differences were that non-mothered cows had a longer latency to first licking and licked their calves longer than the mothered ones (Table 18.2). Two other experiments have been done on that topic. Broom and Leaver (1977) observed 'some signs of inadequacy in calves reared in isolation for 6 months', and Donaldson (1970) found that group reared cows were more aggressive towards their newborn calves than the isolated ones. Effects of isolation on the subsequent maternal

Table 18.2. Effects of breed and rearing condition on cow–newborn calf relationships (4 h period after birth; Mann–Whitney test).

	Breed			Rearing condition			Total mean ± standard deviation
	Friesian		Salers	Non-mothered		Mothered	
Number of animals	6		7	7		6	13
Number of attempts before the first suckling	2.2	***	8.0	3.7	*	7.2	5.3 ± 5.0
Birth–first suckling interval (min)	69.5		75.3	73.7		71.3	72.6 ± 24.0
Latency between calving and first licking (min)	6.6		5.0	7.7	*	3.7	6.0 ± 4.2
Total licking time (min)	44.3		41.9	51.4	**	33.2	43.0 ± 17.7
Number of kicks at the calf	0.5		1.7	0.9		1.5	1.2 ± 1.9

Source: After Le Neindre, 1989a.
*$P < 0.05$, ** $P < 0.01$; *** $P < 0.001$.

behaviour are still unclear and are probably dependent on the specific conditions of isolation and group rearing.

Sex-related Behaviours

Several experiments described the effects of the environment on sex-related behaviours but a lot of them are on rodents.

Prenatal stress of the mother can significantly alter the sexual physiology and behaviour of its offspring. For example, stressful crowding during pregnancy delays puberty and alters the oestrous cycles of female offspring in mice (Harvey and Chevins, 1987). In domestic ungulates, ewes born from mothers treated with implants of testosterone during their gestation showed a permanent male like pattern of sexual behaviour as adults which intact females never presented with (Fabre-Nys and Venier, 1989). After ovariectomy, none of the androgenized animals showed sexual receptivity in response to either the threshold or high oestrogen treatment (Fig. 18.2).

Postnatal stimulation in rodents affects their sexual behaviour as adults. For example, male rats reared in isolation exhibited an aberrant mounting behaviour after puberty (Folman and Drori, 1965). In sheep during sexual tests, young males reared by their dams and/or with females displayed a shorter delay in sexual activities and copulated more than animals reared artificially without females (Lachaux *et al.*, 1983). Hemsworth *et al.* (1986) found that young boars initially handled pleasantly achieved a co ordinated mating response sooner than did their peers which were handled unpleasantly (see Chapter 7).

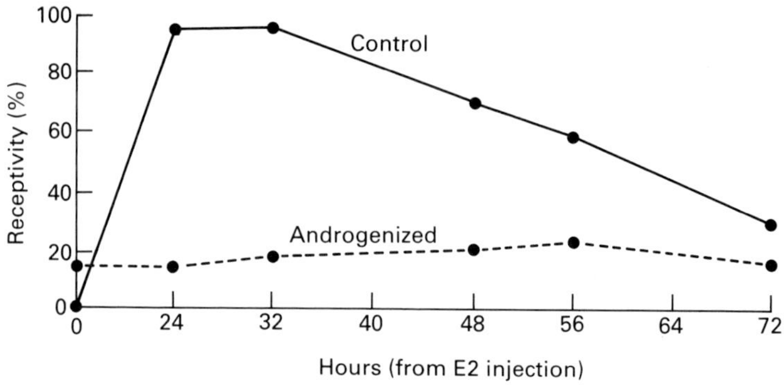

Fig. 18.2. Female sexual behaviour during an artificial oestrous cycle (high oestradiol E_2) (after Fabre-Nys and Venier, 1989).

Human–Animal Relationships

Humans are very important features of the animals' surrounding. They are associated with food and often also with unpleasant handling. The exact place the animals attribute to humans in their environment is not well known. People can be viewed as social partners. For instance, in Africa, Fulani cowmen became 'chiefs' and 'friends' of their cattle by having agonistic (threat, charge, . . .) or non-agonistic (grooming) encounters with them when they are young (Lott and Hart, 1979).

It is known in several species that early contact with humans can be very important for their subsequent relationships with humans (fowl: Hughes and Black, 1976; Murphy and Duncan, 1978; pig: Hemsworth *et al.*, 1986; dog: Scott, 1970). Red partridge that see humans during the two days after hatching will not be afraid of them afterwards (Csermely *et al.*, 1983–84). However in cattle, Sato *et al.* (1984) and Boissy and Bouissou (1988) did not find any long lasting effect of early handling, i.e. during the first months after birth, of dairy calves on their subsequent behaviour. Dairy calves are artificially reared and have quite a lot of contact with humans who at least feed them daily. It is, therefore, possible that the additional contact that was given to these animals did not have any additional effects.

Experiments were then designed with calves from suckling herds to study the effects of early contacts with humans on the subsequent human–animal relationship. In a first experiment (Boivin and Le Neindre, unpublished data), behaviour of calves from two beef breeds (Salers and Limousin) reared by their mothers in two environments were compared. In the first system, calves were indoors during winter and suckled their mothers twice a day under human supervision and they then had a lot of contact with humans and a lot of handling (traditional). In the second system calves were outdoors with their mothers from birth and had little contact with humans (range). When nine months old and after four months together at pasture, heifers reared indoors were easier to handle than the others (Fig. 18.3). When they were 21 months old, heart rate of heifers reared indoors increased less than the others in response to a contact with humans (+16 beats/minute vs. +34, $P < 0.05$).

In another experiment, effects of short periods of contact (20 minutes a day for

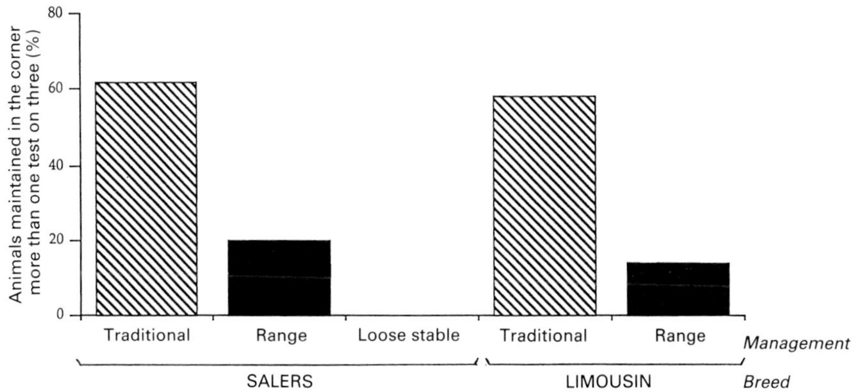

Fig. 18.3. Behaviour of 9-month-old heifers during restraint tests (two tests/animal). The handler has to maintain the animal in a corner of the pen for 30 consecutive seconds during a 2 minute test, three tests/animal (after Boivin and Le Neindre, unpublished data).

10 or 13 days) experienced either just after birth or weaning at 6 weeks were studied on heifers from a beef breed (Aubrac) reared by their mothers outdoors (Boivin *et al.*, 1991). When observed at 3.5 months of age, calves which had received contacts with humans after birth or 1.5 months later were easier to handle than those which had not received such contacts. Moreover, contacts received 1.5 months after birth were more efficient than contacts received immediately after birth. Similarly, in another experiment heifers observed seven months after weaning, which had received human contacts just after weaning or 1.5 months later were easier to handle than those which had received no contact. In this case, contact experienced immediately after weaning reduced fear against people to a greater extent than contact experienced 1.5 months later.

In both experiments, animals having had contact with humans, either as part of the rearing routines or for short periods, were never aggressive towards them, whereas the others attacked their caretakers quite frequently.

As for the relationships with peers a period of contact when animals are young has a stronger effect than later. However, the period immediately after birth does not appear to be the optimum period.

Feeding Behaviour

The early social environment has an impact not only on further social behaviours but also on the other behaviours: feeding behaviour, fear reactions and learning abilities. Warnick *et al.* (1977) compared calves that have been isolated to calves that have been reared in groups. They found that the latter commenced eating concentrates at an earlier age (Table 18.3).

The physical environment is also known to influence further foraging behaviour in some species. Foraging seems to be more efficient in gerbils having experienced an outdoor desert compared to gerbils reared only in cages (Cheal, 1987).

Moreover, experience can also change the intake of specific foods. It is well

Table 18.3. Solid feed consumed and age at which 0.45 kg day^{-1} was eaten by artificially reared calves in three rearing treatments.

	Treatment		
	Group	Individual	Isolated
Concentrate consumed, birth to 10 weeks (kg/day)	0.24	0.21	0.20
Age at which 0.45 kg concentrate consumed daily (days)	22.2[a]	29.2[b]	36.6[c]

Source: After Warnick *et al.*, 1977.
Means in rows with different superscripts differ significantly ($P < 0.05$).

known that food aversion establishes very quickly and a single ingestion of a detrimental food is often sufficient to make it non-consumed later. In a more positive way farmers have known for a long time that habituating an animal to a special flavour when it is young promotes eating foods having this flavour afterwards. Ensminger (1976) in Craig (1981) reported that Chinese farmers used to feed piglets with sow's milk to which hyacinth extract was added to make them eat hyacinths later in life. Lynch *et al.* (1983) compared the ingestion of wheat grain after weaning by lambs which had no previous experience of that food to others which had had contact with it in groups of young or in a group with their mothers. They found that young lambs ate wheat grain more easily when they were exposed to it with their mothers. When tested after weaning, only the latter ate wheat grain. Here again the social context of an experience has an effect. The tendencies of the young to habituate to food items provided by the early environment and to eat the same items as their dams results in feeding preferences when they are older. This was reported in goats by Biquand and Biquand-Guyot (1987).

Reactions of Fear and Anxiety

Numerous studies have shown that the prenatal environment significantly influences the fear and defence reactions of animals. Usually, rough treatment of the mothers during pregnancy is detrimental or 'stressful' for their offsprings' reactivity. For example, in mice, stressful crowding during pregnancy increases the responses of the offspring to threat (Harvey and Chevins, 1985).

Another group of investigations concerns the effect of extra stimulation in the period between birth and weaning. In rodents, whereas prenatal stimulation usually results in increased fearfulness, postnatal environmental handling of pups often reduces their fear responses when adults (removal for a few minutes from the home cage): they defecate less, move and feed more in a novel area, explore more readily and display a lower adrenocortical reactivity in fearful situations (Levine *et al.*, 1967; Ader, 1970; Bodnoff *et al.*, 1987). The type and the intensity of stimulations are not important and handling is as effective as electric shocks. Even stimulation that is not aversive, such as rearing in a complex or enriched environment, has much the same effect (review by Chevins, 1990).

There have been many studies on this topic in domestic animals too. Authors have compared the reactions of animals reared in more or less restrictive conditions. For example Dellmeier *et al.* (1985) and Le Neindre *et al.* (1991) compared calves that had been housed individually or in a group; they found that the former exhibited a higher locomotor activity in an open-field test. However, animals may compensate for their previous inactivity when in an open-field test. For example, the physiological reactions of calves to handling and transport are greater when they have been reared in crates than in groups (Trunkfield *et al.*, 1991).

Environment during the juvenile period can have long lasting effects. For example, rams reared artificially are less reluctant to enter a novel room than males reared by their dams (Lachaux *et al.*, 1983). In the same way, heifers reared artificially were less active in an open-field test than mothered ones (Le Neindre, 1989a). Several studies demonstrate that early regular gentling decreases fear reactions to later encounters with humans: in horses (McCann *et al.*, 1988); pigs (Hemsworth *et al.*, 1986; Gonyou *et al.*, 1986); goats (Lyons and Price, 1987; Lyons *et al.*, 1988); and cattle (Sato *et al.*, 1984; Boissy and Bouissou, 1988). Early contact with man has beneficial effects not only with respect to the reactions of animals towards humans but also in respect to later fear reactions towards frightful events. Subjects handled for several months exhibited fewer reactions to novel objects or to a new environment than non-handled controls in rabbits (Kersten *et al.*, 1989), horses (Heird *et al.*, 1986), pigs (Hemsworth *et al.*, 1986) and cattle (Boissy and Bouissou, 1988).

Exposure to stress or to gentling in adulthood has also been demonstrated to have a wide variety of consequences for the subsequent reactivity of animals towards fearful events. For example, handling rats after weaning decreases fearfulness: handled subjects exhibit less reluctance to eat and only a short latency period before entering a novel arena (Becker and Flaherty, 1966). Nevertheless, these influences do not seem to be as persistent as the ones resulting from neonatal experience.

Learning

Most experiments on previous experience and learning abilities have been carried out on rodents, and learning abilities are measured most of the time by using conditioning (see Delacour, 1981, for general descriptions). These studies aim mainly to analyse human behaviour. However, they can help us to understand how domestic animals are affected by previous rearing conditions.

The effects of handling rodents in infancy are well documented. The procedure means removing a pup from the maternity cage and placing it in a small container for a few minutes. This seems to enhance later associative learning abilities and, for example, handling has been reported to improve acquisition of visual discrimination tasks (Wong and Jamieson, 1968) (Fig. 18.4) and operant tasks (Wong and Scott Wilson, 1976). Moreover, rats that have been handled in infancy appear better in adapting to a change in associative cues: they performed better than non-handled rats when the cues of a visual discrimination task were reversed (Wong and Jamieson,

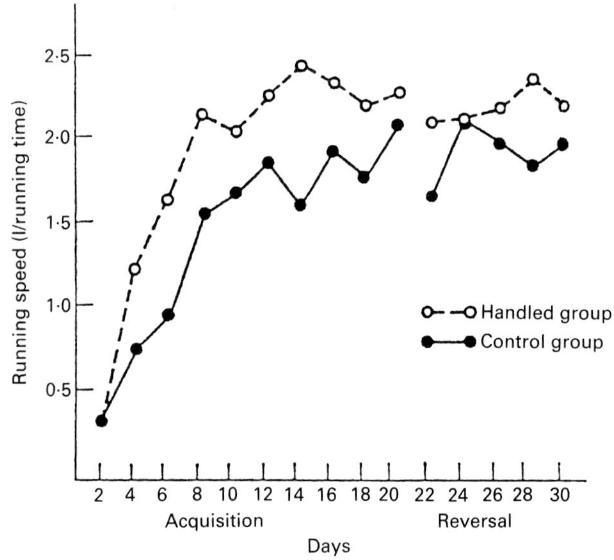

Fig. 18.4. Running speed during acquisition and reversal learning as a function of infant treatment (after Wong and Jamieson, 1968).

1968; Wong and Judd, 1973) and, after punishment, they took a longer time to resume a previous rate of an operant response (Wong, 1972).

It seems that the procedure of handling stresses the pups. Hence its effect on learning might be due to a consequence of stressors experienced early in life, whatever these stressors are, since, e.g. for fear responses, electric shocks have the same impact (Denenberg *et al.*, 1967; Levine, 1969). These experiments could be related to the effects of weaning on the reactions of animals to handling, where we concluded that this period of distress modifies the behaviour of the animals towards association with humans. Weaning may act as a stressor that helps to associate humans with a positive reinforcer.

Studies on later experience and learning in rodents are mainly concerned with physical or social enrichment of rearing conditions. Physical enrichment is obtained by adding several objects (tunnels, ramps, seesaws, etc.) in a standardized rats' cage. Such an enrichment is likely to improve problem solving. This has been observed in mazes where it reduces the number of errors; that is, the frequency of non-rewarding routes taken (Denenberg *et al.*, 1968; Bouchon and Will, 1982). Social impoverishment is obtained by total isolation of rats. Such rats differed from rats reared in groups in several ways:

1. Isolated rats habituated less rapidly to a novel environment: when exposed to an open-field test for the first time, they did not differ from controls but on subsequent sessions, they still moved a lot in the testing cage while the controls habituated rapidly (Dalrymple-Alford and Benton, 1981) (Fig. 18.5).
2. They are slower to abandon a previously rewarding strategy: Morgan (1973) trained rats to pull a ball in an alley in order to be allowed to run to a goal box containing food. Isolated rats learnt as quickly as controls. Then, the rats had to

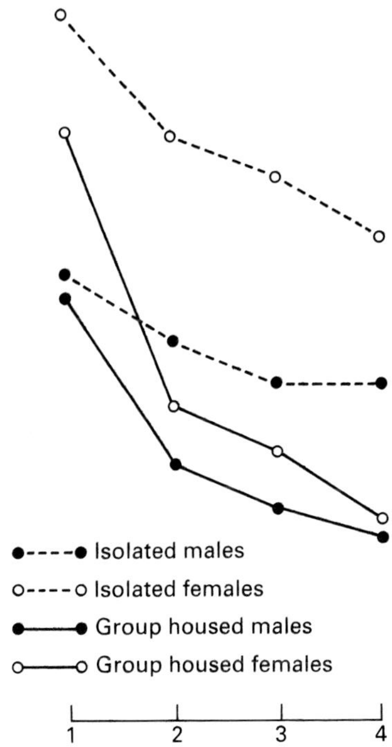

Fig. 18.5. Number of squares entered in an open field by isolated and group housed rats. Each point is the mean activity of 10 animals for each minute of a 4 min daily trial totalled over the 5 days (after Dalrymple-Alford and Benton, 1981)

push the ball instead of pulling it. Isolated rats showed great difficulties in learning this reversed task.

3. They exhibit a slower decline of an operant response under extinction (— no reward) or satiation (Gluck and Pearce, 1977; Morgan *et al.*, 1977).

It might be concluded that social isolation renders a rat less able to change its behaviour in response to change from one situation to another. This assumption fits well with point 2. Repeated exposures to a novel environment make this environment less and less novel. So normal rats respond to it by habituation. Absence of habituation in isolated rats (point 1) can be interpreted as absence of behavioural readjustment when the environment becomes less novel. Under extinction, a normal rat performs the operant response less and less frequently since this does not produce the reward any more. Under satiation, a group reared rat again responds less, since the response is less rewarding as it is not hungry. In both cases (point 3), the higher rate of response of isolated rats can be interpreted as perseverance in a behaviour whose rewarding value has been changed.

On the whole, enriching the environment either by short treatments such as handling in infancy or by modifying the whole rearing environment seems, at least in rodents, to improve behavioural flexibility.

Very little information on this topic is available in farm animals. Early handling, i.e. haltering, brushing or leading an animal, is likely to improve maze learning in horses (Heird *et al.*, 1986). Mother reared animals have been compared to artificially reared ones, since it is the most frequent variation of early rearing conditions for a farm animal. According to Veissier (1987), non-mothered Friesian heifers appeared worse at associative learning than mothered Friesian heifers, since their conditioned fear responses are lower in spite of greater spontaneous fear responses. However, Houpt *et al.* (1982) found no difference between foals artificially reared and foals with their mothers on their later maze-learning ability.

Discussion

First of all, from these results, it is proposed that events experienced during the neonatal and juvenile periods can have long lasting effects on an animal's behaviour. The social environment appears as a major factor in domestic animals which belong to gregarious species. Social contacts during the early life affect not only future social relationships but also other behaviours.

The mother is for the young a very important feature of the environment. The lack of a mother results in altered social activities between peers. In sheep and cattle, animals reared artificially have enhanced spontaneous fear reactions and lower conditioned fear reactions. The latter finding can be interpreted as an impairment of learning. In addition, the dam influences food preference of its offspring. The effects of the presence of the dam are probably more important in calves from beef breeds and probably in the young of other species of domestic animals than in dairy calves where the young–dam relationship seems to be not as strong as in other animals. As a matter of fact, in contrast to other animals, isolation of Friesian calves from their mothers does not have a major detrimental effect on their subsequent social, sexual and maternal behaviour. This genetic variability of behaviour is very important to keep in mind when studying the effects of environment on animals.

Rearing an animal without peers is also likely to alter further relationships. Social isolation has some detrimental effects on social relationships, as it is observed in dairy calves, and the absence of partners from the opposite gender can affect the sexual behaviour. In addition, rodents reared in a poor social environment adapt less easily to environmental changes. This may also be true for farm animals.

Man has a particular place in the animals' group structure. In domestic animals, it seems that man can be considered as a social partner. As with other social partners, young animals need to have contacts with him when they are young to accept him easily afterwards.

The results reviewed in this chapter suggest that the different features of the environment do not all act on the later behaviour at the same time. The dam is probably important in a very young animal. However, when growing older, young animals have more contacts with other individuals. These in turn may be of greater importance. If man is seen as a social partner, the period when his presence has a major effect should be the same as that of the major period of socialization. As a matter of fact, social imprinting in birds occurs just after hatching and the taming of wild birds is possibly only at that time. Similarly, removing the dam for abrupt weaning strengthened relationships among calves (Veissier and Le Neindre, 1989) and contacts with humans at this time are very efficient in helping further human–animal relationships.

In husbandry systems, it is sometime difficult to provide the animals with all these social stimuli. For instance, dairy calves are isolated from their dams for these to be milked. In our opinion, it is better to isolate the young from birth than to leave it with its dam for some days. In the latter case a very strong bond is created and separation brings a high level of unnecessary distress (Le Neindre, 1984; Veissier *et al.*, 1989).

To compensate for the lack of the dam, it is thought that calves should be reared in groups. However, to obtain an acceptable result it is necessary to adjust the environment. Without the dam, social interactions are much more frequent and calves must have a lot of space to be able to escape. Another possibility could be to separate veal calves by open partitions in order to maintain some visual and physical contacts. In addition, veal calves perform non-nutritive suckling very often. It is likely that such oral activities act as a compensation for the lack of external stimuli due to physical and social restraint (Le Neindre *et al.*, 1991). The veal calves should be provided with a substitute for the normal suckling activity. Another idea is that of the caretaker spending more time with his animals. A 'good' bond between animals and caretakers might regulate the social behaviour of the former. In addition, this should make further handling procedures easier.

To study the importance of the major features of the environment it is necessary to define the behaviour of each species in an environment with minimal constraints. Specific characteristics of the species and even of a population can then be known. The effects of experience of a given environment are to be compared to this ethogram. Moreover, as reactivity and learning are general characteristics that influence the animals' behaviour in any aspects of their life, they must be taken into consideration when explanatory mechanisms are looked for.

References

Ader, R. (1970) The effects of early experience on the adrenocortical response to different magnitudes of stimulation. *Physiology and Behaviour* **5**, 837–839.

Becker, G. and Flaherty, T.B. (1966) Effects of postweaning tactual stimulation on emotionality and social dominance in the rat. *Psychological Reports* **19**, 363–366.

Biquand, S. and Biquand-Guyot, V. (1987) Approche Ecoethologique du pâturage de la chèvre créole. Thèse de Doctorat, Université de Rennes I, 302pp.

Bodnoff, S.R., Suranyi-Cadotte, B., Quirion, R. and Meaney, M.J. (1987) Postnatal handling reduces novelty-induced fear and increases [3H] flunitrazepam binding in rat brain. *European Journal of Pharmacology* **144**, 105–107.

Boissy, A. and Bouissou, M.F. (1988) Effects of early handling on heifers' subsequent reactivity to humans and to unfamiliar situations. *Applied Animal Behaviour Science* **20**, 259–273.

Boivin, X., Le Neindre, P. and Chupin, J.M. (1992) Establishment of cattle/human relationships. *Applied Animal Behaviour Science* **32**, 325–335.

Bouchon, R. and Will, B. (1982) Effets des conditions d'élevage après le sevrage sur les performances d'apprentissage des souris 'Dwarf'. *Physiology and Behaviour* **28**, 971–978.

Bouissou, M.F. (1985) Contribution à l'étude des relations interindividuelles chez les bovins domestiques femelles (*Bos Taurus* L.). Thèse Doc. ès sciences, Univ. Paris VI, 365pp.

Bouissou, M.F. and Andrieu, S. (1977) Etablissement des relations préférentielles chez les bovins domestiques. *Biologie du Comportement* **2**, 213–221.

Broom, D.M. and Leaver, J.D. (1977) Mother–young interactions in dairy cattle. *British Veterinary Journal* **133**, 192.

Cheal, M. (1987) Environmental enrichment facilitates foraging behavior. *Physiology and Behaviour* **39**, 281–283.

Chevins, P.F.D. (1990) Early environmental influences on fear and defence in rodents. In: Brain P.F., Parmigiani S., Blanchard R., Mainardi D. (eds), *Fear and Defence*. Harwood Academic Publishers, pp. 269–288.

Craig, J.V. (1981) *Domestic Animal Behaviour: Causes and implications for animal care and management*. Prentice-Hall, Englewood Cliffs New Jersey, 364pp.

Csermely, D., Mainardi, D. and Spano, S. (1983–84) Escape-reaction of captive young partridges (*Alectoris rufa*) reared with or without visual contacts with man. *Applied Animal Ethology* **11**, 177–182.

Dalrymple-Alford, J.C. and Benton, D. (1981) Activity differences of individually and group-housed male and female rats. *Animal Learning Behaviour* **9**, 50–55.

Delacour, J. (1981) *Conditionnement et Biologie*. Masson Paris/New York/Barcelona/Milan, 190pp.

Dellmeier, G.R., Friend, T.H. and Gbur, E.E. (1985) Comparison of four methods of calf confinement. II. Behavior. *Journal of Animal Science* **60**, 1102–1109.

Deneberg, V.H., Brumaghim, J.T., Haltmeyer, G.C. and Zarrow, M.X. (1967) Increased adrenocortical activity in the neonatal rat following handling. *Endocrinology* **81**, 1047–1052.

Deneberg, V.H., Woodcock, J.M. and Rosenberg, K.M. (1968) Long term effects, of pre-weaning and postweaning free environment experience on rats' problem solving behavior. *Journal of Comparative Physiological Psychology* **66**, 533–535.

Donaldson, S.L. (1970) The effects of early feeding and rearing experiences on social, maternal and milking parlor behavior in dairy cattle. PhD Thesis, Purdue University.

Fabre-Nys, C. and Venier, G. (1989) Hormonal control of male and female sexual behaviour in ewes: effects of treatments during prenatal life and adulthood. ESCPB Conference on Hormone, Brain and Behaviour. Liège, Belgium, pp. 61–62.

Fallon, R.J. (1978) The effect of immunoglobulin levels on calf performance and methods of artificially feeding colostrum to the new-born calf. *Annales des Recherches Vétérinaires* **9**, 347–352.

Folman, Y. and Drori, D. (1965) Normal and aberrant copulatory behavior in male rats (*R. Norvegicus*) reared in isolation. *Animal Behaviour* **13**, 427–429.

Gluck, J.P. and Pearce, H.E. (1977) Acquisition and extinction of an operant response in differentially reared rats. *Developmental Psychology* **10**, 143–149.

Gonyou, H.W., Hemsworth, P.H. and Barnett, J.L. (1986) Effects of frequent interactions with humans on growing pigs. *Applied Animal Behaviour Science* **16**, 269–278.

Harvey, P.W. and Chevins, P.F.D. (1985) Crowding pregnant mice affects attack and threat behavior of male offspring. *Hormones and Behaviour* **19**, 80–97.

Harvey, P.W. and Chevins, P.F.D. (1987) Crowding during pregnancy delays puberty and alters oestrus cycles of female offspring in mice. *Experientia* **43**, 306–307.

Heird, J.C., Whitaker, D.D., Bell, R.W., Ramsey, C.B. and Lokey, C.E. (1986) The effects of handling at different ages on the subsequent learning ability of 2-year-old horses. *Applied Animal Behaviour Science* **15**, 15–25.

Hemsworth, P.H., Barnett, J.L., Hansen, C. and Gonyou, H.W. (1986) The influence of early contact with humans on subsequent behavioural response of pigs to humans. *Applied Animal Behaviour Science* **15**, 55–63.

Houpt, K.A., Parsons M.S. and Hintz H.F. (1982) Learning ability of orphan foals, of normal foals and of their mothers. *Journal of Animal Science* **55**, 1027–1032.

Hughes, B.O. and Black, A.J. (1976) The influence of handling on egg production, egg shell quality and avoidance behaviour of hens. *British Poultry Science* **17**, 135–144.

Imanishi, I. (1960) Social organization of subhuman Primates in their natural habits. *Current Anthropology* **1**, 393–407.

Kersten, A.M.P., Meijsser, F.M. and Metz, J.H.M. (1989) Effects of early handling on later open-field behaviour in rabbits. *Applied Animal Behaviour Science* **24**, 157–167.

Lachaux, M., Bouissou, M.F., Berges, J.C. and Orgeur, P. (1983) Etude du comportement en open-field de béliers Ile-de-France soumis à différentes conditions d'élevage. *Biology of Behaviour* **3**, 2577–2690.

Le Neindre, P. (1984) La relation mère–jeune chez les bovins: Influences de l'environnement et de la race. Thèse de l'Université de Rennes. 273pp.

Le Neindre, P. (1989a) Influence of rearing conditions and breed on social behaviour and activity of cattle in novel environments. *Applied Animal Behaviour Science* **23**, 129–140.

Le Neindre, P. (1989b) Influence of rearing conditions and breed on social relationship of mother and young. *Applied Animal Behaviour Science* **23**, 117–127.

Le Neindre, P., Gesmier, V. and Trillat G. (1991) Le confort du veau de boucherie dans les conditions de l'élevage intensif. Document de travail, *Contrat Interbev-Interveaux*, 65pp.

Le Neindre, P. and Signoret, J.P. (1987) Ethological results of welfare research in veal calves. In: Schlichting, M.C. and Smidt, D. (eds), *Welfare Aspects of Housing Systems for Veal Calves and Fattening Bulls*. CEC., Luxembourg, EUR 1077 EN, pp. 77–88.

Levine, S. (1969) An endocrine theory of infantile stimulation. In: Ambrose A. (ed.), *Stimulation in Infancy*. Academic Press, London/New York, pp. 45–55.

Levine, S., Haltmeyer, G.C., Karas, G.G. and Denenberg, V.H. (1967) Physiological and behavioral effects of infantile stimulation. *Physiology and Behaviour* **2**, 55–59.

Lott, D.F. and Hart, B.L. (1979) Applied ethology in a nomadic cattle culture. *Applied Animal Ethology* **5**, 309–319.

Lynch, J.J., Keogh, R.G., Elwin, R.L., Green, G.C. and Mottershead, B.E. (1983) Effects of early experience on the post-weaning acceptance of whole grain wheat by fine-wool Merino lambs. *Animal Production* **36**, 175–183.

Lyons, D.M. and Price, E.O. (1987) Relationships between heart rates and behavior of goats in encounters with people. *Applied Animal Behaviour Science* **18**, 363–369.

Lyons, D.M., Price, E.O. and Moberg, G.P. (1988) Individual difference in temperament of domestic dairy goats: constancy and change. *Animal Behaviour* **36**, 1323–1333.

McCann, J.S., Heird, J.C., Bell, R.W. and Lutherer, L.O. (1988) Normal and more highly reactive horses. II. The effect of handling and reserpine on the cardiac response to stimuli. *Applied Animal Behaviour Science* **19**, 215–226.

Morgan, M.J. (1973) Effects of post-weaning environment on learning in the rat. *Animal Behaviour* **21**, 429–442.

Morgan, M.J., Einon, D. and Morris, R.G.M. (1977) Inhibition and isolation rearing in the rat: extinction and satiation. *Physiology and Behaviour* **18**, 1–6.

Murphy, L.B. and Duncan, J.H. (1978) Attempts to modify the responses of domestic fowl towards human beings. II. The effects of early experience. *Applied Animal Behaviour* **4**, 5–12.

Sato, S., Shiki, H. and Yamasaki, F. (1984) The effects of early caressing on later tractability of calves. *Japanese Journal of Zootechnical Science* **55**, 332–338.

Scott, J.P. (1970) Critical periods for the development of social behaviour in dogs. In Kazda, S. and Denenberg, V.H. (eds), *The Postnatal Development of Phenotype*. Academia, pp. 21–31.

Stephens, D.B. (1974) Studies on the effects of social environment on the behaviour and growth rates of artificially reared British Friesian male calves. *Animal Production* **18**, 23–34.

Trunkfield, H.R., Broom, D.M., Maatje, K., Wierenga, H.K., Lambooy, E. and Kooijman, J. (1991) Effects of housing on responses of veal calves to handling and transport. In: Metz,

J.H.M. and Groenestein, N.C.M. (eds), *New Trends in Veal Calf Production.* EAAP Publication, 52, Pudoc, Wageningen, pp. 40–43.

Veissier I. (1987) Capacités comportementales d'adaptation des Bovins à de nouvelles conditions de vie. Thèse pour l'obtention du doctorat de l'université de Rennes I, 146pp.

Veissier, I. and Le Neindre, P. (1989) Weaning in calves: its effects on social organization. *Applied Animal Behaviour Science* **29**, 43–54.

Veissier, I., Le Neindre, P. and Trillat, G. (1989) The use of circadian behaviour to measure adaptation of calves to changes in their environment. *Applied Animal Behaviour Science* **22**, 1–12.

Veissier I., Lamy D. and Le Neindre, P. (1990) Social behaviour in domestic beef cattle when yearling calves are left with the cows for the next calving. *Applied Animal Behaviour Science* **27**, 193–200.

Vos de, A., Broks, P. and Geist, V. (1967) A review of social behavior of the North American cervids during the reproductive period. *American Midlands National* **77**, 390–417.

Warnick, V.D., Arave, C.W. and Mickelsen, C.H. (1977) Effects of group, individual and isolated rearing of calves on weight gain and behavior. *Journal of Dairy Science* **60**, 947–953.

Waterhouse A. (1978) The effects of pen conditions on the development of calf behaviour. *Applied Animal Ethology* **4**, 286.

Wong, R. (1972) Infantile learning and associative processes of rats. *British Journal of Psychology* **63**, 101–108.

Wong, R. and Jamieson, J.L. (1968) Infantile handling and the facilitation of discrimination and reversal learning. *Quarterly Journal of Experimental Psychology* **20**, 197–199.

Wong, R. and Judd, M. (1973) Infantile handling and successive spatial reversal learning in rats. *Behaviour and Biology* **8**, 391–397.

Wong, R. and Scott Wilson, C. (1976) Infantile handling effects on mixed sessions of DRL 10-sec and DRL 15-sec schedules. *Behaviour and Biology* **16**, 73–78.

Section IV
Effects of Farm Animals on the Human Environment

Apart from the obvious nutritional consequences, farm animals have a range of beneficial and detrimental effects on the human environment. It could be argued that it is selfish for humans to consider only the effects on *their* environment and that they should altruistically maintain a balance between the human and animal environments. However, the reality is far from that and there is little altruism in human activities, as evidenced by our scant regard for pollution of the environment by our own activities.

Detrimental effects on the environment arise largely from many of the practices adopted during the industrialization and mechanization of animal production during the last few decades. Increases in output have been pursued with little regard for, or knowledge of, the effects on the environment. Excessive concentrations of livestock result in uncontainable liquid emissions that have the potential for direct contamination – in particular of our water supplies and more generally of our aquatic and terrestrial environments.

Gaseous emissions are a small but significant contributor to global atmospheric changes and may have a damaging long-term effect. However, by far the most potentially destructive effect of the industrialization process is the evolution of pathogens with mass destruction potential when they are transferred to their final host: man. This could produce epidemics paralleled only by the plagues associated with the increase in the human population density in the Middle Ages and Victorian epochs. The emergence of new pathogens is an inevitable consequence of our recent eradication or suppression of most of the harmful pathogens in our farm animal and human populations. However, it is the concentration of animal populations and the cyclicity of the feeding practices which make the spread of diseases potentially very rapid and uncontrollable. Governments may be tempted into short-term damage limitation exercises to protect farmers but scientists must be heeded even if the potential magnitude of the risks is uncertain.

On the positive side farm animals, as traditionally kept on grazing lands, enrich our countryside by their direct effect on the habitat and by fulfilling our need to coexist with other species, rather than becoming so dominant that our compatibility with other animals is threatened.

19 Liquid Wastes from Farm Animal Enterprises

J.R. ARCHER[1] and R.J. NICHOLSON[2]

[1] *Agricultural Development and Advisory Service, Ministry of Agriculture, Fisheries and Food, London;* [2] *Agricultural and Development Advisory Service, Bristol, UK*

Summary

The main liquid wastes produced on farms in England and Wales are slurry from pigs and cows, dirty water, particularly yard runoff on dairy farms, and silage effluent. All these wastes have a high biochemical oxygen demand (BOD), particularly undiluted slurries and silage effluent. They also contain considerable nitrogen and other nutrients. Dirty water is very variable in concentration but rarely falls below 1000 mg l^{-1} BOD.

The quantity of waste produced varies with animal number for slurry, rainfall and farm layout for dirty water and method of silage production for silage effluent. The adoption of baled silage production has limited the increase in effluent being produced on farms in England and Wales.

Direct runoff of these wastes to surface water either direct from the farm or as a result of land spreading is mainly responsible for the 3000–4000 reported farm pollution incidents which the National Rivers Authority handles each year. Livestock wastes are also a significant factor contributing to nitrate in waters abstracted for drinking.

The recently published MAFF Code of Good Agricultural Practice for the Protection of Water gives guidelines on how farmers can minimize water pollution. It provides guidance on safe application of these wastes to land and how to assess the need for storage on farms. This guidance includes a section on reducing nitrate loss from agricultural land.

Various safe systems of storing liquid wastes on farms are available. New installations must now conform to the standards set out in the Control of Pollution Regulations 1991.

Future limitations on nitrate leaching will be governed by the recently agreed EC Directive. Vulnerable zones will be established where water sources need protection.

The major challenge for the future is determining the right balance between agricultural production and pollution control.

Introduction

Nearly all dairy and pig farms produce substantial quantities of liquid wastes. Dairy farms produce liquid manure (slurry) and parlour washings/yard runoff (dirty water). Most will also produce silage effluent. Beef farms may also produce all three liquid wastes. Pig farms, both rearing and fattening, will commonly produce slurry. Any livestock farm, including sheep and poultry, may produce liquid effluent which has drained from solid manure stored in the open.

Virtually all livestock wastes, whether slurry, dirty water or solid manure effluent, and most silage effluent produced by livestock farms in the United Kingdom are returned to land. The aim is to make good use of the nutrient content through crop uptake, while minimizing the risk of direct and diffuse pollution of water by paying careful attention to the amount and timing of applications. We strive to find the right balance between agricultural production and maintaining the quality of our rivers and drinking water.

The risks of air pollution from livestock farming are dealt with in Chapter 20. The risk of soil pollution from liquid wastes is restricted to the addition of the heavy metals, zinc and copper. The small amounts added to animal diets today have virtually eliminated this as a practical problem. The other main concern is excessive soil nutrient levels. The application of large amounts of livestock waste annually over many years can result in high nitrogen, phosphorus and potassium levels in soils. These may lead to both pollution and agricultural production problems.

Other liquid wastes from livestock farming can cause pollution of water, especially unwanted milk or milk products and used sheep dip. These problems are outside the scope of this chapter. For further information on these wastes, consult the Ministry of Agriculture, Fisheries and Food (MAFF) Code of Good Agricultural Practice for the Protection of Water (MAFF, 1991).

Properties of Liquid Wastes

The physical characteristics of livestock wastes depend on the type of livestock, how they are housed, the amount and type of bedding material and whether the waste is diluted with water.

Slurry contains dung, urine and water with only small amounts of bedding. It flows by gravity and can be collected in slatted-floor systems, below ground tanks, or reception pits. Semi-solid slurry is a mixture of dung, urine and bedding. It is very commonly produced from cubicles on dairy farms. Solid manure may contain large amounts of bedding and can usually be stacked. Nearly all poultry manures fall into this last category. Fresh battery manure can behave as a semi-solid.

All liquid wastes from livestock farms have a high biochemical oxygen demand (BOD) compared to untreated domestic sewage (Table 19.1). Some, such as undiluted slurry and silage effluent, are extremely high and only a small leak into a water course can cause major pollution.

The other main chemical characteristic of liquid wastes which influences their risk of causing pollution is nutrient content, particularly nitrogen and phosphorus.

Table 19.1. Examples of typical BOD levels (mg l^{-1}).

	BOD
Treated domestic sewage	20–60
Raw domestic sewage	300–400
Vegetable washings	500–3 000
Dilute dairy parlour and yard washings (dirty water)	1 000–2 000
Liquid wastes draining from slurry stores	1 000–12 000
Liquid sewage sludge	10 000–20 000
Cattle slurry	10 000–20 000
Pig slurry	20 000–30 000
Silage effluent	30 000–80 000
Milk	140 000

Source: After MAFF, 1991.

The nutrient content of wastes depends primarily on dilution. The range of nutrient content of actual slurry on farms is very large and, for a particular species, is generally well correlated with dry matter content (Smith and Unwin, 1983). The figures for the NPK content of different manures and undiluted slurries are given in Table 19.2 (Archer, 1988).

The nitrogen content of slurry applied to land is less than that of excreta. Volatile loss of nitrogen as ammonia takes place from the housing area, from the slurry store and from the field (Jarvis and Pain, 1990). Whenever slurry is moved either into store, while in store or during spreading, ammonia is lost. For this reason, the nitrogen content of slurry applied to land can vary considerably even if derived from one production unit.

Dirty water on farms is notoriously variable in composition depending on what goes into it and how much it is diluted by rainfall. Liquid manures on farms are often

Table 19.2. Nitrogen, phosphate and potash contents of animal manures and undiluted slurries (on fresh weight basis).

Type	Dry matter (%)	Nitrogen (% N)	Phosphate (% P_2O_5)	Potash (% K_2O)
Cattle manure	25	0.6	0.3	0.7
Pig manure	25	0.6	0.6	0.4
Poultry – battery	70	4.2	2.8	1.9
Poultry – broiler	70	2.4	2.2	1.4
Cattle slurry	10	0.5	0.2	0.5
Pig slurry				
dry meal fed	10	0.6	0.4	0.3
pipeline fed	6–10	0.5	0.2	0.2
whey fed	2–4	0.3	0.2	0.2

Source: After Archer, 1988.

Table 19.3. Amount of excreta produced by livestock.

Type of livestock	Body weight (kg)	Moisture content (%)	Typical volumes (l/day)
1 dairy cow	450–650	90	57
1 beef bullock	200–450	90	27
1 dry sow	90–120	90	4.0
1 lactating sow + litter	90–120	90	12
1 pig, dry meal fed	47–75	90	4.0
1 pig, liquid fed (water : meal ratio 2.5 : 1)	45–75	90	4.0
1 pig, liquid fed (water : meal ratio 4 : 1)	45–75	94	7.0
1 pig, whey fed	45–75	97	14
1 fattening lamb	45	89	2.2
1 mature sheep	60–80	89	4.0
1000 laying hens	2000	70	115
1000 laying hens (air dried)	2000	30	49
1000 broilers + litter	1000	30	36
1000 turkeys + litter	5000	30	124

Souce: After MAFF, 1991.

described as dirty water although they may contain effluent from a weeping wall store or earth banked store strainer box. Equally they may include liquid from a slurry separator. Typical values for the BOD of dirty water are given in Table 19.1. The range of dry matter varies from 0.1 to 1% and the nitrogen content from 0.02 to 2 $kg\,m^{-3}$.

Silage effluent contains sugars, organic acids and products of cell breakdown. It is highly polluting with a BOD of up to 80 000 $mg\,l^{-1}$ and because its pH can be as low as 3.8, it is highly corrosive especially to concrete. Effluent dry matter content is commonly in the range 2–8%, with a nitrogen content of 1–5%.

Quantities of Liquid Wastes Produced

The quantity and chemical composition of animal excreta depends mainly on diet. In preparing the recently published MAFF Code of Good Agricultural Practice for the Protection of Water (MAFF, 1991), ADAS reviewed the available data on quantities of excreta produced by individual animals. Table 19.3 is taken from the code. It gives the quantity and dry matter content for the main livestock systems used in the United Kingdom. Inevitably, these average figures hide a wide variation. The alternative approach is to predict the output for a particular livestock enterprise, using a detailed knowledge of the effect of individual variables. This may be justified for planning new units. For pigs in particular, variation in water intake due to type of feeding will make a large difference to the quantity and dry matter content of the excreta.

Unlike slurry, the quantity of dirty water produced on a dairy unit varies considerably. It depends mainly on the open area of yards, silo and manure stores and the rainfall. Parlour washings and other cleaning wastes are less variable in quantity. A typical figure for a 100 cow dairy unit for the full year, assuming 1000 m^2 of concrete, 1000 mm of average annual rainfall and 2 m^3 of parlour washings per day is 1730 m^3. This is in addition to, say, 1116 m^3 of slurry.

Current research is examining ways of reducing the amount of nitrogen and phosphorus excreted by animals, particularly non-ruminants. The scope for reducing excretion by grazing ruminants is less. Dutch work suggests that synthetic amino acid supplementation gives considerable scope for improved nitrogen utilization by non-ruminants but feed costs will inevitably be higher. Research with pigs is currently assessing the scope for this approach in the United Kingdom.

Silage production in England and Wales has increased from around 10 million tonnes in 1976 to over 35 million tonnes in 1988. Even though latest MAFF census data show 12–15% of the silage is now made in bales, the amount of silage effluent being produced from silo-made silage has increased very substantially. The amount of hay produced has fallen over this period. Baled and bagged or wrapped silage produces much less effluent due to the need for a higher dry matter crop and because it is not subjected to high pressure as in a silo. Any effluent that is produced is contained, so the pollution risk is very small as long as the bales are stored and opened more than 10 metres from a watercourse.

Peak flow rates of silage effluent from unwilted grass in a silo can be as high as 30 litres per tonne ensiled per day. Peak flow occurs within approximately 5 days of ensiling. Total effluent produced can exceed 200 litres/tonne. Up to 1980, it was common practice to wilt grass to at least 25% dry matter before putting it into the silo. At this dry matter content effluent production is minimal (Bastiman and Altman, 1985). In the last decade, farmers have generally wilted less, as research showed that this practice improves feed value and milk production (Gordon, 1981). This has encouraged farmers to ensile grass at lower dry matter contents of 20–22%, particularly in years when the weather is unfavourable for wilting. This has inevitably resulted in more silage effluent being generated on farms, particularly in wet seasons.

Impact on Water Quality

There are two main water quality problems related to liquid wastes in England and Wales at present. The first is direct pollution of surface waters by runoff from farms. The second is the contribution that organic manures make to the nutrient enrichment of waters, particularly nitrate in drinking water abstracted from both surface and, more particularly, groundwater sources.

The number of reported point source, direct pollution incidents on farms in England and Wales exceeded 4000 in the wet year of 1988. The number fell below 3000 in the dry year of 1989. Generally there has been an increase in reported incidents of point source pollution in recent years. This coincides with more slurry and silage being produced on farms. The annual variation in the number is strongly influenced by rainfall, particularly that falling during the main autumn slurry

Table 19.4. Farm pollution incidents in England and Wales, 1987–1989.

	1987	1988	1989	1987–1989 (% of total)
Land runoff	286	434	380	10%
Slurry stores	1027	1148	841	28%
Dirty water	875	895	734	23%
Silage effluent	1003	815	245	19%
Solid stores	148	194	121	4%
Others	551	655	567	16%
TOTAL	3890	4141	2888	–

Source: After NRA, 1990b.

spreading period and the early summer silage cutting season. The numbers include all farm derived pollution incidents, but liquid wastes dominate the numbers as seen in Table 19.4 (NRA, 1990b). To put these numbers in perspective, during 1987–89 agriculture was responsible for 17% of the total reported pollution incidents. Industry and sewage accounted for most of the non-agricultural contribution.

Most of these agricultural incidents are point source in origin from one building, store or field. When waste with a high BOD gets into a watercourse, it is broken down by micro-organisms and oxygen is consumed. This reduced oxygen concentration in the water is primarily responsible for fish kills. Other river flora and fauna may be affected both by low oxygen and by direct effects, such as ammonium toxicity. The actual amount of damage to river life depends on many factors, not least the quantity of high BOD material getting into the water and the resulting concentration in the watercourse after it has been diluted by water flow. Major leaks from stores of slurry or silage effluent invariably cause severe damage because the BOD of these liquids is so high. Dirty water pollution incidents vary much more, depending on the flow in the watercourse at the time. Schofield *et al.* (1990) have carried out a recent study examining the impact of dairy farming practices on water quality in a grassland catchment in West Wales. Detailed study was able to relate diurnal fluctuations in water quality to time of cleaning after milking.

Nitrate in drinking water has become a major problem in the EC in the last few years as water supplies have to conform to the EC drinking water standard of $50\,\text{mg}\,\text{l}^{-1}$ nitrate. In the United Kingdom many borehole sources and some river abstractions exceed this value for at least part of the year and abstraction has to cease, the source has to be blended with lower nitrate water or, in a few cases, the water is treated to reduce its nitrate concentration (HMSO, 1989).

Because of the variability in both total content and form of nitrogen in animal manure sources, it is inevitable that they cause an increase in nitrate leaching from soils to which they are applied. In any catchment where attempts are being made to reduce nitrate loss, management of animal manure applications, both amount and timing, must be a high priority.

A major consideration with groundwaters is the long time scale for abstracted water to reach a new equilibrium nitrate level following a change in the amount being lost from the agricultural soil. Sandstone aquifers typically take 20 years and soft chalks up to 40 years before the full benefit of any change is reflected in abstracted water concentration. Fissured limestone on the other hand will show the impact of changes within a few years, although taking considerably longer to achieve a new equilibrium (Foster *et al.*, 1986).

Eutrophication of inland and coastal waters can be influenced by nutrient levels in the water as well as many other factors, particularly temperature and flow. In recent years, toxic blue-green algae have been a problem in several slow moving lakes and reservoirs in England, particularly during the hot summers of 1989 and 1990 (NRA, 1990a). There is general agreement that phosphorus is the limiting nutrient for inland waters (Birch and Moss, 1990). These waters will naturally contain a nitrogen concentration high enough to support algal growth, if other factors are favourable. The majority of phosphorus in river waters comes from sewage works. The agricultural contribution is poorly researched in England and Wales but is thought to be around 20–30% of the total. Phosphorus is rarely leached from soils. The main pathway of loss is by surface runoff in particulate or suspended forms. Soil particles and organic manures are the main sources. It is difficult to predict loadings as numerous factors influence when runoff occurs and how much phosphorus will be transported by it. If animal manure is applied on fields when the runoff risk is high, the likelihood of this source contributing a considerable input of phosphorus to the river is high. Prevention follows the same rules as those for minimizing high BOD material runoff risk.

By contrast, marine eutrophication can be nitrogen limited. The nitrogen content of the open sea is very low. To date, no significant eutrophication problems have been detected around the UK coast, although transient problems in some estuaries are considered likely to occur. The problem is very limited compared with those on the other side of the North Sea and in the Baltic Sea around Sweden and Denmark where the movement of water is much less and mixing much slower (National Agency of Environmental Protection, Denmark, 1991).

Problems of cryptosporidium, a protozoan parasite, in drinking water have become an increasing concern for water supply companies in recent years. Circumstantial evidence suggests that the particular organism, *Cryptosporidium parvum*, may get into water through contamination with animal manures. Oocysts of cryptosporidium are prevalent in livestock. This suggests that agriculture is the most likely source (HMSO, 1990). It is another reason for the need to minimize the direct pollution of surface waters by livestock wastes.

Application to Land

The principle of application of animal manure and other effluents from livestock farming back to land is sound. The recycling of plant nutrients in this way reduces the need for using inorganic fertilizers. But the risk of land application causing pollution from runoff and leaching must also be considered. Problems can result from excessive or poorly timed nutrient and/or hydraulic loadings.

Soil type, rainfall and slope are the main factors that determine site runoff risk. Proximity to a watercourse is of major importance in determining whether pollution can occur from a particular field or store. Major incidents involving store failure only require a favourable slope between the store and a watercourse. Pollution by runoff from fields on which slurry or other liquid waste has been applied depends on several factors. It is most likely to occur when hydraulic loading exceeds the surface infiltration rate. This may be due either to the quantity of liquid waste itself applied or to heavy rain following waste application. Either can result in polluted water reaching a watercourse. The amount of liquid waste applied is usually only a problem with dirty water disposal. Failure to move fixed sprinklers regularly is a common cause of dirty water runoff.

Soils containing pipe drains present special risks, particularly on clayey soils. These soils are strongly structured and most water flow, even when the soil is fully wetted, is down preferred cracks and pathways between structural units. For this reason, water can move rapidly from the surface to the drains, before it has been held in the soil long enough for the micro-organisms to break down and utilize its organic loading. Pollution of watercourses can occur when these soils are fully wetted or when dry and cracked. The hydraulic loading which results in contaminated water reaching the drains is often very low.

Nutrient loading can be based on nitrogen or phosphorus, on total or available plant nutrient, or on stocking density. The Dutch use phosphorus; other EC countries, including the United Kingdom, prefer to use nitrogen as it is the greater problem nutrient, albeit the more difficult to quantify. While nutrient loading based on animal number per hectare is simple to calculate, it provides no incentive to reduce the nutrient excreted. It is an indirect measure of the real problem. Available nutrient level on first consideration seems appropriate, being most closely related to crop utilization. But it allows higher amounts of nitrogen to be applied at times of the year when availability is low. This could encourage increased autumn application and nitrate leaching, the opposite of the intention behind any controls.

For these reasons, total nitrogen in manure is the best of the three approaches for general use. This principle has been incorporated in the MAFF Code of Good Agricultural Practice (MAFF, 1991). The figure of $250\,kg\,ha^{-1}\,year^{-1}$ of total nitrogen in organic manures, excluding that deposited while grazing, is given as a guide to minimize the risk of both direct pollution and nutrient enrichment of waters in England and Wales. Lower amounts may be appropriate in sensitive catchments such as Nitrate Sensitive Areas (MAFF, 1990). In addition, it is important that the available nitrogen in organic sources applied to the land does not exceed that needed by the crop to be grown in a particular year.

To provide a basis on which farmers can plan their application of all organic wastes on the farm, Section Two of the Code of Good Agricultural Practice sets out five stages that should be applied on individual farms.

Stage 1

Any land on the farm where waste should not be spread at any time is selected. An untreated strip at least 10 metres wide is left next to all watercourses. Irrigation systems should be operated so there is no chance of their spray coming within 10 metres of a watercourse or wind blowing material into a watercourse. To reduce the risk of polluting groundwater, livestock wastes and other organic wastes should not be applied within at least 50 metres of a spring, well or borehole that supplies water for human consumption, or is to be used in farm dairies.

It may not be possible to spread waste on some areas of a farm because of the nuisance that could be caused by smell or because of management agreements.

Stage 2

The amount of land needed to take the total nitrogen in all the waste that has to be spread on the farm is calculated. If there is not enough land, arrangement must be made to spread the excess material on suitable land elsewhere. This calculation should be carried out using the 250 kg ha^{-1} $year^{-1}$ nitrogen application maximum, which includes the total nitrogen in all organic wastes as discussed above. The Code suggests that for new units, the amount of land needed to achieve this should be available from the outset. Existing units currently applying organic wastes at rates which exceed this amount should take steps to change their practice so that they can meet this limit.

Stage 3

Land where waste should not be spread at certain times or where the spreading rate should be limited is selected. This involves assessing the fields on the farm to judge the risk of causing pollution when waste is applied and the number of months in the year when this risk exists. Crop damage may limit the time when waste can be applied.

The risk of damaging the soil might also prevent waste from being spread during the winter months by machinery that is pulled by a tractor.

The Code provides guidance on separating land into three risk categories. High risk areas should not have waste applied to them when certain conditions exist, such as a high flooding tendency, poor surface condition and close proximity to watercourse, spring or borehole. Slope, soil type and field capacity period should also be taken into account, as well as the presence of field drains.

On intermediate risk areas, wastes may be applied but the recommended amount should not be exceeded. These take into account the same principles as in high risk areas. On fields in this condition, it is recommended that no more than 50 m^3 per hectare of slurry or 50 t per hectare of manure should be applied at any one time and that at least three weeks should be left between individual applications. For liquid effluent applied by pumped irrigation systems, a maximum application rate of 50 m^3 per hectare for travelling irrigators and a maximum precipitation rate of 5 mm per hour for sprinklers, moved regularly to suit the conditions, is recommended.

For fields falling into the low risk category, the amount and timing of applications is less critical. On all sites it is important that the nutrient loading is not exceeded and that spreading does not cause ponding or runoff.

When applying silage effluent to land, it should be diluted at least one-to-one with water before application. The amounts recommended for slurry and dirty water should be used for this diluted silage effluent when working out maximum application rates to land.

Stage 4

The largest amount of waste that will need to be stored before it is spread is calculated. If storage facilities on the farm are too small, more will need to be provided. Storage of slurry and dirty water is necessary when safe application to land is not possible all year as assessed at Stage 3. To assess the size of storage facilities needed, for each month the amount of waste produced and the amount that can safely be spread is determined. Monthly surpluses which cannot be spread are then added and the maximum amount of waste which cannot be spread gives the size of store needed. In many cases, dirty water and slurry may be estimated and handled separately. If the type of store chosen cannot be emptied in stages, the animal housing period will also have to be taken into account.

Stage 5

If extra storage is needed, the final stage is to choose and design a suitable storage system to meet the needs of the farm. The best way of handling and storing livestock waste depends on its consistency. This depends on the type of livestock, how they are housed, the amount and type of bedding material and whether the waste is diluted with dirty water. The system chosen will also depend on winter rainfall. In areas with a high rainfall, storage systems that have a large surface area will need extra storage capacity. For storage systems that produce liquid waste continuously, the extra amount needed will depend on rainfall. The type of store chosen also depends on other factors. An earth banked storage system can only be built on a suitable site. Machinery available to empty the store, labour required and costs will also affect the decision.

Judging when a field is suitable for slurry or dirty water application can be difficult. The guidelines given in the Code will undoubtedly need revision in the light of experience of field use and results from current research and development. Two major areas need further refinement.

The first concerns surface condition and runoff risk. In particular, better characterization of the threshold for allowing winter application of slurry at low rates is needed, especially in high risk areas in the Code. Perhaps the more urgent need is to improve understanding of sites on which daily disposal of dirty water is acceptable. At present this practice is carried out on a wide range of soils. Recent field evidence and research results have raised doubts over the practice on pipe drained sites, particularly when at field capacity. Better control of hydraulic loading may be the answer to this problem, using mobile irrigators rather than manually moved sprinklers. Manually moved sprinklers can apply at a low precipitation rate but the application rate depends entirely on how often, in practice, they are actually moved to a new site in the field.

There are increasing reasons why farmers should work out a farm waste plan for their farm following the MAFF Code criteria. Undoubtedly more management time as well as capital expenditure will be needed on many livestock farms in future to ensure pollution risk is minimized. The positive side of land application is maximizing crop uptake of the nutrient content, particularly nitrogen, phosphorus and potassium, leading to a reduction in expenditure on inorganic fertilizer (HMSO,

1988). The amount of readily available nutrients in animal manures is considerable. Dirty water will usually contain very small amounts unless it includes concentrated effluent from manure or slurry storage or treatment.

Phosphorus and potassium content will be well utilized whenever the slurry is applied. Timing is only important to ensure that the potassium is not taken up excessively by grass, inducing low magnesium in the herbage. It is, however, difficult to use the nitrogen in slurry effectively (Voorburg, 1991; Smith and Unwin, 1983). Unless applied in the spring or after silage cutting, nitrogen utilization will be poor. Ideally slurry should be applied in spring and incorporated before sowing spring crops or surface applied before grass growth accelerates in March/April. Considerable agronomic and mechanization research in recent years has been put into top dressing of autumn sown cereals with various livestock manures, particularly pig slurry (Unwin, 1987). This has been the most effective spring top dressed material. The practice suits mainly arable farms with pig units.

The ability of machinery to spread low amounts evenly is still a problem. In the past field machinery for application of liquid wastes on farms has not been capable of applying evenly at 20 m^3 per hectare. New machines are more capable of meeting this standard, particularly for liquids that have no straw or other coarse material in them.

Reducing Nitrate Leaching

Agriculture is normally the major contributor of nitrate to water, particularly groundwater in predominantly rural catchments. All soils, however managed, lose some nitrate by leaching over winter. If the input of nitrogen is high, the leaching loss is correspondingly high but also variable depending on numerous aspects of soil and crop management and their interaction with the weather (Davies *et al.*, 1990). Any application of animal manure, however well timed and balanced by an appropriate reduction in inorganic nitrogen fertilizer, will result in an increased loss of nitrogen from the soil.

For soil types with predominantly vertical water movement, either to depth or to land drains, much of this loss will be leached as nitrate. The quantity of nitrate leached from a soil over winter depends on the nitrate accumulated in the profile at the start of the winter leaching period and the net mineralization of further nitrate over winter. Crop uptake is low even where a crop is present during the winter period. The other main variables in determining the quantity of nitrate lost are soil type and the amount of excess winter rainfall which controls the quantity of drainage water passing through the profile. Up to about 200 mm of excess winter rainfall on sandy soils, there is a direct positive relationship between amount of drainage and the proportion of the potential nitrate loss that occurs. Above 200 mm the relationship changes as the actual loss approaches the potential in the profile. The situation is more complex in structured soils. Models have been produced in recent years to enable predictions of nitrate leaching to be made (Whitmore and Addiscott, 1986). The other main mechanisms of loss are volatilization as ammonia and denitrification as nitrous oxide or nitrogen gas.

Two characteristics of animal manures are important in assessing their contribution to nitrate leaching. Firstly, the total nitrogen concentration in the material and, secondly, the proportion of that nitrogen which is present in readily available form (mainly ammonium and uric). Total nitrogen determines how much will be released in the long term and available nitrogen how much is at risk from leaching in the weeks rather than the months following application.

Traditional farmyard manures have a low proportion of their nitrogen in an available form. By contrast, pig slurry and poultry manures contain a high proportion of available nitrogen in ammonium or uric forms. Cow slurry is intermediate between farmyard manure and pig slurry. No livestock manure contains significant amounts of nitrate-nitrogen but ammonium and uric-nitrogen is quickly converted to nitrate in the soil.

Both the quantity and timing of an application for a particular type of livestock waste will influence the quantity of nitrate lost by leaching. However, as all contain organic nitrogen which is released slowly, the total nitrogen loading in kilograms per hectare per year must be limited if nitrate leaching is to be controlled.

There is a strong case for banning the autumn/early winter application of livestock wastes with a high available nitrogen content to arable land. Work at Gleadthorpe Experimental Husbandry Farm, UK, over the last two winters has shown a marked difference in nitrate leaching from farmyard manure and battery manure (Unwin *et al.*, 1991). Even where crop cover is maintained during this period, crop uptake will only reduce this loss by a small percentage unless the application rate is very low.

Nitrate leaching following autumn application of cow slurry to grassland is less of a problem. Much depends on the mineral nitrogen level already in the soil at that time of the year. Some additional loss is likely from September/October applications to light soils which are intensively grazed on dairy farms.

Where a closed season is necessary to reduce nitrate leaching, the critical period is from when crop uptake ceases or slows down until the soil is cold enough to limit the conversion of added ammonium to nitrate. For arable fields this means from harvest onwards. For grassland this probably means from 1 September onwards in the UK. The end point of the closed season is more difficult to judge but should probably be 1 November or 1 December. Current research is addressing this issue.

Storage and Handling of Liquid Wastes on Farms

Storage of slurry on dairy and pig farms is often necessary to meet both agronomic and environmental needs. As discussed earlier, adequate land may not be available at all times of the year to enable the farm to dispose of slurry without storage. On many dairy farms, it is necessary to store slurry for the full winter housing period. In all cases storage requirements must be assessed for the individual farm circumstances.

Three main types of slurry store are used on UK farms. The choice depends on numerous factors, not least the cost and the suitability of the site for a particular type. Full construction standards are given in CIRIA (1992). The Code of Good Agricultural Practice for the Protection of Water provides guidance on their use (MAFF, 1991).

Above ground circular stores

These stores are normally made from curved steel panels or concrete sections. Storage depth can be up to 6 m so they generally take up less space than other storage systems. A typical system has a reception pit next to the main store. A grid covers the reception pit so that slurry can drop through and long fibres are scraped to one side. Slurry may also get to the reception pit through underground channels. A purpose-built pump moves slurry from the reception pit to the store, a tank or irrigation system for spreading. These pumps are often fitted with an outlet so that the contents of the reception pit can be recirculated. The contents of the store can be mixed by recirculating them through movable, single or multiple outlets in the store, by using the filling pump.

These stores provide a secure and effective method of total containment providing they are sized, maintained and operated correctly. The first stores were erected in the late 1960s. A number of 20-year-old stores are now showing serious signs of corrosion and a limited number of incidents occur each year because of panel failure. These incidents often result in serious water pollution.

This method of storage may well increase in the United Kingdom due to the effect of the Regulations discussed later in the chapter.

Weeping wall stores

Weeping wall stores are built above ground level on a concrete base. Excess liquid drains through narrow gaps in the walls, and is collected and spread on the land. This liquid has a high risk of causing pollution although it is often mixed with yard runoff and termed dirty water. The contents of the store gradually dry out and become similar in consistency to farmyard manure. The side panels of the store can be taken out to empty these solids. The contents are not emptied during the winter housing period, because they have not usually dried out sufficiently until early summer onwards. The size of store needed will usually be worked out from the winter housing period, taking into account the volume of slurry and bedding but not rainfall as this drains into a collection tank. Rain that falls on to the store plus about 10% of the slurry going into it will drain out during storage.

These stores are commonly used for cattle slurry including the straw bedding. They allow emptying using farm equipment. They are not suitable for pig slurry. The successful operation of this type of store usually requires that the liquid draining from it can be disposed of on a regular basis. This may present problems under the new Regulations on some UK farms in future where suitable land is not available for regular disposal of the liquid over winter.

Earth bank stores

Earth bank structures can be used to store slurry that contains bedding, dilute slurry, separated liquids from mechanical separators or dirty water. Depending on the consistency of their contents, they can either have the liquid taken out and be emptied as a solid or be mixed and emptied as a liquid. Such stores are commonly called ‘compounds’ when they contain solids or semi-solids and ‘lagoons’ if they contain more liquid waste. A series of lagoons may be used to settle dirty water before it is pumped through an irrigation system. Stores can be built below, above, or part-below/part-above ground. Typical depth from the base to the top of the embankment is 3–4 m.

This type of store has been regarded as a cheap method of storage by many farmers and unsatisfactory earth bank structures have been erected because no effective control on their siting, sizing and construction existed. A large number of serious pollution incidents are caused by earth bank stores that are too small, badly built or in an unsuitable place. Such stores cannot be built properly in a lot of places, because of unsuitable soil type. In future, under the new Regulations, such structures will only be allowed in the United Kingdom if they have an impermeable base and walls. To achieve this, the structure will have to be lined in some way on permeable soil types.

Correctly constructed, this method of storage can provide a cost-effective solution to many farm problems. Earth bank compounds commonly have a strainer box system to remove excess liquid from the store. This liquid will have a high BOD and nutrient content, and must be considered as highly contaminated when spread on land.

Treatment of Slurry

Treatment of manures and slurries by anaerobic digestion or by aerobic treatment has been used in the United Kingdom to achieve odour control on a limited number of farms. These processes have not been used to treat wastes to meet existing standards for discharge to watercourses, as generally speaking they are not capable of reaching the required standards without several stages of treatment. This would be totally uneconomic.

Mechanical separation

Mechanical separation of slurries into solid and liquid fractions offers a number of advantages. Coarse solids are removed giving an easily pumped liquid which presents fewer storage problems, can be applied by relatively simple irrigation equipment and may produce fewer problems of runoff during application, compared to unseparated slurry. Various workers have compared the different types of slurry separation available for both pig and cattle slurries (Pain *et al.*, 1978).

Treatment of dilute effluents

In situations where winter application of dilute effluent poses a pollution risk and storage is impractical or prohibitively expensive because of the large volumes involved, treatment appears to offer an attractive solution to enable the effluent to be discharged to a watercourse.

In the 1960s and 1970s, barrier ditches which achieved BOD reduction by settlement followed by an aerobic free-flowing section were installed on many farms. In general this technique has now been shown to be unsuccessful in meeting the discharge standards required in the United Kingdom by the National Rivers Authority (NRA). Root zone treatment of dilute effluents by reed beds is currently attracting publicity as a possible alternative (Job *et al.*, 1991). Current work indicates that a combination of down-flow and horizontal-flow beds can achieve BOD reductions of around 95% when input BOD is approximately 1000 mg l^{-1}, with some denitrification of the effluent. It remains to be seen whether such a capability can be employed on farms to reliably meet NRA discharge consent standards.

Field Application Machinery

Conventional tankers commonly in use for pig slurry or dilute cow slurry are vacuum or pump filled and discharge to a splash plate dispersal mechanism which gives a spread pattern 8–15 m wide. Application rates achieved typically range from 20 to 100 m^3 per hectare, but vacuum tankers in particular are not easily controlled, except by varying the forward speed of the tractor.

More sophisticated band spreading methods involving a full width distribution manifold and applying slurry through a series of outlets at ground level can achieve a better distribution pattern than a conventional tanker. This technique is becoming more widely available in the United Kingdom and offers the advantage of reducing odour emissions as well as more accurate spreading.

Tankers adapted or built for sub-surface injection will achieve a similar range of application rates. This technique may reduce the likelihood of surface runoff but can increase the risk of pollution via field drainage systems if soils are cracked or backfill is intercepted by the injection tines.

Travelling irrigators offer advantages over manually moved sprinklers for land application of dirty water and other dilute effluents. A range of purpose-built irrigators is now available in the United Kingdom. These typically have working widths of around 20 m and a range of application rates ranging from 5 to 50 m^3 per hectare. Some machines can be set to apply as little as 2 m^3 per hectare but this may involve a rapid winch rate, which will involve frequent moving and resetting of the machine. If operators are tempted to use higher application rates to reduce the need to move and reset the machine, the risk of runoff is increased. The majority of machines are fitted with automatic shut down devices to cut off the pump when the irrigator reaches the end of its run, thereby avoiding the risk of runoff due to continued operation while static.

Small travelling irrigators are frequently used for the application of separated slurries to growing crops in the spring. Tanker attachments have been devised to achieve a spread pattern and width which matches the 12 m spaced tramlines used for agrochemical applications to cereal crops.

It is generally accepted that fertilizer distributors for granular or prilled inorganic materials are capable of achieving application rates within plus or minus 5% of target and a coefficient of variation in spread pattern of 15% or less when well set up. Distribution of slurry to enable maximum use to be made of its nutrient content involves a number of uncertainties associated with the machinery used.

There is poor control over the application rate with the majority of tankers in current use. New technology may overcome this by automatic control of flow rate. This is available on some machines now. On mobile irrigators, variable application rate is a problem because the drive mechanism is not positive and the wind-in rate may vary from one end to the other end of the run. The best liquid waste applicators available in the United Kingdom will only achieve a coefficient of variation of 15–20%. Field machines used on commercial farms are likely to perform less well in many cases, particularly under variable wind conditions.

Water Protection Legislation

Nearly all water protection legislation in England and Wale stems from the 1989 Water Act. This provides the powers for the Control of Pollution (Silage, Slurry and Agricultural Fuel Oil) Regulations 1991 (DoE, 1991). It also provided a statutory basis for the setting up of the National Rivers Authority (NRA), the provision of the MAFF Code of Good Agricultural Practice for the Protection of Water 1991 and the setting up of Nitrate Sensitive Areas.

Code of Good Agricultural Practice for the Protection of Water

This code was issued by MAFF in July 1991 (MAFF, 1991). It replaces the previous Code published in 1985. Abiding by this code is not a defence against a charge of causing pollution, but it could be taken into account in any legal action.

The Code is a statutory Code in law but compliance with the Code is voluntary. However, a number of the requirements discussed in the code are covered by separate legislation.

An important principle introduced in this Code is the use of a guide figure of 250 kg ha^{-1} $year^{-1}$ of total nitrogen in all organic manures for nitrogen loading of manures to land in England and Wales. This figure as used in the code excludes manure deposited while grazing. The aim of introducing this limit is to provide a general guide to the level of application compatible with minimizing pollution risk, both direct contamination and nutrient enrichment.

The Control of Pollution (Silage, Slurry and Agricultural Fuel Oil) Regulations 1991

These Regulations came into full effect in the United Kingdom on 1 September 1991. The National Rivers Authority is responsible for enforcing them. The Regulations set minimum legal standards for new, substantially enlarged and substantially reconstructed facilities for the storage of silage effluent, slurry and agricultural fuel oil on farms.

Existing facilities on farms are exempt from the Regulations unless they cause or are likely to cause a pollution problem. In this situation the NRA can serve a notice requiring some or all of the Regulation requirements to be satisfied.

For silage, the Regulations allow farmers to make silage in existing silos, in new silos conforming to the Regulations, in wrapped or bagged bales or in a tower silo that meets the appropriate British Standard. Field heaps are only allowed if this technique has been used for the majority of silage made on the farm over the last three years and the NRA has been notified.

Where construction is carried out to conform to the Regulations, the main requirement is that the silo must have an impermeable base and collection channels to collect effluent. The collection channels must lead to a tank which is able to resist acid and to hold at least 20 litres of effluent for each 1 m^3 of silo space. If the collection tank is below ground, it must be able to resist acid attack for 20 years without maintenance.

For slurry storage, the Regulations lay down rules for new, substantially enlarged or substantially reconstructed stores, reception pits and channels. These require that no part of the storage facility may be within 10 m of a watercourse or field drain that slurry could get into if it escaped. Floors and walls must be impermeable. The store must have a life of at least 20 years with proper maintenance. Reception pits must be able to hold at least 2 days' slurry production and the store itself must be big enough to hold at least 4 months' slurry, unless safe year round disposal is possible on the farm.

Numerous other detailed points are covered in the Regulations. A guidance note for farmers on their implementation is available from the Department of the Environment (DoE, 1991).

Pilot Nitrate Sensitive Areas

During 1989–1991, 10 pilot Nitrate Sensitive Areas (NSAs) were set up in England to study the practical reduction of nitrate leaching from agricultural soils in groundwater catchments, where the abstracted nitrate level in drinking water was above or approaching 50 mg l^{-1} nitrate. Under the basic scheme (MAFF, 1990) for which farmers were able to apply on a voluntary basis and receive payment under a five-year management agreement, the following measures relate to organic manures:

1. Organic manure should be limited to 175 kg ha^{-1} total N per year (excluding manure deposited while grazing).
2. No slurry, poultry manure or liquid sewage sludge should be applied between 31 August and 1 November to grass fields, and between 30 June and 1 November to fields not in grass.
3. Slurry and liquid sewage sludge stored in the NSA must be contained in a way that will avoid leaching of nitrate.
4. Application of organic manures to the NSA land on the farm should not be increased by more than 25% of current application, unless the manure is produced within the NSA.

Results from research and NSA monitoring have shown that animal manure use is a major factor in determining nitrate leaching from agricultural fields. Very high levels of nitrate have been found in water draining from the agricultural zone of the soil on fields with a history of substantial animal manure applications.

EC Nitrate Directive

The above Directive was notified on 19 December 1991 and will result in controls on nitrate leaching from agricultural soils in EC vulnerable zones. Under the Directive as agreed, member states will have up to two years to identify and designate vulnerable zones where nitrate levels in drinking water from ground and surface water sources exceed or are likely to exceed 50 mg l^{-1} nitrate. There will then be up to two further years to decide on the agricultural measures which will be set up for each zone. These measures will be compulsory for farmers in these areas. While there is considerable national discretion on the details of these measures, one requirement is fixed by the Directive itself. This is the maximum total nitrogen loading in organic manures which has been agreed at 170 kg ha^{-1} $year^{-1}$ of total nitrogen, including deposition while grazing. For the first four years of the programme a higher figure of 210 kg ha^{-1} $year^{-1}$ may be used. Higher amounts are allowed under the Directive as long as they can be justified. It is likely that substantial areas of lowland England will be included in vulnerable zones, particularly areas with low excess winter rainfall.

Balancing Agronomic and Environmental Considerations

No individual pollution control measure can be looked at in isolation. Indeed there is conflict between the requirements of reducing nitrate leaching from soils and minimizing the risk of runoff from slurry application during the winter. The ideal time for slurry application to avoid runoff risk is often in the autumn on both grass and arable farms. Application at this time will maximize the amount of nitrate leaching. If both environmental constraints are applied to the same field, the practical opportunity for application becomes very restricted or indeed absent in some cases. While the need to apply both constraints to the same field may be uncommon, the practical consequences are all too apparent. For this reason it is essential that constraints on time of application, for whatever reason, are only required in well justified situations.

It is clear that spring applications of liquid wastes are highly desirable. Further development work is needed to develop practical ways of applying slurry and other liquid wastes to growing crops in both arable and grassland farming systems. Otherwise it will become increasingly difficult to overcome the combination of agricultural and environmental limitations on land application, which is generally the best way to recycle the nutrients in livestock wastes.

References

Archer, J.R. (1988) *Crop Nutrition and Fertiliser Use*, 2nd edition. Farming Press Limited, Ipswich, Suffolk.

Bastiman, B. and Altman, J.F.B. (1985) Losses at various stages in silage making. *Research and Development in Agriculture* **2**, 19–25.

Birch, S.P. and Moss, B. (1990) *Nitrogen and eutrophication in the UK*. Report to the Fertiliser Manufacturers Association, Peterborough, England.

Construction Industry Research and Information Association (CIRIA) (1992) *Farm Waste Storage – guidelines on construction. Technical Report No. 126*, CIRIA, London.

Davies, D.B., Archer, J.R. and Lord, E.I. (1990) Management of nitrate in sensitive catchments within England. In: Calvet, R. (ed.), *Nitrates – Agriculture – Eau*, INRA, Paris, pp. 511–526.

Department of the Environment (DoE) (1991) Control of Pollution (Silage, Slurry and Agricultural Fuel Oil) Regulations 1991. Guidance Notes for Farmers, DoE, London.

Foster, S.S.D. *et al.* (1986) The groundwater nitrate problem. *British Geological Survey Hydrogeological Report 86/2.*

Gordon, F.J. (1981) The effect of wilting of herbage on silage composition and its feeding value for milk production. *Animal Production* **32**, 171–178.

HMSO (1988) Fertiliser recommendations for agricultural and horticultural crops. *ADAS Reference Book 209*, MAFF, London.

HMSO (1989) Nitrate in water. *House of Lords Select Committee on the European Communities 16th Report.* Session 1988–89.

HMSO (1990) Cryptosporidium in water supplies. *Report of Group of Experts, DoE and Department of Health.*

Jarvis, S.C. and Pain, B.F. (1990) Ammonia volatilisation from agricultural land. *Proceedings of the Fertiliser Society*, No. 298, London.

Job, G.D., Biddlestone, A.J. and Grey, K. (1991) *Treatment of High Strength Agricultural and Industrial Effluents Using Reed Bed Treatment Systems.* Institute of Chemical Engineering Research, Cambridge.

Ministry of Agriculture, Fisheries and Food (MAFF) (1990) Pilot Nitrate Scheme. Guidelines for farmers in nitrate sensitive areas. MAFF, London.

Ministry of Agriculture, Fisheries and Food (MAFF) (1991) Code of Good Agricultural Practice for the Protection of Water. MAFF, London.

National Agency of Environmental Protection, Denmark (1991) *Environmental Impacts of Nutrient Emissions in Denmark. Report No. 1.*

National Rivers Authority (NRA) (1990a) *Toxic Blue Green Algae. Water Quality Series*, No. 2, NRA, London.

National Rivers Authority (NRA) (1990b) *Water Pollution from Farm Waste 1989. Joint report by NRA and MAFF.* NRA, Exeter.

Pain, B.F. *et al.* (1978) Factors affecting the performance of four slurry separating machines. *Journal of Agricultural Engineering Research* **23**, 231–242.

Schofield, K., Seager, J. and Merriman, R.P. (1990) The impact of intensive dairy farming activities on river quality: the Eastern Cleddau Catchment Study. *Journal of the Institute of Water and Environmental Management* **4**, 176–186.

Smith, K.A. and Unwin, R.J. (1983) Fertiliser value of organic manures in the UK. *Proceedings of the Fertiliser Society*, No. 221, London.

Unwin, R.J. (1987) Application of dilute animal slurries to cereal crops in spring. In: *Agricultural Waste Management and Environmental Protection. Proceedings of 4th International CIEC Conference*, Braunschweig, Germany.

Unwin, R.J., Shepherd, M. and Smith, K.A. (1991) Treatment and use of sewage sludge and liquid agricultural wastes. *Proceedings of Symposium*, CEC, Athens, October 1990.

Voorburg, J.H. (1991) Utilisation of animal manure to avoid pollution. *Proceedings of the Fertiliser Society*, No. 304, London.

Whitmore, A.P. and Addiscott, T.M. (1986) Computer simulation of winter leaching losses of nitrate from soils cropped with winter wheat. *Soil Use and Management* **2**, 26–30.

20 Gaseous Pollutants Produced by Farm Animal Enterprises

S. TAMMINGA

Department of Animal Nutrition, Agricultural University, Wageningen, Netherlands

Summary

Production of milk, meat and wool by farm animals results in the loss of volatiles, most of which are undesirable from an animal health as well as an environmental point of view. Gases produced directly by the animal are carbon dioxide and methane. Indirectly, through aerobic and anaerobic microbial conversions, ammonia, nitrous oxide, dinitrogen oxide, nitrogen dioxide and a wide variety of noxious odours are lost from animal waste (faeces, urine) to the environment.

Despite its possible role as contributor to the greenhouse effect, carbon dioxide produced by farm animals is considered harmless, because it is produced from renewable rather than from fossil energy. Methane, particularly produced by ruminants, causes concern because of its contribution to the greenhouse effect as well as its possible contribution to the impairment of the ozone layer. This is also true for nitrous oxide and dinitrogen oxide. When exposed to high levels, ammonia may cause detrimental effects to the respiratory organs of farm animals, but the main concern is with respect to its contribution to acid deposition on soil or surface water. Other volatiles are not directly hazardous to health or environment, but because their smell is unwelcome, they may cause problems.

Animal waste management and feeding management can contribute significantly to solving the problems and methods for both are discussed in this chapter.

Introduction

Animal production means the conversion of feeds into valuable products like meat, milk, eggs and wool and into less desired but unavoidable waste products. Animal wastes (faeces, urine, respiration and fermentation gases) are excreted in solid, liquid or gaseous form. Respiration and fermentation gases are lost to the environment at the moment they are produced in the animal. After being excreted, solids and liquids in animal waste become subject to (largely anaerobic) microbial conversions which convert solid or liquid organic substrates into microbial biomass and microbial soluble or gaseous waste products.

Until recently most attention given to gases from animal waste was because of their (mal)odour. This was caused by intensification of animal numbers on the one hand and urbanization of areas with an intensive animal population on the other. In recent years, however, attention to gases has shifted from their odour to their damaging impact on the environment. Their contribution to environmental hazards like acid deposition, the impairment of the ozone layer and the greenhouse effect has received much attention.

Excessive levels of gases in the environment may result from too high a concentration of animals in a small area, from less efficient feeding, or from inadequate storage or way of application of animal waste as fertilizer. In this chapter attention will be paid to sources of undesired gases, the problems which are caused by them and some ways to reduce or eliminate them.

Sources and Problems Caused by Gaseous Pollutants from Farm Animals

In high concentrations some of the gaseous pollutants are a direct health risk, but in sufficiently ventilated farm buildings dangerous levels are rarely reached in practice. An exception is waste storage pits with concentrations of carbon dioxide (CO_2), ammonia (NH_3), hydrogen sulphide (H_2S) and methane (CH_4) which are so high that oxygen (O_2) levels drop to almost zero, and if subjected to such a mixture, which may occur when stirring the pit's contents, intoxication and suffocation is a real danger. Although most of the gases resulting from animal production are not directly a hazard to animal and human health, their smell is repulsive, which still makes them undesirable. Besides, a category of gases is becoming more and more recognized as detrimental to the environment, to the soil and its vegetation as well as to the atmosphere.

Groups of gases to be discussed in more detail in this chapter are (i) methane, (ii) ammonia and other nitrogenous gases and (iii) (mal)odours.

Methane production in farm animals

Anaerobic microbial conversions contribute substantially to digestion in farm animals, particularly in ruminants (cattle, sheep) in their forestomachs and to a lesser extent in monogastric animals (swine, poultry) in their hindgut. Organic matter in animal feed, structural polysaccharides in fibrous material in particular, is converted into volatile fatty acids (acetic acid, propionic acid, butyric acid), microbial biomass, fermentation gases (CO_2, CH_4) and heat. This conversion process, which converts organic material that would otherwise remain undigested, requires the presence of an 'electron sink' and CO_2 plays such a role and in that capacity it becomes reduced to CH_4 as shown in Table 20.1.

Volatile fatty acids (VFAs) are absorbed into the blood, transported to organs and tissues and subsequently metabolized and utilized as a source of energy; CO_2 and

Table 20.1. Microbial conversions of carbohydrates in forestomachs and hindgut.

$(C_6H_{10}O_5)_n$	$\longrightarrow$	$nC_6H_{12}O_6$
$C_6H_{12}O_6 + 2H_2O$	$\longrightarrow$	$2C_2H_4O_2 + 2CO_2 + 8[H]$
$C_6H_{12}O_6 + 4[H]$	$\longrightarrow$	$2C_3H_6O_2 + 2H_2O$
$C_6H_{12}O_6$	$\longrightarrow$	$C_4H_8O_2 + 2CO_2 + 4[H]$
$CO_2 + 8[H]$	$\longrightarrow$	$CH_4 + 2H_2O$

CH_4 are removed from the rumen by respiration or eructation. As a result between 5% and 10% of the energy present in the feed is lost in CH_4 (Johnson, 1991). From the scheme in Table 20.1 it is easy to understand that the loss of CH_4 is largely determined by the fermentation pattern (VFA ratio), and that an inverse relation exists between the production of CH_4 and propionic acid in the rumen (Van Nevel and Demeyer, 1988).

It has become clear now that CH_4 contributes significantly to the greenhouse effect and its sources of production have therefore become of interest. Recently, worldwide CH_4 production was estimated to be 400 million metric tonnes (Crutzen *et al.*, 1986). Animals are held responsible for between 15% and 25% of this production and some 90% of total animal production is suggested to come from cattle (74%), sheep (9%) and buffaloes (5%). Although farm animals contribute only around 20% to the total production, there are strong indications of an increase in atmospheric CH_4 levels during the last centuries from 1000 ppb in 1700 to 1700 ppb in 1986 (RIVM, 1988) and that cattle are held responsible for a significant part of this increase.

Losses of NH_3 and other nitrogenous gases

Farm animals consume considerable amounts of protein and other nitrogen-containing substances with their feed. Part of the protein is deposited in the body or excreted in milk, but usually a large part is lost in faeces and urine. Table 20.2 gives an

Table 20.2. Fate of nitrogen consumed with the feed by different categories of livestock.

	Feed protein ($g\,kg^{-1}$ DM)	Body (%)	Milk (%)	Faeces (%)	Urine (%)
Lactating cow					
grazing	250	2	17	25	56
stall fed	175	2	25	34	39
Beef cattle	150	22	–	30	48
Veal calf	180	54	–	5	41
Lactating sow	160	5	20	20	55
Piglet	184	40	–	10	50
Growing pig	170	32	–	15	53
Laying hen	170	32	–	12	56
Broiler	217	42	–	10	48

Source: Adapted from Coppoolse *et al.*, 1990.

indication of protein losses in different classes of livestock (adapted from Coppoolse *et al.*, 1990). The results show that a substantial proportion of the consumed N is excreted in faeces and urine. Nitrogen excreted in faeces originates from the feed, from endogenous sources and from microbes. Undigested feed N and endogenous N will mainly be excreted as true protein (amino acids). Microbial N will partly (15–20%) be present in nucleic acids. Faecal N may be partly excreted as NH_3. Nitrogen in urine is mainly excreted as urea in cattle, sheep and swine and as uric acid in poultry. Other N-containing compounds excreted in urine are allantoin, hippuric acid and creatinine.

After excretion, faeces and urine normally get mixed and, due to the presence of large numbers of microbes with microbial enzymes in faeces, degradation of N-containing compounds occurs. An important end product of this degradation is NH_3, which, during storage or after application to the soil as a fertilizer, may become converted further to nitrate (nitrification) or gaseous nitrogen (denitrification). As a side effect of nitrification and denitrification, oxides like nitrous oxide (N_2O), nitrogen oxide (NO) and nitrogen dioxide (NO_2) are produced and escape in the air. These gaseous intermediates of nitrification and denitrification are believed to contribute to the impairment of the ozone layer and are therefore considered undesirable. Besides, at high levels they may be poisonous and threshold limit values (TLV) for humans at a daily subjection time of 8 hours are 25, 25 and 5 ppm for NH_3, NO and NO_2 respectively (Muir, 1977). Farm animals are subjected to these gases almost continuously and therefore for NH_3 lower threshold levels of between 2 and 10 ppm are considered safe. Contrary to animals, plants appear to be much more sensitive to NH_3 and recommended maximum levels are 30 to 40 ppb for 'wild' vegetation on poor soils to around 100 ppb for cultivated plants (RIVM, 1988).

Noxious odours

Over 10 years ago over 60 volatile components had already been identified in animal waste (Spoelstra, 1978), of which about a dozen were considered important contributors to malodour. According to Coleman *et al.* (1991), noxious odours lost from animal manure can be classified as carboxylic acids (acetic acid, butyric acid), phenolics (*p*-cresol, phenol), aliphatic and nitrogen-containing compounds (ammonia) or sulphur-containing compounds (hydrogen sulphide, dimethyl sulphide, ethyl mercaptan, methyl mercaptan). The odour threshold value (OTV) and the threshold limit value (TLV) of these compounds in air are given in Table 20.3. With regard to OTVs, some discrepancy exists between the two sources, but TLVs are generally at least a factor 500 higher than the OTVs; hence, these odours can easily be recognized, long before their concentrations become a health risk. 'Normal' values in farm buildings (pig houses) are usually far below the TLV, except NH_3 and H_2S, which may occasionally reach dangerous levels.

The results show that the lowest OTVs are for sulphur-containing compounds, which are expected to originate at least partly from the degradation of S-containing amino acids. Sulphur-containing compounds like hydrogen sulphide (H_2S), have been indicated as contributing most to malodours (Coleman *et al.*, 1991), a view challenged by others (Spoelstra, 1978), because part of the H_2S is derived from the reduction of inorganic sulphate rather than from the degradation of S-containing

Table 20.3. Odour threshold values (OTVs), lowest toxic values (LTVs) and concentration normally found ('normal') of different organic compounds found in air in animal houses.

Compound	OTV* (ppb)	LTV*† (ppb)
Ammonia	4700	25 000
Acetic acid	1000	10 000
Phenol	5	5 000
Methyl mercaptan	2	500
Butyric acid	1	–
p-Cresol	1	5 000
Ethyl mercaptan	1	500
Dimethyl sulphide	1	1 000
Hydrogen sulphide	0.5	10 000

Sources: After Coleman *et al.*, 1991*; Muir, 1977†.

amino acids. Most of the other odours also originate from the anaerobic degradation of protein through deamination, dehydroxylation and decarboxylation of amino acids. An important reason for this degradation is that the ratio between nitrogenous and non-nitrogenous compounds in animal waste is between 0.10 and 0.16 (Table 20.5), which is much higher than the optimal ratio of between 0.025 and 0.035 needed for anaerobic microbial growth. Hence, only part of the nitrogenous compounds can become incorporated in microbial biomass. Besides, after exhausting the energy-yielding substrate, microbes show lysis and the released nitrogenous compounds also become degraded, giving rise to even more degradation products and more odours.

Measures to Prevent or Reduce Gaseous Pollutants from Farm Animals

As indicated earlier gaseous losses from animal enterprises originate partly directly from the animals and partly indirectly after microbial conversions in stored waste. Total losses for different classes of farm animals are summarized in Table 20.4. The results show that only ruminants give a significant contribution to CH_4 production from farm animals but that all categories contribute significantly to the loss of NH_3.

Methane

Methane must be considered an integrated part of fermentation in the rumen (Demeyer and Van Nevel, 1975), and although inhibition of CH_4 production is possible, this cannot be achieved as a single event. When hydrogen is an important substrate, like in the rumen, methanogenic bacteria are inhibited by ammonia and propionate. When VFAs like acetate or propionate are important substrates, as in

Table 20.4. Loss of gaseous pollutants (kg per animal per year) from farm animals.

	Body weight (kg)	CH_4	NH_3	VFA	Others
Lactating cow	650	100	8.8	n.d.	n.d.
Beef cattle	450	50	5.7	n.d.	n.d.
Veal calf	150	–	1.5	n.d.	n.d.
Lactating sow	200	–	8	n.d.	n.d.
Growing pig	75	0.3	3.0	0.14	0.012
Laying hen	5	–	0.2	n.d.	n.d.
Broiler	0.5	–	0.1	n.d.	n.d.

Sources: After Hartung, 1988; Klarenbeek and Bruins, 1988.

anaerobic waste digesters, hydrogen, acetate and propionate are inhibitors too (Zeeman, 1991).

Production of CH_4 in the rumen can be manipulated by altering the fermentation pattern, which in turn can be achieved by altering the composition of the diet or by adding chemical compounds to the diet which selectively inhibit methanogenic bacteria (Van Nevel and Demeyer, 1988). Reductions in CH_4 production can result from a decreased extent of fermentation in the rumen or from a shift in the VFA pattern towards more propionate and less acetate.

A shift in site of digestion from the forestomachs to the intestines will result in a reduction in CH_4 production. A decreased degradation in the rumen, when compensated by an increased digestion in the small intestine will limit CH_4 losses, because digestion in the small intestine is not affected by microbial fermentation, and hence this digestion does not result in CH_4 losses. From the point of view of CH_4 losses this can be considered an advantage. If the shift in rumen digestion is compensated by an increased digestion in the hindgut, it depends on the fermentation pattern in the hindgut, but CH_4 production in hindgut fermentation is often lower than in rumen fermentation. Hence this shift seems advantageous too. If, however, the shift in rumen digestion is not compensated by an increase in small intestine or hindgut digestion, the resulting reduced CH_4 formation is no advantage, because, to meet requirements, the animal has to consume a larger quantity of such feed and the total CH_4 loss per animal or per kilogram of animal product (meat, milk) is likely to increase.

Level of feeding can also be used to manipulate CH_4 production. Increasing the

Table 20.5. Dry matter (DM), organic matter (OM), nitrogen (N) and ammonia (NH_4–N)/N contents of animals' waste after storage.

Animal species	DM (%)	OM/DM	N/OM	NH_4-N/total-N
Cattle	6.8	0.685	0.095	0.76
Swine	6.3	0.673	0.111	0.70
Poultry	10.7	0.691	0.157	0.62

Source: After Klarenbeek and Bruins, 1988.

level of feed intake with a constant dietary composition results in a shift in the VFA pattern towards more propionate and less acetate and will also result in a reduced loss of CH_4 (Robinson *et al.*, 1986; Sutton, 1986). Changes in rumen fermentation resulting in a reduced loss of CH_4 can be achieved by improving the quality of forage, by replacing forages with concentrates, or by changing the composition of concentrates. The last is possible by replacing structural carbohydrates (fibre) with starch and sugars or by the addition of fat, particularly unsaturated fatty acids (Van Nevel and Demeyer, 1988). The feeding of starch usually results in more propionate and therefore less CH_4 (Murphy *et al.*, 1982), whereas the inclusion of fat in diets for ruminants often prevents degradation of part of the structural carbohydrates in fibre in the rumen and shifts the site of digestion to the hindgut. In addition the inclusion of fat causes a shift in the VFA pattern towards propionate. Sheep and beef cattle appear to be more sensitive to fat than dairy cattle (Tamminga, 1989).

In estimating CH_4 production from respiration trials in sheep Blaxter and Clapperton (1965) found a positive relationship with digestibility, but at the same time a negative relationship with level of feeding. In dairy cows Moe and Tyrrell (1980) observed a positive relationship between CH_4 production and fibre components (hemicellulose, cellulose), with a specially high influence of cellulose. In sheep Giger-Reverdin *et al.* (1992) found a negative influence of fat in the diet on CH_4 production. This was also observed in dairy cows (Tamminga, 1989). An interesting observation was made with regard to CH_4 losses in grass-fed dairy cows. Data sets from research performed in New Zealand and in The Netherlands showed a decrease in CH_4 losses (expressed as %/unit of digestible energy) with an increasing N content in grass (Maas, unpublished results). Because of an inverse relationship between N and fibre contents in grass, it is not entirely clear if this reduction resulted from a specific effect of NH_3 in the rumen or if it was caused by a shift in VFA pattern towards propionate.

Reduction of methanogenesis in the rumen is also possible with chemical agents like ionophores, halogenated compounds and antibiotics (Van Nevel and Demeyer, 1988). Their application, however, is only permitted in a few species of farm animals because of the danger of residues appearing in milk and meat. Besides, adaptation of the rumen microflora to such compounds has been observed quite frequently (Van Nevel and Demeyer, 1988). To what extent CH_4 production can be or should be reduced depends on a number of factors. Based on the figures of Moe and Tyrrell (1980), shifting the composition of the diet by replacing 25% structural carbohydrates (cellulose + hemicellulose) by non-structural carbohydrates would suggest a reduction in CH_4 production of almost 20%. The inclusion of 10% fat in diets for dairy cows resulted in reductions of CH_4 production of 20–25% (Tamminga, 1989). At the same time milk protein content is usually and milk fat content is occasionally negatively affected. Increasing N content in fresh grass from between 1.5% to around 4.5% resulted in a reduction of CH_4 production of between 15% and 25%. One may assume a significant increase in NH_3 loss at the same time, so this does not seem a very promising way to reduce CH_4 production.

In conclusion, reduction of CH_4 seems best achieved by replacing structural carbohydrates (fibre) with non-structural carbohydrates (starch) or by increasing the level of feed intake. The latter often results in a higher and more efficient production as well.

Table 20.6. Composition of rations fed to dairy cows and of freshly voided faeces.

	OM/DM	NDF/OM*	N/OM	SN/NDS†
Diet 1	0.90	0.46	0.037	0.066
Faeces	0.80	0.45	0.048	0.072
Diet 2	0.91	0.54	0.034	0.067
Faeces	0.85	0.57	0.038	0.061
Diet 3	0.91	0.55	0.035	0.074
Faeces	0.84	0.52	0.040	0.059
Diet 4	0.92	0.62	0.022	0.048
Faeces	0.88	0.69	0.026	0.037

Source: After Bosch, 1991.
*NDF: neutral detergent fibre.
†SN/NDS: soluble nitrogen/neutral detergent solubles.

Ammonia and other N-containing compounds

Freshly voided faeces and urine contain little ammonia. They do, however, contain soluble and easily degradable organic matter, including urea and easily degradable protein. The latter gives rise to a rapid breakdown and the formation of (partly branched chained) VFAs and other degradation products of protein. Because of the presence of ureolytic bacteria in faeces and large amounts of urea in urine, ammonia is also produced very rapidly during storage. Table 20.5 shows the composition of slurry from cattle, pigs and poultry after storage.

Compared to freshly voided faeces (Table 20.6) some striking differences with regard to the N/OM ratio and the contribution of NH_3-N to total N become apparent. Similar differences exist with regard to the S/OM ratio. These differences result from the contamination of faeces with urinary N as well as from the disappearance of N-free organic matter due to microbial fermentation.

Methods of reducing losses in NH_3 are:

- reducing the excretion of compounds from which NH_3 can be formed;
- preventing the conversion of urea into NH_3;
- preventing the escape of NH_3 from stored animal waste.

Reduction of the excretion of N in animal waste is possible by reducing the amount of N ingested with the feed. Ways of achieving this were reviewed recently (Lenis, 1989; Tamminga and Verstegen, 1992; Tamminga, 1992). In dairy cows mainly fed with forage (wilted grass silage) a good relationship appears to exist between the composition and quality of the diet and the composition of freshly voided faeces (Table 20.6) and clearly possibilities exist to reduce N excretion in faeces through feeding. At the same time a reduction of N excretion in urine was observed. Reducing N intake to below the level required by rumen microbes for their activity (below $24\,g\,kg^{-1}$ of DM), may impair digestion in ruminants and is not recommended. Reduction of N intake in monogastric animals like swine and poultry is also possible, but, to prevent a reduction in productivity, upgrading the quality of a small amount of protein may be required either by feeding some protein of a better quality or by supplementing the protein with limiting amino acids. Particularly the amino acids lysine and methionine seem to have possibilities.

As important sites of N losses in farm animals, the forestomachs in ruminants and the small intestine and several organs and tissues in all species were identified. Losses from the forestomachs can be prevented by reducing N input or by inhibiting the degradation of feed protein (Tamminga, 1992). Losses from the small intestine are due to an incomplete digestion or an excessive production of endogenous protein. Endogenous protein is added as enzymes, mucus and epithelial cells to the digesta when passing along the digestive tract (Swanson, 1982). Although sensitive to proteolytic digestive enzymes, their digestion and reabsorption is incomplete and they contribute to N in animal waste, either directly or after microbial conversion. Technological treatment (heat, moisture, pressure) appears a promising way to reduce losses due to low digestibility as well as to excessive endogenous losses (Tamminga and Verstegen, 1991).

Losses from organs and tissues occur because of a lack of other sources of energy or because amino acids are not offered to organs and tissues in the right ratio. Plant proteins, the main source of amino acids for farm animals, are often low in lysin and methionine and as a result the utilization of the other amino acids becomes poor, because deamination followed by oxidation takes place.

Proposed ways to prevent NH_3 from escaping from animal waste during storage have been summarized by Tamminga and Verstegen (1991) and include the addition of clay minerals to the feed, the addition of urease inhibitors to the feed or to the waste, and (airtight) covering of the storage pit or acidification. The latter seems a promising way, because animal waste is enriched with N and becomes more valuable as N fertilizer (Esteban Turzo *et al.*, 1988; Pain *et al.*, 1990). Nitrification may, however, convert part of the trapped N into N_2, which escapes in the air and gets lost.

A further important time of NH_3 loss is during spreading. Here again recommendations have been made, like the application of manure during the growing season only and soil injection (see Chapter 19).

One of the situations with the highest N losses in animal waste is that of the grazing animal. The application of high levels of fertilizer N results in high quality grass, with a (crude) protein content of up to $250\,g\,kg^{-1}$ of DM, but this is about twice as much as the grazing animal's needs and the amount which it can efficiently utilize. It is not only that high levels of fertilizer N result in too high levels of N in grass, they will also almost completely eliminate leguminous forages like clover, with a capacity to bind N through N-fixation.

It has been estimated (Vertregt and Rutgers, 1988) that of the N excreted in the urine of grazing animals some 15% is lost as NH_3, but at least an equally large proportion could not be recovered from the soil or the grazed herbage, suggesting other important losses too. Possible solutions to improve the undesirable large urinary N excretions are supplementation of grass with forages or concentrates with a deficit of N, like corn silage, beet pulp or grains. The surplus of N in the herbage can make up the deficit in the supplement and overall N utilization has been shown to improve considerably (Valk *et al.*, 1990).

Preventing the development of noxious odours

Noxious odours are largely end products of the degradation of protein either excreted in faeces or after lysis of microbes grown during the early stages of storage of animal waste. A positive relationship was found between the excretion of NH_3 and the excretion of noxious odours (Pain *et al.*, 1988). This finding is somewhat difficult to explain, because there is not always a direct relation between N excreted in faeces and N excreted in urine. It is, however, an interesting observation, because NH_3 excretion can possibly be used as a marker for the release of odour, which itself is very difficult to measure.

Various ways exist to prevent the escape of noxious odours. Most widely practised are biofilters (Zeisig, 1988), but acidification seems a good alternative (Esteban Turzo *et al.*, 1988; Pain *et al.*, 1988). Oligolysis (formation of ferro-sulphate through electrolysis) seems also a promising alternative (Coleman *et al.*, 1991) and finally fermentation has been recommended as a means to eliminate noxious odours (Zeeman, 1991).

Legislation to Prevent Escape of Gaseous Compounds from Farm Animals

Most measures to reduce the release of undesired components from animal waste are not (yet) economically attractive. Legislation to limit and reduce environmental pollution through farm animals may be necessary to enforce such reductions. In The Netherlands, legislation aimed at reducing the application of too high levels of phosphorus (P) through manure as a fertilizer was introduced a number of years ago. Maximum application rates of manure, depending on the type of crop produced, were introduced, and are at present 200, 200 and 125 kg P ha^{-1} for land used for the production of grass, corn and other crops respectively. It is expected that before the end of the century, this will be further reduced to 110, 75 and 70 kg ha^{-1} respectively. For surplus manure a tax has been introduced and distribution (as fertilizer or as raw material for a manure refinery) is through a manure bank.

Regulations to limit emissions of gaseous pollutants by farm animals have also been or soon will be introduced. Two groups of regulations can be distinguished: first the expansion of farm capacity and storage of manure is becoming subject to severe restrictions and second the application of manure as a fertilizer will become limited. Regulations to reduce NH_3 emissions can be summarized as follows:

1. Possibilities to expand animal production farms have been restricted in the vicinity of forests and nature reservations. The introduction of a maximum stocking density is being considered.
2. Manure storage pits need to be covered in such a way that emission of ammonia is reduced by at least 75%.
3. Application of manure as a fertilizer is not permitted between 1 October and 1 February and is never permitted on snow covered land.
4. On sandy soils application of manure as a fertilizer between 15 June and 1 October may only take place through soil injection.

These regulations are expected to reduce emission of NH_3 by 70% before the end of this decade.

Conclusions

Three groups of gaseous pollutants originating from farm animals can be distinguished: first fermentation and respiration gases, second ammonia and nitrogen oxides and third noxious odours. Their origin differs in that the first group is produced directly by farm animals, whereas the second and third groups result from microbial conversions during storage. Gaseous emissions can be severely reduced through proper feeding management, proper storage and proper ways of application of manure as a fertilizer. Because such measures may not be economically profitable, legislation may be necessary to enforce them.

References

Blaxter, K.L. and Clapperton, J.L. (1965) Prediction of the amount of methane produced by ruminants. *British Journal of Nutrition* **19**, 511–522.

Bosch, M. (1991) Influence of stage of maturity of grass silages on digestion processes in dairy cows. PhD Thesis, Agricultural University, Wageningen.

Bryant, M.P. (1979) Microbial methane production – theoretical aspects. *Journal of Animal Science* **48**, 193–201.

Coleman, R.N., Feddes, J.J.R. and West, B.S. (1991) What is odour and the potential for its control? In: *Proceedings Western Branch Meeting*. Canadian Society of Animal Production, Chilliwack, 27–28 May (Abstract).

Coppoolse, J., Van Vuuren, A.M., Huisman, J., Janssen, W.M.M., Jongbloed, A., Lenis, N.P. and Simons, P.C.M. (1990) De uitscheiding van stikstof, forsfor en kalium door landbouwhuisdieren. *IVVO Mededeling no. 3 (Update 1990)*. Institute for Livestock Feeding and Nutrition (IVVO), Lelystad, Netherlands.

Crutzen, P.J., Aselmann, I. and Seiler, W. (1986) Methane production by domestic animals, wild ruminants, other herbivorous fauna and humans. *Tellus* **38B**, 271–281.

Curtis, S.E. (1983) The air environment. In: Curtis, S.E. (ed.), *Environmental Management in Animal Agriculture*. The Iowa State University Press, Ames, USA.

Demeyer, D.I. and Van Nevel, C.J. (1975) Methanogenesis, an integrated part of carbohydrate fermentation and its control. In: McDonald, I.W. and Warner, A.C.I. (eds), *Digestion and Metabolism in the Ruminant*. The University of New England Publishing Unit, Armidale, Australia, pp. 366–382.

Esteban Turzo, P., Gonzalez Fernandez, A. and Garcia de Bustos, J.M. (1988) Improved nitrogen fixation by acidification and dehydration of slurry. In: Nielsen, V.C., Voorburg, J.H. and L'Hermite, P. (eds), *Volatile Emissions from Livestock Farming and Sewage Operations*. Elsevier Applied Science, London, pp. 170–177.

Giger-Reverdin, S., Vermorel, M. and Sauvant, D. (1992) Facteurs de variation de la production de methane au cours de la digestion des aliments composes chez les ruminants. *Reproduction, Nutrition et Development* (in press).

Hartung, J. (1988) Tentative calculations of gaseous emissions from pig houses by way of the exhaust air. In: Nielsen, V.C., Voorburg, J.H. and L'Hermite, P. (eds), *Volatile Emissions from Livestock Farming and Sewage Operations*. Elsevier Applied Science, London, pp. 54–58.

Johnson, K.A. (1991) The effect of dietary regime on methane production in ruminants. In: *Proceedings Western Branch Meeting*, 27–28 May. Canadian Society of Animal Production, Chilliwack.

Klarenbeek, J.V. and Bruins, M.A. (1988) Ammonia emissions from livestock buildings and slurry spreading in the Netherlands. In: Nielsen, V.C., Voorburg, J.H. and L'Hermite, P. (eds), *Volatile Emissions from Livestock Farming and Sewage Operations*. Elsevier Applied Science, London, pp. 73–84.

Lenis, N.P. (1989) Lower nitrogen excretion in pig husbandry by feeding: current and future possibilities. *Netherlands Journal of Agricultural Science* **37**, 61–70.

Moe, P.W. and Tyrrell, H.F. (1980) Methane production in dairy cows. In: Mount, L.E. (ed.), *Energy Metabolism*. Butterworths, London, pp. 59–62.

Muir, G.D. (1977) Hazardous chemicals. In: Muir, G.D. (ed.), *Hazards in the Chemical Laboratory*. The Chemical Society, London, pp. 111–436.

Murphy, M.R., Baldwin, L.R. and Koong, L.J. (1982) Estimation of stoichiometric parameters for rumen fermentation of roughage and concentrate diets. *Journal of Animal Science* **55**, 411–421.

Pain, B.F., Rees, Y.J. and Lockyer, D.R. (1988) Odour and ammonia emission following the application of pig and cattle slurry to land. In: Nielsen, V.C., Voorburg, J.H. and L'Hermite, P. (eds), *Volatile Emissions from Livestock Farming and Sewage Operations*. Elsevier Applied Science, London, pp. 2–11.

Pain, B.F., Thomson, R.B., Rees, Y.J. and Skinner, J.H. (1990) Reducing gaseous losses of nitrogen from cattle slurry applied to grassland by the use of additives. *Journal of the Science of Food and Agriculture* **50**, 141–153.

RIVM (1988) Continentale milieuproblemen in Europa. In: Langeweg, F. (ed.), *Zorgen voor morgen. Nationale milieuverkenning 1985–2000*. Samson, Alphen a/d Rijn, pp. 84–155.

Robinson, P.H., Tamminga, S. and Van Vuuren, A.M. (1986) Influence of declining level of feed intake and varying the proportion of starch in the concentrate on rumen fermentation in dairy cows. *Livestock Production Science* **15**, 173–189.

Spoelstra, S.F. (1978) Microbial aspects of the formation of malodorous compounds in anaerobically stored piggery waste. PhD Thesis, Agricultural University, Wageningen.

Sutton, J.D. (1986) Rumen fermentation and gastro-intestinal absorption. In: Neimann-Sorensen, A. (ed.), *Agriculture: New developments and future perspectives in research on rumen function*. European Economic Community, Brussels, pp. 21–38.

Swanson, E.W. (1982) Estimation of metabolic protein requirements to cover unavoidable losses of endogenous nitrogen in maintenance of cattle. In: Owens, F.N. (ed.), *Protein Requirements for Cattle*. Oklahoma State University, Stillwater, USA, pp. 183–197.

Tamminga, S. (1989) Forage digestion as influenced by feeding fats. In: Martin, J. (ed.), *Advances in Dairy Technology*. University of Alberta, Edmonton, Canada, pp. 9–20.

Tamminga, S. (1992) Nutrition management of dairy cows as a contribution to pollution control. *Journal of Dairy Science* **75**, 345–357.

Tamminga, S. and Verstegen, M.W.A. (1991) Protein nutrition and animal production. Consequences for environments and some possible recommendations. In: Eggum, B.O., Boisen, S., Borsting, C., Danfaer, A. and Hvelplund, T. (eds), *Protein Metabolism and Nutrition*, vol. I. National Institute of Animal Science, Foulum, Denmark, pp. 23–36.

Valk, H., Klein Poelhuis, H.W. and Wentink, H.J. (1990) Effect of fibrous and starchy carbohydrates in concentrate supplements in a herbage-based diet for high-yielding dairy cows. *Netherlands Journal of Agricultural Science* **38**, 475–486.

Van Nevel, C.J. and Demeyer, D.I. (1988) Manipulation of rumen fermentation. In: Hobson, P.N. (ed.), *The Rumen Microbial Ecosystem*. Elsevier Applied Science, London, pp. 387–443.

Vertregt, N. and Rutgers, B. (1988) Ammonia volatilization from urine patches in grassland.

In: Nielsen, V.C., Voorburg, J.H. and L'Hermite, P. (eds), *Volatile Emissions from Livestock Farming and Sewage Operations*. Elsevier Applied Science, London, pp. 85–91.

Zeeman, G. (1991) Mesophylic and psychrophylic digestion of liquid manure. PhD Thesis, Agricultural University, Wageningen.

Zeisig, H.D. (1988) Experiments with the use of biofilters to remove odours from piggeries and hen houses. In: Nielsen, V.C., Voorburg, J.H. and L'Hermite, P. (eds), *Volatile Emissions from Livestock Farming and Sewage Operations*. Elsevier Applied Science, London, pp. 209–216.

21 Disease Transfer

R.W. Lacey

Department of Microbiology, University of Leeds, UK

Summary

Intensive farming of fish, poultry and animals requires the provision of adequate artificial feed and the disposal of the resulting unwanted components of the food. In general, both the provision of the feed and the disposal of the remains are geared to each other with the feed in part being derived from the rendering plants. Because the main commercial criterion of intensive farming relates to weight gain, aspects such as safety and nutrition have, until recently, received scant attention. The salmonella problem in eggs is almost certainly partly related to intensive farming and the BSE epidemic entirely so.

While there is no certainty of the source of the infectious agent that caused bovine spongiform encephalopathy there is evident risk of its subsequent transfer to further mammalian hosts. There is already evidence of infection of cats, mice, pigs and some zoo animals such as antelopes from this source. The current annual mortality from Creutzfeldt–Jakob disease counts for about 1% of deaths in the United Kingdom and this figure could be substantially increased next century as a result of consumption of BSE contaminated beef.

The agents that cause transmissible spongiform encephalopathies are in practice more resistant to heat and physical agents than food protein. This indicates that the derivatives from rendering plants should not be returned to the same or similar species from which they were derived.

Intensive methods of rearing fish, broilers, egg layers and pigs can reduce the proportion of *cis* polyunsaturated fatty acids (PUFAs) to saturated fatty acids in the food. Irradiation, and deep freezing and cook–chill systems may further reduce PUFAs. It is not known whether microwave exposure converts *cis* fatty acids to *trans*. We request that all food items produced by intensive farming and/or processing be regularly analysed for PUFAs and labelled accordingly. Changes in the methods of feeding food animals, poultry and fish, and their processing, may be partly responsible for the continued high incidence of coronary heart disease.

Current concern exists about the microbial safety and nutritional quality of much of our food as a result of the cumulative effects of the different stages of the lengthening food chain.

Introduction

The intensive rearing of animals, birds and fish to provide food for human consumption is associated with a number of problems. Indeed the closed systems employed do enable a careful analysis of these problems. It is likely that intensive rearing represents an amplification of problems inherent with all such food production be it free range, partly or fully controlled. Nevertheless intensive rearing probably does provide many additional and unnecessary hazards for the human population, both in the acquisition of infections and also in the provision of unsatisfactory nutrients particularly unsaturated fatty acids. In this chapter, both these issues will be considered.

It is an interesting paradox that in the United Kingdom the urban city dweller now commands five times as much air in his dwellings as in the last century. In contrast, the majority of our meat is obtained from animals and birds reared very much closer together. The factors underlying the health hazards for man from such intensive rearing are on the whole self-evident. No one surely will disagree with the principle that the closer animals or birds are housed, the more chance there is of infective agents spreading between them. It is well known that two major factors involved in the spread of infections through the air involve distance that causes risk according to the inverse square law, and the humidity of the atmosphere. It must also be obvious that if we rear animals and birds of a uniform genetic constitution, then if any one animal or bird is vulnerable to a particular type of infection then all of them, or at least a great majority of them, will be vulnerable.

As an example of intensive rearing, let us consider a typical large broiler farm. This will consist of perhaps 10 houses each harbouring 40 000 developing broiler chickens which will be cropped, say, eight times a year. The sheds will be located within farmland that is likely to be arable. Such intensive rearing produces perhaps 3 million live chickens for human consumption and about 300 000 dead. Because the growing birds do not have access to adjacent land, the first requirement is provision of the feed. The feed is artificial and centrally produced and compounded to produce maximum weight gain for the minimum cost. During rearing there will be a requirement to prevent and control a number of infections that include bacteria, fungi and viruses. The bacterial and fungal causes can generally be controlled by prophylactic or therapeutic drugs that are often added to the drinking water. The viral infections tend to be incapable of therapeutic control apart from vaccination. On occasion slaughtering practices have to be employed.

When the broilers are slaughtered and prepared for sale, any contaminating or colonizing bacteria present within their intestines or associated with the birds are dispersed widely among the chickens because of the automated nature of the slaughtering system. This uses wet conditions and a conveyor belt approach. For this reason, it is often difficult to identify the exact contribution that either the intensive rearing or the slaughtering methods make towards the final levels of carcass contamination. The unwanted remains of the birds – e.g. feathers, heads and intestines – have traditionally been treated in the rendering plants to provide two usable products: tallow and the protein-rich bonemeal. The latter has been returned after central pooling to feed that same species and also to feed other intensively reared animals.

The advantages of the resultant food are its uniform and predictable quality and that it is inexpensive. The price that we pay for these is an increased risk of contamination by, for example, *Salmonella* and *Campylobacter* in addition to deficiencies in nutrients.

Evidently intensively reared animals and birds do require a large number of antibiotics and other drugs to ensure their satisfactory health. All such drugs have been carefully assessed prior to their general use and, in theory, sufficient time should lapse between the completion of their use and the presentation of the animal/bird for slaughter to ensure that no drug residues persist. In an ideal world, this would occur, but there are substantial reasons to think that, because the system is inadequately monitored and controlled, the drug residues can sometimes enter into our food chain without the knowledge of the consumer.

In practical terms we can identify that the problems from intensive rearing result first, from the nature of the artificial feed, and BSE is by far the most important example of this; second, from the conditions of crowding during growth; third, from the methods of slaughter. The food poisoning organisms *Salmonella* and *Campylobacter* are probably caused by all these factors.

At the time of writing, considerable publicity has centred on so-called blue ear disease of pigs. The disease is referred to as blue ear because in a fairly small proportion of infected animals, the ears do take on a bluish hue. This results from the severity of the pneumonia that can cause anoxaemia as a result of involvement of the lungs. The resultant haemoglobin takes on a bluish brown tinge particularly where light passes through it, as in the ear. The disease is essentially an acute pneumonia or influenza that is frequently mild or even subclinical in non-pregnant sows or pigs, but in the pregnant sow it would appear to possess a specific tissue tropism with the infective agent frequently causing damage to the unborn piglets. Miscarriage or stillbirth usually follows. After the infection the sow becomes immune but there seem to be doubts as to the extent to which the sow is able to produce normal litters in the future. The present epidemic began in November 1990 in Germany and has spread westwards and southwards with the disease being described in Belgium, Holland, France, the United Kingdom and Spain. The exact numbers of pigs infected is not known because many of the infections will be subclinical but the figure is likely to exceed hundreds of thousands.

The similarity to human influenza is striking. Human influenza typically spreads westwards epidemically during the winter, and large numbers of the human population who are not immune to the particular strain are affected. Fortunately human influenza does not seem to have a particular tropism for the pregnant woman. Human influenza is thought to have an origin in the Far East and there have been many claims of a reservoir in pigs, particularly in Mongolia. It is thought that the epidemics are initiated in the human population when the influenza virus has changed sufficiently for the existing immunity – either natural or vaccine induced – not to protect the human population. A great deal of research has studied the way in which the human influenza virus can change. Major alterations in its genetic make-up are referred to as shift and minor ones as drift (Jawetz *et al.*, 1984). Pig tissues are very close to human tissues as is shown by the fact that various connective tissue components from the pig have been used as human grafts. The similarities between blue ear pig disease and human influenza have raised the question as to whether the human population might be vulnerable to the pig disease. In June 1991 Dutch workers reported the isolation of a virus from infected pigs in Holland (Wensvoort, *et al.*, 1991). So far this virus has not been characterized. However, a recent report (Walkland, 1991) reinforces the concern of the possible similarities between the pig and human infections. The report describes a trial whereby pigs who had not suffered blue ear disease previously were given injections of the human influenza vaccine. Control groups housed elsewhere were not given any of the vaccine. Subsequently the vaccinated piglets showed a very low incidence (2 out of 100) of the disease, whereas the controls suffered infection at a rate of about 80%. If these findings are confirmed, then the implications may be important because vaccines only protect against a very narrow range of infective agents, and this must mean that not only are the pig and human influenzas within the same broad class, but they may well be very close indeed. If this is the case, there is a distinct possibility that the pig influenza virus will have a specific tissue tropism for the pregnant human. Further research is urgently needed but the occurrence of this disease would seem to be yet another illustration of risks that result from modern methods of animal husbandry which involve animals of uniform genetic type, and intensive rearing in cramped spaces.

Salmonella Food Poisoning: the Egg Problem

Food poisoning from salmonellae usually results within 12–24 hours of eating contaminated food in a moderate to severe symptom complex of diarrhoea, abdominal pain, vomiting with dehydration and subsequent debility lasting for between a few days and a few weeks. While otherwise healthy adults die exceedingly rarely, vulnerable groups – babies, hospital patients, the elderly – may succumb. The mortality is typically between 1 in 500 and 1 in 3000 of reported cases (Hook, 1961). The effect of salmonella food poisoning on British society is very considerable. The last two large outbreaks occurred in 1984 with 455 cases at Stanley Royd Hospital and 766 on British Airways with 19 and 2 deaths respectively.

During 1988–1990, about 25 000 annual notifications of salmonella food poisoning have been made to the Communicable Disease Surveillance Centre, London. Because of the failure to notify all cases, the actual annual number of people who suffer from salmonella food poisoning could be about 250 000.

The explanation for this recent dramatic increase cannot be accounted for by the following:

1. Changes in isolating and reporting procedures.
2. Change of vulnerability of the patient or pathogenicity of the bacterium.
3. Major changes in kichen hygiene or catering practice.
4. The occurrence of one or a few large outbreaks that artificially 'inflate' the figures. Indeed, *the number of incidents has recently declined.*

Similarly, there has been a decline in the last few years of hospital salmonella outbreaks associated with catering errors (Communicable Disease Reports, 1982–

1988). This suggests that hygiene control may be improving, perhaps stimulated by the publicity over the Stanley Royd episode in 1984.

Eggs are the source of that additional contamination

Many species of salmonella, notably *typhimurium* and *enteritidis* have been associated with poultry and eggs. There has been an acceptance that the contamination of egg shell by droppings, associated with washing and cracking can implant small numbers of salmonellae into the egg albumen. This problem has been containable because:

1. the egg white is relatively hostile to the multiplication of salmonella due to the presence of inhibitory substances, including lysozyme;
2. eggs are now rarely washed in the United Kingdom;
3. battery produced eggs permit the ready removal of droppings through the wire floors with gentle rolling of the laid eggs for collection.

Salmonella enteritidis

The major change in the epidemiology of salmonella concerns the increase of *Salmonella enteritidis*, the source of which is essentially poultry (Humphrey *et al.*, 1988). Between 1982 and 1987, the number of reported cases in England and Wales increased sixfold from 1101 to 6858 (Editorial, 1988). During this time, the notifications of salmonella other than *enteritidis* have remained fairly constant (Communicable Disease Reports).

Salmonella enteritidis type 4 has only rarely been associated with foods other than those containing egg ingredients. It is certainly possible that some of the food poisoning from *S. enteritidis* could be from chicken meat, but there would appear to be no reason for such an increase. No data have been presented that show contamination of chicken meat by salmonella to have increased recently. Furthermore, the contamination is usually 'light', and 'normal' cooking should destroy salmonellae associated with surfaces of chicken (see below for effect of 'normal' cooking on salmonellae in egg yolks). However, it is accepted that some *S. enteritidis* type 4 may be derived from chicken meat, rather than eggs. It also follows that types of *S. enteritidis* other than type 4, and *S. typhimurium* may also contribute to egg-derived food poisoning (e.g. the House of Lords outbreak, May 1988).

It is often difficult to establish the cause of food poisoning because while the bacteria can be isolated from the patient, the food responsible may have been discarded. Eggs, and particualrly egg yolks, present a particular problem because individual eggs are unlabelled, making tracing the source difficult. The egg, particularly the yolk, may have been completely consumed, and there is often no yolk adhering to the inside of the shell after cracking.

Despite these problems, the association of *S. enteritidis* type 4 (and some other types) food poisoning and eggs is well established (Coyle *et al.*, 1988; Perales and Audicana, 1988; St Louis *et al.*, 1988). Epidemiological studies using food histories and matched controls have described four episodes in Wales (Coyle *et al.*, 1988). In other studies, culture evidence has been described. There have been suggestions by the Department of Health that as many as 1000 patients, in about 50 incidents, have been identified as suffering from *S. enteritidis* type 4 derived from eggs. These reflect many 'outbreaks' and, as has been stated above, most cases are necessarily sporadic.

Salmonella enteritidis type 4 in internal organs of laying birds and in eggs

Three independent reports have shown that *S. enteritidis* type 4 is present in a high percentage of the internal organs including ovaries and oviducts of laying birds (Hopper and Mawer, 1988; Lister, 1988; O'Brien, 1988). The findings of Hopper and Mawer (1988) are particularly significant. These authors state 'despite the evident presence of *S. enteritidis* type 4 infection in this flock of 60 000 birds, mortality and egg production have remained at expected levels'. These authors studied the livers, caeca, oviduct and ovule and found that 13 (26%) of 50 birds contained *S. enteritidis* type 4. This organism was also found in some of the egg contents, including yolks. Other workers have also found salmonellae, including *enteritidis* in the internal contents of shell eggs (Chapman *et al.*, 1988; Perales and Audicana, 1988).

The evidence linking *Salmonella enteritidis* type 4 food poisoning to eggs

Given the nature of the problems discussed above, the evidence for such a link is as strong as can be obtained, and is based partly on epidemiology, and partly on direct culture of the chicken's organs and eggs. However, it is the *only* plausible explanation for the dramatic rise in *S. enteritidis* food poisoning. It is, of course, often not possible to establish a certain link in sporadic cases for the reasons given, and it is not reasonable to expect such proof. A useful analogy may illustrate the problem. Epidemiological and experimental data have established that smoking causes lung cancer in general, although in the individual patient who succumbs from the disease, such causal proof is lacking.

It is the opinion of the author of this chapter that the origin of *S. enteritidis* type 4 (and probably some other species) food poisoning is usually hens' eggs.

The scale of the problem

As has been stated above, the numbers of reported isolations of *S. enteritidis* type 4 have been about 300 weekly over the last three years, with peaks in late summer. While it is possible that not all these are derived from eggs, and some other species may also contaminate eggs, the figure of 300 reports may be taken to be representative of egg-derived bacteria. The true incidence of the disease can only be arrived at by conjecture. Because of the passive nature of our surveillance system and the failure of patients to offer themselves for diagnosis, or to provide specimens, or of laboratories to isolate and report, the numbers officially presented are obviously less than the actual figures. Rowe (personal communication) believes that we should assume that only 1% of cases are reported; in contrast the Editorial (1988) considered 10% was nearer the mark. If the latter conservative figure is adopted, then the current incidence of *S. enteritidis* food poisoning is at a rate of 150 000 per year. Calculations of the numbers of eggs contaminated must consider the infectivity of the salmonella, another most uncertain factor. If, again conservatively, it is assumed that only 1 in 10 contaminated eggs cause food poisoning, and that 30 million eggs have been consumed daily, then 1 egg in 7000 is infected. If this is the case, then

Table 21.1. Positive salmonella 1988. Isolations in Leeds City Council residents.

Serotype	Phage type	Infected in UK	Infected abroad
Salmonella enteriditis	1	2	1
	4	244*	33
	5	–	1
	6	–	16
	8	53†	–
	13a	8	–
	23	1	–
	24	6	6
	unknown	16	2
	Total	330	59
Salmonella typhimurium	8	3	–
	10	8	2
	12	16	1
	15a	2	–
	18	2	–
	35	–	1
	44	1	–
	49	10	1
	96	–	2
	104	3	1
	110	2	–
	124	4	–
	136		–
	141	1	–
	194	1	–
	193	12	2
	208	4	–
	204a	2	1
	u285	–	1
	u296	1	–
	unknown	4	2
	Total	76	14

*Includes 41 in one prison and 38 in other outbreaks.
†Includes 32 in restaurant outbreaks.

surveys showing freedom of infection in numbers of eggs of this order or less have little meaning.

Of salmonellae isolated in the United Kingdom, about 20% are from patients returning ill from holidays abroad. The figures for Leeds are typical: of 595 isolates reported in 1988, 112 were from patients who acquired the infection in Spain, Portugal, France and Holland (Table 21.1).

The most significant finding is that among 277 isolates of the 'egg strain', *S.*

enteritidis phage type 4, 33 were acquired by patients when overseas. The isolation of bacteria that appear genotypically identical from several different sources suggests either a common factor or vector.

It is highly pertinent to this issue that the novelty of egg contamination results from transovarian (or vertical) spread and the breeding of élite flocks for much of Western European egg production is found in just two localities in France and Holland. The only plausible explanation for the scale of this problem is the contamination of these élite flocks with vertical transmission to grandparent and parents of the laying chickens throughout Western Europe. There are, however, no published data on this matter.

One significant factor in the relationship of salmonella to poultry, is that while the incidence of human infection rose dramatically during the 1980s, that of overt disease in laying chickens declined (see above). The only major natural host of *S. enteritidis* phage type 4 appears to be poultry, with man infected as a secondary host. Hopefully, further transfer from man should be uncommon. Other salmonella strains, e.g. *S. pullorum* tended to have produced much clinical disease in the chicken. The essential attribute of *S. enteritidis* type 4 is that its usual biological status in poultry is commensal, but in man it is parasitic. This would appear to be the key to understanding and solving the problem.

It is pertinent to consider the means by which this salmonella achieved its essential commensal status with the chicken. Simple evolutionary pressures probably apply, since it is not in the 'interest' of a bacterium or its host to result in an extreme parasitic relationship whereby the bacterium destroys its host.

Perhaps the pressure for this evolutionary change towards commensalism was derived from the repeated refeeding of the bacterium to the élite flocks? It is certainly known experimentally that chickens fed *S. enteritidis* type 4 can excrete the bacterium in their eggs.

However, the origins of the epidemic (a word used carefully) must remain speculative because of insufficient research, monitoring or reporting of the élite flocks. The problem will not be solved until élite flocks are established that are *S. enteritidis* free.

Laying flocks in the United Kingdom still from time to time excrete the *S. enteritidis* type 4, and presumably the conditions of rearing sometimes cause the commensal status of the bacterium to convert to parasitic.

The control of salmonella in eggs requires a three-pronged approach:

1. The development of a salmonella-free élite flock.
2. Rearing of egg layers in such a way as to discourage the return to a parasitic relationship between bacterium and egg (e.g. good husbandry, adequate space, control of diseases).
3. Food hygiene that discourages the multiplication of salmonellae were they to be present at all.

Even with these measures, not even the most optimistic observer will anticipate that salmonella infection from eggs will be reduced to zero.

The control of salmonella in poultry and meat is amenable to traditional hygiene measures, notably thorough cooking. Unfortunately, for cooking to be relied on to achieve control in eggs, every egg must be cooked until the yolk is completely hard throughout.

Bovine Spongiform Encephalopathy (BSE)

The first cases of the current epidemic of bovine spongiform encephalopathy (BSE) probably occurred in late 1985 and were confirmed during 1986 (Dealler and Lacey, 1990). By the summer of 1992 the number of notified cases had risen to a total of around 60 000. Infected dairy cows have been reported much more often than beef cattle because cows live typically to be 6–8 years old before slaughter and beef cattle live to typically between one and three years. The incubation period of this disease is thought to be around five years in cows and cattle. This explains why many cattle will be slaughtered before the infectious agent they are carrying has had time to produce the final clinical disease.

Perhaps the most significant feature of BSE has been its wide geographical spread, with farms involved in many parts of England, Wales, Scotland, Northern Ireland, Eire and even in the Isle of Man. However, a substantial number of herds have not reported any cases at all and it is by no means certain that all herds are infected. That the agent has appeared so widely and in many different breeds, and that the incubation period is several years indicate that the feed was the vector for the infection (Wilesmith *et al.*, 1988). This is extremely important because with this group of infectious diseases (transmissible spongiform encephalopathies or TSEs) there has been uncertainty over which is the predominant means of acquisition of the disease.

While the figure of 60 000 reported cases among perhaps 12 million cows and cattle reared annually is an indication of the extent of the problem, the real numbers must be even greater than this since many animals will be slaughtered before the time required for the clinical disease to occur. Moreover it is not known how many animals carry the infectious agent but never become clinically ill. The reason for this uncertainty is that no research has been done by the Government in this area.

The contaminated feed was used between 1981 and 1988, when feeding of rendered offal to ruminants was prohibited by the BSE Order. The feed had been derived from the addition of protein supplements from the rendering plants. This protein concentrate was also added to feeds for poultry, sheep and pigs. For this reason the dissemination of the infectious agent causing BSE has been extremely widespread, although the disease has so far been manifest primarily in dairy cows and to a lesser extent in beef cattle. One of the implications is that any attempt to eliminate the infection from the pastures and buildings of this country would be a truly monumental task. This does, of course, explain the attitude currently taken by the Government that plays down the scale of the problem, as it in reality knows that it cannot deal effectively with it because the resources needed are too great.

The disease BSE results from the damaged brain that shows histological spongiform change. The disease agent must, however, pass through a number of organs between its entry into the animal's body through the guts and the final multiplication in the brain (Table 21.2). Most of the work on the distribution of TSEs has been performed in sheep and the Government seems to rely on such evidence to identify

Table 21.2. Tissues containing infectious agent for TSEs[1].

Tissue	Sheep	Goat	Mouse	Mink	Man	Cow
Brain	+	+	+	+	+	+
Spinal cord	+	+	+	+	+	
Peripheral nerve	+	+	+			
Eye					+	
Adrenal		+				
Lymph node	+	+	+	+		
Tonsil		+				
Salivary gland		+	+	+		
Spleen	+	+	+	+	+	
Gut	+	+	+	+		
Liver		+	+	+		
Kidney			+	+		
Bladder				+		
Lung			+	+		
Thymus			+	+		
Uterus			+			
Blood/serum				+	+	
Bone marrow			+			
CSF		+			+	
Faeces				+		
Muscle		+		+		

+ Denotes presence of infectious agent.
[1]TSEs have not been detected in pancreas, heart, thyroid, testis, ovary, urine, saliva, milk or mammary gland from one or more species.

which organs are likely to contain the infectious agent in cows. Thus the spleen, the lymphatics, the thymus, the spinal cord and the brain have been designated high risk organs and are not permitted to enter the food chain. However, certain knowledge of the distribution of the infectious agent in cattle will take very many years, because the identification of each agent is still extremely lengthy. While tissue proteins associated with infection, such as prion protein (PrP) and scrapie-asssociated fibrils are amenable to detection through tissue staining and biochemical methods, the actual certain identification of the presence of infectious agent can still only be made by injection into test animals, such as mice. Most of these experiments take one year to be completed.

Indeed, one uncertainty about these experiments is that even if a group of mice is not infected by an injection of potentially infected material, it does not establish that there is no such infected material present, since it could mean that the material in the donor animal has no specific affinity for mice. One of the features of these agents is that once they have moved from one host to another the *subsequent* host range can change. Thus, in the natural experiment in the 1950s and 1960s in North America the infective agent from sheep scrapie passed to mink, and when the agents from sheep and mink were then studied for the host range, major differences were apparent. For example, sheep scrapie had no affinity for rhesus monkeys (consistent with a view that sheep scrapie does not affect man) but the agent from mink did produce disease in this host.

This means that it is extremely difficult to perform experiments relating to possible infectivity in man. Of course, it is not ethical to deliberately challenge members of the human population with such infectious agents (and it is gratifying to see that the Health and Safety Executive is treating all such tissues from infective animals as suspect and requires very careful containment facilities to be provided before any laboratory experiments are considered). Consider, for example, the possibility of a person already infected with this agent which may perhaps be present in the spleen. By this time the agent may have undergone some host modification as a result of its presence in the human body for some time. If that spleen is injected into mice, and then if infection is transferred, it establishes that the infection in the human spleen is now in a state capable of infecting mice, but it does not prove that the agent is certain to damage the human brain. A negative result in mice similarly cannot guarantee that the human brain will not be involved subsequently.

One disturbing feature of these agents is that there is no evidence of an immunological response to their presence. As a consequence of this, there is no serological test for their presence. This does raise the disturbing possibility that infectivity is related directly to the total amount of agent ingested over the years, in the same way that the adverse effects of smoking or exposure to ionizing radiation are cumulative. This means that the repetitive consumption of a small amount of agent in contaminated meat, due for example to its persistence in the nervous tissue, could well provide a danger comparable to ingestion of a large amount of brain material on a single occason. It is for this reason that advice not to consume beef products must stand. (Incidentally, these infectious agents have not been detected in infected mammals' milk.)

Nature of infectious agent

None of the theories relating to the identity of this agent can accommodate all the known facts. Either a substantial amount of experimental work is in error or has been misinterpreted, or the actual mode of infectivity, once it has been elucidated, will reveal new insights into fundamental cellular biochemistry. Certainly a number of tissue proteins as mentioned above have been associated with the infectious agent. It is, however, not known whether these are the actual infectious agents themselves or the result of an infectious agent reacting with existing tissue structures or indeed in various ways promoting the formation of these substances. What is known about the pathogenesis of TSEs is that there is an infectious basis in at least the majority of the animals involved. The agent must be capable of securing its self-replication and there must be a means for it to spread between tissues.

It is also known that the agent is extremely small and resistant to damage by physical agents, notably heat. Persistence of some infectivity can occur after heating to 360°C for one hour (Brown *et al.*, 1990). It is also resistant to most routinely used antiseptics and ionizing radiation. TSEs can survive for three years in soil. This suggests that it will be virtually impossible to eliminate these agents from our country for many decades if not centuries.

As far as the chemical composition of the agent, Prusiner suggests that it is made

Table 21.3. Mammals suffering 'naturally' from transmissible spongiform encephalopathies (TSEs).

Man (CJD, Kuru, Gerstmann–Straussler syndrome)
Sheep (Scrapie)
Goats
Deer
Cattle
Cats
Mink
Oryx
Antelope

of a tightly packed protein known as a prion, and contains very little else (Prusiner, 1989). It is known that protease K is able to inactivate the agent. Other workers, such as the Edinburgh group, suggested that there is a small piece of DNA involved that somehow enables the infective agent to replicate. However, the presence of any DNA is difficult to reconcile with its extraordinary physical toughness.

It is worth mentioning briefly the other human diseases which are due to TSEs (Table 21.3). Kuru was endemic in the Fore tribe in New Guinea in the first part of the century and was due to the cannibalistic practices, particularly among the women who tended to eat human brain tissues. Education has now stopped this practice. The women typically developed the disease 15–20 years after the first exposure. The main clinical features of Kuru were musculo-skeletal disorders rather than problems of intellect. Creutzfeldt–Jakob disease (CJD), which has been known since the early 1920s, typically occurs in middle-age, and while perhaps 10% of cases are familial, the great majority are sporadic and the main effect is seen as damage to the intellect, with less evidence of musculo-skeletal disorders. For this reason, many patients with CJD endure this terminal illness in mental institutions. In these patients, post-mortems are rarely performed, so many cases of CJD will be missed. It has been estimated that the real number of CJD cases in the United Kindom is of the order of 1500 to 9000 per year (Roberts, 1990) whereas the official notification figures are about one case per million per year.

The important and unsolved questions are as follows:

1. Is BSE capable of spreading to man?
2. What is the origin of CJD?

The most pessimistic hypothesis gives the same answer to these questions, namely that CJD is due to the consumption some 20–30 years ago of the BSE agent. This would be before it had become amplified to the extent it has now. Rare cases or subclinical infection in cows and cattle could well have been missed (Dealler and Lacey, 1990).

Generally about 60% of the transmissible spongiform encephalopathies are capable of experimental transfer from one mammal to another. Some authorities believe that this understates the potential infectivity because some test mammals may not live long enough to express the disease. Because each agent is modified on its subsequent host range by passage in the most recently infected animal, it is not particularly important whether or not BSE came from sheep scrapie because the crucial question is whether the infectious agent is able to infect the human host. Many cats, zoo animals such as antelopes and oryx, and, experimentally, mice and pigs, have seemingly been infected with the BSE agent. The ability of the pig to sustain infection with BSE must provide pessimistic implications for man, since pig and human tissues are similar in many ways.

The question of the origin of the sporadic cases of CJD suggests that the dairy cow is the most likely source because it is the main mammal that is eaten regularly when it is elderly. Other mammals are a less probable source. There appears to be no association between sheep scrapie and CJD (Southwood, 1989). Consumption of pig products usually occurs when the animal is only five or six months old, and CJD is most prevalent in Jewish populations which do not eat pork.

Knowledge from accidents whereby human growth hormone and various transplants have been transferred to patients, and the knowledge of CJD and Kuru suggest that the incubation period for TSEs in man will rarely be less than 10 years when acquired by the oral route, and could well be 40 years or more. For this reason there will be no reassurance that man is not liable to be infected by these agents from BSE-infected meat until 1996 at the earliest and probably not until the first part of the next century.

Meanwhile, there must be a massive investment into research of the disease, and a search for chemotherapeutic agents that might modify the illness.

The dilemma our politicians face is that to undertake such a programme would admit the gravity of the problem, which itself would have a devastating effect on farming and other aspects of our community. For this reason the Government has relied on hoping that the effect on man will be minor and delayed and has attempted to reassure members of the public by stating that BSE has the same infectivity as sheep scrapie.

Government action

The Southwood Report

In April 1988, Sir Richard Southwood was asked to set up a working party to advise on BSE. In June 1988, Southwood recommended that carcasses of affected animals should be destroyed. This was never implemented adequately in that most carcasses were dumped in land-fill sites, a most unsatisfactory procedure because of the extreme hardiness of the infective agent (Brown and Gajdusek, 1991). Of course, such carcasses should be incinerated. Unfortunately, there were inadequate facilities, and little effort has been made to build new incinerators.

In November 1988, Southwood recommended that the ban on the use of ruminant-based protein for ruminants should continue indefinitely. This was an extraordinarily narrow recommendation as it permitted the continued feeding of the remains of rendered pigs and poultry to the same species. What is more, Southwood maintained that they 'could not be wholly sure that rendering as currently practised would eliminate the agents causing BSE'. The extreme toughness of these agents is responsible for these doubts. Surely the logical approach would have been to curtail the whole unsafe system of feeding all animals with pooled rendered protein? This was the first of many examples where the problems associated with BSE were recognized but the action taken was incomplete, inadequate or unenforceable.

The full text of the Southwood report was published in February 1989 (Southwood, 1989). The key conclusion (page 21) was

> From present evidence, it is likely that cattle will prove to be a 'dead-end host' for the disease agent and most unlikely that BSE will have any implications for human health. Nevertheless, if our assessments of these likelihoods are incorrect, the implications would be extremely serious.

The first statement is evidently tautological in that if cattle are indeed a dead-end host for the infectious agent, this means it will not transfer to any other host, including man. However, the real worry with this statement is that by February 1989, insufficient time had elapsed to know whether BSE might spread spontaneously or experimentally to other mammals. *There was not any evidence* as to whether cattle would prove a dead-end host. Since then, BSE has now been successfully transmitted to mice and a pig by experiment (Dawson *et al.*, 1990) and also has probably infected cats and zoo animals such as oryx and antelopes and their offspring. So, we now know that cattle are not a 'dead-end' host, and we agree fully with Southwood's first conditional comment, that is now an operational statement that 'the implications would be extremely serious'.

This has largely been ignored by much of the UK's Ministry of Agriculture, Fisheries and Food (MAFF). In particular, ministers have continued to quote that it is 'most unlikely that BSE will have any implications for human health', out of context.

The final and most decisive recommendation of Southwood was 'that this [i.e. the rendering plant] method of disposing of animal waste should be changed so as to eliminate these novel pathways for pathogens'.

The Tyrell Report

Southwood recommended the establishment of a second committee to look at certain aspects of BSE. The chairman was Dr David Tyrell, a retired virologist, and members did include those with particular expertise in TSEs. Their report was presented for publication in June 1989, but for some inexplicable reason the report was not published until January 1990 (Tyrell, 1989). This caused a substantial delay of several months in the prohibition of certain infected tissues entering the food chain. Moreover, these tissues were never removed from items already available at retail outlets – and presumably some dried, canned and frozen foods that contain substantial amounts of beef offal are still being purchased and consumed.

The most disturbing paragraph in the Tyrell Report is as follows:

> Many extensive epidemiological studies around the world have contributed to the current consensus view that scrapie is not causally linked with Creutzfeldt–Jakob Disease (CJD). It is urgent that the same reassurance can be given about the lack of effect of BSE on human health. The best way of doing this is to monitor all UK cases of CJD over the next two decades.

This section begins with an accurate statement. It is true that there is no evidence linking sheep scrapie with CJD (there is also no evidence linking BSE with scrapie). It is also true that it would be desirable for reassurance to be given that BSE will not infect man. But surely this intent should only be considered after the reassuring evidence has been obtained. This illustrates much of the Government's attitude to food safety: the prime intent is to reassure the public; that is, to maintain the commercial *status quo*. Very much a secondary consideration, is to take any curative action, particularly if it is costly.

The last sentence of this paragraph of the Tyrell Report quite amazingly states that such reassurance is to be achieved by seeing how many more people die of CJD in the next 20 years. Already CJD accounts for about 1% of deaths (Roberts, 1990) and was probably acquired from eating beef many years ago (see above). So the experiment has started to see what happens! Most of the members of the Tyrell committee will not be expected to survive 20 years anyway!

Recommendations for the Elimination of BSE and Related Organisms

1. *Rendering (processing) plants.* There is a clear case for the total prohibition of the use of the bone meal fraction from rendered animal and poultry remains in those species. Rather, this product should be used as a fertilizer for cereal crops. This would interrupt the potentially dangerous cycle animal → remains → animal. Protein supplements for animal and poultry feeds should be obtained from fish or vegetable sources, e.g. soya.

It may be difficult to ensure that such a recommendation be enforced. One way might be to put all the rendering plants under direct government control.

There will also need to be legislation and surveillance procedures that ensure imported products comply with these stipulations.

2. *Replacement of BSE-infected herds with BSE-free stock.* Since it may be 1992 or beyond before the frequency of vertical transmission of the infectious agent will be capable of definition and since there is no way of identifying symptomless but infected animals, there is pressure to develop BSE-free herds now. BSE-free herds would presumably have to be kept on new pastures and in thoroughly cleaned buildings. These detailed proposals should eliminate BSE, but will be extremely costly to implement.

(a) There should be an intent to rear no further calves from infected herds.
(b) Any calf born in the near future from a cow belonging to an infected herd should be destroyed.
(c) Milk-producing cows at the end of their useful lactation should be slaughtered, and incinerated. The carcass should not enter any part of the food chain.
(d) Beef cattle should be slaughtered as soon as possible – preferably under 18 months of age. Careful removal of meat from the carcass might produce some relatively safe product, say, for people over 50. The brain, spinal cord and body cavities containing 'high risk' material should not be manipulated. After

removal of meat, the remainder of the carcass should be incinerated.

(e) A system of introduction and documentation of BSE-free stock will be required. Every BSE-free animal should be required to be identified according to a pedigree indicating source, and subsequent movements.

(f) Urgent research must be done to quantify the extent to which herds are infected, but not showing signs of disease. Subsequently, any infected herds should be managed by procedures as outlined above.

Intensive Farming: Implications for Polyunsaturated Fats

There is now substantial evidence that the amount and type of fatty acids in animals, birds and fish is determined by the nature of their diet, and indeed by other rearing conditions. Because of the conversions of many farming systems from 'free-range' to intensive over the last few decades, the fatty acid profile of our food will be altered as the intensively reared animals and fish will be fed novel diets. These changes will not be expected to be confined to macroscopic fat. It follows from this that recommended diets may not now reflect the anticipated fatty acid composition, particularly that of *n*-3 polyunsaturated fatty acids (PUFAs).

Farmed fish: the loss of essential fatty acids

Fish are an excellent source of omega-3 (*n*-3) PUFAs including linolenic acid (LN). The major *n*-3 PUFAs of phytoplankton, the fundamental diet of marine life, are eicosapentaenoic acid (EPA) and docosahexaenoic acid (DHA) and most wild marine life will eventually become enriched with these (Denton and Lacey, 1991). EPA and DHA, in much smaller amounts, may also be acquired from endogenous synthesis of ingested LN. The major PUFA of typical Western diets is linoleic acid (LA), an *n*-6 PUFA, found in many vegetable and seed oils. In humans, LA by further metabolic conversion is desaturated and elongated to arachidonic acid (AA).

Interest in the role of *n*-3 PUFAs in human disease was first aroused by the observation that Eskimos, with diets consisting mainly of marine life, had a low incidence of atherosclerotic disease in comparison to those consuming typical Western diets (Kromann and Green, 1980). Further studies have supported the view that diets rich in *n*-3 PUFAs are effective in reducing the incidence of subsequent coronary artery disease. Consumption of fish oils in man leads to the accumulation of EPA and DHA in plasma and platelet and tissue phospholipids at the expense of AA. This change appers to have a number of biochemical and physiological effects. There is a reduction in plasma triglyceride levels and effects on total plasma cholesterol and that of high density lipoprotein cholesterol. There is a marked reduction in platelet aggregation associated with a lengthened bleeding and an increase in erythrocyte deformability (see Denton and Lacey, 1991).

As fishing grounds have become depleted and the demand for fish continues, there has been an increase in the production of farm-reared fish for human consumption. Such fish will be fed artificial feeds. The practice of feeding rendered offal to farmed fish has been claimed to be uncommon (Richards, 1983); ideally most commercial feed should consist of a mixture of marine and vegetable material. However, we believe that much of the diet of farmed fish in the United Kingdom may not comprise these ingredients exclusively.

As with mammals the fatty acid content of fish can be substantially altered by making changes to their diet. Commercial fish feeds appear to contain fewer *n*-3 PUFAs and more *n*-6 PUFAs than natural fish diets (Agren *et al.*, 1987). There is a reduction of *n*-3 PUFA content of 15–33% and an increase in *n*-6 PUFA content of 50–100% in comparison to wild counterparts. Farm-reared fish also have a total lipid content expressed per gram of flesh that is up to 50% higher (Agren *et al.*, 1987).

The effect that these changes may have on the desirable biochemical and physiological properties of these fish, when consumed, has not been studied in clinical trials. However, it is known that coronary artery atherosclerosis runs a more rapid course in farmed salmon in comparison to wild salmon (Saunders and Farrell, 1988). It must be concluded that fish oils derived from farmed fish may not produce the beneficial effects in man seen with those from wild fish. Further research is needed in this area.

Chickens and eggs

In the United Kingdom, chicken meat comprises the greatest single source of meat and about 30 million eggs are consumed daily. One factor responsible for the popularity of chicken meat was the Royal College of Physicians of London and the British Cardiac Society's advice in 1976 (Report, 1976) that chicken meat, being low in fat, was preferable to red meat. However, Crawford and Marsh (1989) have pointed out that at the end of the last century the carcass fat on chicken was in the order of 2.4%, but it had risen to 22% by the early 1980s – evidently much greater than that for some red meats. The amount and nature of fatty acids in chicken meat is influenced by the diet and rearing conditions. Such variables include the degree of unsaturation of the dietary fatty acids, the proportion of *cis* to *trans* isomers, the amount of biotin, and the temperature of rearing.

The relationship between dietary fatty acids and their concentration in tissues may not correlate closely because there is evidence of synthesis *de novo* of some fatty acids (even unsaturated ones) in chicken tissues (Phetteplace and Watkins, 1989). However, the following dietary ingredients rich in omega-3 fatty acids, particularly alpha-LN, appear to result in enhanced deposition of such fatty acids in chicken tissues: corn oil and partially hydrogenated soyabean oil (Rogel and Watkins, 1987), linseed, menhaden and natural soyabean oil, canola, flax oil (Nwokolo and Sim, 1989) and redfish oil (Hulan *et al.*, 1988). In these experiments control diets were rich in barley, wheat, spent restaurant grease or chicken fat itself.

The composition of the fatty acids in egg yolks is also markedly influenced by the diet of the laying chicken. Nwokola and Sim (1989) found that the addition of 20 g canola seeds (a type of oil seed rape) daily to the diet increased the percentage of LN to total fat in egg yolk fatty acids from 0.49 to 1.63. Even more striking were the findings of Simopoulos and Salem (1989) who compared the effect of rearing laying chickens on a diet rich in the plant purslane with a 'standard' diet that yielded 'supermarket' eggs. Purslane is claimed to be the richest source of *n*-3 fatty acids of any green, leafy vegetable (Simopoulos and Salem, 1986). The average total *n*-3 fatty acid content of the 'supermarket' eggs was 1.73 mg g^{-1} and that in the Greek purslane-fed eggs was 17.66 mg g^{-1}. The corresponding figures for alpha-LN were 0.52 and 6.9.

In general, the chicken diets rich in *n*-3 fatty acids increased the proportion of *n*-3 fatty acids by two- to fivefold (see above). The relationship of the nature of the diet to the commercially decisive parameter of weight gain is not predictable. However, the addition of redfish meal reduced weight gain (Hulan *et al.*, 1988) and in one report, Al-Athari and Watkins (1988) found that chickens fed either hydrogenated soyabean oil or saturated fat had lower gains than those fed natural soyabean oil or spent restaurant grease. But in another series of experiments, chickens fed spent restaurant grease had a lower weight gain than those fed hydrogenated soyabean oil (Rogel and Watkins, 1987).

Biotin is an essential co-factor for the formation of malonyl-CoA, itself required for *de novo* fatty acid synthesis and elongation of long-chain PUFAs (Watkins, 1988). Watkins and Kratzer (1987) therefore undertook experiments to study the effect of biotin deficiency: major changes were noted, particularly in the elongation of LN. The detailed changes are complex, but the rearing of chickens under conditions of limited dietary biotin could adversely affect the fatty acid content of chicken meat produced for human consumption.

While in mammals, including man, the 'unnatural' *trans* isomers of oleic acid (OA) and LA are thought to be utilizable for energy through β-oxidation, there is strong experimental evidence that such isomers are incapable of adequate metabolic conversion through elongation and further desaturation in rats (Al-Athari and Watkins, 1988). These authors have studied the effect of feeding chicks diets supplemented with soyabean oil, chemically defined saturated fat, hydrogenated soyabean oil (chemical hydrogenation yields both *cis* and *trans* derivatives) or spent restaurant grease for three weeks. The *trans* isomer of OA was incorporated into many chick tissues after feeding hydrogenated soyabean. In these tissues were reduced levels of both tissue LA and the metabolite, AA. These chicks also exhibited reduced weight relative to the other diets. The addition of spent restaurant grease also increased *trans* fatty acids in liver microsomes.

The synthesis of polyunsaturated fatty acids in life forms in general is favoured by relatively low temperatures (Crawford and Marsh, 1989). This has been studied in chickens by Sonaiya (1988). In these experiments temperatures were either maintained constantly at 30°C (high temperature) or fluctuated between 30°C over 18 h and 21°C for a 6 h period (low temperature). When the depot fats were analysed after 54 days, the proportions of OA, LA and LN were all greater with the lower temperature than with the higher. Interestingly, at the low temperature the proportion of LA to total fatty acids was higher in females than males. There is clearly a need for further research in this area.

Cattle

In 1968, Crawford published the observation that the nature of fatty acids varied dramatically between animals reared under free-living or domestic conditions. Notably the percentage fat content of total weight of wild buffalo was 2.8% whereas that of domestic cattle (i.e. butcher's meat) was 25.0 (Crawford, 1968). Within the total fat, the proportions of polyunsaturated to mono- and saturated were also variable: domestic beef contained just 2.2% of its fatty acid as polyunsaturated, and buffalo between 10% and 30% polyunsaturated according to herd. The lower figure was from animals confined to open parkland and the higher from woodland. In recent years, the effect of different cattle diets on milk and meat have been studied under experimental conditions, and generally the variations are much less than with chickens. The reason for this is mainly the bacterial activity within the rumen that tends to convert any dietary fat towards saturation. White *et al.* (1987) showed that following a diet enriched by sunflower seeds, the rumen of Holstein steers decreased the amount of LA and increased that of the saturated acids, stearic and palmitic. Work in France has shown that this change in the rumen was due directly to hydrogenation (Doreau *et al.*, 1989). Thus when lactating cows are fed diets high in unsaturated oils, the effects on milk composition may be small. Another reason for this could be the decrease of fatty acid synthesis in the udder (Banks *et al.*, 1990). However, some effects do occur since these authors showed that the composition of 18:1 *cis* (i.e. oleic) acid in milk increased from 118 to 177 mmol mol^{-1} of milk fats with dietary supplement with soya oil.

To a small degree differences in dietary lipid can also influence the amount and species of fatty acids in beef. Thus in steers fed a diet supplemented with canola seeds there was a resultant increase in both stearic and alpha-LN in their adipose tissue.

This work has been confirmed by the finding that cows eating feed supplemented with oats had an increased proportion of OA in their milk at the expense of some saturated fatty acids (Martin and Thomas, 1988). However, there were few changes in either LA or LN and it remains questionable at present whether such dietary manipulations do produce any useful change in fatty acid profile. Indeed some supplements might be positively deleterious. Kankare *et al.* (1989) found that the addition of calcium salts of fatty acids to the feed of dairy cows increased both *trans* isomers and the proportion of saturated acids in milk, largely at the expense of OA.

A series of papers from Japan has explored the relationship between diet and temperature and the nature of fatty acids in Japanese Black or Shorthorn steers. During fattening, the animals were fed additional 'concentrates' (details not provided) and the proportion of unsaturated fatty acids rose by about 10%. Interestingly the main increase in palmitoleic (*c*.16.1) and oleic occurred in the cool season (Ishada *et al.*, 1988).

However, more emphatic changes were sometimes seen with diets based on barley or corn. Those fed corn diet showed an increase in LA as a percentage of total fatty acids of around 10–20% (Mitsuhashi *et al.*, 1988). The effect of feeding 'concentrates' has been shown to produce a detrimental effect on fatty acids. The term 'concentrates' presumably refers to the protein rich meal derived from the rendering plants. In these experiments, the addition of 'concentrates' caused a reduction in omega-3 fatty acid (Tsuneishi *et al.*, 1989).

The genetic variations of cattle are much greater than those of intensely reared chickens, and subsequently some caution must be exercised before assuming that such dietary effects would be anticipated with all species.

Pigs

In their natural state pigs roam and forage in forests. In medieval times, man controlled wild pigs and fed them mainly acorns in winter. These are rich in LA and pigs fed in this way would be expected to have a ratio of polyunsaturated to saturated fatty acids of 2:1. In contrast, the ratio for the modern pig is reversed to 0.2:1 (Crawford and Marsh, 1989).

The relationship between the fatty acid content of the pig's feed and that of the animal's carcass is well understood. For example, Rammell *et al.* (1988) studied the effect of a chemically defined diet on the various compartments of the pig. The proportion of polyunsaturated to total fatty acids in the feed was 38%, in serum 36%, and in liver 39%.

Feeding pigs supplements of sunflower oil produces softer and oilier carcass fat and does dramatically increase (by about 75%) the proportion of polyunsaturated to saturated fatty acid ratio in both muscle and adipose tissue (Rhee *et al.*, 1988). Indeed the flavour of the meat, in addition to biochemical parameters, is influenced by the nature of the pig's diet. Thus while canola oil is effective in increasing the carcass content of PUFAs, it produces meat with a greater tendency to 'off flavours' (Miller *et al.*, 1990). These authors fed pigs a control diet of corn and soyabean meal and test diets containing 10% animal fat, safflower oil, sunflower oil or canola oil. The pigs were slaughtered after 90 days at 100 kg weight. There were no obvious differences in growth with each regime, except that there was somewhat more fat with each of the fat supplements. All the oils increased the proportion of OA and reduced that of palmitic.

Discussion

It has been assumed here that either the parent *n*-6 18:2 LA or *n*-3 18:3 LNs, or some of their metabolic derivatives are essential or at least desirable for human health. Some immediate biochemical effects are required from these substances; in addition, there is good evidence for their value in preventing future vascular disease. Current uncertainties over these essential fatty acids concern mainly the amount of their desirable intake in the long term. Moreover, it is also not known whether some of their value results from the possible associated reduction of intake of saturated acids. Since polyunsaturated acids are capable of ready oxidation, it is unlikely that a relatively high intake, within a total ideal calorific content of the diet, will be harmful.

If this argument is correct, the continued high incidence of vascular disease, particularly in the north of Britain, could in part be due to loss of PUFAs in food as a result of modern intensive methods of farming.

Thus, as a result of the nature of the feed, intensely reared fish (i.e. farmed), poultry and eggs and pigs are all expected to contain a lower proportion of PUFAs/total fat compared to their wild counterparts. For these species, the feed is in part derived from meal generated by the rendering plants. Such material must be expected to be low in PUFAs, as a result of the heat treatment and prolonged storage. Furthermore, fish, poultry and pigs reared under intensive conditions do not have access to natural vegetation rich in PUFAs. It would seem reasonable for products from this range of livestock to be regularly analysed and labelled as to their nutritional content. This is particularly important for poultry where the belief that 'white meat' is preferable to 'red meat' as far as fatty acids are concerned, lingers on. In addition, because of the many other differences between farmed and wild fish, these two types should be clearly marked.

One reason for the high proportion of saturated fat in pig products is that the food industry, and indeed the consumer, can consider oily meat difficult to handle, and there may be generation of unpleasant tastes and smells during storage. Much supermarket bacon is packaged and preserved with sodium nitrite and its shelf-life could well be reduced if the proportion of PUFAs in it were to be increased.

Generally, under conditions of intensive rearing, the proportions of PUFAs in pigs and poultry could well be increased if the current feeds were reformulated to produce this effect. The prospects for changing the fatty acid content of beef and dairy products are actively being researched at present, e.g. by Klusmeyer *et al.* (1991).

References

Agren, J., Muje, P., Hanninen, O., Herranen, J. and Penttila, I. (1987) Seasonal variations of lipid fatty acids of boreal freshwater fish species. *Comparative Biochemistry and Physiology (B)* **88**, 905–909.

Al-Athari, A.K. and Watkins, B.A. (1988) Distribution of *trans* and *cis* 18 : 1 fatty acid isomers in chicks fed different diets. *Poultry Science* **67**, 778–786.

Banks, W., Clapperton, J.L. and Girdler, A.K. (1990) Effect of dietary unsaturated fatty acids in various farms on the *de novo* synthesis of fatty acids in the bovine mammary gland. *Journal of Dietary Research* **57**, 179–185.

Brown, P. and Gajdusek, D.G. (1991) Survival of scrapie virus after 3 years' interment. *The Lancet* **337**, 269–270.

Brown, P., Liberski, P.P., Wolff, A. and Gajdusek, D.L. (1990) Resistance of scrapie infectivity to steam autoclaving after formaldehyde fixation and limited survival after ashing at 360°: practical and theoretical implications. *Journal of Infectious Diseases* **115**, 393–399.

Chapman, P.A., Rhodes, P. and Rylands, W. (1988) *Salmonella typhimurium* phage type 141 infections in Sheffield during 1984 and 1985: association with hens' eggs. *Epidemiology and Infection* **101**, 75–82.

Communicable Disease Reports (CDR) (1982–1988) PHLS Communicable Disease Surveillance Centre, 61 Colindale Avenue, London NW9 5EQ.

Coyle, E.F., Ribeiro, C.D., Howard, A.J., Palmer, S.R., Jones, H.I., Ward, L. and Rowe, B. (1988) *Salmonella enteritidis* phage type 4 infection in association with hens' eggs. *Lancet* **2**, 1295.

Crawford, M. and Marsh, D. (1989) *The Driving Force. Food evolution and the future.* Heinemann, London.

Crawford, M.A. (1968) Fatty-acid ratios in free-living and domestic animals. *Lancet* **1**, 1329–1332.

Dawson, M., Wells, G.A.H., Parker, B.N.J. and Scott, A.C. (1990) Primary parenteral transmission of bovine spongiform encephalopathy to the pig. *The Veterinary Record*, 338.

Dealler, S.F. and Lacey, R.W. (1990) Transmissible spongiform encephalopathies: the threat of BSE to man. *Food Microbiology* **7**, 253–279.

Denton, M. and Lacey, R.W. (1991) Intensive farming and food processing: Implications for polyunsaturated fats. *Journal of Nutritional Medicine* **2**, 179–189.

Doreau, M., Batisse, V. and Bauchart, D. (1989) Appreciation de l'hydrogénation des acides gras alimentaires dans le rumen de la vache: étude méthodologique préliminaire. *Annals Zootech* **38**, 139–144.

Editorial (1988) *Salmonella enteritidis* phage type 4: Chicken and Egg. *Lancet* **2**, 720–722.

Hook, E.W. (1961) Salmonellosis: certain factors influencing the interaction of *Salmonella* and the human host. *Bulletin of the New York Academy of Medicine* **37**, 499–512.

Hopper, S.A. and Mawer, S. (1988) *Salmonella enteritidis* in a commercial laying flock. *Veterinary Record* **123**, 351.

Hulan, H.W., Ackman, R.G., Ratnayake, W.M.N. and Proudfoot, F.G. (1988) Omega-3 fatty acid levels and performance of broiler chickens fed red fish meal or red fish oil. *Canadian Journal of Animal Science* **68**, 533–547.

Humphrey, T.J., Mead, G.C. and Rowe, B. (1988) Poultry meat as a source of human salmonellosis in England and Wales. *Epidemiology and Infection* **100**, 175–84.

Ishida, M., Takeda, T., Saito, T., Kano, Y., Matsumoto, T. and Takahashi, I. (1988) Fatty acid composition of the subcutaneous fat in Japanese Black steers during fattening. *Japanese Journal of Zootechnical Society* **59**, 496–501.

Jawetz, E., Melnick, J.L. and Adelberg, E.A. (1984) *Review of Medical Microbiology*, 16th edition. Lange, California.

Kankare, V., Antila, V., Vaatainen, H., Miettinen, H. and Setala, J. (1989) The effect of calcium salts of fatty acids added to the feed of dairy cows on the fatty acid composition of mik fat. *Finnish Journal of Dairy Science* **1**, 1–9.

Klusmeyer, T.H., Lynch, G.L., Clark, J.H. and Nelson, D.R. (1991) Effects of calcium salts of fatty acids and protein source on ruminal fermentation and nutrient flow to duodenum of cows. *Journal of Dairy Science* **74**, 2206–2219.

Kromann, N. and Green, A. (1980) Epidemiological studies in the Upernavik district, Greenland. *Acta Medica Scandanavica* **208**, 401–406.

Lister, S.A. (1988) *Salmonella enteritidis* infection in broilers and broiler breeders. *Veterinary Record* **123**, 350.

Martin, P.A. and Thomas, P.C. (1988) Dietary manipulation of the yield and composition of milk: effects of dietary inclusions of barley and oats in untreated or formaldehyde-treated forms on milk fatty acid composition. *Journal of Science, Food and Agriculture* **43**, 145–154.

Miller, M.F., Shackleford, S.D., Hayden, K.D. and Reagan, J.O. (1990) Determination of the alteration in fatty acid profiles, sensory characteristics and carcass traits of swine fed elevated levels of monounsaturated fats in the diet. *Journal of Animal Science* **68**, 1624–1631.

Mitsuhashi, T., Kitamura, Y., Mitsumoto, M., Yamashita, Y. and Ozawa, S. (1988) Effect of barley and corn feeding on fatty acid composition and color values of adipose tissue from Japanese black steers. *Bulletin of the Chugoku National Agricultural Experimental Station* **3**, 71–79.

Nwokolo, E. and Sim, J. (1989) Influence of canola seed consumption on omega-3 fatty acid concentration in plasma, tissues and egg yolk of hens. *Agriculture and Forestry Bulletin* **68**, 52–54.

O'Brien, J.D.P. (1988) *Salmonella enteritidis* infection in broiler chickens. *Veterinary Record* **122**, 214.

Perales, I. and Audicana, A. (1988) *Salmonella enteritidis* and eggs. *Lancet* **2**, 1133.

Phetteplace, H.W. and Watkins, B.A. (1989) Effects of various n-3 lipid sources on fatty compositions in chicken tissues. *Journal of Food and Comparative Analysis* **2**, 104–117.

Prusiner, S.B. (1989) Scrapie prions. *Annual Review of Microbiology* **43**, 345–374.

Rammel, C.G., Pearson, A.B. and Bentley, G.R. (1988) Vitamin E, selenium and polyunsaturated fatty acids in clinically normal growers (9–16 week pigs) and their feed: their relationship to the vitamin E/selenium deficiency ('VESD') syndrome. *New Zealand Veterinary Journal* **36**, 133–137.

Report (1976) Royal College of Physicians and British Cardiac Society. Prevention of coronary heart disease. *Journal of the Royal College of Physicians* 213–275.

Rhee, K.S., Davidson, T.L., Knabe, D.A. Cross, H.R., Ziprin, Y.A. and Rhee, K.C. (1988) Effect of dietary high-oleic sunflower oil on pork carcass traits and fatty acid profiles of raw tissues. *Meat Science* **24**, 249–260.

Richards, R.H. (1983) Diseases of farmed fish: salmonids. *Veterinary Record* **112**, 124–126.

Roberts, G.W. (1990) Memorandum submitted to the House of Commons Committee on Agriculture 10 July. HMSO.

Rogel, A.M. and Watkins, B.A. (1987) Liver subcellular fatty acid profiles of chicks fed diets containing hydrogenated fats and varying LA levels. *Lipids* **22**, 637–642.

Saunders, R.L. and Farrell, A.P. (1988) Coronary arteriosclerosis in Atlantic salmon. No regression of lesions after spawning. *Arteriosclerosis* **8**, 378–384.

Simopoulos, A.P. and Salem, N. (1986) Purslane: a terrestial source of omega-3 fatty acids. *New England Journal of Medicine* **315**, 833.

Simopoulos, A.P. and Salem, N. (1989) N-3 fatty acids in eggs from range-fed Greek chickens. *New England Journal of Medicine* **11**, 1412.

Sonaiya, E.B. (1988) Fatty acid composition of broiler abdominal fat as influenced by temperature, age and sex. *British Poultry Science* **29**, 589–595.

Southwood, R. (1989) Report of the Working Party on Bovine Spongiform Encephalopathy, MAFF.

St Louis, M.E., Dale, L.M., Potter, M.E., Thomas, M.D., Guzewich, J.J., Tauxe, R.V. and Blake, P.A. (1988) The emergence of Grade A eggs as a major source of *Salmonella enteritidis* infections. *Journal of the American Medical Association* **259**, 2103–2107.

Tsuneishi, E., Takimoto, Y., Mishimurak, Watanabe, A. and Takeda, H. (1989) Fatty acid composition of depot fat and muscle from grazing cattle. *Japanese Journal of Zootechnical Society* **59**, 614–618.

Tyrell, D.A.J. (1989) *Consultative Committee on Research into Spongiform Encephalopathies Interim Report.* MAFF and Department of Health, HMSO.

Walkland, C. (1991) Human vaccine cuts abortion. *Pig Farming*, 5.

Watkins, B.A. (1988) Influences of biotin deficiency and dietary *trans*-fatty acids on tissue lipids in chickens. *British Journal of Nutrition* **67**, 778–786.

Watkins, B.A. and Kratzer, F.H. (1987) Dietary biotin effects on polyunsaturated fatty acids in chick tissue lipids and prostaglandin E_2 in freeze-clamped hearts. *Poultry Science* **66**, 1818–1828.

White, B.G., Ingalls, J.R., Sharma, H.R. and McKirdy, J.A. (1987) The effect of whole sunflower seeds on the flow of fat and fatty acids through the gastrointestinal tract of cannulated holstein steers. *Canadian Journal of Animal Science* **67**, 447–459.

Wensvoort, G., Terpstra, C., Pol, J. (1991) Letter – 'Blue ear' disease in pigs. *Veterinary Record* 15 June, 574.

Wilesmith, J.W., Wells, G.A., Cranwell, M.P. and Ryan, J.B. (1988) Bovine Spongiform Encephalopathy: Epidemiological Studies. *Veterinary Record* **123**(25).

22 Role in Landscape and Wildlife Conservation

M.B. Alcock

School of Agricultural and Forest Sciences, University of Wales, Bangor, UK

Summary

The importance of landscape and wildlife conservation on agricultural land and particularly in livestock areas is emphasized.

Account is given of the 12.8 million hectares of grassland and rough grazing in the UK and the decline that has occurred in wildlife habitats. The causes of this decline are discussed. The current status of agriculture together with future predictions of yield increases suggest that future food supplies could be met on a much reduced land area. The latter suggests that more attention could be paid to enhancing wildlife resources on the surplus areas. The potential roles of livestock and livestock systems in conservation are outlined. Particular attention is paid to the role of livestock species in managing vegetation. Examples of specific habitat management are given including calcicolous grasslands, hay meadows and heather moorland.

The development of relevant environmental policies is reviewed and suggestions are made for their future improvement.

Introduction

Of the 23 million hectares of land in Britain, 71% are in agricultural use of which almost three-quarters is grassland or grazing land of one sort of another. Collectively, rough and hill grazings, permanent grassland and short-lived grass completely dominate the landscape of the north and west of the country. The economic significance of this is reflected in the animal output from these areas, which in 1987 accounted for 56% of all agricultural output (Owen, 1987). Conversely only about 1% of the total land area is solely devoted to wildlife conservation, although in total some 15–20% of the land area in England and Wales is officially designated for its wildlife and landscape values. Much of this is inevitably situated on farmed land, primarily in livestock areas. Dominant designations include National Parks, Areas

Table 22.1. The conservation estate in England and Wales.

Designation[1]	Area (ha)
National Parks	2 360 000
Areas of Outstanding Natural Beauty	1 708 500
Environmentally Sensitive Areas[2]	110 495
National Nature Reserves	133 840
Sites of Special Scientific Interest[3]	1 364 404
Local Nature Reserves	7 919
County Trust Reserves	32 410
Royal Society for the Protection of Birds Reserves	34 484
National Trust	262 783
Country Parks	21 296

Source: Potter *et al.*, 1991.
[1]Not mutually exclusive categories.
[2]Land enrolled in ESA agreements, June 1989.
[3]As at 31 March 1987.

of Outstanding Natural Beauty, Sites of Special Scientific Interest and Environmentally Sensitive Areas – Table 22.1 (Potter *et al.*, 1991).

While the link between livestock farming systems and landscape is dramatically obvious, the link with wildlife needs further examination. Agriculture has changed significantly since the Second World War and more particularly in the last two decades. There has been a process of expansion and intensification which has resulted, and continues to result, in considerable loss of wildlife habitat, either by the elimination of areas of semi-natural vegetation or by the transformation of areas to uniform monoculture. While this has been most noticeable in arable areas, notably in the south and east of the country, the livestock based systems have not escaped entirely.

It is generally accepted among the scientific community that efforts to protect the last remaining remnants of high-value wildlife habitats will be less effective if they continue to remain in an inhospitable matrix of intensively farmed land. Indeed the survival of a diverse wildlife in Britain is dependent on its fortunes in agricultural areas.

Concern for environmental issues has rapidly grown since the early 1960s, as witnessed by the large and firmly established membership of environmental pressure groups (Green, 1981). Given this concern and the well developed arguments from the scientific community, the UK Government has gradually responded, notably with the Wildlife and Countryside Act 1981/85, the Agriculture Act of 1986, which identifies the wider role of agriculture in wildlife and landscape conservation, and the White Paper on the environment – *This Common Inheritance* – which identifies the vital contribution farmers and other land managers can make to the environment by adopting management practices which protect wildlife and their habitats and which enhance the rural landscape (HMSO, 1990).

A significant stage was reached in the fortunes of agriculture with the attainment of self-sufficiency in the main temperate foods, and subsequent over-production in relation to Europe's internal market has resulted in the introduction of measures to curb production and in attempts to reform the Common Agricultural Policy (CAP). It is, perhaps, understandable that environmentalists now see this as an opportunity for increasing conservation measures on farms, while many farmers' organizations support such a move if it presents a potential basis for generating additional income at a time of serious financial difficulty.

There is much to be done if these opportunities are to be utilized wisely to everyone's benefit. This chapter examines some of the issues and opportunities with particular relevance for the livestock sector of the farming industry.

Basic Considerations

Why conservation?

The arguments for conservation in general have been widely debated and are now selectively illustrated. In the World Conservation Strategy (IUCN, 1980) it is stated that human beings, in their quest for economic development and enjoyment of the riches of nature, must come to terms with the reality of resource limitation and the carrying capacity of ecosystems. In the context of living resource conservation the three main objectives should be;

1. To maintain essential ecological processes and life-support systems.
2. To preserve genetic diversity.
3. To ensure the sustainable utilization of species and ecosystems.

Norton (1987) has discussed reasons for preserving natural variety, while Green (1981) has presented arguments why the countryside should be conserved, including the following bases or values: (i) ethical values; (ii) aesthetic values; (iii) cultural values, i.e. part of our heritage; (iv) scientific – for its information content and understanding of natural systems; (v) for material benefits arising from the use of natural resources both now and in the future; (vi) the need to maintain global equilibrium or ecological balance – an example of which is illustrated by current concern over global warming. O'Riordan (1990) eloquently concludes 'Human survival depends on a healthy natural World. Conservation is the recogniton of this relationship. The loss of a species of butterfly or the desecration of a beautiful rural scene, are symbols of our dangerous denial of this ultimate truth'.

While all of the above are generally accepted, if not always fully understood, the reasons why the farmer is concerned with conservation are frequently pragmatic and based on economics. Farmers are in the business of making a livelihood and readily recognize the significance of satisfying their customers, who, increasingly, demand a conservation ethic. In addition the steps of Government to produce stable and realistic financial incentives for conservation activities offer increasing opportunities for broadening their source of income. The economic argument can also be extended to support the case for landscape conservation. In England, for example, tourism in the countryside, which depends on the attractiveness of this resource, generates sufficient income to support 530 000 jobs, which are equivalent to the total number of people employed in agriculture (Countryside Commission, 1991a).

Scale of perception and organization

The one characteristic of farmland which, above all others, poses problems for conservation is its patchiness (MacDonald and Smith, 1990). Thus, while attention is frequently concentrated specifically on individual sites, habitats, or land units such as, for example, hedges, field boundaries and hay meadows, their role for wildlife is also dependent on their interrelationships and it is the implication of the latter which constitutes a major challenge for scientific investigation. The growing subject of landscape ecology is recognition of the importance and the need for such an integrative approach (Forman and Godron, 1986).

A central issue is how wildlife species survive in remnants of semi-natural vegetation and how, and to what extent, they utilize, or are affected by, the larger matrix of intensively managed fields. An associated issue involves the dispersibility of species within the agricultural landscape (Bunce and Howard, 1990).

When planning the conservation of species it is necessary, therefore, to consider the feasibilty of remnant habitat improvement (maintenance or enhancement), the extension of patches or sites and the simultaneous construction of wildlife corridors, e.g. hedgerows (Woodmansee, 1989). It is important to remember that, from the farming point of view, land use decisions are made within the context of the farm boundary, and, therefore, whole-farm conservation planning is both logical and necessary.

Landscape is a hierarchical level whose main component parts are site clusters, or, from an administrative point of view, farm or forest units. Whereas farms in the United Kingdom are predominantly in private ownership and hence subjected to individual farmers' decisions, concern for landscape is necessarily the concern of statutory bodies and agencies who, in developing policies, have to take a wider view than the individual farmer.

Given consideration of a hierarchy of sites, farms and landscapes it is also necessary, when making conservation plans, to take account of the national and international context (Fig. 22.1). Thus the wildlife value of a site may be determined by whole-farm activities, which in turn contribute to the local agricultural landscape. Significantly the site may also support populations of a nationally or internationally important species.

What is currently there?

If rough grazing and commons are included there are no less than 12.8 million hectares of grassland. Most farming statistics deal with broad categories such as rough grazing or permanent pasture (Holmes, 1989), however Bunce *et al.* (1984) have produced survey data for 1978 (Table 22.2) while Fuller (1987) has reviewed the changes that have occurred in lowland grassland in England and Wales. Since 1982, for example, there has been a 92% loss of unimproved and rough grasslands in lowland England and Wales with only 0.6 million hectares remaining.

The most extensive areas of semi-natural vegetation in Britain are found in the hills and uplands where they represent the largest areas of land with a high wildlife and landscape conservation value (Ratcliffe and Thompson, 1988). The vegetation is diverse ranging from expanses of blanket-bog, mosaics of dwarf shrub heath and

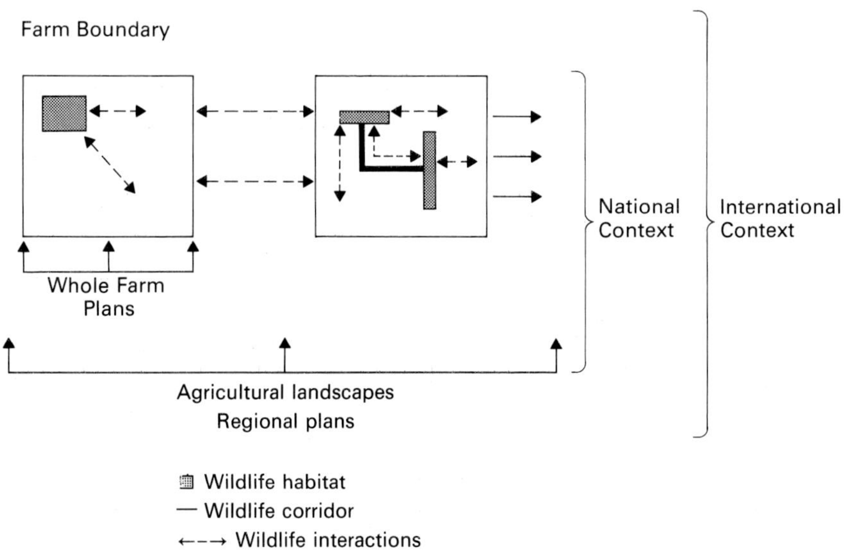

Fig. 22.1. Scale of perception for conservation management.

Table 22.2. Allocation of grassland and moorland as a percentage of the land surface of Britain.

Land use	Percentage of total area
Temporary grass	
Short term leys	10.4
Other leys	4.8
Permanent grass	
Generally re-seeded	6.4
Older grassland	6.3
Rough grazing	
Rough grassland	
Rough grass	2.2
Mixed rough grass	1.4
Bracken	1.6
Rushes	1.6
Mountain grass	2.7
Moorland	
Purple moor grass	3.3
Cotton grass	2.9
Heather	5.5
General moorland	2.8
Total	51.8

Source: Bunce *et al.*, 1984.

moss heaths on high mountain plateaux, together with significant areas of acidic grassland and semi-natural woodland. Of particular note is the international importance of a number of breeding birds such as the peregrine falcon (*Falco peregrinus*), golden eagle (*Aquila chrysaetos*), red grouse (*Lagopus lagopus*), ptarmigan (*Lagopus mutus*) and raven (*Corvus corax*) with at least 13 other species which breed in higher densities than elsewhere in Europe.

The lowlands are extremely varied in their farming systems, which, in the ruminant livestock sector, can range from very intensive dairy units to traditional, low intensity systems of livestock rearing or from specialist all-grass farms to mixed farming systems. Wildlife habitats are, in contrast to those in upland areas, thinly dispersed and highly fragmented and which include herb-rich grassland, grazing marshes, hedgerows, shelterbelts and small fragments of undisturbed vegetation (NCC, 1990).

The implication for a conservation strategy is that attention should primarily be paid to protecting, expanding or diversifying linear features and relatively small areas of species-diverse vegetation on lowland areas, whereas in upland areas attention should be focused on the whole landscape.

How and why has it changed?

Most of the grazed areas of Britain represent a plagioclimax vegetation which, in the absence of some form of management, would revert to forest through a process of natural succession – a point apparently not always appreciated by those who advocate wholesale removal of grazing sheep from certain hill areas! The origins and relevant changes in the vegetation up to the 19th century have been reviewed by Green (1989) and will not be discussed here.

Farming practices prior to the Second World War were essentially characterized by a slow rate of technical development and land use change which provided opportunities for organisms to adapt to the changing landscape. Of considerable importance was the existence, in the lowlands, of a rich matrix of remnant habitats while field systems were generally on a small scale with a more varied, frequently mixed arable–grassland system. This heterogeneity and relatively stable land use provided an ideal wildlife resource (Fry, 1990).

In the last 40 years there have been dramatic developments in agriculture which have considerably eroded this wildlife and landscape resource. Principally these have resulted in the loss of habitat and in the specialization and intensification of farming practice. In the uplands moorland conversion through drainage, fertilizing and re-seeding has led to major losses in habitat. In Wales alone, 100 000 ha of unimproved grazing land was reclaimed in the 20 years up to 1986 (NCC, 1990). The accompanying increased intensity of use, involving increased stocking rates, particularly of sheep, has contributed in a large measure to the decline in heather moorland as well as to localized problems of soil erosion and deterioration of grazed woodlands.

In the lowlands the pace of change has been even greater. In the grassland–livestock sector this has involved increased use of drainage, fertilizers and herbicides (1 million tonnes of fertilizer nitrogen applied annually to UK grassland (Owen, 1987)) increased intensity of utilization of grassland with an ever-growing dominance of ryegrass, changes to silage systems and earlier cutting, increases in the use of slurry, field enlargement and hedge removal. There have also been associated changes in farming systems with increased specialization and a decline in mixed farming. Investment in modern farm buildings has often produced adverse effects on landscape (Soper and Carter, 1991).

The consequences for wildlife in the United Kingdom have been considerable and these are not always fully revealed in published statistics. For example, while hedgerow removal in western livestock areas has not been so marked as in the arable east of the country many surviving hedges are neglected and suffer from repeated annual flailing, and deposition of fertilizers and slurry in the hedge bottom. This results in the dominance of a few vigorous plant species and, in some areas, poor hedge structure as a consequence of severe browsing and mechanical damage. While these hedges still contribute to landscape their wildlife value is very low.

There have been major losses of semi-natural vegetation which between 1947 and 1980 amounted to an overall loss of 25% (Countryside Commission, 1986). The Nature Conservancy Council (NCC, 1984) has recorded that in about the same time interval 80% of limestone grasslands and 50% of wetlands were lost. The impact on wildlife has been the extinction of some species and a serious decline in others, which conservatively includes well over 200 plant and animal species (Green, 1989). An example of how intensive use of grassland areas can affect wildlife is seen in the Somerset Levels and Moors where land drainage coupled with lowered water tables, re-seeding, higher use of fertilizers, increased stocking rates and a switch to early cut silage systems have caused a serious loss of wader populations including those of international significance (see Fig. 22.2; Taylor and Dixon, 1990; Robins *et al.*, 1991). The extent of wildlife decline has been frequently reviewed (see Jenkins, 1984; O'Connor and Shrubb, 1986; Bell and Bunce, 1987; Usher and Thompson, 1988; NCC, 1990). The causes of these changes are largely embodied in agricultural policies since 1947. These policies concentrated on increasing farm production through the provision of financial incentives and in the support of research, development and extension. There is no doubt that this particular policy was successful in achieving its aims but at some considerable environmental cost. It has also created, for many farmers, a technological treadmill which is proving very difficult to get off. Details of the progress of the farm policy and its results for the ruminant livestock sector are reviewed in Jollans (1981).

Currently British agriculture is faced with economic decline arising from the inevitable problems of surplus production, rising production costs and relatively low farm-gate prices. The problems of the CAP and its potential reform will continue to have an overwhelming influence on the fortunes of agriculture, wildlife and landscape. Although it is difficult to predict future political decisions, the future of technical development and its implications is more obvious.

An analysis of potential primary production throughout Europe reveals that this still exceeds actual production by a considerable amount and consequently it is reasonable to assume that output per hectare will continue to increase, at least at the present rate (de Wit *et al.*, 1987). North (1991), has pointed out that new technologies, such as genetic engineering, more effective disease control in crops and animals, improved nutritional efficiency, improvements in the basic photosynthetic process and the extension of nitrogen fixation to a wide range of non-leguminous

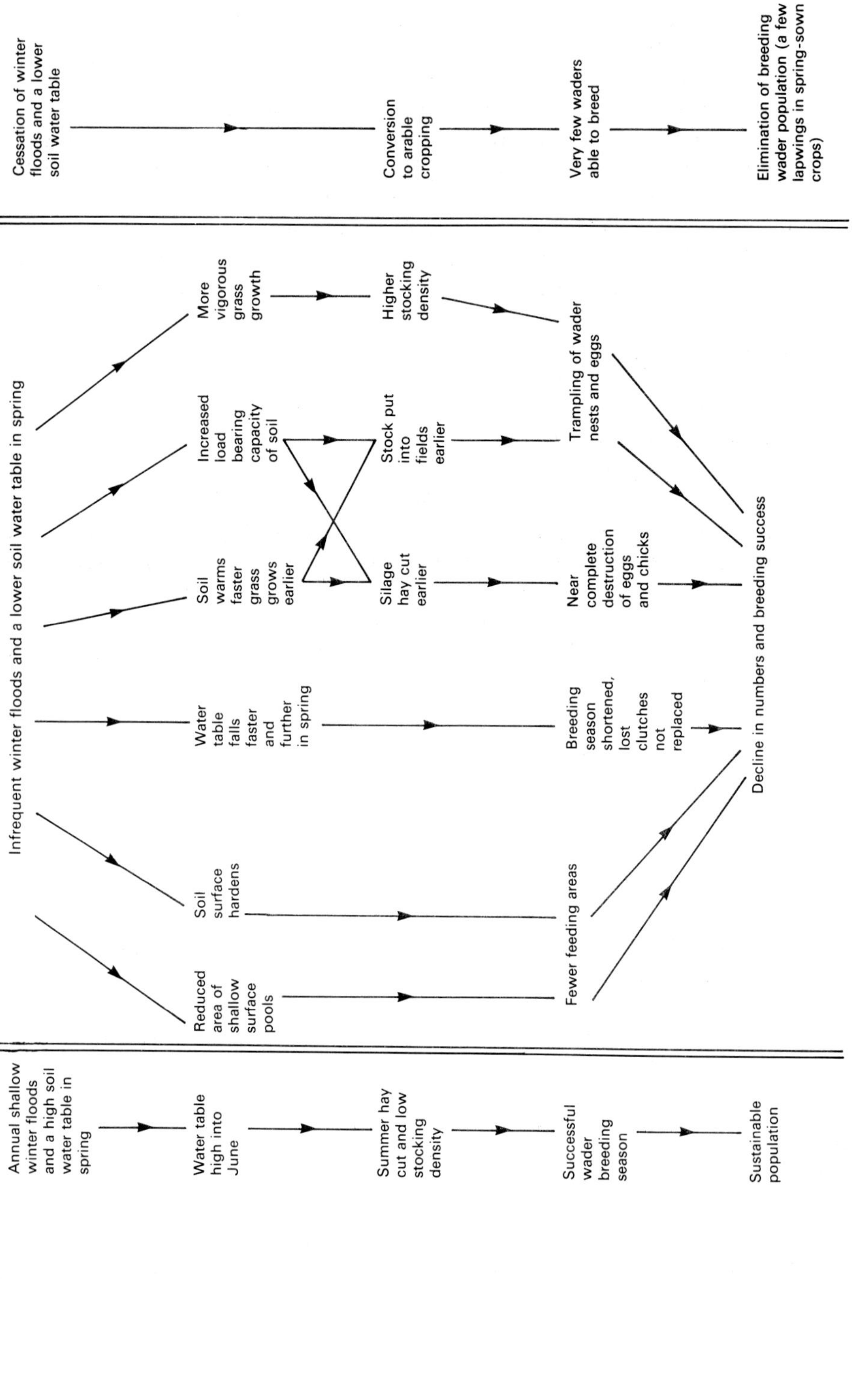

Fig. 22.2. The effect of water level management on breeding waders (Robins *et al.*, 1991).

species could all enhance the opportunities for higher production from land. Given the demographic changes in the human population (UK), and income elasticities of demand for food, North has estimated that the demand at the farm gate for increased food production will only be in the order of between 3% and 5% higher in the year 2015 than in 1985. The implications of this are that the expected food production targets could be met from a much smaller land resource.

Under various scenarios of production, for example ranging from maximum use of grassland forage in intensive dairy systems to systems based more on cereals as a major source of livestock feed, the conclusion is that in all cases about 5 million hectares could be released from agricultural use by the year 2015 and with the probability that this could come from the hills and lowlands in about equal proportions.

That this has significant implications for wildlife and landscape is obvious. In the next section the potential direction that this could take is considered.

How do we want it to change?

This rather depends on who 'we' are. In 1985 only 2.5% of the population was active economically in agriculture and only 8% was classed as rural (Robinson, 1990). Increasingly, therefore, 'we' are the large urban populations particularly when considered in the context of their influence on markets and land use policies. In a multi-purpose countryside there should be a place for all legitimate uses of land. The important question is how can this be achieved in a way which optimizes land use and satisfies the competing demands on its resources? The Countryside Commission (1991b), in a consultative paper emphasized the need for a clear vision of the future of the countryside to which all can subscribe and for a coherent national strategy for the use of rural land. Key objectives would include the following.

1. An environmentally healthy countryside, which embraces the main resources of land, air and water and of diverse species and habitats.
2. A beautiful countryside.
3. A diverse countryside.
4. A countryside of quality.
5. An accessible countryside.
6. A thriving countryside.

It is predictable, therefore, that in the future there will be increasing pressure on the farming population to meet these objectives while at the same time striving to maintain an efficient agriculture. The opportunities that exist for incorporating wildlife and landscape conservation objectives within the livestock farming sector will now be considered.

The Role of Livestock

Livestock management, wildlife and landscape conservation

Ruminant livestock systems have major influences on landscape and wildlife ranging from land impoverishment, erosion and loss of species (in the case of over-grazing) to, at the other extreme, a benign or positive influence where stable low intensity grazing systems and vegetation are co-adapted to produce species-diverse habitats.

The optimal management of land initially requires a choice together with a clear statement of objectives. These must be followed by appropriate prescriptions and guidelines for practical application. Also of importance is the need for simple and speedy methods of monitoring or appraisal – have the objectives been achieved? While this is normally straightforward in agricultural terms, the appraisal of wildlife and landscape is far more complex. Finally, in the context of the farm business, there should be financial reward for conservation work. In the first instance this should allow for a rational business decision to be made as to the balance of activity between food and fibre production and conservation. In addition the possibility may eventually exist to link performance determined payments to conservation work, an aspect which will be referred to later.

Management, especially of natural resources, is based on scientific knowledge, common sense, rule-of-thumb knowledge, untested theories and hunches. Managers must plan with the best knowledge and thought at hand, regardless of their true value in situations where scientific evidence is sparse or incomplete (Provenza, 1991).

In the case of the conservation of wildlife on specific sites, careful examination of associated historic systems of management, as for example found on heather moorland, herb-rich hay meadows or marshland grazing, provides a pragmatic solution for the instant framing of management prescriptions. Taken in a wide context, traditional or heritage systems of farming are, however, 'producing' wildlife as a passive by-product that is without any objectivity in this direction.

Many of these systems are inherently of low agricultural productivity and the question arises as to whether more productive systems, in keeping with modern requirements, can be devised while still retaining a high wildlife content. The development of objective management of wildlife on farmland requires detailed ecological understanding of, at least, key organisms and habitats and of spatial and temporal relationships determining the survival of species populations in largely patchy landscapes. Progress in acquiring this understanding is being made although it suffers from the lack of attention paid to agricultural areas by ecologists in the past. Examples will be given subsequently of some of the more significant areas of development.

Farm livestock: implications for conservation

Farm livestock have a role to play in conservation in that (i) they replace large wild herbivores in the task of managing plagioclimax vegetation and (ii) the associated livestock systems determine the form and pattern of land use which in turn is intimately associated with wildlife and landscape.

Foraging characteristics

While sheep and cattle are the most commonly available species there are local opportunities for extending the range of foraging strategies with horses, goats and deer. Dietary profiles compiled from world literature indicate species differences (Van Dyne *et al.*, 1980; Table 22.3).

Table 22.3. Summary of average dietary botanical composition for all seasons.

	Diet		
Species	Grass (%)	Forb (%)	Shrub (%)
Cattle	72	15	13
Sheep	50	30	20
Goats	29	12	59
Horses	69	15	16
Red deer	40	21	39

Source: Adapted from Van Dyne *et al.*, 1980.

Cattle and horses are grazers with a particular preference for grasses while goats are browsers and show a preference for shrubs. Sheep and red deer have a more balanced dietary profile. The ability to select discrete food items from a heterogeneous array of plant material is associated with the size and shape of the animals' mouths (Gordon and Jason, 1989). Cattle are generally least selective, horses and deer are more selective, while sheep and goats exhibit a high degree of selectivity. In addition differences related to body size, digestive tract morphology and function are linked with contrasting abilities to exploit different vegetation types (Hodgson *et al.*, 1991). Cattle, for example, are better than sheep at utilizing material of low digestibility and goats are more capable of utilizing browse such as willow and gorse than sheep.

These variations give rise to differences in habitat exploitation under free range conditions and are of considerable importance where vegetation management is one of the objectives.

Detailed studies of the utilization of plant communities under free range conditions have been carried out by Hunter (1962) on sheep in the Cheviot hills, Hughes *et al.* (1975) on sheep in Snowdonia, Welch (1984) on heather moorland, which involved cattle, sheep and red deer, Gordon (1989) on the Isle of Rhum, which involved cattle, deer, goats and ponies and Putman *et al.* (1991) on ancient grassland involving cattle and horses.

Hunter (1964) demonstrated the existence of family groups within a flock of sheep which maintained individual home-ranges associated with specific vegetation types. Ellis-Williams (1989) has suggested that modern flock management may disrupt such spatial variation in the evolved utilization of hill vegetation to the detriment of the long-term stability of plant communities.

Provenza and Balph (1990) have stressed the need to consider such intraspecific plasticity with respect to diet and habitat selection, and Provenza (1989) has suggested that habitat use could be manipulated by selectively culling individual animals.

Hall (1988), in a study of the Chillingham herd of free-ranging white cattle, concluded that the intensity of grazing is more likely to be spread across different vegetation communities of conservation importance if a mixed-age herd of cows and young cattle are used.

Mixed-species grazing may be beneficial for livestock production where complementary use of heterogeneous vegetation occurs or where parasitic infestations of livestock need to be reduced. Gordon (1988), refers to grazing facilitation where, on the Isle of Rhum, the use of cattle on pasture dominated by *Molinia caerulea* (L.) improved conditions for grazing deer, particularly in the spring. Clearly a high level of understanding of foraging strategies is a necessary prerequisite to effective use of farm livestock in vegetation utilization and management.

Impact on vegetation

That grazing animals can influence the composition and productivity of vegetation has long been appreciated by agriculturalists (Jones, 1933; Milton, 1940; Jones, 1967). Apart from choice of animal species, management control can be exerted through:

1. the time of year at which grazing occurs;
2. the duration of grazing and method of control, i.e. continuous or rotational;
3. the intensity of grazing.

Prescriptions for the management of productive swards are well developed (Hodgson, 1990) and the introduction of sward height, as a simple measurement reflecting sward state, has provided a basis for objective regulation of grazing. In this situation the primary objective is that the needs of the individual animal and sward productivity are balanced.

Where vegetation is to be managed for conservation the objectives are different. Livestock performance is not necessarily an immediate concern, while the diversity and structure of the vegetation is paramount.

Plant species diversity is strongly affected by the level of soil fertility. Nitrogen fertilizer application, for example, stimulates the growth of responsive species, such as ryegrass, which by a process of competitive exclusion, results in a decline in species variation. Grazing, by constraining growth can reduce this influence and so maintain or increase species diversity.

Grime (1979) has suggested that where the objective is to create a species-rich herbaceous vegetation, a target in management should be to maintain the seasonal maximum in standing crop and litter within the range 350–750 g DM m^{-2}.

Structural heterogeneity within plant communities is important for species diversity – both flora and fauna – where, for example, short turf provides wide diversity of plant species while patches of taller vegetation have favourable micro-climates for invertebrates (Bakker, 1989). Moderately intensive grazing creates an ideal mosaic vegetation. The exact intensity of grazing which represents ‘moderate’ may vary with the evolutionary history of the grazing–vegetation complex, being higher for longer established associations (Milchunas *et al.*, 1988). Grazing animals may introduce diversity through excretory behaviour causing spatial variations in soil nutrients – an extreme example is the development of latrines in horse paddocks – or through disturbance caused by cattle poaching the ground under wet conditions and in so doing provide an opening for plant invasion.

Changes in vegetation structure caused by selective grazing, trampling and manuring are followed by changes in plant species composition, and attempts to predict this sequence are being made by Thalen *et al.* (1987) using a modelling approach. Other broad guidelines relating to the prediction of vegetation change are being developed in conjunction with the National Vegetation Classification (NVC), an example of which will be discussed later, while Miles (1987) has reviewed the successional transitions that occur under low or high grazing pressure on semi-natural hill vegetation.

Management of Habitats and Landscape – Examples

Habitats

Calcicolous grasslands

During the 18th and 19th centuries extensive, open unfertilized sheep walks existed on downland in southern Britain. Management of these areas was characterized by a stable system of sheep grazing based on a daily cycle, with the sheep grazing the downland during the day and returning to lowland fields at night.

Remnants of the characteristic downland vegetation that developed under this system still exist, although in a much fragmented and depleted state following the ploughing up campaigns of two world wars and more recently the swing to arable production. Many of the remnants are found on the steeper less accessible slopes where currently considerable effort is being directed to their conservation (NCC, 1982). The old grazed chalk grassland is particularly notable for the highly intricate mixture of plant species (as many as 40 species per square metre) and an equally diverse invertebrate population, and therefore management objectives pay attention to these two key areas.

The national vegetation classification (NVC) can be used as a basis for determining management strategies (Jones-Walters, 1990). The fixed influence of soil type, underlying bedrock, climate and topography determines the NVC sub-community while the influence of grazing determines the major community type. Suggested relationships of vegetation types showing the influence of grazing are illustrated (Fig. 22.3).

Individual invertebrate communities can be related directly to particular plant communities (Jones-Walters, 1990), and therefore NVC classes can be used as indicators of whole community performance.

An important step forward is the use of interaction matrices in assessing and formulating conservation objectives and the development of expert-systems as an aid to determining specific management prescriptions.

Structual heterogeneity is aimed for with a range of short through to long swards. Grazing intensity ranges from 1–7 sheep ha^{-1} $year^{-1}$ or 0.2–1.5 cattle ha^{-1} $year^{-1}$ with stocking rates at any instant of time being determined by the season, the inherent fertility of the unfertilized soil and the overall NVC objective. Management of these relatively small areas is most effective when carried out in conjunction with improved grassland so that conservation needs and the grazing animals' needs can be adequately met in a two-pasture system (Bacon, 1990).

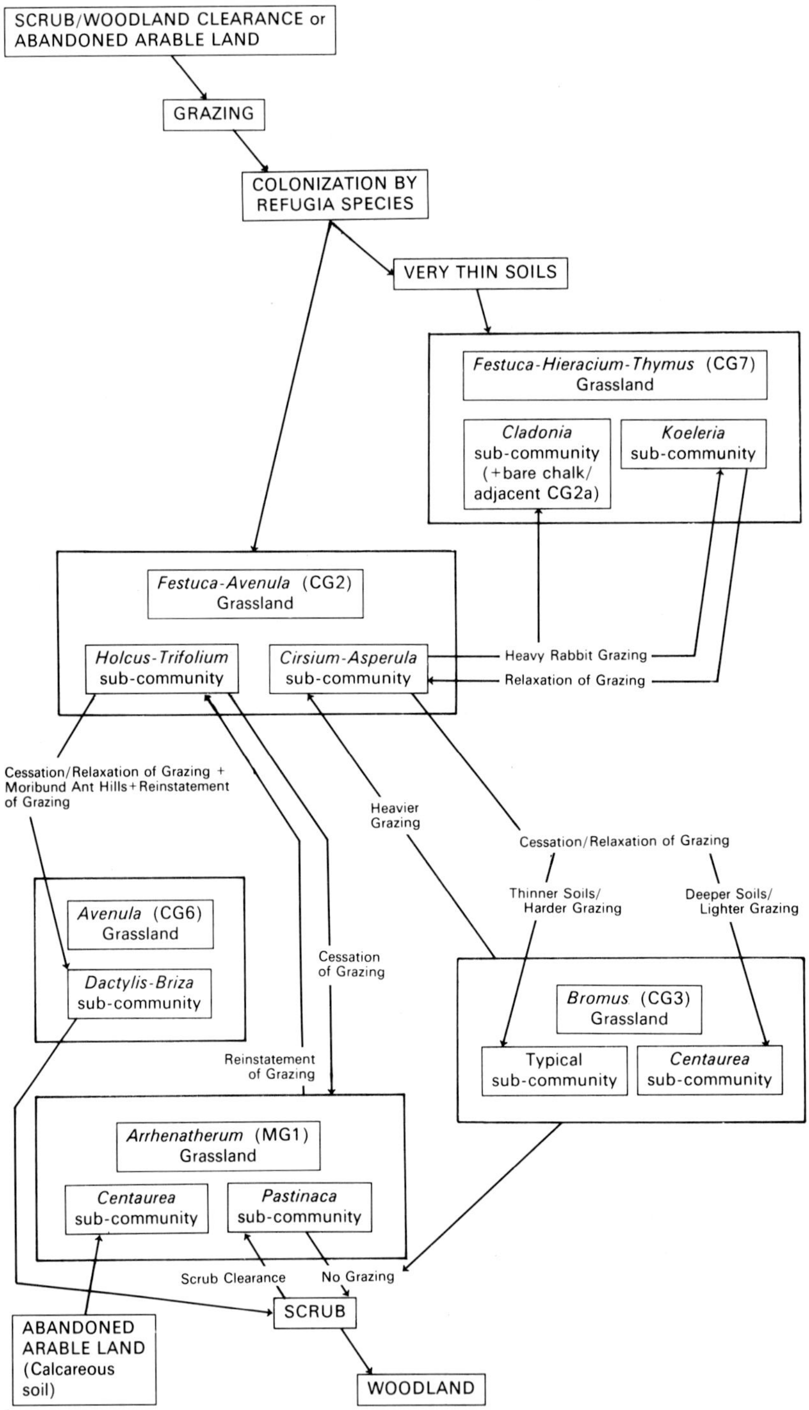

Fig. 22.3. Suggested relationship of vegetation types (communities and sub-communities) for Aston Rowant and Martin Down in combination (Jones-Walters, 1990).

Mesotrophic grasslands: traditional hay meadows

Traditional hay meadows can be defined by their management regime and by their floristic composition. They are one of Britain's rarest and most fragile habitats. They contain a rich variety of flowering plants, some of local and some of national rarity. Some of the wetter meadows provide important breeding sites for waders such as snipe, curlew, lapwing and golden plover. While varying in detail, the dominant features of traditional management, which is part of a low intensity system of livestock production, include: overwintering of cattle in barns with the ensuing farmyard manure being spread on the hay meadows in the spring. At this time the meadows were grazed by ewes prior to and during lambing time. In mid-May the ewes and lambs were removed and the meadows left until hay harvest in late June to early August, depending on site and weather, when they were subsequently once more grazed with cattle and sheep and during the winter by sheep at low intensity (Smith, 1985).

The introduction of Environmentally Sensitive Area (ESA) designation is accompanied by management guidelines based on the above but allowing restricted use of fertilizers, herbicides and lime.

Two features are of particular importance and call for improved guidelines. Firstly, the application of nitrogen fertilizer is detrimental from a floristic point of view. The traditional use of manure with its slow and relatively low release of nitrogen provides a nutrient status which is not damaging to plant diversity. In addition the inclusion of organic material benefits invertebrate populations.

Secondly, a factor giving rise to concern in the ESAs, is the fixed prescription for the earliest date of hay harvest which, depending on location, ranges, for example, from 1 July to 15 July, in the North Pennines and Yorkshire Dales (Smith and Jones, 1991). The logic of this is based on the need for meadow species to shed seed before the hay harvest in order to ensure continuity. Operational difficulties arise for farmers when weather conditions are suitable for harvesting just before the target date. The possibility of improved and more flexible guidelines being developed arises from detailed studies of the phenology of the main meadow species (Smith and Jones, 1991; Fig. 22.4). The proposed alternative strategy is partly based on the fact that fertilizer nitrogen use is most damaging to floristic diversity. In meadows identified for their presence of late flowering indicator species the prescription would be no fertilizer nitrogen followed by cutting dates later than mid-August every other year. In other situations a restriction in nitrogen use would, by virtue of slower growth, lead to relatively late harvest dates but in this case the farmer could make hay or silage whenever wished.

The counsel of perfection, when the ideal conservation management plan is based upon a complete understanding of the ecological factors that influence the conservation interest of a site, is hard to achieve. However, as information accumulates, as above, more effective management can be prescribed which is beneficial to both agricultural and conservation interests. Other areas that would benefit from attention include examining the importance of spatial heterogeneity caused by dung and urine patches deposited by grazing livestock and the role of grazing and treading in maintaining species diversity even under moderate fertilizer inputs (Smith, 1988).

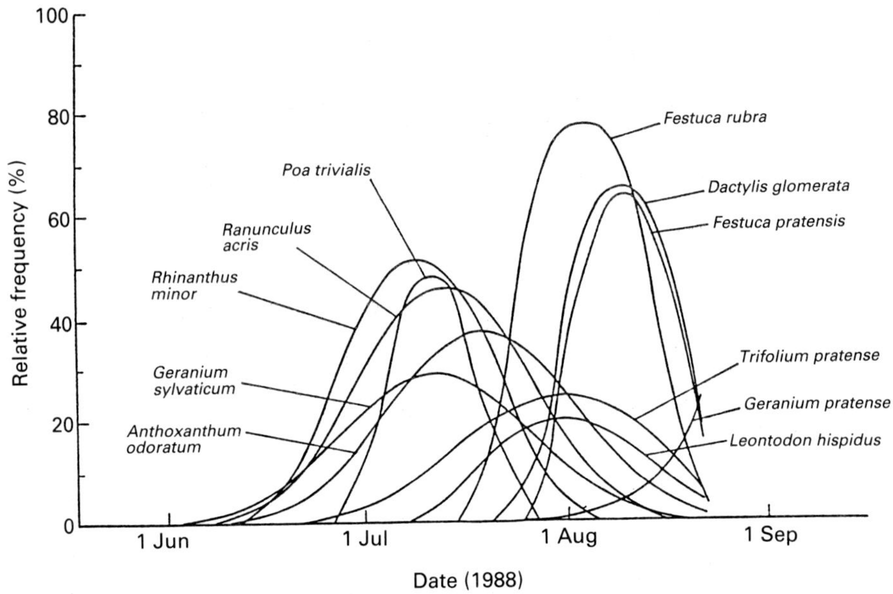

Fig. 22.4. The development of ripe seed in 11 species (after Smith and Jones, 1991).

Upland communities: heather moorland

The most widespread economic use of upland moorland with heather communities is livestock agriculture, principally sheep but with cattle as a common subsidiary activity. Heather moors, if managed appropriately, also can support viable grouse populations. These areas potentially support a characteristic and internationally important wildlife assemblage and their landscapes are highly valued representing for many people the epitome of natural beauty (Felton and Marsden, 1990). Currently, considerable attention is being directed to developing improved management prescriptions, especially for multiple use (Whitby and Grant, 1990).

The main uses of heather moorland are to a certain extent compatible but with agricultural use presenting the main area of conflict. Increased stocking of sheep, coupled at times with over-burning, has led to heather decline and conversion to grassland. The practice of supplementary winter feeding causes localized concentrations of stock with consequent severe over-grazing. The general rise in sheep numbers is reflected in statistics for the broad category of rough grazing in England and Wales, where in 1977, 71% of the area was stocked at below 2 ewes ha^{-1}, whereas in 1987 71% of the area was stocked at between 2 and 6 ewes ha^{-1} (Felton and Marsden, 1990). In addition to ecological over-grazing, losses of heather moorland have been due to reclamation for improved grassland and forestry although this has now almost ceased.

Characteristic moorland requirements for each of the main uses are given in Table 22.4.

The fundamental issue to be discussed here is that of assessing the sheep carrying capacity or stocking rate of the moor. In a restricted context this may require grazing

Table 22.4. Features of heather moors according to their main use.

	Conservation		Main use		
Characteristic	Plant	Animal	Shooting	Grazing	Recreation
1. Heather age range	High proportion of old	Wide range	Wide range	Mainly young	Any combinations – too much old is dangerous
2. Size of heather age units	Medium/large	Small elongated	Small elongated	Medium/large	Variable
3. Wet areas	Encouraged	Encouraged	Encouraged	Discouraged	Discouraged
4. Intensity of management:					
(i) Burning/cutting	Low	High	High	Medium	Variable
(ii) Predator control	Low	Medium	High	Medium/high	Low
5. Bracken control	High priority	Low priority	High priority	High priority	Variable should be high priority
6. Grazing level	Low	Low/medium	Low/medium	High	Low/medium or high
7. Level of access	Low	Low	Low	Low	High
8. Wildlife diversity	High	High	High	Medium/low	Medium/low
9. Seral development to scrub	Discouraged	Discouraged	Discouraged	Discouraged	Discouraged
10. % moor as heather	50–75%	75–85%	50	50–75	

Source: Whitby and Grant, 1990.

to be limited in order to ensure continuity of heather as well as ensuring that sheep achieve a satisfactory diet. In this it is necessary to take into account rotational burning practices which have an important influence on the area of young palatable heather. For sheep to achieve adequate nutritional use of heather there is a need to balance heather intake with an equal amount of grass which is an essential source of available nitrogen. Sibbald *et al.* (1987) have developed a computer model which accounts for vegetation productivity, the limits of sustainable use of heather and grass communities, and the dietary constraints for sheep. Using areas of heather with recognition of different age classes and grassland, together with information of the area of heather burning and its frequency, it is possible, through the model, to predict a sustainable carrying capacity of the moor. This represents one of the first practical approaches to solving multidimensional problems affecting livestock and vegetation use in the field. It is currently being extended to account for regional variations in heather productivity and the activities of other herbivores in the system, including those of grouse (Armstrong, 1990). An important area where knowledge is insufficient concerns the use sheep and wildlife make of vegetation mosaics. Prediction of the foraging behaviour of sheep in a patchy environment, for example, is currently the objective of investigations in a number of research centres in Britain.

If a complete understanding of essential ecological details of a system is a counsel of perfection for conservation management (*sensu* Smith) then the development of a mechanistic model of the system is the ultimate vehicle for its application in guiding land use decisions. This is currently closer to reality for heather moorland than in any other system.

Farm and landscape

Holistic assessments are required for farms and for landscapes – which may involve many farms – where the aim is that of integrated farming, wildlife and landscape conservation. Methodologies used in developing integrated land management plans on farms are well developed in certain areas including those for the initial assessment of resources (Heymann and Tittensor, 1981; Willis and Young, 1983; Cobham *et al.*, 1984; Laurie *et al.*, 1984; Tait *et al.*, 1988).

Wildlife surveys can be rapid, being largely based on vegetation characteristics and the identification of the area and location of remnant habitats, or lengthy and detailed, including year-round surveys of both flora and fauna.

Landscape assessment still remains the most contentious area although largely qualitative methods have gained some standardization (Countryside Commission, 1987, 1990; Tait *et al.*, 1988). Agreement about outstanding or representative landscapes has resulted in or has contributed to the introduction in the United Kingdom of a number of land designations such as National Park status, Areas of Outstanding Natural Beauty and more recently Environmentally Sensitive Areas. Many of these designated areas are in livestock farming regions where the landscape status is dependent on the maintenance of the more traditional systems of management.

The more obvious opportunities for the management of wildlife and landscape features are listed in Table 22.5 (FWAG, 1991a,b).

Coupled with these approaches is the need to practise environmentally responsible farming. The Code of Good Agricultural Practice for the Protection of Water

Table 22.5. Wildlife and landscape features on farms requiring positive management.

Hedgerows	Woodland
Dry stone walls	Scrub
Headland and field margins	Unimproved grassland
Wetland	Moorland and heath
Watercourses	Public rights of way
Ponds	Historic features
	Farm buildings

Source: Adapted from FWAG, 1991a, b.

issued by MAFF (1991) is a good example of how this may be achieved and is very timely following a large number of damaging incidents of pollution from livestock wastes and from silage effluents in recent years (NCC, 1991).

While pragmatic solutions are being pursued in current integrated land management plans and progress is being made to arrest the more environmentally damaging aspects of modern agriculture, there is no doubt that further scientific investigation is needed on all fronts.

The emergence of landscape ecology (Forman and Godron, 1986) as a scientific discipline is giving stimulus to those seeking to optimize the multiple-objective use of land. In Wales, studies on integrated land management of Bardsey Island are illustrating ways in which new technologies such as Geographic Information Systems (GIS) and computer models can be effective tools in planning, while at the Pwllpeiran Experimental Husbandry Farm environmental problems of hill land are being approached with vigour. The forthcoming establishment of an organic dairy farm on the Institute of Grassland and Environmental Research farm at Trawsgoed, Wales, will provide further opportunities for detailed studies of the interrelationship between wildlife and the introduction of a completely new system of farming.

In parallel with such scientific and technical investigations there is a need to develop policies which embrace national and European controls, guidelines and incentives if farmers are to respond to these emerging environmental imperatives.

Policy

A basic concern is the need for farmers to make a satisfactory financial return in order to remain viable. In turn a viable farming community is necessary if an attractive and ecologically diverse countryside is to be achieved and maintained. However, conservation measures cause costs which can only be justified in economic terms if the value of the output equals or exceeds the costs. The real difficulty arises in evaluating the output of environmental goods which contain largely non-marketable elements. Weinschenck (1991) points out that while economists have developed sophisticated methods to overcome the problems of evaluating those non-marketable elements, their operational value is zero partly because of neglecting to include all external costs of the consumption activities, for social and other reasons, and mainly on account of a limited anthropocentric point of view. In the absence of market forces effectively operating it is necessary for political intervention on behalf of the public to safeguard wildlife and landscape. Real difficulties arise where policies directed at conservation come into conflict with policies supporting conventional farm production and it is in the area of reconciliation of opposing views that attention is now focused.

Early solutions to the problem of providing an adequate return to farmers for environmental goods were via management agreements. Farmers wishing to develop designated areas such as SSSIs were offered management agreements in which profits forgone, as a result of maintaining the *status quo*, were paid as compensation. The cost of this approach is high both in the context of the profit base in the calculation of compensation and in the cost of administration where each case is negotiated separately. This, however, has received some simplification, notably in the Berwyn Mountain Scheme in Wales, where agreement has been reached on a standard payment and a streamlined procedure for negotiation.

A sequence of schemes has recently been introduced by the Ministry of Agriculture, Fisheries and Foods in England and Wales (MAFF) and the Countryside Commission such as Set-aside, Premium and Stewardship schemes and experimental extensification schemes. Their impact, however, is still restricted, particularly in the livestock sector.

The introduction of Environmentally Sensitive Area designation, however, has been generally seen to be a most significant and welcome development in agricultural policy. The scheme, which evolved from an experimental Broads Grazing Marshes Scheme, was introduced in late 1986. Areas designated had to satisfy the following criteria.

1. They should be of national environmental significance.
2. Their conservation must depend on the adoption, maintenance or extension of a particular form of farming practice.
3. They should be areas where encouragement of traditional farming practices would help to prevent damage to the environment.
4. They should comprise a discrete and coherent unit of environmental interest.

A large proportion of the areas subsequently designated were farmed with livestock systems (Table 22.6; see Potter *et al.*, 1991). Farmers in the scheme receive annual area payments for agreeing voluntarily to manage eligible land in a defined environmentally acceptable way. Payments range from £10 to £300 per hectare, depending on the degree of involvement and reflecting the economic status of the farming system. Farmers' response has generally been good with an average in England, for example, of 87% of targeted land entering into the scheme. In some areas, however, the response has been slow or poor, and Brotherton (1991) has identified that in the Lleyn Peninsula and Somerset Levels ESAs the participation is depressed by the relatively small total payment that entry would bring to many of the farmers. No doubt other factors have also contributed, including the attitude of farmers and the inflexibility of single schemes operating in diverse areas (Evans, 1989).

The first five years of designation have been considered experimental and future policy has yet to be announced*. The issues are obviously complex. Agriculture is in a depression, the problem of the CAP and over-production dominate the political

*At the time of going to press the government has announced a major extension of the ESA designated areas.

Table 22.6. Landscape, habitat and farms targeted by ESAs in England.

ESA	Total area (ha)	Landscape characteristics	Habitats	Target farms/ farming systems
Breckland	94 032	Broad and open heathland landscape with distinctive shelterbelts of Scots pine; river valley landscape	Grazed heath; calcareous grassland; wet grazing meadows	Remaining livestock farms in river valleys; arable farms with characteristic landscape features
Broads	32 160	Open, homogeneous grazing marsh landscape: flat valleys and fen and fen carr	Marshland; dikes; fen/carr woodland	Traditional livestock farms and graziers
North Peak	50 250	Wild moorland plateaux incented by deep valleys	Peat moorland; oak woods; hay meadows	Hill sheep farms on high moors and surrounding valleys
Pennine Dales	15 600	Characteristic limestone scenery of grassland and valley bottom meadows, drystone walls and field barns	Wet pasture, hay meadows and woodland	Hill sheep farms and some sheep and cattle farms (for overwintered 'store' animals)
Somerset Levels and moors	26 970	Flat landscape criss-crossed by rivers, ditches and rhines	Wetland	Traditional dairy farms and mixed farms combining dairy, beef and sheep
South Downs	54 343	Chalk grassland landscape with scattered trees and shrubs	Species rich chalk grassland; river valley mosaic of fields, ditches and dikes	Traditional downland farms plus mixed farms
Suffolk river valleys	32 149	Pastoral landscape of river valleys with grassland, hedges and reedbeds	Wet grasslands, reedbeds, rivers and ditches	Mixed farms with livestock on river valley grassland
Test valley	2 690	Pastoral mosaic of river, trees, carr, reedbed and pasture	Chalk stream communities and chalk grassland	Mixed farms
Shropshire borders	21 000	Characteristic intricate pastoral landscape of small fields and unimproved pasture	Rough grazing, valley woodlands and hedgerows	Cattle and sheep farms
West Penrith	7 210	Rugged coastline, moors, small fields	Coastal to inland heathland; scrub and wetland	Traditional dairy farms

Source: Potter *et al.*, 1991.

scene. It can be argued that much of the political incentive for environmental legislation, such as the introduction of ESAs, was initially motivated by the need to regulate farm production. Nevertheless the opportunities that may exist in switching support from conventional farm production to environmental management have not escaped the interest of a growing sector of the farming population as well as environmentalists.

Haines *et al.* (1991) argue for a more market orientated agriculture coupled with a redeployment of funds released from commodity support to the encouragement of a sustainable agriculture which provides food and the environmental services that society demands. A series of measures is advocated including the adoption of the principle of cross-compliance in a modified Hill Livestock Compensatory Allowances scheme which accounts for the need to control maximum stocking rates; emphasis on extensification rather than set-aside, which is particularly relevant in Wales; individual five-year farm plans and a tiering system (as in the current ESA scheme), which should involve a standard farm management payment and a menu of conservation management payments which would give high flexibility in their application.

An interesting suggestion is that farmers should be paid to monitor wildlife and make returns on wildlife observed each year. Espie (1990) has described a system of paying farmers for results being developed by Leiden University. Currently there are no incentives for doing conservation work well and the introduction of a scheme could appeal to the farmer as an entrepreneur. Criteria for a scheme would include the following.

1. There must be a clear definition of the 'product' and how to measure it.
2. The 'product' must be easily measurable; the farmer must be able to count his own 'product' and send in his returns.
3. The system must be easily checked.
4. The scheme should provide high enough payments to encourage the farmer to take part.

The RSPB (Taylor and Dixon, 1990; Dixon 1991) and NCC (1990) advocate extending the ESA scheme over the whole country. The RSPB suggests geographical tiering to target areas of highest conservation value. They propose a menu of options, including sensitive, extensive and diversion categories. Cross-compliance, so that the receipt of CAP commodity support measures becomes dependent on farmers involving themselves in some degree of environmental management, is an important requirement in the proposal.

Potter *et al.* (1991) advocate setting up a conservation reserve in the United Kingdom. The criteria for enrolment in this reserve would include the following.

1. Areas of mismatch – unsuitability of cropping systems.
2. Vulnerability of land from degrading uses.
3. Conservation potential.

Creating a conservation reserve would involve farmers in the target areas, taking land out of production of surplus crops and putting it into a range of conservation uses tailored to their individual circumstances and interests. This would include conversion to grassland or the creation of specialist habitats all of which could be managed with grazing livestock.

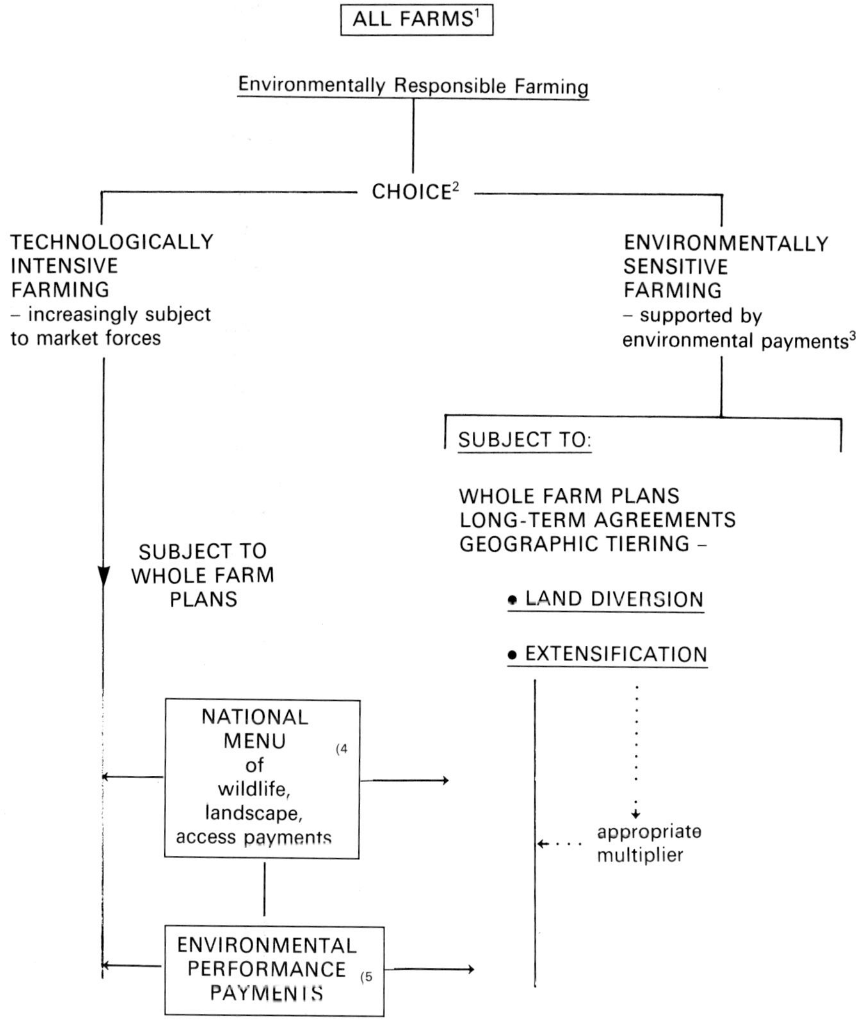

Fig. 22.5. A framework for a future environmental policy for farming?

[1]All farms should adhere to environmental codes of practice and work within limits imposed by environmental legislation (see FWAG, 1991a,b).

[2]The voluntary principle should allow a choice between conventional intensive farming and environmentally sensitive farming (ESF) based on a development of ESA procedures.

[3]ESF payments would be subject to agreed whole farm plans on a long-term basis (10–20 years). The basic approach would involve land diversion (particularly in the lowlands) or extensification (particularly in the hills and uplands). Depending on geographic location and length of agreement each farm entered would receive a multiplier which would be applied to enhance payments from a national menu of land-use activities ranging from hedges to heather moorland and to woodland.

[4]The 'national menu' would be comprehensive and would give, in relation to each payment, an equivalent to an expected gross margin for that activity so that farm economic plans can be prepared. Specific items would be restricted to ESF farmers whilst others would be widely available.

[5]Environmental performance payments would be developed and issued on the basis of environmental returns, related to specific indicators which are easy to monitor and check.

The National Sheep Association (1991), while recognizing the implications for the upland farmers if sheep production is increased in lowland areas, also recognizes that if stocking rates are deliberately reduced in the uplands for environmental reasons, there is a case for accepting increased use and stocking of new lowland grass areas. Opportunities in the lowlands also exist for the non-competitive development of fine wool and dairy production from sheep.

The Nature Conservancy Council (NCC, 1990) emphasizes the need for:

- whole farm involvement in conservation schemes, such as extended ESAs
- longer term agreements of up to 20 years to provide greater stability where needed
- flexibility to ensure regional and local priorities can be targeted.

The NCC also advocates that all farms should comply with a set of basic environmental standards.

The National Farmers' Union (NFU, 1991), reflecting farmers' attitudes, accepts the need for a comprehensive conservation scheme for agriculture which (i) assists and encourages farmers to retain and improve their role as stewards of the countryside and (ii) is sufficiently comprehensive to be applicable to any farmer whether upland or lowland, small or large, within a designated area or outside.

Conclusions

The impetus towards the adoption of conservation management objectives within a broad environmental imperative has achieved a self-sustaining momentum and the necessity for this to be closely integrated with agriculture in a sound holistic use of land is now generally accepted.

Whatever final solutions emerge to current difficulties there is no doubt that the livestock sector of farming will be at the forefront of enlightened systems of land management. A possible solution in the form of a framework for an environmental policy is presented in Fig. 22.5.

References

Armstrong, H. (1990) Modelling the effects of vertebrate herbivore populations on heather moorland vegetation. In: Whitby, M. and Grant, S. (eds), *Modelling Heather Management* – a workshop report. Department of Agricultural Economics and Food Marketing, University of Newcastle upon Tyne, pp. 56–67.

Bacon, J.C. (1990) The use of livestock in calcareous grassland management. In: Hillier, S.H., Walton, D.W.H. and Wells, D.A. (eds), *Calcareous Grasslands – Ecology and Management. Proceedings of British Ecological Society/NCC symposium at University of Sheffield*, 14–16 Sept. 1987. Bluntisham Books, Bluntisham, pp. 121–127.

Bakker, J.P. (1989) *Nature Management by Grazing and Cutting.* Kluwer Academic Publishers.

Bell, M. and Bunce, R.G.H. (eds) (1987) *Agriculture and Conservation in the Hills and Uplands.* Institute of Terrestrial Ecology, Cumbria.

Brotherton, I. (1991) What limits participation in ESAs? *Journal of Environmental Management* **32**, 241–249.

Bunce, R.G.H. and Howard, D.C. (eds) (1990) *Species Dispersal in Agricultural Habitats.* Institute of Terrestrial Ecology with Belhaven Press.

Bunce, R.G.H., Tranter, R.B., Thompson, A.M.N., Mitchell, C.P. and Barr, C.J. (1984) Models for predicting changes in rural land use in Great Britain. In: Jenkins, D. (ed.), *Agriculture and the Environment.* Institute of Terrestrial Ecology, Cambridge.

Cobham, R.O., Matthews, J.R., McNab, A., Stephenson, E. and Slatter, M. (1984) *Agricultural Landscapes: Demonstration farms.* Countryside Commission, Cheltenham, Gloucester.

Countryside Commission (1986) *Monitoring Landscape Change.* Countryside Commission, Cheltenham.

Countryside Commission (1987) *Landscape Assessment.* Countryside Commission, Cheltenham.

Countryside Commission (1990) *The Cambrian Mountains Landscape.* Countryside Commission, Cheltenham.

Countryside Commission (1991a) The seven pound trip. *Countryside Commission News* No. 50, Jul/Aug.

Countryside Commission (1991b) An agenda for the countryside. Countryside Commission, Manchester.

Dixon, J. (1991) *A Future for Environmentally Sensitive Farming.* RSPB, Sandy, Bedfordshire.

Ellis-Williams, I. (1989) Plant–grazier relationships in upland grasslands. In: *Environmentally Responsible Grassland Management.* British Grassland Society Winter Meeting 6 Dec. 1989. British Grassland Society and AFRC Institute for Grassland and Animal Production.

Espie, A. (1990) Ways of modifying agriculture to the benefit of the environment – a report. Nuffield Farming Scholarship Trust.

Evans, S.J. (1989) A preliminary investigation of the Lleyn Environmentally Sensitive Area Scheme. MSc Thesis, University of Wales, Bangor.

Felton, M. and Marsden, J. (1990) Heather regeneration in England and Wales – a feasibility study for the Department of the Environment. Nature Conservancy Council.

Forman, R.T.T. and Godron, M. (1986) *Landscape Ecology.* John Wiley and Sons.

Fry, G.L.A. (1990) Conservation in agricultural ecosystems. In: Spellerberg, I.F., Goldsmith, F.B. and Morris, M.G. (eds), *The Scientific Management of Temperate Communities for Conservation. 31st Symposium of the British Ecological Society*, Southampton 1989. Blackwell Scientific Publishers.

Fuller, R.M. (1987) The changing extent and conservation interest of lowland grasslands in England and Wales; A review of grassland surveys 1939–84. *Biological Conservation* **40**, 281–300.

FWAG (1991a) *Guidelines for Environmentally Responsible Farming.* Farming and Wildlife Advisory Group and Rhone-Poulenc.

FWAG (1991b) *Handbook for Environmentally Responsible Farming.* Farming and Wildlife Advisory Group and Rhone-Poulenc.

Gordon, I.J. (1988) Facilitation of red deer grazing by cattle and its impact on red deer performance. *Journal of Applied Ecology* **25**, 1–10.

Gordon, I.J. (1989a) Vegetation community selection by ungulates on the Isle of Rhum. I Food Supply. *Journal of Applied Ecology* **26**, 35–51.

Gordon, I.J. (1989b) Vegetation community selection by ungulates on the Isle of Rhum. II Vegetation community selection. *Journal of Applied Ecology* **26**, 53–64.

Gordon, I.J. (1989c) Vegetation community selection by ungulates on the Isle of Rhum. III Determinants of vegetation community selection. *Journal of Applied Ecology* **26**, 65–79.

Gordon, I.J. and Iason, G.R. (1989) Foraging strategy of ruminants: its significance to vegetation utilization and management. In: Macaulay Land Use Research Institute Annual Report, 1988. 31–41.

Green, Bryn (1981) *Countryside Conservation.* Unwin Hyman, London.

Green, Bryn (1989) The impact of agricultural management practices on the ecology of grasslands. In: *Environmentally Responsible Grassland Management.* British Grassland Society Winter Meeting, London 6 Dec. 1989, pp. 1.1–1.13.

Grime, J.P. (1979) *Plant Strategies and Vegetation Processes.* Wiley, Chichester.

Haines, M., Hooson, J., Hopkins, W., Jones, T., MacCarthy, B., Tamplin, J., Trant, I. and Caldwell, N. (1991) *Agriculture and Rural Wales – A report to the Campaign for the Protection of Rural Wales* (CPRW).

Hall, S.J.G. (1988) Chillingham park and its herd of white cattle: relationships between vegetation classes and patterns of range use. *Journal of Applied Ecology* **25**, 777–789.

Heymann, M. and Titensor, R. (1981) Ecological appraisal of an agricultural estate. *Chartered Surveyor* **113**, 424–425.

Her Majesty's Stationery Office (HMSO) (1990) *This Common Inheritance.* Cm 1200, HMSO, London.

Hodgson, J. (1990) *Grazing management – Science into Practice.* Longman Scientific & Technical Publications, Harlow.

Hodgson, J., Forbes, T.D.A., Armstrong, R.H., Beattie, M.M. and Hunter, E.A. (1991) Comparative studies of the ingestive behaviour and herbage intake of sheep and cattle grazing indigenous hill plant communities. *Journal of Applied Ecology* **28**, 205–227.

Holmes, W. (ed.) (1989) *Grass – its Production and Utilization.* Blackwell Scientific Publishers.

Hughes, R.E., Dale, J., Mountford, M.D. and Ellis-Williams, I. (1975) Contemporary distribution of sheep populations and environment. *Journal of Applied Ecology* **12**, 165–178.

Hunter, R.F. (1962) Hill sheep and their pasture, a study of sheep grazing in South East Scotland. *Journal of Ecology* **50**, 651–680.

Hunter, R.F. (1964) Home range behaviour in hill sheep. In: Crisp, D.J. (ed.), *Grazing in Terrestrial and Marine Environments.* Symposium of the British Ecological Society, April 1962. Blackwell Scientific Publishers.

IUCN (1980) *World Conservation Strategy.* IUCN-UNEP-WWF.

Jenkins, D. (ed.) (1984) *Agriculture and the Environment* ITE Lavenham Press, Suffolk.

Jollans, J.L. (ed.) (1981) *Grassland in the British Economy.* CAS paper 10. Centre for Agricultural Strategy, Reading University.

Jones, L.J. (1967) *Studies on Hill Land in Wales.* Welsh Plant Breeding Station.

Jones, M. (1933) (a) I Factors influencing the growth of pasture plants, 45–57; (b) II The management of 'clovery' swards and its effects, 122–128; (c) III The management of a 'grassy' sward and its effects, 224–234; (d) IV The management of poor pastures, 361–366; (d) V Edaphic and biotic influences on pasture 366–367 *The Empire Journal of Experimental Agriculture* **1**.

Jones-Walters, L.M. (1990) A new approach to the management of chalk grassland with particular reference to the integration of conservation measures for invertebrates. In: Hillier, S.H., Walton, D.W.H. and Wells, D.A. (eds), *Calcareous Grasslands – Ecology and Management. Proceedings of a British Ecological Society/NCC Symposium at University of Sheffield*, 14–16 September 1987. Bluntisham Books, Bluntisham, pp. 140–148.

Laurie, I., Cook, L.M., Hounsome, M. and Anderson, P. (1984) *Eaton Estate – Landscape Conservation Plan.* Manchester University, Commercial Centre, Oldham.

MacDonald, D.W. and Smith, H. (1990) Dispersal, dispersion and conservation in the agricultural ecosystems. In: Bunce, R.G.H. and Howard, D.C. (eds), *Species Dispersal in Agricultural Habitats.* Belhaven Press, London, pp. 18–64.

Milchunas, D.G., Sala, O.E. and Lavenroth, W.K. (1988) A generalised model of the effects of grazing by large herbivores on grassland community structure. *American Naturalist* **132**, 87–106.

Miles, J. (1987) Effects of man on upland vegetation. In: Bell, M. and Bunce, R.G.H. (eds), *Agriculture and Conservation in the Hills and Uplands.* Institute of Terrestrial Ecology, Huntingdon, pp. 7–18.

Milton, W.E.J. (1940) The effects of manuring, grazing and cutting on the yield, botanical and chemical composition of natural hill pastures. 1. Yield and botanical section. *Journal of Ecology* **28**, 326–356.

National Sheep Association (1991) *Conservation and Sheep Farming – Environment and Conservation Report.* NSA, Aldine Press, Malvern.

Nature Conservancy Council (NCC) (1982) *Chalk Grassland – its Conservation and Management.* Nature Conservancy Council, London.

NCC (1984) *Nature Conservation in Great Britain.* Nature Conservancy Council, Peterborough.

NCC (1990) *Nature Conservation and Agricultural Change. Focus on Nature Conservation* No. 25. Nature Conservancy Council, Peterborough.

NCC (1991) *Nature Conservation and Pollution from Farm Wastes.* Nature Conservancy Council, Peterborough.

National Farmers' Union (NFU) (1991) Environmental strategy – interim statement of the Parliamentary Land Use and Environment Committee. NFU London.

North, J. (1991) Technology and future land use patterns. In: Britton, D. (ed.), *Agriculture in Britain: Changing pressures and policies.* CAB International, Wallingford, pp. 45–93.

Norton, B.G. (1987) *Why Preserve Natural Variety?* Princeton University Press, New Jersey.

O'Connor, R.J. and Shrubb, M. (1986) *Farming and Birds.* Cambridge University Press.

O'Riordan, T. (1990) *Conserving Wildlife and the Countryside – Key Environment Issues* No. 7. British Gas.

Owen, D. (1987) Report to the Priorities Board, Grassland and Forage Research Consultative Committee. (MAFF, DAFS, AFRC).

Potter, C., Burnham, P., Edwards, A., Gasson, R. and Green, B. (1991) *The Diversion of Land – Conservation in a Period of Farming Contraction.* Routledge, London.

Provenza, F.D. (1989) The evaluation of range science and its implications for grazing management. Radunautafundur, Iceland.

Provenza, F.D. (1991) Viewpoint: Range science and range management are complementary but distinct endeavours. *Journal of Range Management* **44**, 181–183.

Provenza, F.D. and Balph, D.F. (1990) Applicability of five diet-selection models to various foraging challenges ruminants encounter. In: Hughes, R.N. (ed.), *Behavioural Mechanisms of Food Selection.* Springer-Verlag. Berlin.

Putman, R.J., Fowler, A.D. and Tout, S. (1991) Patterns of use of ancient grassland by cattle and horses and effects on vegetational composition and structure. *Biological Conservation* **56**, 329–347.

Ratcliffe, D.A. and Thompson, D.B.A. (1988) The British uplands: their ecological character and international significance. In: Usher, M.B. and Thompson, D.B.A. (eds), *Ecological Changes in the Uplands.* Blackwell Scientific Publishers.

Robins, M., Davies, S.G.F. and Buisson, R.S.K. (1991) *An Internationally Important Wetland in Crisis.* RSPB, Sandy, Bedfordshire, UK.

Robinson, G.M. (1990) *Conflict and Change in the Countryside.* Belhaven Press, London.

Sibbald, A.R., Grant, S.A., Milne, J.A. and Maxwell, T.J. (1987) Heather moorland management – a model. In: Bell, M. and Bunce, R.G.H. (eds), *Agriculture and Conservation in the Hills and Uplands.* ITE, Grange-Over-Sands, pp. 107–108.

Smith, R.S. (1985) *Conservation of Northern Upland Meadows.* Yorkshire Dales National Park.

Smith, R.S. (1988) Farming and the conservation of traditional meadowlands in the Pennine Dales' environmentally sensitive area. In: Usher, M.B. and Thompson, D.B.A. (eds), *Ecological Change in the Uplands.* British Ecological Society. Blackwell Scientific Publishers, Oxford, pp. 183–199.

Smith, R.S. and Jones, L. (1991) The phenology of mesotrophic grassland in the Pennine Dales, northern England: Historic hay cutting dates, vegetation variation and plant species phenologies. *Journal of Applied Ecology* **28**, 42–59.

Soper, M.H.R. and Carter, E.S. (1991) *Farming and the Countryside.* Farming Press, Ipswich.

Tait, J., Lane, A., and Carr, S. (1988) *Practical Conservation – Site Assessment and Management Planning*. The Open University. Hodder and Stoughton Educational.

Taylor, J.P., and Dixon, J.B. (1990) *Agriculture and the Environment. Towards Integration*. RSPB, Sandy, Bedfordshire, UK.

Thalen, D.C.P., Poorter, H., Lotz, L.A.P. and Oosterveld, P. (1987) Modelling the structural changes in vegetation under different grazing regimes. In: Van Andel, J. *et al.*, *Disturbance in Grasslands*. Dr W. Junk Publishers, The Netherlands.

Usher, M.B. and Thompson, D.B.A. (eds) (1988) *Ecological Change in the Uplands*. Special Publication No. 7, British Ecological Society. Blackwell Scientific Publishers, Oxford.

Van Dyne, G.M., Brockington, N.R., Szocs, Z., Duck, J. and Ribic, C.A. (1980) Large herbivore subsystem. In: Breymeyer, A.I. and van Dyne, G.M. (eds), *Grasslands, Systems Analysis and Man*. IBP 19. Cambridge University Press, pp. 269–537.

Weinschenick, G. (1991) Environmental economics and harmony with nature. In: Burger, K., de Groot, M., Post, J. and Zachariasse, V. (eds), *Agricultural Economics and Policy: International challenges for the nineties*. Elsevier, Amsterdam, pp. 166–173.

Welch, D. (1984) Studies in the grazing of heather moorland in north-east Scotland. I Site descriptions and patterns of utilization. *Journal of Applied Ecology* **21**, 179–195.

Whitby, M. and Grant, S. (eds) (1990) Modelling heather management – a workshop report. Department of Agricultural Economics and Food Marketing, University of Newcastle upon Tyne.

Willis, M.J. and Young, S.W. (1983) Cheshire College of Agriculture wildlife and landscape management plan review 1976–1981. Cheshire County Education Committee Publication.

de Wit, C.T., Huisman, H. and Rabbinge, R. (1987) Agriculture and its environment – Are there other ways? *Agricultural Systems* **23**, 211–236.

Woodmansee, R.G. (1989) Biogeochemical cycles and ecological hierarchies. In: Zonneveld, I.S. and Forman, R.T.T. (eds), *Changing Landscapes: an ecological perspective*. Springer-Verlag, Berlin, pp. 57–71.

Appendix

Abstracts of Posters presented at the Conference

Part 1
Environmental Pollution and Farm Animals

Changes in the Radioactivity of Welsh Mountain Sheep Grazing Contaminated and Uncontaminated Pastures

I. Ap Dewi and J.B. Owen
School of Agricultural and Forest Sciences, University of Wales, Bangor LL57 2UW, UK

Following the accident at the Chernobyl nuclear power plant in April 1986, high levels of radioactivity were detected in ewes and lambs grazing upland areas of North Wales. Restrictions were imposed on the movement of sheep on 5100 farms from 20 June 1986. Experiments were instigated at the University College of North Wales' College Farm and another farm in upland Gwynedd, both of which are within the 'restricted' area. In both experiments ewes and their lambs were moved from 'low-level' pastures to contaminated pastures. Changes in the radioactivity of animals grazing contaminated pastures were compared to those of animals grazing 'low-level' pastures by live-animal monitoring. The decline in the level of radioactivity following movement of sheep from contaminated pastures was also monitored. It was observed that levels of radioactivity changed rapidly when animals were moved either to or from contaminated pasture and a biological half-life for $^{137}Cs/^{134}Cs$ of 10–14 days was calculated. It was also evident that there existed considerable variability in the level of radioactivity recorded for ewes from the same flock grazing contaminated pasture, and it is suggested that this reflects the differential grazing pattern of hill sheep. Those with the highest radioactivity levels grazed areas of mountain where contamination levels were high. There was a significant relationship between the radioactivity level of ewes and their lambs when both grazed together in mountain areas. Since a similar relationship was not observed on lowland pasture, despite similar initial levels, it is suggested that the association between the radioactivity levels in ewes and lambs is the result of their grazing in the same environment. For ewes and lambs grazing mountain pasture the effect of common environment was maintained as lambs became less dependent on their mothers' milk because the correlation between ewe and lamb radioactivity increased during the period up to weaning.

The Effect of Backyard and Large-scale Farming on the Bacterial Quality of Groundwater

G. Sinkovics
University of Agricultural Science Faculty of Animal Husbandry, H6800 Hodmezovasarhely, Andrassy U.15. Hungary

Groundwater samples have been collected for three months from old wells that had been dug in several parts of a Hungarian city (population 60 000) for keeping farm animals – mainly swine – in backyard systems. In these districts the dungwater is not conducted into the recently built pipe systems. Groundwater samples were also collected simultaneously from three new wells, bored near a large-scale swine farm of 1200 sows. The treatment of liquid manure here meets the requirements of Hungarian standards. Further groundwater samples – as controls – were collected from garden suburb areas also, where the breeding of farm animals is prohibited and this area has a suitable pipe drainage system. Samples were examined bacteriologically for determination of total live germ counts (faecal streptococci, coli-coliformes and clostridia) by the method of the International Organization for Standardization (ISO) standards. The results were compared in order to establish the effect of the different ways of animal farming on the bacterial contamination of groundwater.

Effects of Supplementation of some Heavy Metals (Cadmium and/or Zinc and Mercury and/or Selenium) on Growth Performance and Feed Utilization of Tilapia (*Oreochromis niloticus*)

A.M. Nour,[1] E. Omar,[1] A. Mohamed,[1] M.H.-El Sokkary[1] and I. El-Sokkary[2]
[1] *Department of Animal and Fish Production, Faculty of Agriculture, Alexandria University, Egypt.*
[2] *Department of Soil and Water Science, Faculty of Agriculture, Alexandria University, Egypt*

Two experiments were conducted in order to study the effects of supplementation of some heavy metals on growth performance and feed utilization of tilapia (*Oreochromis niloticus*). In the *first* experiment different levels of zinc (Zn) (10, 20 and 30 μg Zn/g diet) or cadmium (Cd) (5, 10 and 15 μg Cd/g diet) or a mixture of Zn and Cd (30, 5; 20, 10 and 10, 15 μg Zn, Cd/g diet respectively) and in the *second* experiment different levels of selenium (Se) (0.5, 1.0 and 1.5 μg Se/g diet) or mercury (Hg) (0.25, 0.50 and 1.0 μg Hg/g diet) or a mixture of Se and Hg (1.0, 0.5; 0.25, 1.5 and 0.5, 1.0 μg Se and Hg/g diet respectively) were supplemented in tilapia diets and compared with the control diet (without supplementation). Fish were reared in glass jars; each contained 105 l of tap water and stocked with ten fingerlings of tilapia with an initial weight of 4.5 and 2.6 g/fish in experiments *one* and *two* respectively. Forty glass jars were arranged for 20 feeding treatments (10 in each experiment). Fish were fed on the experimental diets twice daily for 112 days. The results showed that supplementation of tilapia diets with Zn or Se increased significantly ($P < 0.05$) growth performance and feed utilization of tilapia. However, supplementation of the diets with Cd or Hg decreased significantly ($P < 0.05$) growth performance and feed utilization of tilapia. Also it was observed that growth performance and feed utilization values were improved when tilapia were fed on diets containing a mixture of Zn and Cd (30 μg Zn and 15 μg Cd/g diet) or a mixture of Se and Hg (1.5 μg Se and 0.5 μg Hg/g diet respectively).

Chemical composition of fish carcasses showed that crude protein and ether extract were decreased with increased levels of supplementation of Cd or Hg and increased with increased levels of Zn and Se.

The results clearly showed a significant role of Zn and Se in elimination of the toxic effects of Cd and Hg respectively.

Effects of Different Levels of Dried Rumen Liquor or Poultry Litter on Growth Performance, Feed Utilization and some Metabolic Parameters of Tilapia (*Oreochromis niloticus*) and Common Carp (*Cyprinus carpio*)

E. Omar, A.M. Nour, A.R. Abou Akkada, M.A. Kassan and M.R. Hegazy
Department of Animal and Fish Production, Faculty of Agriculture, Alexandria University, Egypt

The present work aims to study the effect of feeding different levels (0, 10, 20 and 30%) of dried rumen liquor (DRL) or dried poultry litter (DPL) on the growth performance, feed utilization and some metabolic parameters of tilapia (*Oreochromis niloticus*) and common carp (*Cyprinus carpio* L.). Fingerlings of tilapia and carp (about 20 g each) were reared in separate enclosures and were fed for 105 days on the experimental diets.

The results showed the following.

1. Inclusion of DRL at more than 10% in tilapia diets decreased the growth performance, feed *utilization*, and increased the cost of feed required to produce 1 kg of fish as compared with the control diet. None of the tested levels of DRL improved the performance of carp. Increasing the levels of DRL in the diets did not affect the total protein although total lipids, cholesterol and the activity of lipid peroxidation were increased in the livers of tilapia and carp respectively. Total phospholipids were increased in the livers of tilapia and decreased in carp.
2. Inclusion of 20% DPL in tilapia diet significantly improved ($P < 0.05$) growth performance and feed utilization and the cost of production. However, no significant differences were observed in the growth and nutritional parameters of carp fed different levels of DPL.

Total liver protein in tilapia was significantly increased with 20% DPL, while carp were unaffected. Total lipids and phospholipids were decreased, but total cholesterol values were increased in fish livers with an increased level of DPL in the diet. The activity of lipid peroxidation increased with increasing levels of DPL in the diet of tilapia and carp.

It is concluded that tilapia utilize the tested by-products better than carp. DPL was better utilized than DRL. The recommended levels of inclusion of DPL should not exceed 20% in tilapia diets and 10% in carp diets. Both by-products increased in the liver of tilapia and carp.

Ecological and Economic Benefits of Growth Promoters in Animal Husbandry

J.M. Gropp and Doris R.T. Birzer
Institut für Physiologie, Physiologische Chemie und Ernährungsphysiologie Veterinärstr. 13, D-8000 München 22, Germany

Growth promoters all show a more or less definite antibacterial and/or antiparasitic activity. In small concentrations (2–50 mg kg^{-1} feed) they improve gain and feed efficiency of animals. A recent literature review allows an estimation of average improvements in gain and feed efficiency.

For monogastric animals, there is strong evidence that growth promoters act locally in the gastrointestinal tract. The flora stabilizing effect and reduction of potentially hazardous micro-organisms lead to a reduced waste of nutrients, reduced toxin formation, especially of ammonia and amines and therefore to a higher level of metabolic well-being. As a result the intestinal function (absorption) is improved as is the liver function, both resulting in an increase in liver protein synthesis and nutrient retention. It seems possible that the higher metabolic well-being alters the endocrine regulation systems which secondarily add to the growth response.

The most consistent feature associated with growth promoters for ruminants is an increase in molar proportion of propionate at the expense of acetate and frequently butyrate. They partially inhibit ruminal methane production (about 4–31%). Both factors would increase the amount of dietary energy available for absorption. The ionophore antibiotica also reduce feed intake.

The nutrient utilization of animals is affected by growth promoters. Digestibility of

Table A.1. Improvement (%) by growth promoters.

	ADG	FE		ADG	FE
Piglet	+16	+9	Beef cattle	+6	+6
Growing pig	+9	+5.5	Broiler	+4	+4
Finishing pig	±0	±0	Laying hen	+2*	+1
Veal calf	+7	+4			

ADG: average daily gain; FE: feed efficiency.
*Laying performance

nutrients especially of crude protein, amino acids and probably also that of energy is improved. There is only weak evidence for the same effect in ruminants. Also nitrogen retention is increased. As a result of better digestibility and higher nitrogen retention, a nutrient sparing effect is well demonstrated especially for proteins and amino acids in monogastric animals. Environmental stress by housing density and temperature is also partially compensated by growth promoters.

In two experiments (one on growing/finishing pigs, one on veal calves), the extent to which growth promoters (tylosin and carbadox) led not only to better performance but also to a reduced N-excretion was studied.

Nitrogen Emission

Three groups of 12 replicates (two pigs per pen) were each fed a grower diet from 37 to about 55 kg live weight and a finisher diet up to slaughter weight (97 kg). Both diets were wheat–soya based; they contained 17% and 15% crude protein (CP) for the control group and the (20/15 ppm) tylosin group in the grower and finisher period, respectively. A third group received protein reduced diets (16% CP in the grower diet) with carbadox (50 ppm) until an age of 120 days; the finisher (14% CP) contained 15 ppm tylosin. Due to different growth the experiment lasted 80 days for the control, 77 days for the tylosin group and 78 days for the protein reduced group until they had reached the final body weight. Average daily gain (gain/feed) amounted to 763 (323), 785 (341) and 770 g (334 g kg^{-1}) for the control, the tylosin and the protein reduced group. The digestibility of feed dry matter and of crude protein was improved by tylosin and carbadox in both feeding periods (crude protein digestibility was for the control, the tylosin and the protein reduced group 76.4, 77.3 and 77.6% in the grower phase and 78.6, 80.5 and 79.8% in the finisher phase, respectively). Carcass measurements revealed no differences in the meat : fat ratio between the three groups. Therefore, an equal N-retention per gain could be supposed.

From these data, the nitrogen intake as well as the N-excretion (faecal excretion measured; N-retention estimated from literature) could be calculated. The values are shown in Fig. A.1.

An experiment on a total of 48 veal calves (66–129/155 kg live weight) was performed in four groups for 56 days. Two groups (without and with carbadox) received a 60% skim milk containing milk replacer (24.5% CP), two groups (without and with carbadox) a 0% skim milk replacer (23.7% CP). Gain of the 60% skim milk control group was 1468 g/day, of the 0% skim milk control 1177 g day^{-1}. Average daily gain was improved by carbadox by 9% and 7% for the 60% and 0% skim milk groups, respectively. The corresponding feed conversion data were 708 g and 630 g gain/kg feed, 8% and 5% improvement, respectively (Fig. A.2). Nitrogen excretion was measured and calculated as described above.

Conclusions

As long as growth promoters are known to improve nutrient utilization they evidently reduce N-excretion and contribute to environmental protection. Consequently a withdrawal of growth promoters would increase nitrogen release from animal production to groundwater, rivers and lakes. Furthermore, the nutrient sparing capacity of growth promoters could be

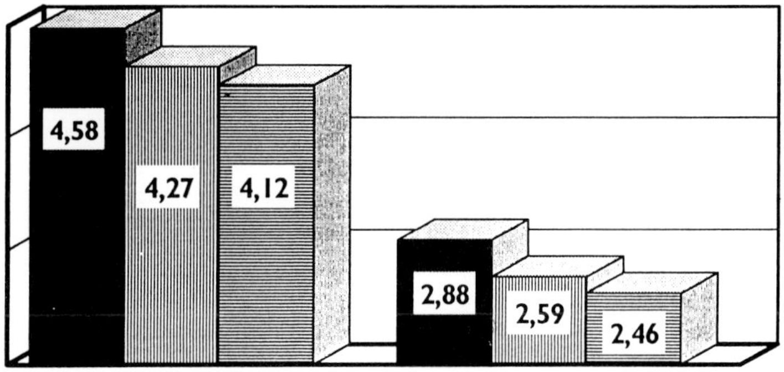

Fig. A.1. Nitrogen intake and excretion of pigs fed diets without and with growth promoters (12 × 2 pigs/group). ■ Control group: grower 17% CP; finisher 15% CP; ▥ Tylosin group: grower 17% CP; finisher 15% CP; ▤ Protein reduced group: grower 16% CP; finisher 14% CP.

used in feed formulation by decreasing dietary protein. As a result an additional reduction of N-excretion would be obtained.

Methane Emission

Methane emission of animal origin is an issue of concern and decrease has been urged by environmental scientists. Nitrogen in animal manure is also of considerable environmental concern.

Ionophore growth promoters reduce methane release by cattle twofold: (i) by decreasing methane production and (ii) by reducing feed intake. It can be shown by model calculations for beef cattle production in the United Kingdom that methane release is probably reduced by the use of growth promoters from 249 to 242 (by reduced feed intake) to a final 219 $m^3 \times 10^6$ (by better energy conversion in the rumen), based on experimental results or in a different scenario from 344 to 335 and final 325 $m^3 \times 10^6$. The withdrawal of growth promoters

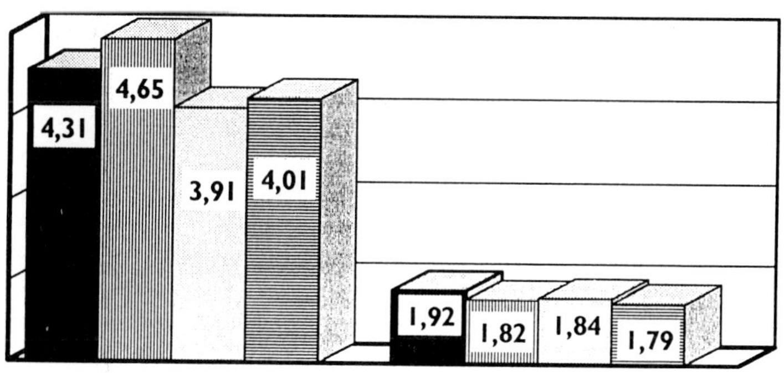

Fig. A.2. Nitrogen intake and excretion of veal calves fed different diets without and with growth promoters (12 calves/group). ■ Control group, 60% skim milk; ▥ Carbadox group, 60% skim milk; ▨ Control group, 0% skim milk; ▤ Carbadox group, 0% skim milk.

approved for cattle would increase methane emission by 11–20% with immediate effect.

Summary

In two experiments on a total of 72 pigs and 48 calves growth promoters not only improved live weight gain, feed conversion and N-digestibility considerably but also decreased nitrogen output. In pigs fed the same basal diet tylosin could be shown to reduce faecal and urinary N-excretion by about 300 g per animal in an 80 day feeding period. Decreasing dietary protein by 1% carbadox followed by tylosin led to a further reduction of 130 g. In calves carbadox reduced N-excretion by 100 g per calf in a 56 day trial.

Assessment of Increase in Milk Yield with Particular Respect to the Ecological Effects in the former FRG

A. Heissenhuber and T. Reitmayr

Lehrstuhl für Wirtschaftslehre des Landbaues, Technische Universität München, D-8050 Freising-Weihenstephan

The relation between agriculture and environment is gaining more attention from the public. Besides nitrate and ammonia, methane and phosphate, which affect the natural resources, are also important.

To reduce animal excreta, yield increases in animal production are discussed, for example by genetic gain or feed additives. This study[1] analyses the ecological effects of yield increase in milk production in the Federal Republic of Germany (FRG) with respect to nitrogen (N), phosphate (P_2O_5) and methane (CH_4). The changes in quantity of animal excreta under yield increase are examined in various scenarios, with respect to different consequences of a reduction of total dairy production.

In the initial situation (Fig. A.3) 6.07 million dairy cows, with 4000 kg milk yield, are needed in the FRG for a total milk production of 24.27 million tonnes. These cows produce 537 000 t nitrogen, 229 000 t phosphate and 642 000 t methane. Only if the total milk production remains constant, and meat production decreases, does the excretion decrease progressively with an increase in milk yield per cow. If the meat production is diminished, caused by production of fewer calves, then the excretion will decrease according to the reduction in beef production. If the milk to meat production ratio per cow remains constant, then the total excretion rises with an increase in milk yield per cow. With an average of 6000 kg milk yield the total excretion is reduced to 425 000 t nitrogen, 183 000 t phosphate and 468 000 t methane, if the milk production is constant and meat production decreases, but increases to 579 000 t nitrogen, 248 000 t phosphate and 653 000 t methane if the milk to meat production ratio per cow remains constant. On the individual farm the number of dairy cows decreases with an increase in milk yield, if the milk production of the farm is kept constant. This means that an increase in milk yield will reduce the excretion on an individal farm basis, but globally the total excretion rises if the milk to meat production ratio remain constant.

The constituents of animal excreta especially nitrogen, phosphate and methane may pollute the environment in different ways. The changes in quantity of animal excreta with yield increase in the FRG are examined in various scenarios, with respect to different consequences of a reduction of dairy husbandry. The data base for the model calculations was taken from the literature. Various strategies to adopt for milk yield increase may result in very different consequences for the quantity of animal excreta. If cows are destocked because of an increase in milk yield under constant milk production, the total animal excretion is reduced. This effect will be greater, if the meat production as bulls or heifers is reduced proportional to the number of dairy cows. With constant meat and milk production there will be a contrary effect concerning animal excreta, because the deficient calves will be replaced by alternative production methods such as suckling cows. Besides the calculations for the entire dairy sector the

[1]Published in full in *Berichte über Landwirtschaft*, No. 4/1991, in press.

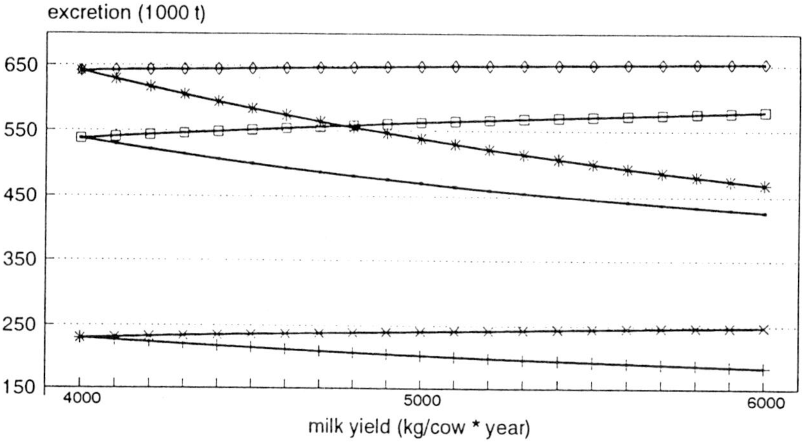

Fig. A.3. Excretion of all dairy and suckling cows without respect to beef production in the FRG according to milk yield per cow and year with constant milk production and with constant milk and meat production. —— nitrogen I; —⊟— nitrogen II; —+— phosphate I; —×— phosphate II; —✱— methane I; —◇— methane II; I = constant milk production without respect to changes in meat production; II = constant milk to meat ratio.

effects of animal excreta and economic parameters for an individual farm are also examined. The excretions of nitrogen and phosphate are a significant threat to the environment, but they may not necessarily pollute the environment.

Part 2
Genotype–Environment Interactions

The Performance of Exotic Cattle Breeds and their Role in the Farm Economy in Bangladesh

D.C. Paul[1] and M. Saadullah[2]
[1] *On-farm Research Division, Bangladesh Agricultural Research Institute, Bangladesh.*
[2] *Department of General Animal Science, Bangladesh Agricultural University, Mymensingh*

The purpose of this investigation was to identify the most promising exotic cattle breeds (referring to their calf mortality, feed efficiency, grazing habits and economic benefits to farmers) in an upland district (Jessore) of Bangladesh during the period from January 1980 to December 1990. The work concerns a study of the perceptions of 12 farmers owning exotic cattle breeds together with some production trials of exotic cattle breeding in the study area. Holstein and Jersey were judged to be the best with the highest yield in the third generation of 16 kg per day compared to Sindhi and Sahiwal at 5 kg per day. Proper care and management were indeed the best criteria for higher yield and less mortality and were more evident in suburban areas than in villages. It was also recorded that cowdung from an adult cow and calf could meet up to 10 months' fuel supply for a five member family and estimated that such a family could easily be maintained by rearing a high yielding cross-bred cow in this environment. The total yield was 2000 kg milk in 300 days. Average net income from exotic cattle was about 15 000.00 Taka per cow per year including yearling calf selling.

It is concluded that exotic cattle breeds can be kept more profitably with slight improvements in the management systems. One cow per household provides for family nutrition as well as income, and the utilization of labour of the family members.

Genotype–Environment Interaction in Milk and Lamb Production

A. Javor and I. Komlosi
University of Debrecen, Faculty of Agriculture, PO Box 36, Debrecen 4015, Hungary

The objective of this study was to compare different genotypes for milk and lamb production in two nutritional environments in Hungary. Four cross-bred ewes: a Merino × Langhe (M × L), Merino × Sarde (M × S), Merino × East Friesian (M × F) and Merino × Pleven (M × P), were compared to the control group Merino × Merino (M × M) ewes in two different nutritional environments over two years. In environment 'A', with summer grazing plus corn silage and hay during winter, and corn supplements during the lactation period, the feed intake was 11.5 MJ/day on average. In environment 'B' with free choice *ad libitum* zero grazed herbage, alfalfa silage, hay and creep fed concentrates during mating and from the last third of pregnancy until the end of lactation, the feed intake was 17 MJ/day on average. Significant genotype–environment and year effects were found at 42 days of age when weaned lambs of M × L weighed 16.8 kg and 17.8 kg; M × S 12.6 kg and 14.3 kg; M × F 16.3 kg and 18.8 kg; M × P 16.1 kg and 16.7 kg; and M × M 16.3 kg and 18.6 kg in environments 'A' and 'B' respectively. The lactation length varied from 50 to 90 days between genotypes over the two years in the two environments. The milk production yields were for M × L 48.7 l, and 92.4 l; M × S 46.3 l, and 72.4 l; M × F 50.8 l, and 86.4 l; M × P 36.1 l and 77.2 l; and M × M 30.7 l and 47.6 l, in environments 'A' and 'B', respectively. The profitabilities in F/head of the genotypes were the following: M × L 2.17 and 15.10; M × S 1.93 and 7.64; M × F 1.87 and 4.62; M × P −1.70 and 6.66; and M × M 0.73 and 2.73 in environments 'A' and 'B', respectively. Selection for milk production should take place in the Merino before cross-breeding for milk. In high (B) and low (A) feeding intensity environments M × S and M × L crosses, respectively, are recommended.

Part 3
Thermoregulation in Farm Animals

In Utero Heat Stressed Fetal Dwarfs are Metabolically Prepared for Neonatal Cold Stress

D.E. Brown,[1] J.O. Westerik,[1] J.N. Mugaas,[1] C.E. Dreiling,[2] and F.S. Carman, III[2]
[1] *WV School of Osteopathic Medicine, 400 North Lee Street, Lewisburg, WV 24901.*
[2] *Department of Biochemistry, University of Nevada, Reno, NV 89557, USA*

Intrauterine growth retardation (IUGR) is a result of late pregnancy maternal heat stress. Pups from rabbits kept at 21.5°C (TN) or at 35.9°C (H) from day 21 to day 30 were used to determine neonatal heat production and gluconeogenic capability at birth and subsequent to a one hour cold challenge at 31°C. Birthweights for H and TN pups were 39.9 + 1.8 and 45.5 + 1.3 kg ($P < 0.05$). Weight reductions varied from 10% at the ovarian end to 25% at the cervical end of a uterine horn. Dwarfs were disproportionate in organ growth and metabolic activity, shifting from protein deposition towards carbohydrate synthesis. Plasma glucose and lactate did not differ between H and TN pups. Liver L-alanine amino transferase (ALT), L-aspartate amino transferase (AST), glutamate dehydrogenase (GDH), γ-glutamyl transferase (GGT), glutaminase and arginase were elevated in H pups. The H liver glycogen stores were increased by 20–30% at birth and decreased with cold exposure. Oxygen uptake was 0.011 ml min^{-1} g^{-1} at birth in both H and TN while uptake was 0.022 vs. 0.026 ml min^{-1} g^{-1} upon 31°C cold stress for TN and H pups ($P < 0.05$). Enzyme activities were altered with one hour of cold stress; altered enzymes reverted to thermoneutral values while enzymes unaltered *in utero* were affected by neonatal cold exposure. IUGR rabbit pups have altered metabolic pathway enzyme activities, but normal heat production at term and increased potential for gluconeogenesis with increased heat production when postnatally cold stressed.

Effect of Chronic Cold Exposure during Late Pregnancy on Perirenal Adipose Tissue Development in the Neonatal Lamb

M.E. Symonds, M.J. Bryant and M.A. Lomax
University of Reading, Department of Biochemistry and Physiology, AMS Building, Whiteknights, PO Box 228, Reading, Berkshire

This study investigates the extent to which altering the maternal metabolic environment by winter shearing affects perirenal adipose tissue (PAT) development over the first month of life. Fourteen twin-bearing Bluefaced Leicester cross Swaledale ewes were housed at ambient temperature (3.5–19°C) and fed daily 200 g barley concentrate and 1 kg chopped hay. At four weeks before lambing, seven ewes were shorn. PAT was taken from eight lambs born from shorn (S) or unshorn (US) ewes within 2 h of birth. Remaining lambs (six S and six US) were artificially reared at a warm ambient temperature (20–25°C) on identical diets of 200 g milk replacer in 2.0 litre volume plus 100 g hay and concentrate on day 14. PAT was sampled at day 32 of age. Thermogenic capacity of the PAT was assessed from guanosine 5′-diphosphate (GDP) binding to mitochondrial protein. Lambs born from S ewes weighed 18% more than those from US controls (S 5.18 $\pm$ 0.18; US 4.40 $\pm$ 0.21 kg; $P < 0.01$), and contained 21% more PAT at birth (S 29 $\pm$ 1; US 24 $\pm$ 2 g; $P < 0.05$) which had a 40% higher level of GDP binding (S 95 $\pm$ 7; US 68 $\pm$ 8 nmol/lobe; $P < 0.05$). On day 32 lambs born from S ewes had twice the amount of PAT (S 35 $\pm$ 3; US 17 $\pm$ 2 g; $P < 0.001$) which possessed 76% more GDP binding (S 6.7 $\pm$ 1.0; US 3.8 $\pm$ 0.3 nmol/lobe; $P < 0.05$). It is concluded that chronic cold exposure of ewes during late pregnancy stimulates the growth of fetal PAT, and over the first month of life its mobilization is prevented while maintaining a higher thermogenic capacity, hence the thermoregulatory ability of lambs born from shorn ewes may be improved.

Effect of the Spacing in an Agroforestry System on Energy Expenditure of Grazing Sheep

P.D. Fenn
Macaulay Land Use Research Institute, Pentlandfield, Roslin, Midlothian EH25 9RF

With the present reliance on imported timber the conversion of grazing land to timber production is actively encouraged. During the establishment stages it is possible to graze animals between trees. These animals will be sheltered from the adverse effects of winter weather which may reduce their energy expenditure. On two occasions during the winter of 1988 CO_2 production rate as an index of energy expenditure was measured in Scottish blackface ewes grazing plots of 15-year-old Sitka spruce planted at 4 m and 8 m spacing on plots at Cloich forest near Edinburgh, compared to an unsheltered plot at the same site (Table A.2). Over a 5 day period the rate of production of CO_2 was measured using continuous infusions of $NaH^{14}CO_3$ and automatic sampling of blood. Animals were penned for one hour per day during which time they received 500 g of a pelleted complete diet, but were otherwise left to graze undisturbed on plots where ground cover was sparse.

The results show that CO_2 production rate was greater (22.7% and 20.7%) in ewes grazing the exposed, compared to the sheltered plots ($P < 0.05$). There appears to be no difference between the CO_2 production rates on the 8 m and 4 m sites suggesting an equal level of protection from the weather.

Table A.2. Carbon dioxide entry rate (grammage of $C\,day^{-1}\,kg^{-0.75}$) of Scottish Blackface ewes grazing under spaced trees.

	Nov. 1988	Dec. 1988
Control	13.5	13.4
4 m	11.0	
8 m		11.1
SED	1.09	1.11

Index